Applied Mathematics to Mechanisms and Machines

Applied Mathematics to Mechanisms and Machines

Editors

Higinio Rubio Alonso
Alejandro Bustos Caballero
Jesus Meneses Alonso
Enrique Soriano-Heras

MDPI • Basel • Beijing • Wuhan • Barcelona • Belgrade • Manchester • Tokyo • Cluj • Tianjin

Editors

Higinio Rubio Alonso
Department of Mechanical
Engineering, Universidad
Carlos III de Madrid,
Leganés, Spain

Alejandro Bustos Caballero
Department of Mechanics,
National University of
Distance Education,
Madrid, Spain

Jesus Meneses Alonso
Department of Mechanical
Engineering, Universidad
Carlos III de Madrid,
Leganés, Spain

Enrique Soriano-Heras
Department of Mechanical
Engineering, Universidad
Carlos III de Madrid,
Leganés, Spain

Editorial Office
MDPI
St. Alban-Anlage 66
4052 Basel, Switzerland

This is a reprint of articles from the Special Issue published online in the open access journal *Mathematics* (ISSN 2227-7390) (available at: https://www.mdpi.com/journal/mathematics/special_issues/Applied_Mathematics_to_Mechanisms_Machines).

For citation purposes, cite each article independently as indicated on the article page online and as indicated below:

LastName, A.A.; LastName, B.B.; LastName, C.C. Article Title. *Journal Name* **Year**, *Volume Number*, Page Range.

ISBN 978-3-0365-5923-0 (Hbk)
ISBN 978-3-0365-5924-7 (PDF)

Contents

Preface to "Applied Mathematics to Mechanisms and Machines"

The science of mechanisms and machines has necessitated mathematics from the very moment of its birth. The much celebrated renaissance figure Leonardo da Vinci already pointed out how necessary mathematics is for the advancement of science, stating that "No human investigation can be called real science if it cannot be demonstrated mathematically". In fact, the first steps of the development of mechanisms occurred hand in hand with geometry. At first, graphic methods, of undoubted pedagogical value, provided results for the motion of mechanisms at a given instant, which involved the arduous task of determining its kinematics and dynamics for the trajectory over time; conversely, analytical methods were limited to simple case resolution. With the advent and development of computing, numerical methods have made their way into all areas of knowledge and technology. In recent decades, these methods have made possible a rapid development of the analysis, optimization and synthesis of mechanical systems, mechanisms and machines in the academic, scientific and industrial fields.

The 16 chapters of this book (all of them accepted and published articles in the Special Issue "Applied Mathematics to Mechanisms and Machines" of the MDPI's *Mathematics* journal) cover a wide variety of topics, including the synthesis, design and optimization of mechanisms, robotics, automotive, Maintenance 4.0, machine vibrations, control, biomechanics and medical devices, among others, that combine mechanisms and machine science with mathematics in synergy.

The Guest Editors of the Special Issue are very grateful to the authors of these works for their high quality contributions, as well as to their reviewers for the sharpness and insightfulness of their comments. We also want to sincerely thank the administrative staff of MDPI publications for their invaluable support throughout the process.

Higinio Rubio Alonso, Alejandro Bustos Caballero, Jesus Meneses Alonso, and Enrique Soriano-Heras

Editors

Article

Hybrid Optimization Based Mathematical Procedure for Dimensional Synthesis of Slider-Crank Linkage

Alfonso Hernández [1], Aitor Muñoyerro [2], Mónica Urízar [1,*] and Enrique Amezua [1]

[1] Faculty of Engineering in Bilbao, University of the Basque Country (UPV/EHU), Plaza Ingeniero Torres Quevedo, 48013 Bilbao, Spain; a.hernandez@ehu.es (A.H.); enrique.amezua@ehu.es (E.A.)

[2] SENER Aeroespacial, Avda. de Zugazarte 56, 48992 Getxo, Spain; aitor.munoyerro@aeroespacial.sener (A.M.)

* Correspondence: monica.urizar@ehu.es (M.U.)

Abstract: In this paper, an optimization procedure for path generation synthesis of the slider-crank mechanism will be presented. The proposed approach is based on a hybrid strategy, mixing local and global optimization techniques. Regarding the local optimization scheme, based on the null gradient condition, a novel methodology to solve the resulting non-linear equations is developed. The solving procedure consists of decoupling two subsystems of equations which can be solved separately and following an iterative process. In relation to the global technique, a multi-start method based on a genetic algorithm is implemented. The fitness function incorporated in the genetic algorithm will take as arguments the set of dimensional parameters of the slider-crank mechanism. Several illustrative examples will prove the validity of the proposed optimization methodology, in some cases achieving an even better result compared to mechanisms with a higher number of dimensional parameters, such as the four-bar mechanism or the Watt's mechanism.

Keywords: path generation; dimensional synthesis; hybrid optimization; slider-crank mechanism

Citation: Hernández, A.; Muñoyerro, A.; Urízar, M.; Amezua, E. Hybrid Optimization Based Mathematical Procedure for Dimensional Synthesis of Slider-Crank Linkage. *Mathematics* **2021**, *9*, 1581. https://doi.org/10.3390/math9131581

Academic Editor: Raimondas Ciegis

Received: 31 May 2021
Accepted: 2 July 2021
Published: 5 July 2021

Publisher's Note: MDPI stays neutral with regard to jurisdictional claims in published maps and institutional affiliations.

1. Introduction

Dimensional synthesis consists of finding a geometry that enables a mechanism to generate certain motion characteristics, such as trajectories or positions of elements. It is tough to solve this problem intuitively and often requires the implementation of specific methods. Depending on the type and amount of prescribed motion characteristics, it is not always possible to obtain an exact solution to this problem, forcing us to use optimization methods to find an approximation with minimal error. The target most commonly addressed in bibliographies is the synthesis type, known as path generation, where a point of a single degree of freedom mechanism is sought to run through a sequence of prescribed positions. This motion may or may not be synchronized with the location of the input element, resulting in prescribed or unprescribed timing problems, respectively. It should be noted that there also exist two other goals that are frequently studied, these being beyond the scope of this paper. These are function generation, where the motion of two elements of the mechanism is synchronized, and motion generation, where a sequence of locations for a certain element of the mechanism is prescribed.

Most of the literature on dimensional synthesis focuses on individual cases, the four-bar hinged mechanism being the most widely studied. In the existing literature, both the exact synthesis, by means of graphic [1] or analytical methods [2], and the approximate synthesis, by means of dimensional optimization [3,4], have been studied. On the other hand, the slider-crank mechanism has been used to solve function generation problems [5–8], for dynamic synthesis [9,10], and to serve as an adjustable mechanism [11–14]. However, no publications on dimensional optimization for path generation have been found. It should be noted that, in addition to the usual kinematic objectives, some papers include more specific characteristics within the error function. For example, in [15] a formulation is developed by using exact differentiation that allows for establishing the position of the instant center of rotation and the centrode.

An often-employed dimensional optimization procedure consists of minimizing the error function, formulated as the sum of squared differences between the points of the discretized prescribed path, and those belonging to the real generated path. The minimization process can be solved with different methods that can be classified into two main groups, these being local and global methods, which are mentioned below. On the other hand, there are also publications focused not so much on studying the mathematical optimization techniques, but on proposing new ways of describing the output generated by the mechanism. This may result in a more advantageous definition for the optimization error function or for the elaboration of atlases and databases. In relation to the path generation problems discussed in this paper, there exist different approaches to describe the trajectories, and probably the most typical ones are based on the Fourier series [16] or Haar Wavelet transform [17]. Similarly, reference [18] describes a unified theory of the harmonic characteristic parameter method for mechanism synthesis. Apart from the dimensional synthesis, other publications focus on the phase that precedes it, i.e., structural synthesis, the first step in the conceptual design of mechanisms. In this sense, there can be found some proposals of automatic algorithms intended for the structural synthesis of robots and closed-loop mechanisms [19].

In relation to the mathematical optimization techniques for dimensional synthesis, the most effective and widely used local methods consist of applying the null gradient condition, which leads to a non-linear system of equations. This system often includes some passive variables that cannot be eliminated. To solve it, the function is linearized and an iterative method, such as Gauss–Newton, is used, starting from an approximate initial solution provided by the designer. The original reference for this type of method is a paper published in 1966 by Chi-Yeh [20], which was dedicated to the four-bar linkage. From then on, several papers related to dimensional synthesis of this mechanism by means of gradient methods have been published, exploring different ways to improve the effectiveness of optimization. As part of these alternative approaches, in [21] the authors proposed modifying the set of variables to be optimized, considering the nodal coordinates instead of the usual dimensional parameters, thereby allowing the elimination of some constraints that were present in the original problem. It is also noteworthy that other publications focus on reformulating the error function, such as [22,23]. The authors of those works proposed to minimize the strain energy originated when the mechanism is forced to run exactly through the prescribed trajectory. On the other hand, some authors have chosen to estimate the error by avoiding its sensitivity to translation and rotation effects, such as [24], where a system of relative coordinates between precision points is used. Following the same idea, the authors proposed to perform a prior and independent phase to optimize the translation, rotation, and scaling parameters [25]. In addition to the different alternatives to characterize the design parameters of the mechanism and to estimate the resulting error, a relevant aspect to achieve good performance in gradient methods is to carry out an exact calculation of the partial derivatives, avoiding the numerical derivation, since it increases the computational cost and results in a lower efficiency. Given the interest in solving this problem, reference [26] presents a general method for calculating the exact partial derivatives from the loop equations previously identified by the designer.

Despite the large number of existing publications devoted to optimal dimensional synthesis by means of gradient methods, and to the improvement of their performance, none of them are capable of solving the main limitation they have. Unfortunately, these methods are highly sensitive to the chosen starting approximation, since they are local in nature and hence converge to the nearest minimum, which will not necessarily be the optimal overall solution. To overcome this drawback, global methods make it possible to explore the entire space where solutions can be found. Metaheuristic methods are the most common ones, and they have been covered in several references. These include genetic algorithms [27–29], differential evolution [30–32], ant search [33], krill herd algorithm [34], imperialist competitive algorithm [35], or neural networks [36]. Nevertheless, the weakness of heuristic methods in comparison with gradient methods is their higher computational cost and a lower convergence rate. Furthermore, there is no guarantee that they will converge to a minimum, neither locally nor globally.

Hybrid optimization algorithms, such as [37], gain greater strength when a global method generating seeds for starting mechanisms is combined with a good local method. Normally, hybrid methods start by running a global method to obtain one or several designs that will later be used as initial approximations in a local method to quickly converge to the nearest relative minimum. In this paper, as described below, a hybrid optimization approach is proposed.

In the case of the slider crank mechanism analyzed in this paper, thanks to the simplicity of its kinematics, it is possible to express the synthesis variables directly as a function of the dimensional and input parameters, and thus completely eliminating the passive variables. A novel aspect of the proposed approach is the way in which the resulting system of equations is solved. Considering the more general case of unprescribed timing synthesis, the final system of equations associated with the null gradient condition can be divided into two subsystems with different characteristics. As will be explained, the procedure described in this article allows each subsystem to be solved separately within an iterative process that connects them together. This makes it possible, in some particularly simple cases, such as the two-parameter slider-crank, to solve all the equations analytically, while in more complex cases, numerical methods must necessarily be adopted. The optimum solution reached will be a relative minimum of the error function and will be influenced by the initial approximation used to solve the system of equations numerically. In this paper, reducing this dependency will be attempted by running the local optimization algorithm from different starting points previously selected by a genetic algorithm. The generation of the starting points could also be done by another type of heuristic method, or through a sweeping process that generates random points within the entire search space. Even so, the latter technique would not achieve the most promising regions as the genetic algorithm does, but it would only generate a wide grid of different starting points. Therefore, the genetic algorithm is the preferred choice.

It is important to highlight that the procedure described in this paper seeks to obtain reliable solutions, not only from a mathematical vision, but also from a practical point of view. For this reason, it will be stated how to avoid the circuit defect in the slider-crank mechanism, relying on the concept of branch index. This concept was introduced in reference [21] to analyze the kinematics of the four-bar hinged linkage. The incorporation of design constraints will be addressed by means of penalty functions included in the error function to be minimized. This is essential to impose the Grashof criterion and thus ensure that the input element is able to fully rotate (crank input).

Finally, the effectiveness of the proposed methodology will be illustrated through different examples. The final solutions obtained in this paper are as accurate as the ones reached in other papers when solving the same problem by using more complex designs, such as four-bar or Watt linkages.

The main novelties and highlights of this paper are:

- Deduction of the equations required for the optimal dimensional synthesis of the slider-crank mechanism, which constitutes an alternative to the hinged four-bar linkage usually used in the literature to solve this type of problem.
- Proposal of an original methodology to solve a non-linear system of equations resulting from the null gradient condition, based on the decoupling of two subsystems of equations. It facilitates the resolution of the system and, in some cases, allows to obtain all the solutions in an analytical way.
- Integration of the local optimization methodology within a hybrid optimization method, which uses a genetic algorithm to search for the best starting approximations. The fitness function has been adapted to solve not only the prescribed timing problem, but also unprescribed timing.

- Solving and comparison of examples proposed by other authors in the literature dealing with the four-bar linkage. Thanks to the effectiveness of the method proposed in this work, the slider-crank mechanism, though being simpler and more limited, is able to provide similar performances (or even better in some cases) in path generation problems.

2. Materials and Methods

In this section, the basis of the optimum synthesis procedure as well as the functioning of the hybrid optimization strategy is presented.

2.1. Bases of the Optimum Synthesis Procedure

Before tackling the optimization process, the loop-closure equations of the mechanism and the position equations of the coupler point that traces the trajectory will be obtained. In addition, the variables that take part in the synthesis process and their functional dependence law will be described. This enables obtaining the partial derivatives that intervene in the minimization process.

2.1.1. Synthesis Equations for a General Design

First, the kinematic problem of the slider-crank mechanism with five dimensional parameters, represented in Figure 1, is obtained.

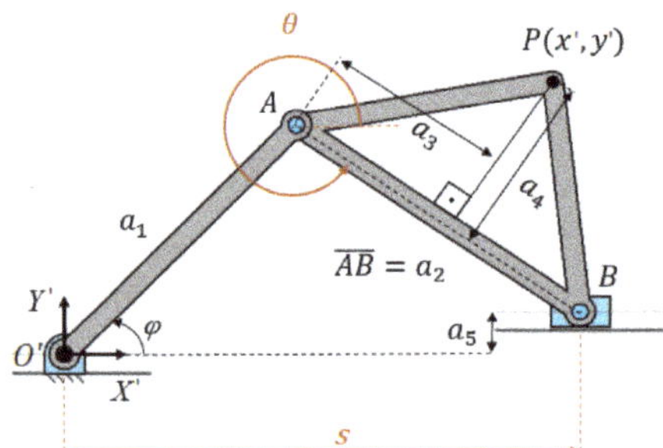

Figure 1. Slider-crank mechanism with 5 dimensional parameters.

The loop-closure equations are the following:

$$a_1 \cdot sin\varphi = a_5 - a_2 \cdot sin\theta \tag{1}$$

$$s = a_1 \cdot cos\varphi + a_2 \cdot cos\theta \tag{2}$$

From Equation (1) yields:

$$sin\theta = \frac{a_5 - a_1 \cdot sin\varphi}{a_2} \tag{3}$$

Then, $cos\theta$ is given by Equation (4), where $K = \pm 1$.

$$cos\theta = K\sqrt{1 - \left(\frac{a_5 - a_1 \cdot sin\varphi}{a_2}\right)^2} \tag{4}$$

The synthesis equations are as follows:

$$x' = s - (a_2 - a_3) \cdot cos\theta + a_4 \cdot cos\left(\theta - \frac{3\pi}{2}\right) \tag{5}$$

$$y' = -(a_2 - a_3) \cdot sin\theta + a_5 + a_4 \cdot cos\theta \tag{6}$$

Using the loop equations to solve the passive variables s (Equation (2)) and θ (Equations (3) and (4)), and substituting them in Equations (5) and (6), the following expressions are obtained for the synthesis variables, referring to the local system O'X'Y':

$$x' = a_1 cos\varphi + a_3 \cdot K\sqrt{1 - \left(\frac{a_5 - a_1 sin\varphi}{a_2}\right)^2} - \frac{a_4 \cdot (a_5 - a_1 sin\varphi)}{a_2} \tag{7}$$

$$y' = \frac{(a_3 - a_2)\cdot(a_5 - a_1 sin\varphi)}{a_2} cos\varphi + a_5 + a_4 \cdot K\sqrt{1 - \left(\frac{a_5 - a_1 sin\varphi}{a_2}\right)^2} \tag{8}$$

Remark Regarding Branches and Circuits

The sign $\pm$ in Equations (7) and (8), which has been substituted for $K = \pm 1$ for simplicity, is related to the two possible positions of the coupler point P for the same input φ. This means that two branches exist, each one associated with the positive or negative value of K. This circumstance is illustrated in Figure 2, in which the two possible configurations of the mechanism, for a given value of the input φ, are represented.

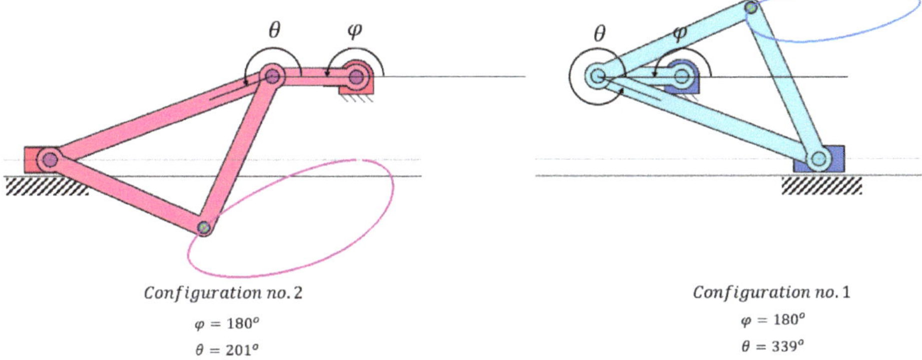

Figure 2. Two possible configurations for the same input.

The two possible trajectories of point P associated with the different configurations of the coupler element, commonly known as *branches*, can be connected or unconnected, resulting in a unique circuit (a unicursal curve), or two circuits (a bicursal curve). In this work, designs where the crank input is able to perform a 360° full rotation are considered, meaning that the Grashof criterion must be fulfilled. Therefore, the two possible branches will be two unconnected circuits. To avoid branch defects, all the selected points must have the same value of K. This value will be the one corresponding to the branch that yields a minimum error with respect to the desired path.

In the most general case, represented in Figure 3, the local reference system O'X'Y' has a rotation relative to the global system OXY, defined by the parameter a_6, and a translation in the plane defined by the parameters a_7 and a_8.

The equations that express the synthesis variables in the global reference system are the following:

$$x_i = x_i' \cdot cos(a_6) - y_i' \cdot sin(a_6) + a_8 \tag{9}$$

$$y_i = x_i' \cdot sin(a_6) + y_i' \cdot cos(a_6) + a_7 \tag{10}$$

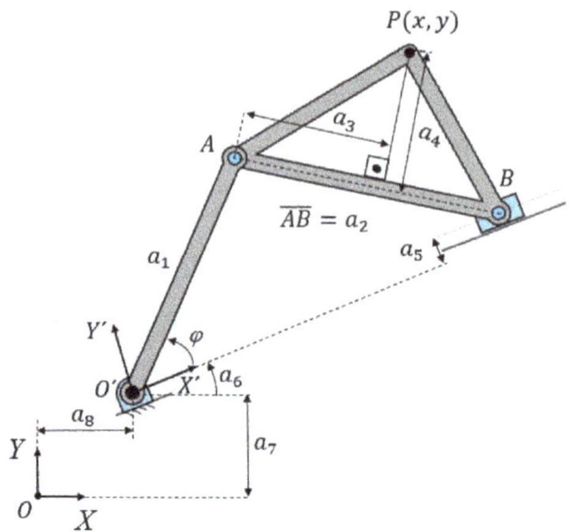

Figure 3. Slider-crank mechanism with 8 dimensional parameters.

Now that all the variables involved in the synthesis problem have been defined, the following classification can be established:

- **Dimensional variables:** $a_1, a_2, \ldots, a_8$. These are variables that define the lengths of the bars and the translation or rotation parameters of the studied mechanism.
- **Input variable:** φ. This is an independent variable corresponding to the degree of freedom of the mechanism under study.
- **Passive variables:** θ, s. These are not independent variables, but rather depend on the input and the dimensional parameters.
- **Output variables or synthesis variables:** x, y. These correspond to the coordinates of the coupler point P. In the case of path generation synthesis, these are indeed the synthesis variables.

2.1.2. Optimal Design Based on the Error Function

The error function E, commonly used in synthesis problems, is defined as the sum of the squared Cartesian distances between the prescribed points and those actually generated:

$$E = \sum_{i=1}^{N} \left[\left(x_i - x_i^d \right)^2 + \left(y_i - y_i^d \right)^2 \right] \tag{11}$$

The error between prescribed and generated trajectory must be minimized to obtain the optimal mechanism. There are two options for carrying out this minimization. In the modality known as prescribed timing, only the dimensional parameters are optimized, requiring solving the system shown in Equation (12). In this case, the input parameters φ_i are not variables to be optimized, but constant values (prescribed values). However, in an optimization known as unprescribed timing, both the dimensional parameters a_j and the set of input parameters φ_i are optimized, requiring solving the systems given by Equations (12) and (13). This last option is more complex but its potential to obtain precise solutions is greater, since the value of the input parameters is not being restricted.

$$\frac{\partial E}{\partial a_j} = 0 \rightarrow \sum_{i=1}^{N} \left[\left(x_i - x_i^d \right) \frac{\partial x_i}{\partial a_j} + \left(y_i - y_i^d \right) \frac{\partial y_i}{\partial a_j} \right] = 0 \qquad \forall j = 1, 2, \ldots, n \tag{12}$$

$$\frac{\partial E}{\partial \varphi_i} = 0 \rightarrow \sum_{i=1}^{N} \left[\left(x_i - x_i^d \right) \frac{\partial x_i}{\partial \varphi_i} + \left(y_i - y_i^d \right) \frac{\partial y_i}{\partial \varphi_i} \right] = 0 \qquad \forall i = 1, 2, \ldots, N \tag{13}$$

2.2. Hybrid Optimization Procedure

Thanks to the simplicity of its kinematics, in the slider-crank mechanism it is possible to eliminate the passive variables. In this way, the synthesis variables (x, y) are expressed as an explicit function of parameters $\{a_j\}$ and φ, facilitating the obtaining of the partial derivatives that appear in Equations (12) and (13). After performing the substitution, it can be verified that the system of Equations (12) and (13) is non-linear. This system can be decomposed into two subsystems of equations. On the one hand, Equation (12) is composed of 8 equations, as many as the dimensional parameters of the mechanism. Each of these equations has the form $g(a_1, a_2 \ldots a_8, \varphi_1, \varphi_2, \ldots \varphi_N) = 0$. On the other hand, Equation (13) is made up of N equations, as many as points of precision has the trajectory. Each of these equations has the form $h(a_1, a_2 \ldots a_8, \varphi_i) = 0$. This circumstance is fundamental when establishing the procedure for solving the system of equations.

In addition, in order to find the optimal initial approximation, the one that will be the starting point of the local optimization process, and that plays a relevant role in the process, a multi-start approach will be implemented. In the following sections, the hybrid procedure proposed by the authors of this work will be explained.

2.2.1. Solving the Equation System for Local Optimization

In this section, the iterative algorithm to solve the non-linear system of equations resulting from the null gradient condition is addressed. The procedure is based on the decoupling of the two subsystems (12) and (13), as described below.

- **First phase:**

This starts from a certain mechanism, coming from the multi-start procedure that will be explained in Section 3.2. The dimensions $(a_1, a_2, \ldots, a_8)$ of this mechanism are assumed constant in this phase. The values of these dimensions are substituted in each of the equations of subsystem (13), resulting in a total of N equations, each with a unique φ_i.

In the simplest case of the slider-crank mechanism, with 2 dimensional parameters $(a_1 = a_2 = a_3$ and $a_5 = a_6 = a_7 = a_8 = 0)$, and using the transformation of half-angle tangent, $t_i = \tan \frac{\varphi_i}{2}$, Equation (13) becomes a fourth degree polynomial:

$$\left(2a_1a_4 + x_i^d a_4\right) t_i^4 + \left(8a_1{}^2 + x_i^d a_1 + 2y_i^d a_4\right) t_i^3 - 12a_1a_4 t_i^2 + \left(-8a_1{}^2 + 4x_i^d a_1 + 2y_i^d a_4\right) t_i + 2a_1a_4 - x_i^d a_4 = 0 \quad (14)$$

From the 4 possible values of $t_i(\varphi_i)$, the one that gives the minimum error is the selected one.

For a design with 3 dimensional parameters, Equation (13) becomes a polynomial of degree 10. Even so, it is easy to obtain the 10 roots and detect the correct one proceeding as in the previous case. However, with 4 or more dimensional parameters, it is no longer as easy to determine its corresponding univariate polynomial, nor is it really worth it. It is more practical to operate as explained next.

The Equation (13) is solved numerically, starting from an initial approximation, and arriving at a unique solution of the parameter φ_i. To guarantee that the global optimal value of φ_i is obtained, it will be necessary to start from an initial approximation obtained as follows: each sum of Equation (11), which represents the error made in each synthesis position i, is evaluated as a function of the input parameter φ_i along the discretized domain $[0, 2\pi)$. In this way, a graph similar to the blue curve represented in Figure 4 will be obtained, where two minima of the error function appear, the one indicated on the right being the one with the lowest value. The latter value of φ_i is taken as a starting approximation to solve Equation (13) (in our case, using the MATLAB *fsolve* command).

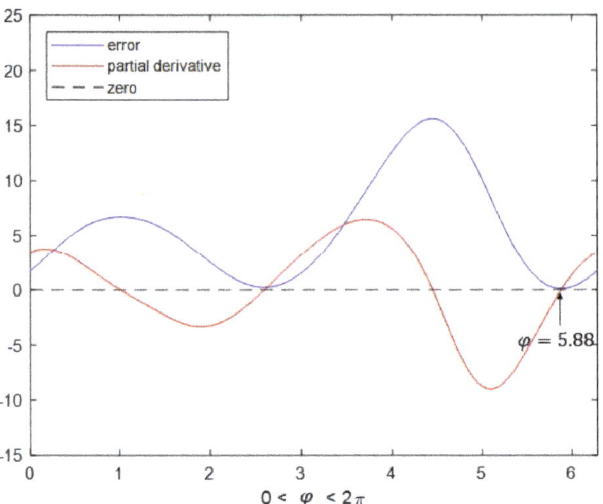

Figure 4. Error function for a point *i* evaluated at [0,2π).

To illustrate this concept, in Figure 5 the prescribed point for a synthesis position *i* and the trajectory generated by the mechanism of dimensions $\{a_j\}$ in the current iteration are indicated. The effect of evaluating the sum of the error function (11) and choosing the absolute minimum is equivalent to traversing the generated trajectory and selecting the point of it (the black point) closest to the prescribed one (red point). In this example, the black point indicated in Figure 5 corresponds to the absolute minimum, $\varphi = 5.88$ rad, which is the one previously indicated in Figure 4.

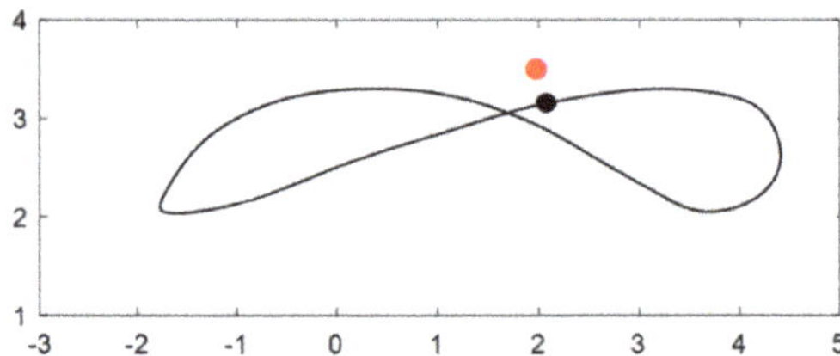

Figure 5. Generated trajectory and prescribed point (red).

- **Second phase:**

In this phase, the values φ_i obtained in the previous phase will be assumed as constants, and the unknowns $(a_1, a_2, \ldots, a_8)$ will be calculated from the subsystem from Equation (12).

As an example, in the simplified particular case of the slider-crank mechanism with 2 parameters, Equation (12) becomes the following linear system:

$$\begin{bmatrix} 4\sum_{i=1}^{N} cos^2\varphi_i & \sum_{i=1}^{N} sin2\varphi_i \\ \sum_{i=1}^{N} sin2\varphi_i & N \end{bmatrix} \left\{ \begin{array}{c} a_1 \\ \\ a_4 \end{array} \right\} = \left\{ \begin{array}{c} 2\sum_{i=1}^{N} cos\varphi_i \cdot x_i^d \\ \sum_{i=1}^{N} \left(sin\varphi_i \cdot x_i^d + cos\varphi_i \cdot y_i^d \right) \end{array} \right\} \tag{15}$$

However, in the design cases with 3 or more dimensional parameters, the subsystem of equations turns out to be non-linear, making it necessary to apply numerical solving methods. As a starting approximation, the values $(a_1, a_2, \ldots, a_8)$ that were assumed as constants in the previous phase will be taken.

- **Next steps:**

With the new values of $\{a_j\}$ obtained in the second phase, we go back to the first phase, and thus continue iteratively until convergence. The chosen stopping criterion consists of comparing the current values of the optimization variables $(a_1, a_2 \ldots a_8, \varphi_1, \varphi_2, \ldots \varphi_N)$ with those of the previous iteration, so that when the difference is less than a specified tolerance, the iterative process will stop.

2.2.2. Implementing a Multi-Start Strategy

The method proposed in Section 3.1 has the disadvantage of being very sensitive to the starting approximation. To avoid this drawback, a multi-start approach based on a genetic algorithm will be used to guide the local optimization method towards a search for the global optimal solution.

The proposed multi-start approach obtains a sufficiently large set of starting approximations, representative of a global sweep of the design space $\{a_j\}$. In this article, a genetic algorithm is used to locate 100 candidate solutions that serve as initial approximations. It is decided to use the genetic algorithm incorporated in MATLAB, assigning as arguments of the *fitness function* the values for the 8 dimensional parameters. The latter function assigns the optimal correspondence of input parameters, according to the procedure outlined in the first phase of Section 3.1, and returns a scalar value that quantifies the individual's fitness as the quadratic sum of the distances between generated and prescribed points.

It is important to bear in mind that, considering the way the fitness function is posed, each starting solution for the unprescribed timing problem will not depend on $8 + N$ variables (8 dimensional + N inputs), but only on the 8 dimensional parameters, which facilitates the exploration of the search space by the genetic algorithm without the need to restrict the input parameters.

The complete scheme of this hybrid procedure is shown in Figure 6. As can be seen, it consists of executing a local optimization from the different starting points obtained in the scan of the space of the dimensional parameters $\{a_j\}$. In this way, different local minimums will be obtained and the best of them will be selected.

To avoid excessive computational cost, the execution time of the genetic algorithm is limited to an acceptable value (i.e., 1 min). Bear in mind that it will not be necessary to obtain the optimal solutions, but that it will be enough to be close to them. On the other hand, the execution time of the local method to be applied later will depend on the number of iterations performed, but it generally consumes a few seconds for each starting point used.

2.2.3. Incorporation of Design Constraints

Searching for an optimal design usually implies adapting the generated trajectory to a prescribed one. In addition, some design requirements must be fulfilled. These design constraints are related to several aspects, such as the maximum size of the mechanism, the maximum and minimum lengths of certain bars, the Grashof criterion, the transmission angle limits, and so on.

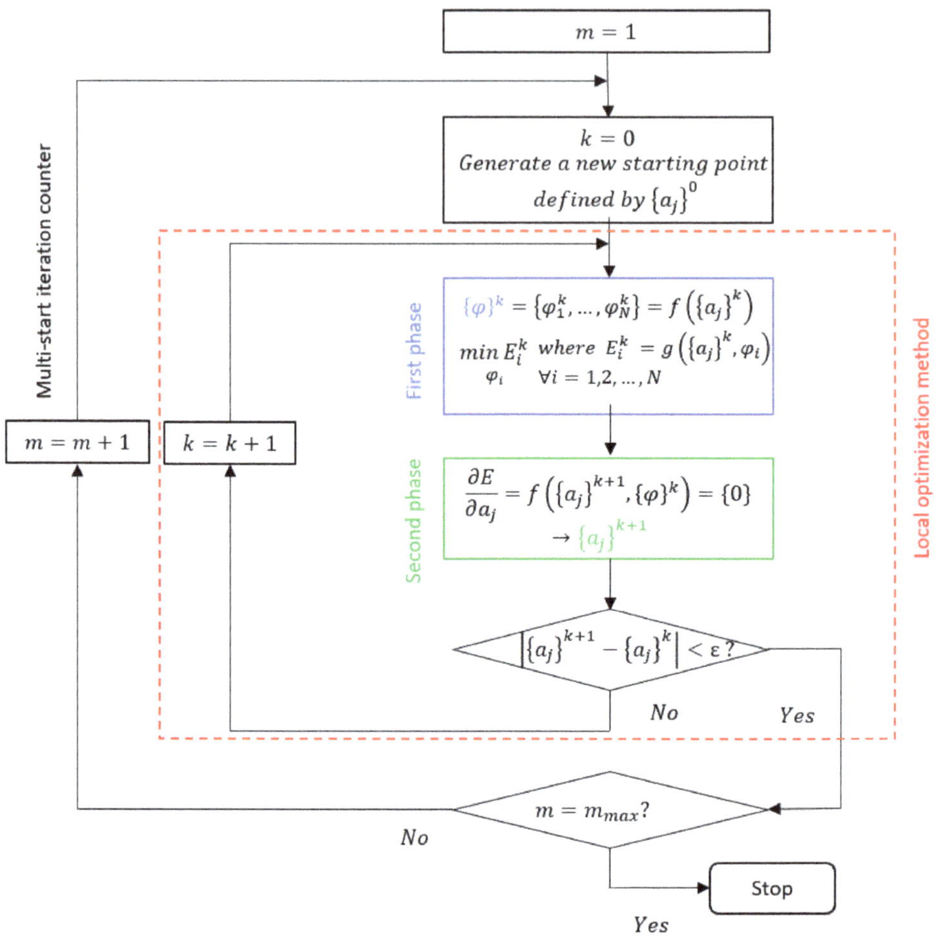

Figure 6. Flowchart of the hybrid optimization scheme.

In general, the optimization problem with certain inequality constraints can be formulated as the minimum error subjected to

$$g_k\{a_1, \ldots, a_n\} \geq 0 \ \forall k = 1, 2, \ldots, l$$

A typical constraint equation, g_k, is expressed as follows:

$$\sum_{j=1}^{m} a_j C_j \geq C_0 \tag{16}$$

When an inequality constraint, such as Equation (16), is included in the process, the error function yields

$$E = \sum_{i=1}^{N} \left(x_i - x_i^d\right)^2 + \sum_{i=1}^{N} \left(y_i - y_i^d\right)^2 + \lambda \cdot \left(\sum_{j=1}^{m} a_j C_j - C_0\right)^2 \tag{17}$$

If, on the one hand, Equation (16) is not fulfilled during the iterations of the optimization process, it is necessary to amplify the value of the error to invalidate the corresponding design $\{a_j\}$. This can be done by means of the λ parameter, which acts as a penalty factor, proceeding as follows:

$$\text{If } \sum_{j=1}^{m} a_j C_j - C_0 < 0 \qquad \text{then} \qquad \lambda = Predefined\ value$$

$$\text{If } \sum_{j=1}^{m} a_j C_j - C_0 \geq 0 \quad \text{then} \quad \lambda = 0$$

If, on the other hand, the condition to be imposed is an equality constraint, then λ will always be a certain predefined value. In this paper, after reviewing similar works in the literature and conducting several trials with different values, the value $\lambda = 50$ was established, insofar as it provides good results for all the analyzed design cases.

3. Results

The following examples will demonstrate the validity of the proposed methodology, showing the results of each design case.

3.1. Demonstrative Example 1

The desired trajectory, shown in Figure 7, is formed by two straight lines and two circumference arcs tangent to these lines. Considering the simplicity of the slider-crank, achieving a design that traces this trajectory is challenging. Moreover, this example is interesting because the trajectory combines both straight lines and arcs, which are trajectories commonly used in the path planning synthesis of planar mechanisms. The coordinates of the desired points, $\left(x_i^d, y_i^d \right)$, are given in Table 1. A total of 16 points are defined.

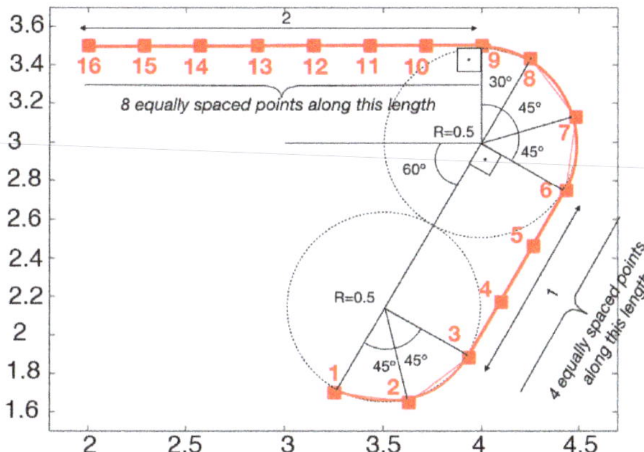

Figure 7. Example 1. Desired trajectory.

The eight dimensional parameters and the 16 input parameters (unprescribed timing) will be optimized, that is, the most complete case possible. Following the previously described technique, this example is solved using the proposed multi-start method, starting from 100 different starting mechanisms. The conditions of the crank-input (based on Grashof's criterion) and a maximum 3:1 ratio between the longest and shortest bar are imposed by means of penalty functions.

Table 1. Example 1. Initial data.

i	x_i^d	y_i^d
1	3.2500	1.7010
2	3.6294	1.6510
3	3.9330	1.8840
4	4.0995	2.1724
5	4.2665	2.4616
6	4.4330	2.7500
7	4.4829	3.1294
8	4.2500	3.4330
9	4.0000	3.5000
10	3.7143	3.5000
11	3.4286	3.5000
12	3.1429	3.5000
13	2.8571	3.5000
14	2.5714	3.5000
15	2.2857	3.5000
16	2.0000	3.5000

The best solution, shown in Figure 8, leads to an error equal to 0.0073. The corresponding design parameters are included in Table 2. In the Supplementary Materials, a video showing the motion of this optimum mechanism can be found.

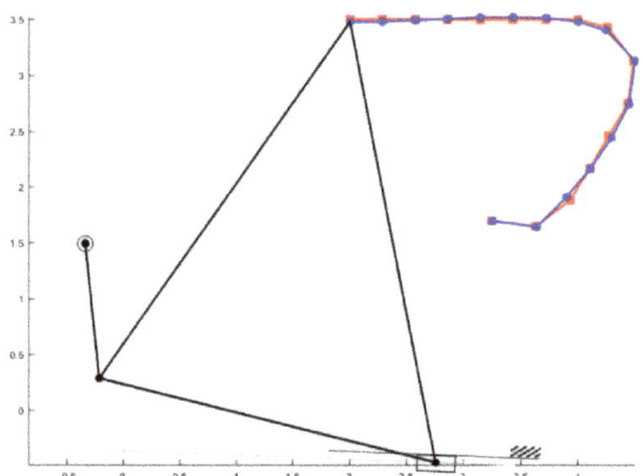

Figure 8. Example 1: best multi-start solution for slider-crank, E = 0.0073.

To illustrate the influence that the starting point has on the final solution obtained, Figure 9 shows, by way of example, the optimal mechanism obtained from another starting point, which does not reach a solution as good as the one shown above, so it would be excluded.

Table 2. Example 1: optimum design parameters and inputs.

Parameters		Inputs			
a_1	1.292	φ_1	4.877	φ_9	2.992
a_2	3.277	φ_2	4.568	φ_{10}	2.804
a_3	1.292	φ_3	4.409	φ_{11}	2.634
a_4	3.875	φ_4	4.267	φ_{12}	2.476
a_5	1.970	φ_5	4.104	φ_{13}	2.323
a_6	3.090	φ_6	3.905	φ_{14}	2.171
a_7	1.356	φ_7	3.579	φ_{15}	2.014
a_8	−0.583	φ_8	3.192	φ_{16}	1.846

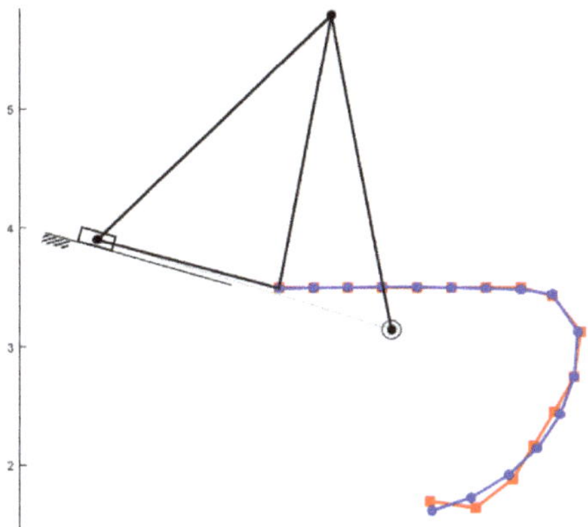

Figure 9. Example 1: comparison to other multi-start solution (multi-start no. 10), E = 0.0227.

This same example has been used by the authors of this work in a recent work, in particular reference [38], making use of another optimization method and applying it to the four-bar mechanism. It should be taken into account that the four-bar mechanism has an additional dimensional parameter with respect to the slider-crank and thus, in principle, a greater potential to adjust to the prescribed trajectories. Nevertheless, comparing the results of both achieved optimal mechanisms (the error in [38] with the four-bar was 0.0089), it can be seen that they are almost of the same value, even having a slightly smaller error in the current case.

3.2. Demonstrative Example 2

Figure 10a shows the results obtained by other authors [30,34,39,40] for an example of path generation synthesis of a closed path formed by 25 points marked in red. Each of the authors uses a different optimization method, but all of them use the four-bar mechanism. On the other hand, Figure 10b shows the results obtained when solving the same example with the slider-crank under the methodology proposed in this paper.

The comparison is made with other studies that analyze the four-bar because, as mentioned in the introduction, the vast majority of path generation synthesis works make use of the four-bar. Even so, as it has been shown in the previous example and it will happen in this one, the method that we propose achieves very similar (or even better) results, despite making use of a more limited mechanism such as the slider-crank.

The optimal solution obtained in this paper with the slider-crank mechanism (Figure 10b) corresponds to an error of 0.103 and is of similar quality to Bulatović et al. [34]. They obtained an error of 0.0392, which is slightly lower than the one we found, but is of the same order of magnitude. On the other hand, in [34] they also indicated the error achieved by Laribi et al. [39] and Smaili and Diab [40], these being 0.9022 and 0.5504, respectively, so they would be considerably higher. On the other hand, the error obtained by Kafash and Nahvi [30] is not directly comparable with the rest because they use a different error function. However, the trajectory generated by the authors in [30] is graphically represented in Figure 10a. The corresponding optimal design parameters for the more general case of slider-crank are shown in Table 3. As in the previous example, a video showing the motion of this optimum mechanism is included as Supplementary Materials.

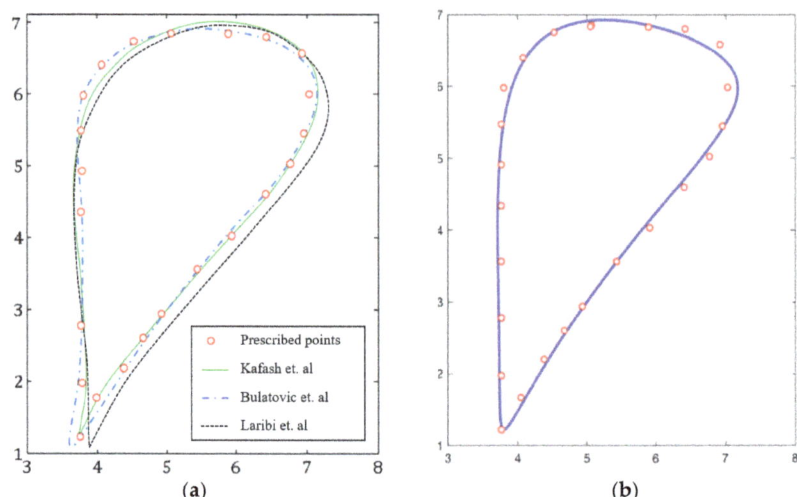

Figure 10. Example 2: optimal solution for: (**a**) four-bar and (**b**) slider-crank mechanism.

Table 3. Example 2: Optimum design parameters and inputs.

Parameters									
a_1	2.309	φ_1	4.703	φ_9	-0.237	φ_{17}	2.715	φ_{25}	4.451
a_2	48.819	φ_2	4.92	φ_{10}	0.049	φ_{18}	2.933		
a_3	-3.304	φ_3	5.041	φ_{11}	0.672	φ_{19}	3.17		
a_4	24.498	φ_4	5.181	φ_{12}	1.331	φ_{20}	3.483		
a_5	46.315	φ_5	5.362	φ_{13}	1.698	φ_{21}	3.762		
a_6	4.726	φ_6	5.528	φ_{14}	1.996	φ_{22}	3.752		
a_7	-18.02	φ_7	5.745	φ_{15}	2.281	φ_{23}	4.091		
a_8	15.293	φ_8	5.875	φ_{16}	2.494	φ_{24}	4.265		

3.3. Demonstrative Example 3

On this occasion, a path generation problem, previously solved in reference [41] for the Watt mechanism, is presented. Their optimal solution is represented in Figure 11a, together with the set of prescribed points represented in red. The optimal solution obtained with the slider-crank mechanism, represented in Figure 11b, has an equal, and, it could be said, even higher precision, despite being a simpler mechanism that has fewer design parameters involved.

The optimal design parameters of the slider-crank mechanism are listed in Table 4. Yet again, a video showing the motion of this optimum mechanism is included as Supplementary Materials.

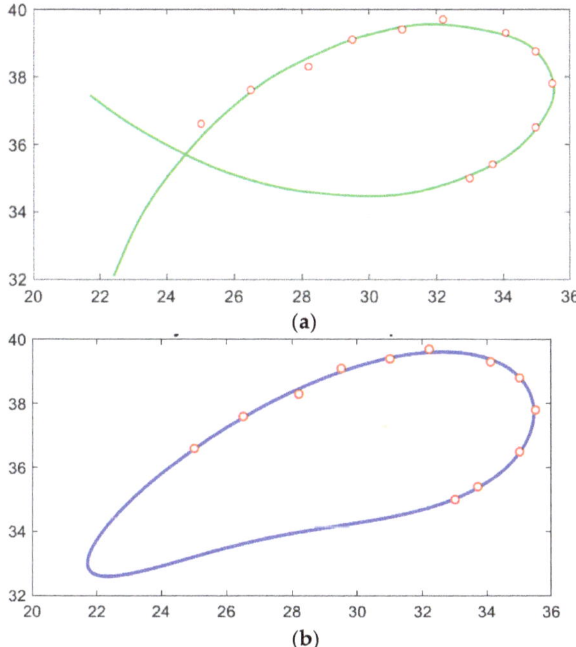

Figure 11. (a) Watt's mechanism optimal solution from [41]; (b) optimal solution for slider-crank mechanism.

Table 4. Example 3: Optimum design parameters and inputs.

Parameters					
a_1	6.719	φ_1	5.580	φ_9	3.502
a_2	15.635	φ_2	5.349	φ_{10}	3.101
a_3	5.217	φ_3	5.116	φ_{11}	2.708
a_4	8.737	φ_4	4.918	φ_{12}	2.549
a_5	−6.710	φ_5	4.703		
a_6	−2.924	φ_6	4.508		
a_7	42.513	φ_7	4.124		
a_8	35.584	φ_8	3.848		

4. Discussion

In this paper, a hybrid optimization procedure has been presented to address the dimensional synthesis of path generation using the slider-crank mechanism. On the one hand, a novel approach has been proposed to solve the non-linear equations corresponding to the null gradient condition, which is based on the decoupling of two subsystems of equations, and which simplifies the resolution procedure without prejudice to the precision obtained. On the other hand, the incorporation of the multi-start method manages to give the local optimization method a global character, carrying out a previous sweep of the total space of the dimensional variables to find the most promising starting approximations. To carry out this search, a genetic algorithm is used where only the eight dimensional parameters are involved, thanks to the fact that the fitness function has been programmed to select the optimal input parameters based on the dimensions.

In addition, it should be noted that this methodology is aimed at achieving designs that are valid from a practical point of view. Therefore, the non-existence of branch defects is ensured, and different design restrictions can be incorporated in a simple way through the use of penalty functions. In this way, it can be ensured that the crank input mechanism does not exceed a maximum ratio between bar lengths or other additional design requirements at the discretion of the designer.

Finally, the effectiveness of the method has been proved by verifying, through various examples, that the slider-crank mechanism allows us to achieve solutions with a precision comparable to other one degree of freedom mechanisms that have a greater number of dimensional parameters, such as the four-bar or the Watt's mechanism, with the additional advantage of having simpler kinematics.

Supplementary Materials: The following are available online at https://www.mdpi.com/2227-7390/9/13/1581/s1. The three videos corresponding to the motion of the optimum mechanisms are available.

Author Contributions: Conceptualization, A.H., A.M., M.U. and E.A.; methodology, A.H., A.M. and M.U.; software, A.M.; validation, A.H., A.M., M.U. and E.A.; formal analysis, A.H.; investigation, A.H., A.M., M.U. and E.A.; resources, A.H. and A.M.; data curation, A.M.; writing—original draft preparation, A.H. and A.M.; writing—review and editing, M.U. and E.A.; visualization, A.M.; supervision, A.H.; project administration, A.H. and M.U.; funding acquisition, A.H., M.U. and E.A. All authors have read and agreed to the published version of the manuscript.

Funding: The authors wish to acknowledge financial support received from the Spanish government through the Ministerio de Economía y Competitividad (Project DPI2015−67626-P (MINECO/FEDER, UE)), the support for the research group through Project Ref. IT949−16, provided by the Departamento de Educación, Política Lingüística y Cultura from the regional Basque Government, and the Program BIKAINTEK 2020 (Ref. 012-B2/2020) provided by the Departamento de Desarrollo Económico, Sostenibilidad y Medio Ambiente from the regional Basque Government.

Institutional Review Board Statement: Not applicable.

Informed Consent Statement: Not applicable.

Data Availability Statement: Not applicable.

Conflicts of Interest: The authors declare no conflict of interest.

References

1. Erdman, A.G.; Sandor, G.N. *Mechanism Design: Analysis and Synthesis*, 4th ed.; Pearson: London, UK, 2001.
2. Wampler, C.; Morgan, A.P.; Sommese, A.J. Complete Solution of the Nine-Point Path Synthesis Problem for Four-Bar Linkages. *J. Mech. Des.* **1992**, *114*, 153–159. [CrossRef]
3. Lee, W.-T.; Russell, K. Developments in quantitative dimensional synthesis (1970-present): Four-bar motion generation. *Inverse Probl. Sci. Eng.* **2017**, *26*, 133–148. [CrossRef]
4. Lee, W.-T.; Russell, K. Developments in quantitative dimensional synthesis (1970–present): Four-bar path and function generation. *Inverse Probl. Sci. Eng.* **2018**, *26*, 1280–1304. [CrossRef]
5. Alizade, R.I.; Mohan Rao, A.V.; Sandor, G.N. Optimum Synthesis of Four-Bar and Offset Slider-Crank Planar and Spatial Mechanisms Using the Penalty Function Approach with Inequality and Equality Constraints. *J. Eng. Ind.* **1975**, *97*, 785–790. [CrossRef]
6. Rao, A.C. Optimum synthesis of a slider-crank mechanism using geometric programming. *Int. J. Numer. Methods Eng.* **1980**, *15*, 1595–1602. [CrossRef]
7. Plecnik, M.M.; McCarthy, J.M. Five position synthesis of a slider-crank function generator. In Proceedings of the ASME International Conference IDETC/CIE 2011, Washington, DC, USA, 28–31 August 2011; pp. 317–324.
8. Almandeel, A.; Murray, A.P.; Myszka, D.H.; Stumph, H.E. A Function Generation Synthesis Methodology for All Defect-Free Slider-Crank Solutions for Four Precision Points. *J. Mech. Robot.* **2015**, *7*, 031020–031021. [CrossRef]
9. Liniecki, A. Synthesis of a slider-crank mechanism with consideration of dynamic effects. *J. Mech.* **1970**, *5*, 337–349. [CrossRef]
10. Davidson, J.K. Analysis and synthesis of a slider-crank mechanism with a flexibly-attached slider. *J. Mech.* **1970**, *5*, 239–247. [CrossRef]
11. Zhou, H.; Ting, K.-L. Adjustable slider–crank linkages for multiple path generation. *Mech. Mach. Theory* **2002**, *37*, 499–509. [CrossRef]
12. Russell, K.; Sodhi, R. On the Design of Slider-Crank Mechanisms. Part I: Multi-Phase Motion Generation. *Mech. Mach. Theory* **2005**, *40*, 285–299. [CrossRef]
13. Russell, K.; Sodhi, R.S. On the design of slider-crank mechanisms. Part II: Multi-phase path and function generation. *Mech. Mach. Theory* **2005**, *40*, 301–317. [CrossRef]

14. Zhou, H. Dimensional synthesis of adjustable path generation linkages using the optimal slider adjustment. *Mech. Mach. Theory* **2009**, *44*, 1866–1876. [CrossRef]
15. Sancibrian, R.; Sarabia, E.G.; Sedano, A.; Blanco, J.M. A general method for the optimal synthesis of mechanisms using prescribed instant center positions. *Appl. Math. Model.* **2016**, *40*, 2206–2222. [CrossRef]
16. Sun, J.; Chu, J. Fourier series representation of the coupler curves of spatial linkages. *Appl. Math. Model.* **2010**, *34*, 1396–1403. [CrossRef]
17. Liu, W.; Sun, J.; Zhang, B.; Chu, J. Wavelet feature parameters representations of open planar curves. *Appl. Math. Model.* **2018**, *57*, 614–624. [CrossRef]
18. Jianwei, S.; Jinkui, C.; Baoyu, S. A unified model of harmonic characteristic parameter method for dimensional synthesis of linkage mechanism. *Appl. Math. Model.* **2012**, *36*, 6001–6010. [CrossRef]
19. Ding, H.; Huang, P.; Zi, B.; Kecskeméthy, A. Automatic synthesis of kinematic structures of mechanisms and robots especially for those with complex structures. *Appl. Math. Model.* **2012**, *36*, 6122–6131. [CrossRef]
20. Chi-Yeh, H. A general method for the optimum design of mechanisms. *J. Mech.* **1967**, *1*, 301–313. [CrossRef]
21. Angeles, J.; Alivizatos, A.; Akhras, A. An unconstrained nonlinear least-square method of optimization of RRRR planar path generators. *Mech. Mach. Theory* **1988**, *23*, 343–353. [CrossRef]
22. Avilés, R.; Navalpotro, S.; Amezua, E.; Hernández, A. An Energy-Based General Method for the Optimum Synthesis of Mechanisms. *J. Mech. Des.* **1994**, *116*, 127–136. [CrossRef]
23. Vallejo, J.; Avilés, R.; Hernández, A.; Amezua, E. Nonlinear optimization of planar linkages for kinematic syntheses. *Mech. Mach. Theory* **1995**, *30*, 501–518. [CrossRef]
24. Sancibrian, R.; Viadero, F.; García, P.; Fernández, A. Gradient-based optimization of path synthesis problems in planar mechanisms. *Mech. Mach. Theory* **2004**, *39*, 839–856. [CrossRef]
25. Sancibrian, R.; De Juan, A.; Sedano, A.; Iglesias, M.; García, P.; Viadero, F.; Fernandez, A. Optimal Dimensional Synthesis of Linkages Using Exact Jacobian Determination in the SQP Algorithm. *Mech. Based Des. Struct. Mach.* **2012**, *40*, 469–486. [CrossRef]
26. Mariappan, J.; Krishnamurty, S. A generalized exact gradient method for mechanism synthesis. *Mech. Mach. Theory* **1996**, *31*, 413–421. [CrossRef]
27. Cabrera, J.; Simon, A.; Prado, M. Optimal synthesis of mechanisms with genetic algorithms. *Mech. Mach. Theory* **2002**, *37*, 1165–1177. [CrossRef]
28. Acharyya, S.; Mandal, M. Performance of EAs for four-bar linkage synthesis. *Mech. Mach. Theory* **2009**, *44*, 1784–1794. [CrossRef]
29. Buśkiewicz, J.; Starosta, R.; Walczak, T. On the application of the curve curvature in path synthesis. *Mech. Mach. Theory* **2009**, *44*, 1223–1239. [CrossRef]
30. Kafash, S.H.; Nahvi, A. Optimal synthesis of four-bar path generator linkages using Circular Proximity Function. *Mech. Mach. Theory* **2017**, *115*, 18–34. [CrossRef]
31. Gogate, G.R.; Matekar, S.B. Optimum synthesis of motion generating four-bar mechanisms using alternate error functions. *Mech. Mach. Theory* **2012**, *54*, 41–61. [CrossRef]
32. Bulatović, R.R.; Đorđević, S.R. Control of the optimum synthesis process of a four-bar linkage whose point on the working member generates the given path. *Appl. Math. Comput.* **2011**, *217*, 9765–9778. [CrossRef]
33. Xiao, R.; Tao, Z. A Swarm Intelligence Approach to Path Synthesis of Mechanism. In Proceedings of the Ninth International Conference on Computer Aided Design and Computer Graphics (CAD-CG'05), Hong Kong, China, 7–10 December 2005; Institute of Electrical and Electronics Engineers (IEEE): Piscataway, NJ, USA, 2005; pp. 451–456.
34. Bulatovic, R.R.; Miodragovic, G.; Boskovic, M.S. Modified Krill Herd (MKH) algorithm and its application in dimensional syn-thesis of a four-bar linkage. *Mech. Mach. Theory* **2016**, *95*, 1–21. [CrossRef]
35. Ebrahimi, S.; Payvandy, P. Efficient constrained synthesis of path generating four-bar mechanisms based on the heuristic opti-mization algorithms. *Mech. Mach. Theory* **2015**, *85*, 189–204. [CrossRef]
36. Vasiliu, A.; Yannou, B. Dimensional synthesis of planar mechanisms using neural networks: Application to path generator linkages. *Mech. Mach. Theory* **2001**, *36*, 299–310. [CrossRef]
37. Sedano, A.; Sancibrian, R.; De-Juan, A.; Viadero, F.; Egaña, F. Hybrid Optimization Approach for the Design of Mechanisms Using a New Error Estimator. *Math. Probl. Eng.* **2012**, *2012*, 1–20. [CrossRef]
38. Hernández, A.; Muñoyerro, A.; Urízar, M.; Amezua, E. Comprehensive approach for the dimensional synthesis of a four-bar linkage based on path assessment and reformulating the error function. *Mech. Mach. Theory* **2021**, *156*, 104126. [CrossRef]
39. Laribi, M.A.; Mlika, A.; Romdhane, L.; Zeghloul, S. A combined genetic algorithm-fuzzy logic method (GA-FL) in mechanism synthesis. *Mech. Mach. Theory* **2004**, *39*, 717–735. [CrossRef]
40. Smaili, A.; Diab, N. Optimum synthesis of hybrid-task mechanisms using ant-gradient search method. *Mech. Mach. Theory* **2007**, *42*, 115–130. [CrossRef]
41. Fernández-Bustos, I.; Aguirrebeitia, J.; Avilés, R.; Angulo, C. Kinematical synthesis of 1-dof mechanisms using finite elements and genetic algorithms. *Finite Elements Anal. Des.* **2005**, *41*, 1441–1463. [CrossRef]

Article

Path Analysis for Hybrid Rigid–Flexible Mechanisms

Oscar Altuzarra, David Manuel Solanillas *, Enrique Amezua and Victor Petuya

Department of Mechanical Engineering, University of the Basque Country UPV/EHU, 48013 Bilbao, Spain;
oscar.altuzarra@ehu.eus (O.A.); enrique.amezua@ehu.eus (E.A.); victor.petuya@ehu.eus (V.P.)
* Correspondence: david.m.solanillas.frances@gmail.com or davidmanuel.solanillas@ehu.eus

Abstract: Hybrid rigid–flexible mechanisms are a type of compliant mechanism that combines rigid and flexible elements, being that their mobility is due to rigid-body joints and the relative flexibility of bendable rods. Two of the modeling methods of flexible rods are the Cosserat rod model and its simplification, the Kirchhoff rod model. Both of them present a system of differential equations that must be solved in conjunction with the boundary constraints of the rod, leading to a boundary value problem (BVP). In this work, two methods to solve this BVP are applied to analyze the influence of external loads in the movement of hybrid compliant mechanisms. First, a shooting method (SM) is used to integrate directly the shape of the flexible rod and the forces that appear in it. Then, an integration with elliptic integrals (EI) is carried out to solve the workspace of the compliant element, considering its buckling mode. Applying both methods, an algorithm that obtains the locus of all possible trajectories of the mechanism's coupler point, and detects the buckling mode change, is developed. This algorithm also allows calculating all possible circuits of the mechanism. Thus, the performance of this method within the path analysis of mechanisms is demonstrated.

Keywords: hybrid compliant mechanisms; path analysis; numerical methods; elliptic integrals; kinematics

Citation: Altuzarra, O.; Solanillas, D.M.; Amezua, E.; Petuya, V. Path Analysis for Hybrid Rigid-Flexible Mechanisms. *Mathematics* **2021**, *9*, 1869. https://doi.org/10.3390/math9161869

Academic Editors: Higinio Rubio Alonso, Alejandro Bustos Caballero, Jesus Meneses Alonso and Enrique Soriano-Heras

Received: 8 July 2021
Accepted: 2 August 2021
Published: 6 August 2021

Publisher's Note: MDPI stays neutral with regard to jurisdictional claims in published maps and institutional affiliations.

1. Introduction

A compliant body is one whose motion depends on its geometry, its material properties and the location and magnitude of the applied forces. If a body of this kind belongs to a mechanism, it is known as a compliant element. When a mechanism is fully composed of compliant elements, e.g., slender rods, it is named a compliant mechanism, and it gains all or part of its mobility, thanks to the relative flexibility of those compliant elements. On another note, if the mechanism combines rigid and compliant elements, it takes the name of a hybrid compliant or hybrid rigid–flexible mechanism [1].

Regarding applications, such compliant—and, especially, hybrid-compliant—mechanisms are an alternative to robotic systems of rigid bodies [2], or even cable robots [3] used in tasks where there is a human–machine interface. Additionally, such hybrid-flexible mechanisms are a plausible alternative to sub-systems subjected to impacts, such as four-link legs in bio-inspired mobile multipod robots [4]. Finally, its elasticity, if appropriately tuned, can play a beneficial role in the dynamic characteristics of systems used in high-speed link motions, reducing the need for balancing [5].

For the complete kinematic characterization of a linkage with rigid bars, all the circuits of the mechanism should be obtained [6]. Each circuit is the set of all possible orientations of links that can be calculated without disconnecting any of the joints [7]. If the linkage needs to be disassembled to move from one position to another, these positions lie on different circuits. These mechanism's circuits do not depend on the input link chosen. In [8,9] is presented the circuit analysis for the Watt and Stephenson –like six-bar mechanisms. It should be noted that, in the case of compliant mechanisms, it is not needed to disassemble any link to obtain the different circuits but to deform a flexible rod through an external load.

To solve the movement of these hybrid rigid–flexible mechanisms, it is needed to obtain the change of shape of compliant elements and the loads that these bodies withstand.

There are a few modeling methods to analyze flexible rods, such as the *Chained Beam-Constraint Model (CBCM)* [10] or *Pseudo-Rigid-Body Model (PRBM)* [11,12]. Another model is the Cosserat rod model, which produces the equations that describe the deformed shape of a slender flexible element and the loads that appear in it [13,14]. This modeling method leads to a set of differential equations that, in conjunction with the boundary constraints of the compliant element, takes the form of a boundary value problem (BVP). This BVP can be solved with the aid of a shooting method (SM) [15], which combines a direct integration method, such as 4th order Runge–Kutta (R-K), and a numerical minimization method, e.g., Newton–Raphson (N–R), which refines the solutions given by the direct integration.

If the extension of the neutral line and the transverse shear strain are neglected in the Cosserat rod model, the Kirchhoff model is obtained. The equations of this new model can also be solved by applying elliptic integrals (EI) that make use of the two parameters known as the modulus of the function (k) and the amplitude of the elliptic integral (ϕ) [16].

When a compliant element deforms, points of null moment can appear along its length. These points of null moment are those where a change of curvature is produced, and they are also known as inflection points [17,18]. Hence, the slender rod deforms changing its curvature in each of these points, leading to what is known as *buckling modes*. These buckling modes are ordered according to the number of inflection points. For instance, if a rod has only one of these points, it is in its first buckling mode. If the same rod has two points, it is in its second buckling mode, and so on. It is worth noting that the higher the buckling mode, the more deformation energy is needed for it to appear. For this reason, the compliant elements tend to deform in lower buckling modes, as they require less energy. These buckling modes can be determined using EI, which can also be used to obtain the rod workspace in each buckling mode [19].

As mentioned above, the motion of the compliant bodies, and hence that of the compliant mechanisms, depends on the applied forces. Therefore, this paper makes use of the SM and EI to check the loads' influence in the movement of a compliant mechanism. In addition, the work here developed shows how these methods allow obtaining the locus of all possible paths of the mechanism's coupler under any applied load.

A mechanism similar to that found in [20] was used to carry out the work in this paper, and it is based on a mechanism of [21] (ch. 12). The mechanisms differ in the boundary constrains of the compliant element. In the case of [20], a clamped-clamped rod is used, but in this paper, a clamped-hinged rod is analyzed.

Thus, in the following section, an introduction to the general equations of flexible rods and the different integration methods is given. Moreover, a way to obtain the workspace of a slender rod using EI is also depicted here. In the third section, a closed-loop hybrid mechanism is proposed, and it is used to analyze the load-dependent paths after explaining its general equations. Finally, the potential of the developed algorithm to detect the buckling mode changes of flexible rods and the possible circuits of the mechanism is also shown.

2. Elastic Rod Fundamentals

2.1. General Equations

Among all of the modeling methods of flexible rods, because of its easy implementation within the control of a continuum robot, a simplification of the Cosserat model, i.e., the Kirchhoff model, was used to develop the algorithms of this work. Recalling the assumptions made in the Kirchhoff model, they are that the extension of the neutral line and the transverse shear strain are neglected. It means that the deformation of the rod is mainly due to bending and torsion. This is more likely to happen in rods such as the one used in the studied mechanism because of its slenderness. All this reduces the complexity of the equations, making more efficient algorithms without greatly increasing calculation errors.

The derivation of the mathematics structure of this section draw on Antman's work [13] (chs. 4 and 8) on nonlinear elasticity. Nevertheless, the nomenclature used here are those used in the robotics community, in the line of the work of Caleb [22].

The framework needed to describe the shape of a slender rod in space includes three main aspects that, coupled, generate the nonlinear system of differential equations that has to be solved in order to obtain the relationship between force and deformation. These are as follows: a kinematic definition of the deformed rod, material constitutive laws and static equilibrium equations.

2.1.1. Kinematics

A three-dimensional parametric Cartesian curve $\mathbf{p}(s) \in \mathbb{R}^3$ that links the centroids of each transverse section and the orthonormal rotation matrix $\mathbf{R}(s) \in \mathrm{SO}(3)$ (SO(3) is the special orthogonal subgroup in three dimensions, $\mathrm{SO}(3) = \{\mathbf{R} \in \mathbb{R}_{3x3} \mid \mathbf{R}^T\mathbf{R} = \mathbf{I},$ and $\det(\mathbf{R}) = 1\}$) that orients a local frame attached to the section are used to define the deformed shape of a rod. The principal axes of each section are usually named the x- and y-axes. The z-axis is perpendicular to the cross-section and tangent to the deformed shape; see Figure 1. A scalar reference arc-length parameter s is used to position and orient each cross-section. This parameter is within the finite interval $s \in [0, L]$, being that L is the length of the rod in the initial state. Due to the assumption that the sections are not distorted, the whole deformation of the flexible bar can be described by mapping from s to a homogeneous rigid-body transformation, $\mathbf{T}(s) \in \mathrm{SE}(3)$ (SE(3) is the special Euclidean subgroup in three dimensions, with $\mathbf{p}(s) \in \mathbb{R}^3$ and $\mathbf{R}(s) \in \mathrm{SO}(3)$).

$$\mathbf{T}(s) = \begin{bmatrix} \mathbf{R}(s) & \mathbf{p}(s) \\ \mathbf{0}^T & 1 \end{bmatrix} \tag{1}$$

where we will often refer to $\mathbf{T}(s)$ as a "frame" hereafter.

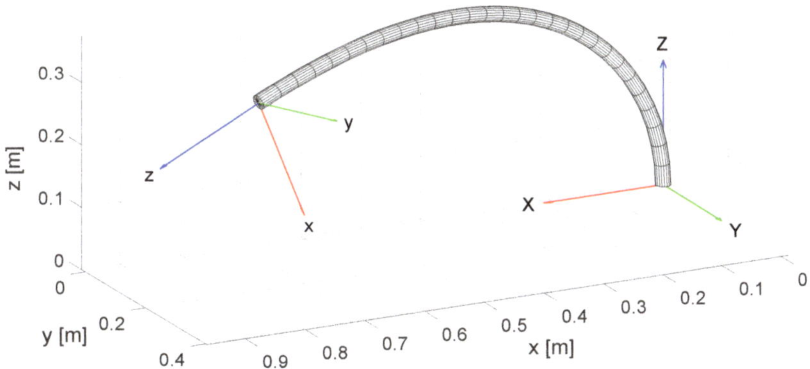

Figure 1. A rod deformed in space, along with frames used.

For the unloaded pre-curved state of the rod, the magnitudes described above are designated with the subscript $_\circ$.

A priori, $\mathbf{R}_\circ(s)$ could be assigned arbitrarily. However, one can establish conventions governing the assignment of reference orientations such that the mapping form $\mathbf{R}_\circ(s)$ to $\mathbf{R}(s)$ has an easily interpretable meaning in terms of material strains. As mentioned above, we have chosen to assign reference orientation such that the z axis of the local frame is parallel to the tangent curve to the reference curve (see Figure 1). Then, the following applies:

$$\mathbf{R}_\circ(s)\mathbf{e}_3 = \frac{\dfrac{d\mathbf{p}_\circ}{ds}}{\left\|\dfrac{d\mathbf{p}_\circ}{ds}\right\|} = \frac{\mathbf{p}_\circ'}{\|\mathbf{p}_\circ'\|} \tag{2}$$

where $\mathbf{e}_1$, $\mathbf{e}_2$ and $\mathbf{e}_3$ are used for the standard basis vectors of the local frame: $[1\ 0\ 0]^T$, $[0\ 1\ 0]^T$ and $[0\ 0\ 1]^T$, respectively. The superscript $'$ is used to indicate differentiation with respect to the arc length s.

Now, let us consider the derivatives of $\mathbf{p}(s)$ and $\mathbf{R}(s)$ with respect to s. The position and orientation evolve along the arc length according to the rates of change, linear $\mathbf{v}(s) \in \mathbb{R}^3$ and angular $\mathbf{u}(s) \in \mathbb{R}^3$. These rates of change are defined in the local frame, and obtained from the above mentioned derivatives, defined in the global frame, with the use of $\mathbf{R}(s)^T$ as follows:

$$\mathbf{v}(s) = \mathbf{R}(s)^T \frac{d\mathbf{p}(s)}{ds} = \mathbf{R}(s)^T \mathbf{p}'(s)$$

$$\mathbf{U}(s) = \mathbf{R}(s)^T \mathbf{R}'(s) \tag{3}$$

where $\mathbf{U}(s)$ is a skew-symmetric matrix of the following form:

$$\mathbf{U}(s) = \begin{bmatrix} 0 & -u_z(s) & u_y(s) \\ u_z(s) & 0 & -u_x(s) \\ -u_y(s) & u_x(s) & 0 \end{bmatrix} \tag{4}$$

Because $\mathbf{U}(s)$ is defined through three independent values, $u_x(s)$, $u_y(s)$ and $u_z(s)$, it can be expressed with a vector $\mathbf{u}(s) = \begin{bmatrix} u_x(s)\ u_y(s)\ u_z(s) \end{bmatrix}^T$ such that $\mathbf{u} = \mathbf{U}^\vee$ and $\mathbf{U} = \hat{\mathbf{u}}$, where the $^\vee$ operator denotes conversion of an element of $\mathfrak{so}(3)$ ($\mathfrak{so}(3)$ is the Lie algebra of SO(3)) to its corresponding element in $\mathbb{R}^3$. The inverse operation, denoted by $\hat{\ }$, maps $\mathbb{R}^3$ to $\mathfrak{so}(3)$, so that $\hat{\mathbf{u}}^\vee = \mathbf{u}$. In the following, $\hat{\mathbf{u}}(s)$ is used whenever a vector form is needed to express $\mathbf{U}$.

When $\mathbf{v}(s)$ and $\mathbf{u}(s)$ are those of the reference, i.e., the unloaded shape of the rod, they are designated as $\mathbf{v}_\circ$ and $\mathbf{u}_\circ$. Because of the convention of the rotation matrix stated in (2), an initially straight rod has $\mathbf{v}_\circ = [0\ 0\ 1]^T$ and $\mathbf{u}_\circ = \mathbf{0}$.

The material strain can be defined by the variation of $\mathbf{v}(s)$, $\Delta\mathbf{v}(s) = \mathbf{v}(s) - \mathbf{v}_\circ(s)$, and $\mathbf{u}(s)$, $\Delta\mathbf{u}(s) = \mathbf{u}(s) - \mathbf{u}_\circ(s)$. The local elongation is represented by Δv_z; values above 1 correspond to extension, values below 1 to compression and values equal to 1 mean that the length of the curve has not changed. The shear deformation along the local axes is depicted by Δv_x and Δv_y. The x and y components of $\Delta\mathbf{u}(s)$, Δu_x and Δu_y, measure the bending along the local x- and y-axes. Concerning Δu_z, it collects the value of the torsion.

2.1.2. Material Constitutive Relationships

The way the material behaves (it can be elastic or plastic, for instance) affects the constitutive relationships between strain and stress. However, the Cosserat rod model only is valid if the cross-sections do not deform.

$\Delta\mathbf{v}(s)$ and $\Delta\mathbf{u}(s)$ can be used to define the stress field since it is obtained from the strain field. Additionally, $\Delta\mathbf{v}(s)$ and $\Delta\mathbf{u}(s)$ can also be used to express internal force $\mathbf{n}(s)$ and moment $\mathbf{m}(s)$, due to the fact that they are calculated from the stress field.

Hereafter, the linear deformation $\mathbf{v}$ is related to internal forces by a linear constitutive law, which is also used to relate the angular deformation $\mathbf{u}$ to the internal moment. Diagonal stiffness matrices, i.e., $\mathbf{K}_{SE}(s)$ and $\mathbf{K}_{BT}(s)$, allow expressing the aforementioned magnitudes in the local frame. It is worth noting that the subscript SE indicates shear and extension, and BT, bending and torsion.

$$\mathbf{n}_{loc}(s) = \mathbf{K}_{SE}(s)\Delta\mathbf{v}(s) \tag{5}$$

$$\mathbf{m}_{loc}(s) = \mathbf{K}_{BT}(s)\Delta\mathbf{u}(s) \tag{6}$$

with

$$\mathbf{K}_{SE} = \begin{bmatrix} GA(s) & 0 & 0 \\ 0 & GA(s) & 0 \\ 0 & 0 & EA(s) \end{bmatrix} \tag{7}$$

$$\mathbf{K}_{BT} = \begin{bmatrix} EI_x(s) & 0 & 0 \\ 0 & EI_y(s) & 0 \\ 0 & 0 & GJ_z(s) \end{bmatrix} \tag{8}$$

where A is the area of cross section, E is Young's modulus, G is the shear modulus, I_x and I_y are the second moment of area of the rod cross section about the principal axes x and y, and J_z is the polar moment about the local axis z. The linear constitutive law applied here is commonly used in mechanics for the small strain range. This happens for most metals and for polymers in the range of small strain.

The rotation matrix is used to change $\mathbf{n}(s)$ and $\mathbf{m}(s)$ from the local axes to the fixed global axes.

$$\mathbf{n}(s) = \mathbf{R}(s)\mathbf{K}_{SE}(s)\Delta\mathbf{v} \tag{9}$$

$$\mathbf{m}(s) = \mathbf{R}(s)\mathbf{K}_{BT}(s)\Delta\mathbf{u} \tag{10}$$

2.1.3. Static Equilibrium Equations

The forces and moments on an infinitesimal piece of the rod, $[s, s + ds]$, are considered to obtain the static equilibrium equations. The internal force and moment exerted on the piece by the material of $[0, s)$ are named $-\mathbf{n}(s)$ and $-\mathbf{m}(s)$. On the other side of the portion, the material of $(s + ds, L]$ exerts a force $\mathbf{n}(s + ds)$ and a moment $\mathbf{m}(s + ds)$. In addition, distributed forces $\mathbf{f}(s)$ and moments $\mathbf{l}(s)$ can also be applied on the interval ds.

On the one hand, the equilibrium of forces yields the following:

$$-\mathbf{n}(s) + \mathbf{n}(s + ds) + \mathbf{f}(s)ds = \mathbf{0} \tag{11}$$

Expanding the magnitudes in $s + ds$ with the Taylor series until the element of first order, simplifying terms, and dividing by ds gives the following:

$$\frac{d\mathbf{n}(s)}{ds} + \mathbf{f}(s) = \mathbf{0} \tag{12}$$

On the other hand, a balance of moments, expanding variables in $s + ds$ with the Taylor series until the element of first order, simplifying terms, and taking into account (12), yields the following:

$$\frac{d\mathbf{m}(s)}{ds} + \frac{d\mathbf{p}(s)}{ds} \times \mathbf{n} + \mathbf{l}(s) = \mathbf{0} \tag{13}$$

So, the nonlinear ordinary differential equations for the equilibrium of a Cosserat rod describing the evolution of the internal force $\mathbf{n}(s)$ and moment $\mathbf{m}(s)$ along arc length are as follows:

$$\mathbf{n}'(s) + \mathbf{f}(s) = \mathbf{0} \tag{14}$$

$$\mathbf{m}'(s) + \mathbf{p}'(s) \times \mathbf{n}(s) + \mathbf{l}(s) = \mathbf{0} \tag{15}$$

2.1.4. Static Model

If the kinematic, constitutive, and equilibrium laws are combined, a static model of the flexible rod can be achieved.

Thus, the Cosserat static model for a flexible rod can be obtained using $\mathbf{m}$ and $\mathbf{n}$ as state variables, and combining (3), (14) and (15) to obtain a set of differential equations. If it is considered that the rod is deformed only by bending and torsion, the Cosserat rod model is simplified, leading to the Kirchhoff model. In the reference rod's configuration, shear and extension deformations do not usually appear, i.e., $\mathbf{v} = \mathbf{v}_\circ = \mathbf{e}_3$, where $\mathbf{e}_3$ is a unit vector tangent to the deformed shape. In this way, a new system of differential equations can be stated:

$$
\begin{aligned}
\mathbf{p}' &= \mathbf{Re}_3 \\
\mathbf{R}' &= \mathbf{R}\hat{\mathbf{u}} \\
\mathbf{n}' &= -\mathbf{f} \\
\mathbf{m}' &= -\mathbf{p}' \times \mathbf{n} - \mathbf{l}
\end{aligned}
\tag{16}
$$

where $\mathbf{u} = \mathbf{K}_{BT}^{-1}\mathbf{R}^T\mathbf{m} + \mathbf{u}_\circ$.

2.2. Kirchhoff Static Model Equations for the Planar Case

The three-dimensional deformation of a rod in space is defined by (16). However, if the rod only deforms within a plane, the model can be further simplified and a shorter formulation is obtained. This has an impact on the computational cost when the resolution method is implemented within an algorithm.

The rod that appears in Figure 2 deforms within the plane OXY. The local frame attached to each cross-section is oriented accordingly, so note that expressions in the previous chapter vary slightly now.

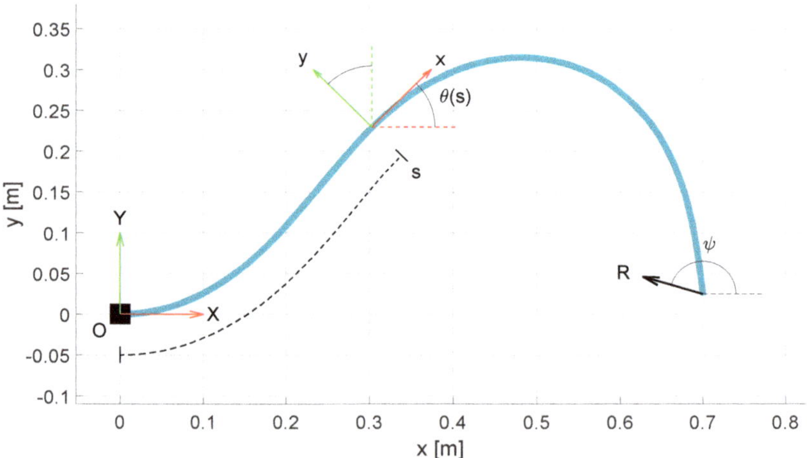

Figure 2. Deformed shape of a clamped-hinged planar rod under load R at extreme.

To orientate each cross-section the rotation matrix is used and it depends on the angle $\theta = \theta(s)$:

$$
\mathbf{R}(s) = \begin{bmatrix} \cos\theta & -\sin\theta & 0 \\ \sin\theta & \cos\theta & 0 \\ 0 & 0 & 1 \end{bmatrix}
\tag{17}
$$

The derivative of $\mathbf{R}$ with respect to arc length s is as follows:

$$\mathbf{R}'(s) = \begin{bmatrix} -\sin\theta & -\cos\theta & 0 \\ \cos\theta & -\sin\theta & 0 \\ 0 & 0 & 0 \end{bmatrix}\theta' \tag{18}$$

and allows finding vector $\mathbf{u}$ in the following:

$$\mathbf{U} = \mathbf{R}^T\mathbf{R}' = \begin{bmatrix} 0 & -\theta' & 0 \\ \theta' & 0 & 0 \\ 0 & 0 & 0 \end{bmatrix} = \begin{bmatrix} 0 & -u_z & 0 \\ u_z & 0 & 0 \\ 0 & 0 & 0 \end{bmatrix} \tag{19}$$

Thus, only the component u_z of vector $\mathbf{u}$ is not null:

$$u_z = \theta' \tag{20}$$

This makes sense, as there is no bending moment with respect to the local y-axis and no torsion (expressed through u_x).

Taking (20) into (16) considering this time $\mathbf{v} = \mathbf{v}_\circ = \mathbf{e}_1$, and taking $\mathbf{n}$ and $\mathbf{m}$ as state variables, the expression of the system of nonlinear ordinary differential equations in (16) yields the following:

$$\begin{Bmatrix} \frac{dx}{ds} \\ \frac{dy}{ds} \\ \frac{d\theta}{ds} \\ \frac{dn_x}{ds} \\ \frac{dn_y}{ds} \\ \frac{dm_z}{ds} \end{Bmatrix} = \begin{Bmatrix} \cos\theta \\ \sin\theta \\ \frac{m_z}{EI} + u_{o,z} \\ -f_x \\ -f_y \\ n_x\sin\theta - n_y\cos\theta - l_z \end{Bmatrix} \tag{21}$$

If no distributed force and moment ($\mathbf{f}$ and $\mathbf{l}$) are applied along the rod, we obtain the following:

$$\begin{Bmatrix} \frac{dx}{ds} \\ \frac{dy}{ds} \\ \frac{d\theta}{ds} \\ \frac{dn_x}{ds} \\ \frac{dn_y}{ds} \\ \frac{dm_z}{ds} \end{Bmatrix} = \begin{Bmatrix} \cos\theta \\ \sin\theta \\ \frac{m_z}{EI} \\ 0 \\ 0 \\ n_x\sin\theta - n_y\cos\theta \end{Bmatrix} \tag{22}$$

Hence, in this particular case, the internal force $\mathbf{n}$ is constant. Such a rod with no distributed loads is quite common in the applications.

The system of differential equations in (22) can be solved by using a direct integration method, such as the 4th order R-K, or a method based on elliptic integrals.

2.3. Direct Integration Using Fourth-Order Runge–Kutta Method

Typical problems in robotics are the forward kinematic problem (FKP) and the inverse kinematic problem (IKP). In the first one, the mechanism inputs are known, and the position and orientation of the end-effector are unknown. In contrast, in the IKP, the end-effector is wanted to describe a given trajectory, so its position is known, but the needed inputs to accomplish the task must be calculated.

When a flexible rod is part of a mechanism, usually, it is under external loads coming from that mechanism. Here, the aim should be to calculate how these loads deform the rod. So, it could be considered as a kind of forward kinematic (FK) position problem because the position and orientation of the extreme in which force and moment are applied are not known.

So focusing on the FKP of a rod that is clamped at the proximal end $s = 0$ and subjected to a known force $\mathbf{n_{ext}}$ and moment $\mathbf{m_{ext}}$ at the distal end $s = L$ (see Figure 2), the rod undergoes a deformation that depends on these boundary conditions and the geometric and mechanical characterization of the rod.

If the effect of gravity is not taken into account, there is not distributed force along the length. Therefore, the shape and the load status of the rod are defined by (22).

This takes the form of a BVP, where the independent variable is s in $[0, L]$, and the vector of dependent variables, $\mathbf{y}$, and the non-linear function, $\mathbf{f}(s, \mathbf{y})$, are the following:

$$\mathbf{y} = \left\{ \begin{array}{c} x \\ y \\ \theta \\ n_x \\ n_y \\ m_z \end{array} \right\} \quad , \quad \mathbf{f}(s, \mathbf{y}) = \left\{ \begin{array}{c} \cos\theta \\ \sin\theta \\ \frac{m_z}{EI} \\ 0 \\ 0 \\ n_x \sin\theta - n_y \cos\theta \end{array} \right\} \tag{23}$$

The boundary conditions are at $s = 0$ of the kinematic kind, and at $s = L$ of the load kind. Regarding the kinematic variables, the position and orientation at $s = 0$ are data, and unknown at $s = L$. In relation to internal forces, and because no distributed force is considered, its value does not change along the length of the rod. At $s = 0$, and because of the static equilibrium, the internal force takes the same value that of the external load, $\mathbf{n_{ext}}$, at $s = L$, but in opposite direction (although the internal forces could be taken out of the integration, in this work, they are included because they are needed to solve the FKP of the whole mechanism). With respect to the internal moment, the value is a datum at $s = L$ (in fact, $\mathbf{m}(L) = \mathbf{m}_{ext}$), and unknown at $s = 0$.

Then, $\mathbf{y}$ at both ends are the following:

$$\mathbf{y}(0) = \left\{ \begin{array}{c} x(0) \\ y(0) \\ \theta(0) \\ n_x(0) \\ n_y(0) \\ m_z(0) \end{array} \right\} = \left\{ \begin{array}{c} x_0 \\ y_0 \\ \theta_0 \\ ? \\ ? \\ ? \end{array} \right\} \quad , \quad \mathbf{y}(L) = \left\{ \begin{array}{c} x(L) \\ y(L) \\ \theta(L) \\ n_x(L) \\ n_y(L) \\ m_z(L) \end{array} \right\} = \left\{ \begin{array}{c} ? \\ ? \\ ? \\ n_{ext,x} \\ n_{ext,y} \\ m_{ext} \end{array} \right\} \tag{24}$$

being Dirichlet conditions.

Since there are three unknowns at the initial of the range of integration $s = 0$, three guess values (one for each variable) are needed: $\mathbf{guess} = \left[n^x_{guess} \; n^y_{guess} \; m^z_{guess} \right]^T$. Once these values are chosen, the initial value problem is solved using the 4th R-K method.

$$\frac{d\mathbf{y}}{ds} = \mathbf{f}(s, \mathbf{y}) \quad , \quad \mathbf{y}(s_0) = \mathbf{y}_0 \tag{25}$$

where $\mathbf{y}$ is a vector that contains the unknown variables at each s, function $\mathbf{f}$ is given, s is the independent variable, and subscript $_0$ indicates values at initio.

2.4. Integration Using Elliptic Integrals

The system of differential equations in (22) can be further developed looking for an analytical integration. If only end-point load is applied (expressed in terms of magnitude R and orientation ψ as in Figure 2), we can express the internal force as the following:

$$n_x = R \cos\psi \tag{26}$$

$$n_y = R \sin\psi \tag{27}$$

For constant E and I, with a stress-free reference straight rod, considering the well-known Bernoulli–Euler law, and upon substitution of (26) and (27) into (22), we obtain the following:

$$\frac{d^2\theta}{d^2s} = \frac{1}{EI}(R\cos\psi\sin\theta - R\sin\psi\cos\theta) = \frac{R}{EI}\sin(\theta - \psi) \tag{28}$$

Therefore, the system of differential Equation (22) can be simplified to the following:

$$\left\{\begin{array}{c} \frac{dx}{ds} \\ \frac{dy}{ds} \\ \frac{d^2\theta}{d^2s} \end{array}\right\} = \left\{\begin{array}{c} \cos\theta \\ \sin\theta \\ \frac{R}{EI}\sin(\theta - \psi) \end{array}\right\} \tag{29}$$

Elliptic integrals is the classical mathematical tool to solve (29) because of its rapid computation. The following equations are used in this work to solve for the coordinates of an arbitrary point along the beam, and they are found in [21] (ch. 4).

$$x(\phi_i) = -\sqrt{\frac{EI}{R}}\cos\psi[2E(k,\phi_i) - 2E(k,\phi_1) - F(k,\phi_i) + F(k,\phi_1)]$$
$$- \sqrt{\frac{EI}{R}}2k\sin\psi[\cos\phi_i - \cos\phi_1] \tag{30}$$

$$y(\phi_i) = -\sqrt{\frac{EI}{R}}\sin\psi[2E(k,\phi_i) - 2E(k,\phi_1) - F(k,\phi_i) + F(k,\phi_1)]$$
$$+ \sqrt{\frac{EI}{R}}2k\cos\psi[\cos\phi_i - \cos\phi_1] \tag{31}$$

where the functions $F(k,\phi)$ and $E(k,\phi)$ are the incomplete elliptic integrals of the first and second kind, respectively. Moreover, the following holds:

$$\sqrt{\frac{EI}{R}} = \frac{L}{[F(k,\phi_2) - F(k,\phi_1)]} \tag{32}$$

Thus, the value of the end-tip force R in terms of k and ϕ can be obtained from (32) as:

$$\sqrt{R} = \frac{\sqrt{EI}}{L}[F(k,\phi_2) - F(k,\phi_1)] \tag{33}$$

In addition, the bending moments along the rod are given by the following:

$$m_i = 2k\sqrt{REI}\cos\phi_i \tag{34}$$

The functions $F(k,\phi)$ and $E(k,\phi)$ depend on two parameters:

- **The non-dimensional parameter k**
  It is known as the **modulus of the function** and it can vary between -1 and 1. In this application, k corresponds roughly but non-linearly to the magnitude of the force R.
- **The variable ϕ**
  It is called the **amplitude of the elliptic integral**. It is measured in radians, and it varies continuously along the beam from ϕ_1 on the left edge to ϕ_2 on the right. These amplitudes, ϕ_1 and ϕ_2, are defined by the boundary conditions of the rod extremes. The variable ϕ is related to the beam angle θ by the following relation:

$$k\sin(\phi) = \cos\left(\frac{\psi - \theta}{2}\right) \tag{35}$$

Hence, (29) is integrated to yield (30), (31), and (33), i.e., a parametric system that uniquely defines a deformed rod in terms of parameters k and ψ for **known boundary conditions** and a **given buckling mode**.

2.4.1. Boundary Conditions for a Clamped-Pinned Rod

The studied mechanism in this work has a clamped-pinned rod (as in Figure 2) that substitutes one of the rigid bars of the four-bar linkage.

The first extreme clamped implies a given slope θ_1 for the rod at $s = 0$; hence, using (35), we can state the following:

$$\phi_1 = \arcsin \frac{1}{k} \cos \left(\frac{\psi - \theta_1}{2} \right) \tag{36}$$

A pinned end-tip implies that no moment is acting at that point. Upon application on (34), we obtain the following:

$$0 = 2k\sqrt{REI} \cos \phi_2 \tag{37}$$

This means that the end-tip is a point of null curvature, then an inflection point of the deflected curve. Additionally, the second part of the equation implies the following:

$$\phi_2 = (2q - 1)\pi/2 \quad \text{for } q = 1,2,\dots \tag{38}$$

being that q is the order of the Buckling Mode shape.

Note that feasible values of k may depend on ψ as, for example, is the case of the limiting value of $k_{min} = ||\cos \left(\frac{\psi - \theta_1}{2} \right)||$ that is necessary to produce real solutions in (36). This condition is indicated with the gridded area in Figure 3a. Therefore k is in the range $k = (k_{min}, 1)$ or $k = (-1, -k_{min})$. Provided that range for k, ϕ_1 will have values in $[-\pi/2, \pi/2]$. On the other hand, ϕ_2 will have discrete values $\pi/2$, $3\pi/2$, and so on, for buckling modes $1, 2 \cdots$, respectively. The variable ϕ will vary continuously from ϕ_1 to ϕ_2. As shown in Figure 3, if we set a duple of values to k and ψ, inside the feasible range (Figure 3a), we will obtain the deformed shape of the rod for every buckling mode chosen (Figure 3b).

From a practical point of view, the data will come from physical quantities that should be used to find those parameters k and ψ. For example, if end-tip force is given, R and ψ, along with boundary conditions, (33) should be used to find k, and then the deformed shape with (30) and (31). If the end-tip position, X_P and Y_P, is given for some boundary conditions, again the objective will be to find the corresponding parameters k and ψ from (30) and (31). In most occasions, this process is numerical.

The different problems that can be stated regarding the shape of a rod accomplishing given conditions may produce a multiplicity of solutions for each buckling mode considered. For clamped-pinned bars those multiple solutions do not have to be symmetrical, but refer to completely different configurations. For example, given a same end-point position, we can obtain different solutions for given buckling modes as shown in Figure 4.

2.4.2. Clamped-Pinned Rod's Workspace Analysis by Modes

The rod's end-tip position can be obtained upon substitution of ϕ_2 from (38) into (30) and (31). These coordinates can be calculated for a range of values of $\psi = [0, 2\pi]$ and $k = (-1, -k_{min})$ or $k = (k_{min}, 1)$. This calculation results in a set of curves that is the workspace of the rod's end-tip, for positive values of k (Figure 5a) and negative values of k (Figure 5b). This workspace is obtained by calculating the rod's end-tip position for given ψ's and varying the value of k in the corresponding interval.

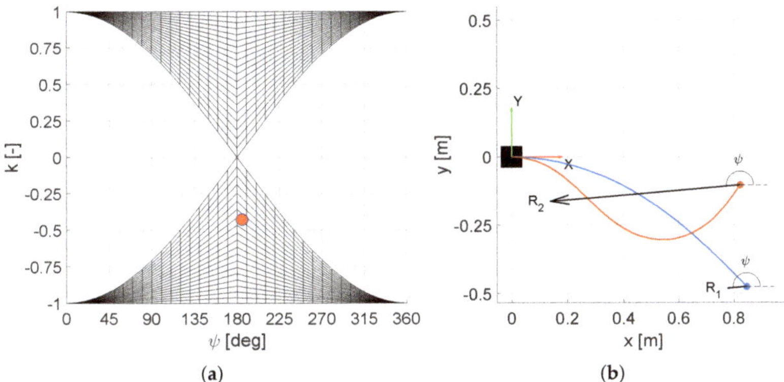

Figure 3. (**a**) Given values of k and ψ. (**b**) Deformed shapes for buckling modes 1 and 2 of a clamped-hinged rod.

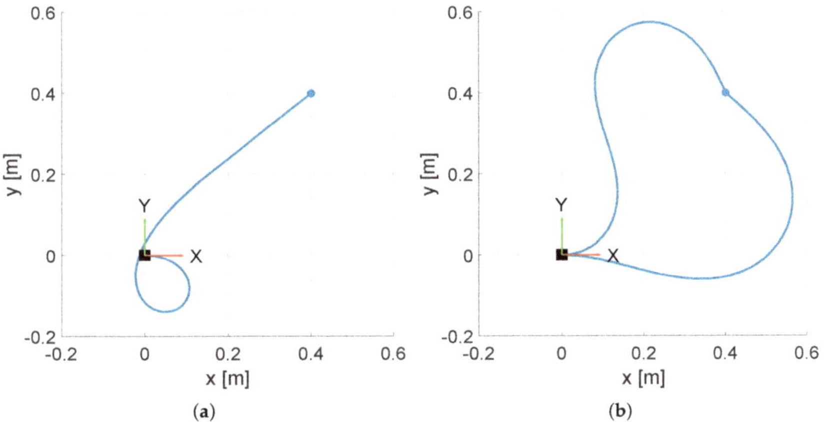

Figure 4. (**a**) Deformed shape for buckling mode 1. (**b**) Deformed shape for buckling mode 2. Both correspond to a clamped-pinned rod for a given end-tip position.

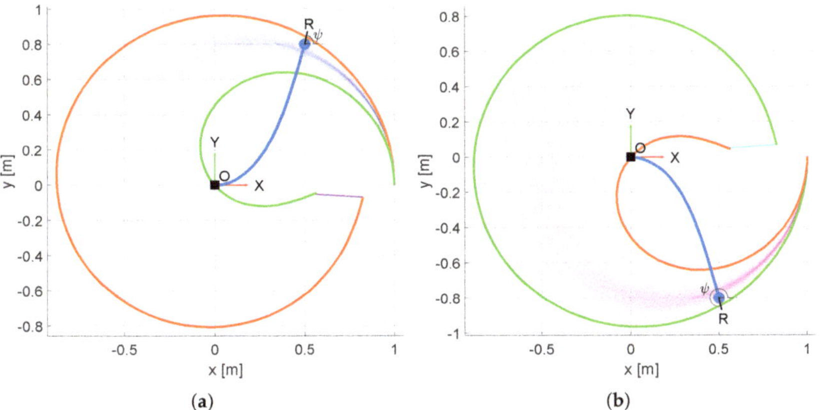

Figure 5. Rod's end-tip workspace for buckling mode 1. (**a**) Positive k. (**b**) Negative k.

The curves plotted in Figure 5 were calculated for the first buckling mode of the rod. To obtain the workspace for other buckling modes, it is necessary to execute the calculation again, changing ϕ_2 in (38), for the desired mode. For instance, Figure 6a shows the workspace for positive k and buckling mode 2 of the rod, and Figure 6b represents that for negative k and the same mode.

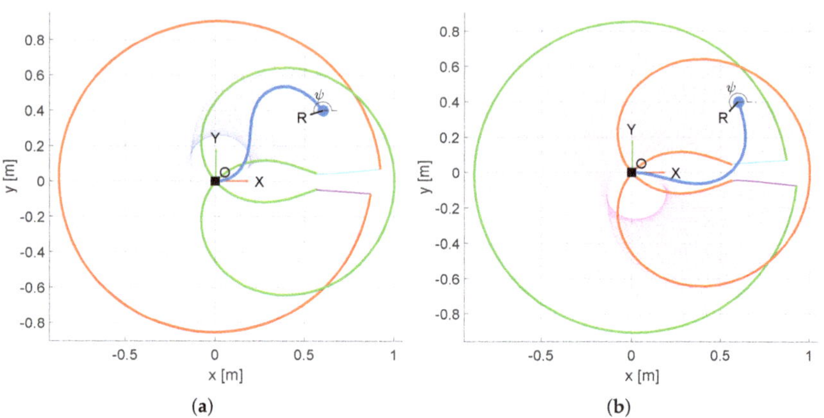

(a)

(b)

Figure 6. Rod's end-tip workspace for buckling mode 2. (**a**) Positive k. (**b**) Negative k.

As for the orange lines in Figures 5 and 6, they come from the calculation of the rod's end-tip position for the higher limit of the range of k and varying ψ. The green lines are calculated with the lower limit of the interval of k and varying ψ. To close the workspace area, the purple lines are plotted. They are related to the calculation with the higher limit of the interval of ψ and varying k. In contrast, the cyan lines derive from the computation with the lower limit of the interval of ψ and varying k.

3. Closed-Loop Hybrid Mechanisms

Once the fundamentals of elastic rods are understood, this type of element can be used in conjunction with rigid bars to create new hybrid mechanisms. These mechanisms are characterized by combining rigid and flexible bars to perform a task, such as describing a trajectory or performing a given force. In this paper, one of the mechanisms proposed in [21] (ch. 12), the four-bar linkage with one flexible clamped-pinned rod (see Figure 7), is the subject of study.

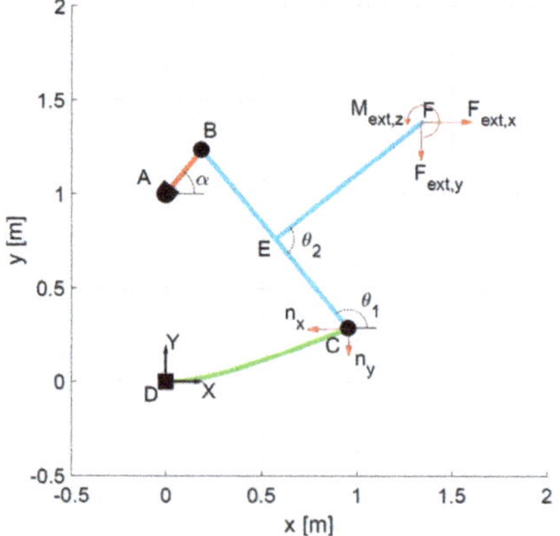

Figure 7. Analysed mechanism in an arbitrary position.

The dimensional data and the description of each bar are given in Table 1. Each angle showed in Figure 7 is explained in Table 2.

Table 1. Dimensional data of the analyzed mechanism.

Segment	Length (m)	Description
$\overline{AB}$	0.3	Input Link (Crank)
$\overline{BC}$	1.2207	Coupler
$\overline{CD}$	1	Output Link (Flexible clamped-pinned rod)
$\overline{BE}$	0.6103	Distance to place the end-effector bar
$\overline{EF}$	1	End-effector bar

Table 2. Angles data of the analyzed mechanism.

Angle	Description
α	Input angle, which positions the crank
θ_1	Orientation of the rigid coupler $\overline{BC}$
θ_2	Relative angle between $\overline{BC}$ and $\overline{EF}$ segments. It has a constant value of 90° through the cycle.

The flexible segment $\overline{CD}$ is a slender rod, made of nitinol, with a circular cross-section; its properties are shown in Table 3.

Table 3. Properties of the flexible rod.

Ø (m)	E (Pa)	I (m^4)
3×10^{-3}	63×10^{9}	3.9761×10^{-12}

3.1. General Equations for the 4-Bar Linkage

The problem to solve is the FKP since the inputs to the mechanism, i.e., the α angle and the external forces, $\mathbf{F_{ext}}$, and moment $\mathbf{M_{ext}}$, are known, but the position and orientation of the coupler $\overline{BC}$ and the shape of the flexible rod $\overline{CD}$ are unknown.

A **shooting method** that combines the **direct integration with a 4th order R-K method** (explained in Section 2.3) and a **N-R scheme** is used.

The resolution begins from a known position of the mechanism for an α angle of 0 degrees.

In the next step, a given α is introduced in the calculation. Thus, the crank $\overline{AB}$ rotates and the coordinates of B point can be calculated. Then, the coordinates of point C must be obtained to position the coupler $\overline{BC}$. However, this calculus requires a deeper analysis because the rod $\overline{CD}$ deforms, so one needs to integrate (22) to obtain the position of C point.

Here, the SM is used to solve the position of the C point and, thus, that of the mechanism, for each step. This method is focused on the minimization, with the aid of the N-R scheme, of a **residue** function that represents all the constraints of the mechanism, and is given by (39).

$$\text{residue} = \left\{ \begin{array}{c} m^*_{z,C} \\ \sum M_z \big|_B \\ \sqrt{(x^*_C - x_B)^2 + (y^*_C - y_B)^2} \end{array} \right\} - \left\{ \begin{array}{c} 0 \\ 0 \\ |\overline{BC}| \end{array} \right\} \qquad (39)$$

where $m^*_{z,C}$ is the bending moment at C, $\sum M_z \big|_B$ the set of all moments about B produced by the external loads, and the superscript * indicates that the variables are not the exact solutions; instead, they are the approximate solutions obtained numerically in each step of the SM.

The first two components of (39) are restrictions of the loads kind, and state that the moments about points B and C must be null because in those points there are hinges. The other restriction is of the geometric kind and imposes that the distance between points B and C is the length of the rigid coupler $\overline{BC}$.

To find the solution of the mechanism's position, the SM needs two inputs: the **residue** function (39) and a vector of guess values, $\mathbf{guess} = \begin{bmatrix} n^x_{guess} & n^y_{guess} & m^z_{guess} \end{bmatrix}^T$, to start the integration of the rod $\overline{CD}$. These guess values are needed to solve the BVP shown in Section 2.3. The 4th order R-K method explained in that section can be particularized for the case of the rod $\overline{CD}$ if $\mathbf{m_{ext}}$ is equal to 0 since there is a hinge in the point C.

For the first α step, these guess variables can take arbitrary values. Despite this, previous calculations were made in this work to obtain guess values that improve the convergence of the SM. For the next α steps, the vector of guess values is composed by the $\begin{bmatrix} n_x & n_y & m_z \end{bmatrix}^T$ values of the solution of the previous α step.

With these two inputs to the SM, the process to obtain the position of point C and the shaped of rod $\overline{CD}$ is as follows:

1. For each **guess**, apply 4th order R-K to obtain $\mathbf{y}^* = \begin{bmatrix} x^*_C & y^*_C & \theta^*_C & n^*_{x,C} & n^*_{y,C} & m^*_{z,C} \end{bmatrix}^T$; see Section 2.3.
2. Evaluate the residue (39). To do this, x^*_C and y^*_C should be used to calculate θ_1, and θ_1, $n^*_{x,C}$ and $n^*_{y,C}$ to solve $\sum M_z \big|_B$. $m^*_{z,C}$ from the direct integration of the rod $\overline{CD}$ is directly used to compare its value with 0.
3. If all terms are below a *tolerance*, exit: the final solution is obtained.
4. If not, calculate variation of **residue** with **guess**: $\mathbf{J} = \frac{d\,\mathbf{residue}}{d\,\mathbf{guess}}$.
5. Modify the guess values, $\mathbf{guess}^{i+1} = \mathbf{guess}^i - \mathbf{J}^{-1}\mathbf{residue}$.
6. Start the loop again.

The results are obtained when the three components of the residue are below a *tolerance*. With this, the position of the mechanism is obtained for the current α step, and the results are the guess values for the next α step.

3.2. Load-Dependent Paths

Applying the aforementioned SM, the trajectory of point F can be calculated, but it depends on external loads because these affect the deformed shape of the flexible rod $\overline{CD}$. Depending on the load applied, point C takes different positions and, in consequence, the position and the orientation of the coupler $\overline{BC}$ varies. This influence of the loads in the path described by point F can be seen in Figure 8 where a set of trajectories for different load cases is plotted.

If these trajectories are plotted as a function of the moment that the external forces and moments generate in point B, it results in Figure 9. In Figure 9a, the trajectories plotted for constant external moments applied to the coupler with no external forces result in level curves. However, if external forces are applied, the relative position between points B and F varies during the movement, so the moment of external loading about B takes different values. For this reason, the trajectories produced are not flat when forces are applied. Despite this, they belong to the same three-dimensional surface that contains the previous level set in Figure 9a, as can be seen in Figure 9b.

Hence, by applying this method to solve the FKP of mechanisms of this kind, it is possible to obtain the trajectories of the end-effector points. Moreover, this method predicts the three-dimensional surface in which the paths under different load cases are.

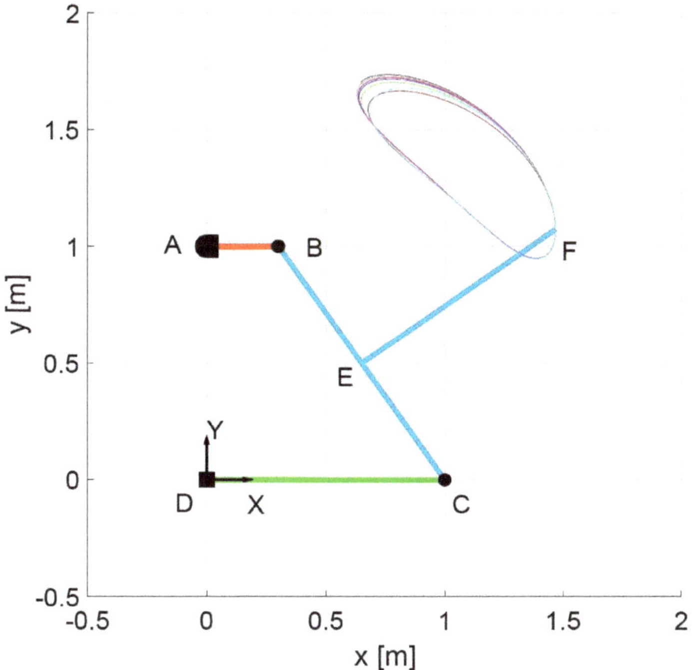

Figure 8. Trajectory of point *F* after a complete turn of the crank $\overline{AB}$ and under different load cases.

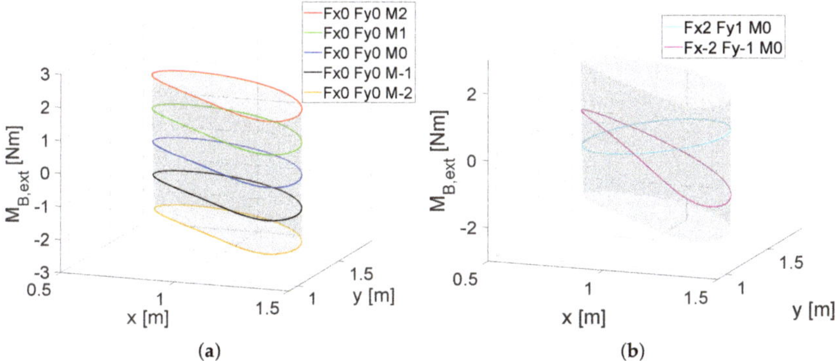

Figure 9. Trajectories for different load cases. (**a**) External constant moments. (**b**) External constant forces.

3.3. Buckling Mode Change

Next, if the SM is combined with the integration using EI, explained in Section 2.4, a powerful algorithm that solves the FKP and shows the workspace of the flexible rod $\overline{CD}$ is obtained. In this way, it is possible to know in which buckling mode rod $\overline{CD}$ is working at each instant of the work cycle. The process should follow these steps:

1. Integrate the deformed shape using the SM explained in Section 3.1. By doing this, the distribution of internal moments along the bar is obtained.
2. Search for sign changes of the internal moments along the bar. Each sign change means a point of curvature change. The number of points of changes of the curvature corresponds to the buckling mode of the flexible rod. For instance, if there are two curvature changes, the rod is in the second buckling mode. It must be borne in mind

that, in the case of this mechanism, there is always a point of null moment, i.e., null curvature, in the distal end of the slender rod $\overline{CD}$ because of the hinge.

3. Once the buckling mode is known, solve the workspace of the rod $\overline{CD}$, applying the concepts of Section 2.4.2, and plot it.

An example of the application of this algorithm is shown in Figures 10–15 (an animation is attached in the Supplementary Materials: Video S1), in which a complete turn of the crank $\overline{AB}$ is simulated while an external horizontal force of 5 N at point F is applied. Taking into account that the points of null moment of the rod $\overline{CD}$ are plotted in red, the flexible rod $\overline{CD}$ is working in its first buckling mode when the cycle starts (see Figure 10). Moreover, k parameter of the EI that allows calculating the workspace is positive. The rod $\overline{CD}$ is working in this mode (see Figure 11) until it reaches an α of 159 degrees, where its buckling mode changes from 1 to 2 (see Figure 12). In addition, k also changes from positive values to negatives ones. The rod $\overline{CD}$ remains in its second buckling mode (see Figure 13) until the input α acquires a value of 285 degrees (see Figure 14). From this pose on, the rod $\overline{CD}$ works in its first buckling mode, with negatives k, until the crank $\overline{AB}$ completes the turn (see Figure 15).

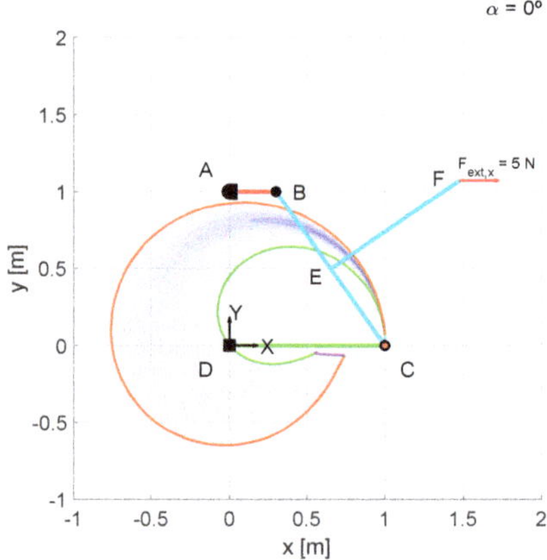

Figure 10. Cycle with buckling mode changes. Initial position.

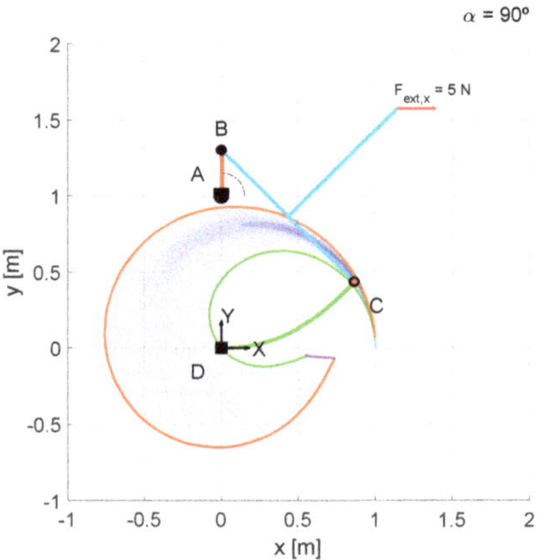

Figure 11. Cycle with buckling mode changes. $\alpha = 90°$.

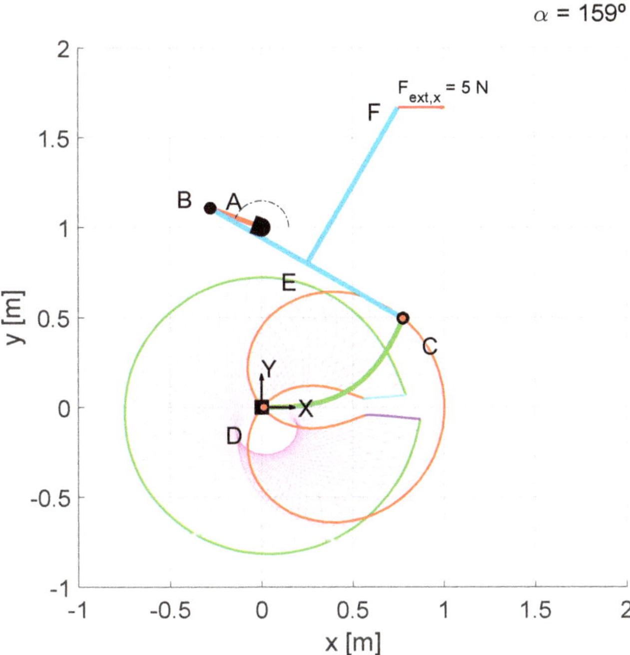

Figure 12. Cycle with buckling mode changes. Mode 1 to Mode 2 change.

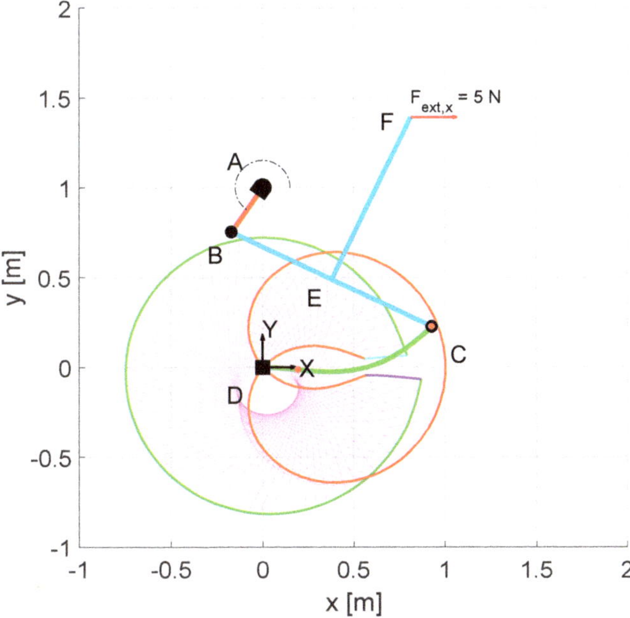

Figure 13. Cycle with buckling mode changes. $\alpha = 235°$.

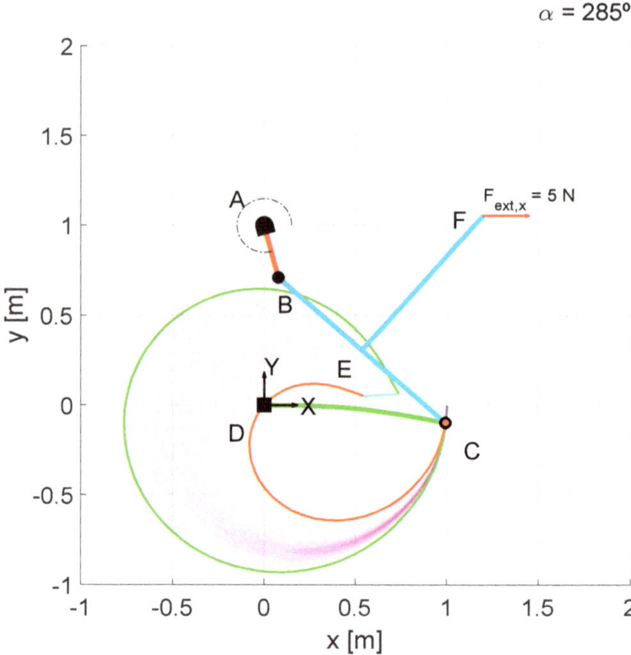

Figure 14. Cycle with buckling mode changes. Mode 2 to Mode 1 change.

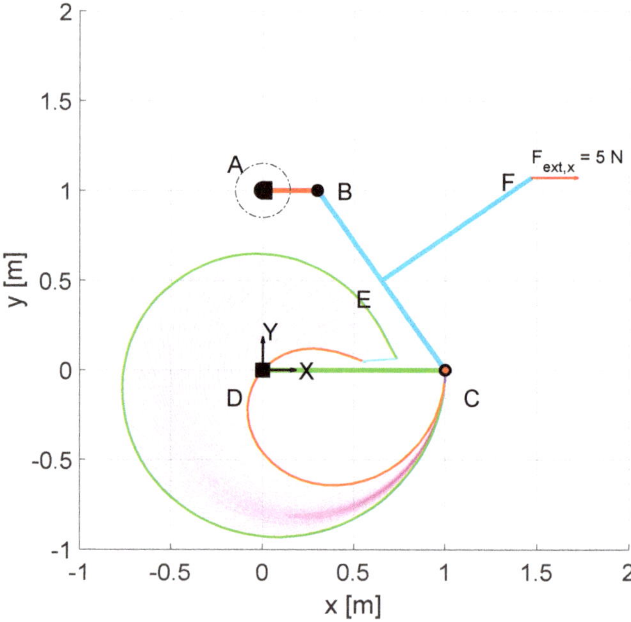

Figure 15. Cycle with buckling mode changes. Final position.

If the trajectory of point F is plotted against the moment of external loading about B for this load case, Figure 16 is obtained. It can be seen that the path is contained in the locus exposed on the previous Section 3.2.

It is worth noting that a cut appears on the surface for moments less than -2 Nm. The potential of this method to obtain the locus in which trajectories can lie for different load cases is highlighted here. Looking at the shadowed surface of Figure 16, it can be detected that load cases that generate moments about B less than -2 Nm cannot produce complete closed trajectories of the point F.

Thus, it is demonstrated that the SM proposed in Section 3.1 can detect and solve buckling mode changes as long as the trajectory of the end-effector point is contained in its corresponding locus.

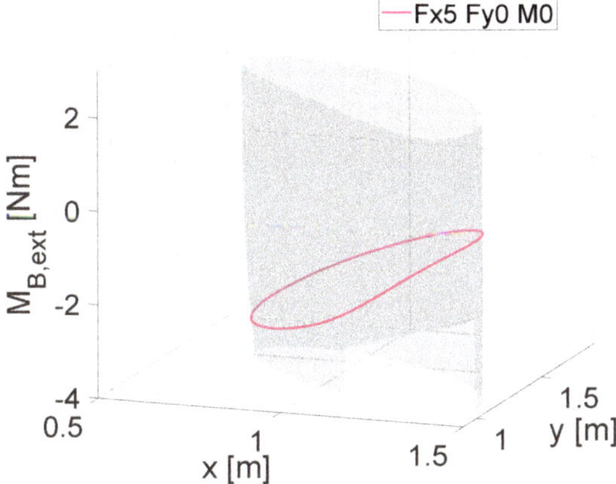

Figure 16. Trajectory for an external applied load of 5 N in the positive x-axis.

3.4. Multiplicity of Mechanism Circuits

Besides buckling mode changes, the method here developed also can find all mechanism circuits. The process to obtain all circuits is the following:

1. Integrate the deformed shape with a 4th order R-K method in points along a circumference with the center at point B and a radius equal to the length of bar $\overline{BC}$. In this way, all potential solutions of rod $\overline{CD}$ that satisfy geometric conditions of the mechanism are calculated. Once the forces in the distal end of the rod $\overline{CD}$ are known, continue with the next step.
2. Find the points of the circumference where there is a sign change in the summation of moments about B. These points are the potential circuits of the mechanism.
3. Apply a SM to refine the potential initial solutions. The function to minimize is (39).
4. Run the SM explained in Section 3.1 to solve the cycle for each calculated circuit.

An example of the application of this process is shown in Figures 17–22. For the case of the mechanism analyzed in this paper, three circuits appear when external forces of $F_{ext,x} = -4$ N and $F_{ext,y} = -4$ N, and a external moment of $M_{ext,z} = 4$ Nm are applied on point F. In the initial position of the first circuit, the rod $\overline{CD}$ is straight (Figure 17), and completes a turn without changing the buckling mode, but with a sign change of EI's parameter k. As for the second circuit, the rod $\overline{CD}$ starts and carries out the whole cycle in its second buckling mode (Figure 19). Finally, if the mechanism moves on its third circuit, the rod $\overline{CD}$ works in its first buckling mode (Figure 21). The complete trajectories for the first, second and third circuits are plotted in Figures 18, 20 and 22, respectively. In these figures, it can be seen that each circuit generates a specific trajectory.

For a better understanding, Videos S2–S4 are available as Supplementary Materials.

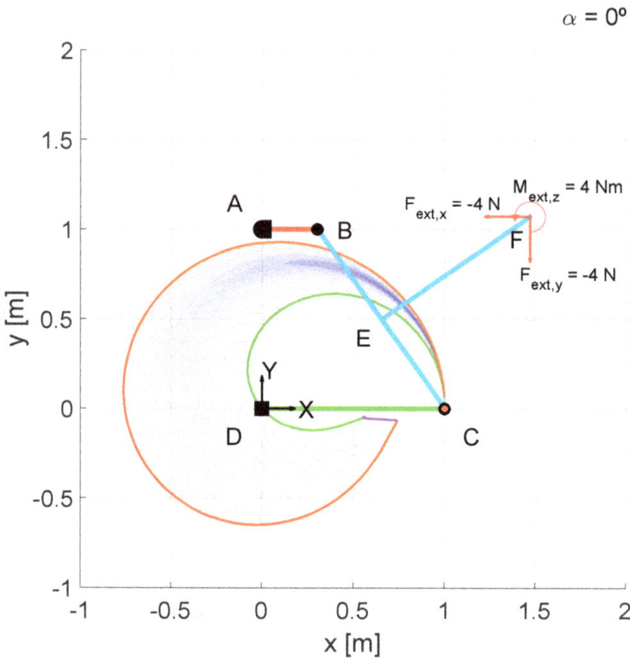

Figure 17. First mechanism circuit under external loading of $F_{ext,x} = -4$ N, $F_{ext,y} = -4$ N, and $M_{ext,z} = 4$ Nm. Initial position.

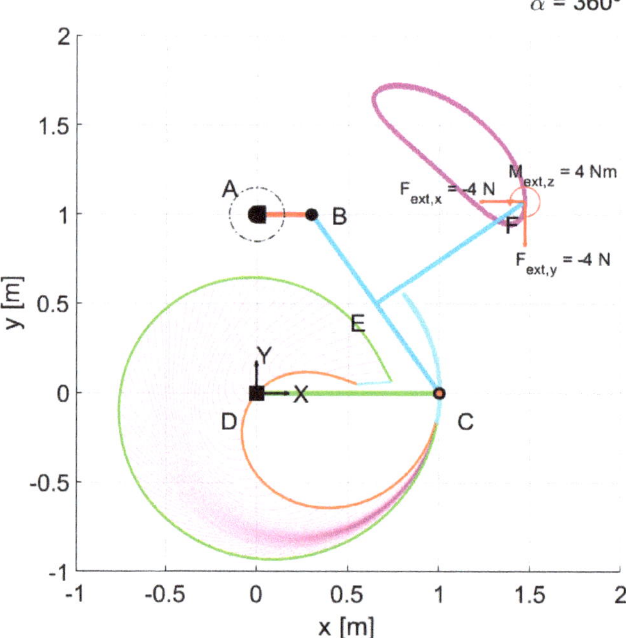

Figure 18. First mechanism circuit under external loading of $F_{ext,x} = -4$ N, $F_{ext,y} = -4$ N, and $M_{ext,z} = 4$ Nm. Final position and trajectory.

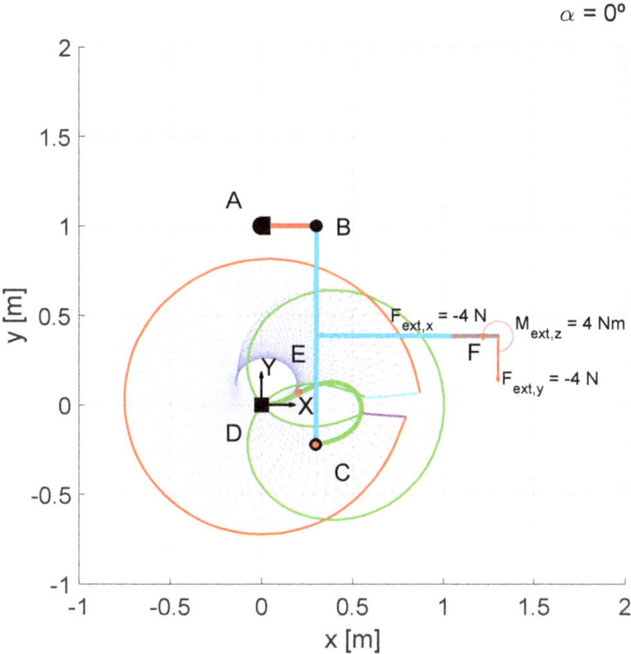

Figure 19. Second mechanism circuit under external loading of $F_{ext,x} = -4$ N, $F_{ext,y} = -4$ N, and $M_{ext,z} = 4$ Nm. Initial position.

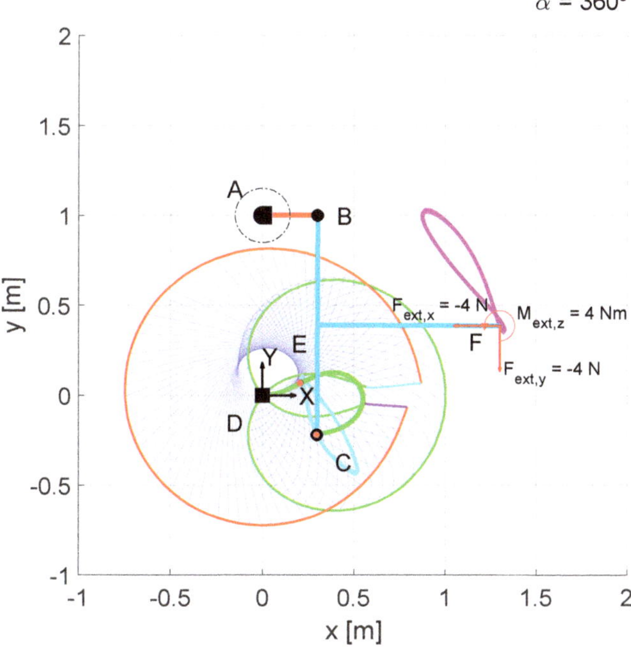

Figure 20. Second mechanism circuit under external loading of $F_{ext,x} = -4$ N, $F_{ext,y} = -4$ N, and $M_{ext,z} = 4$ Nm. Final position and trajectory.

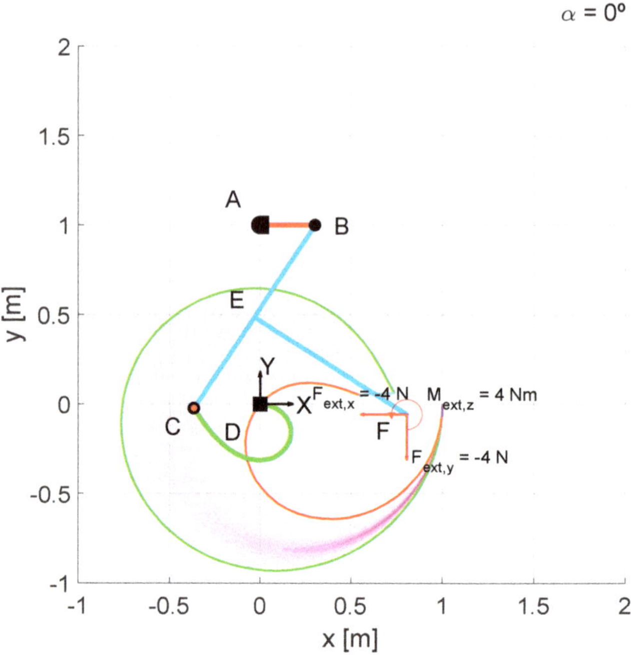

Figure 21. Third mechanism circuit under external loading of $F_{ext,x} = -4\,\text{N}$, $F_{ext,y} = -4\,\text{N}$, and $M_{ext,z} = 4\,\text{Nm}$. Initial position.

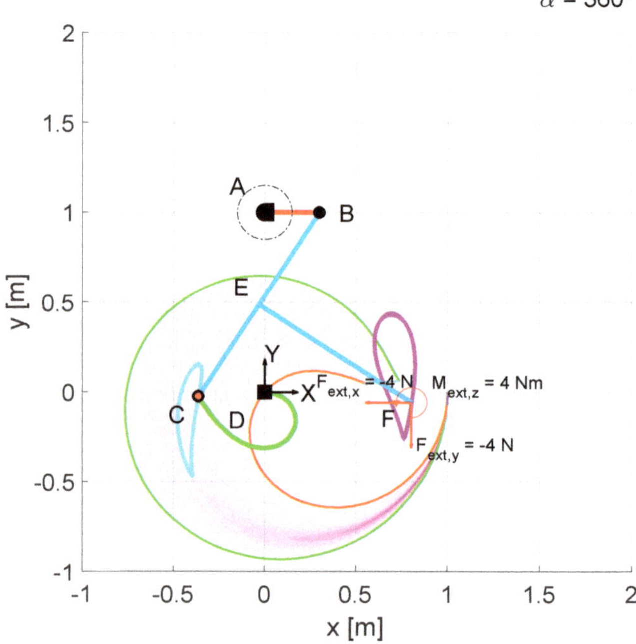

Figure 22. Third mechanism circuit under external loading of $F_{ext,x} = -4\,\text{N}$, $F_{ext,y} = -4\,\text{N}$, and $M_{ext,z} = 4\,\text{Nm}$. Final position and trajectory.

3.5. Algorithm's Flowchart

To summarize all steps of the entire algorithm, its flowchart is given in Figure 23.

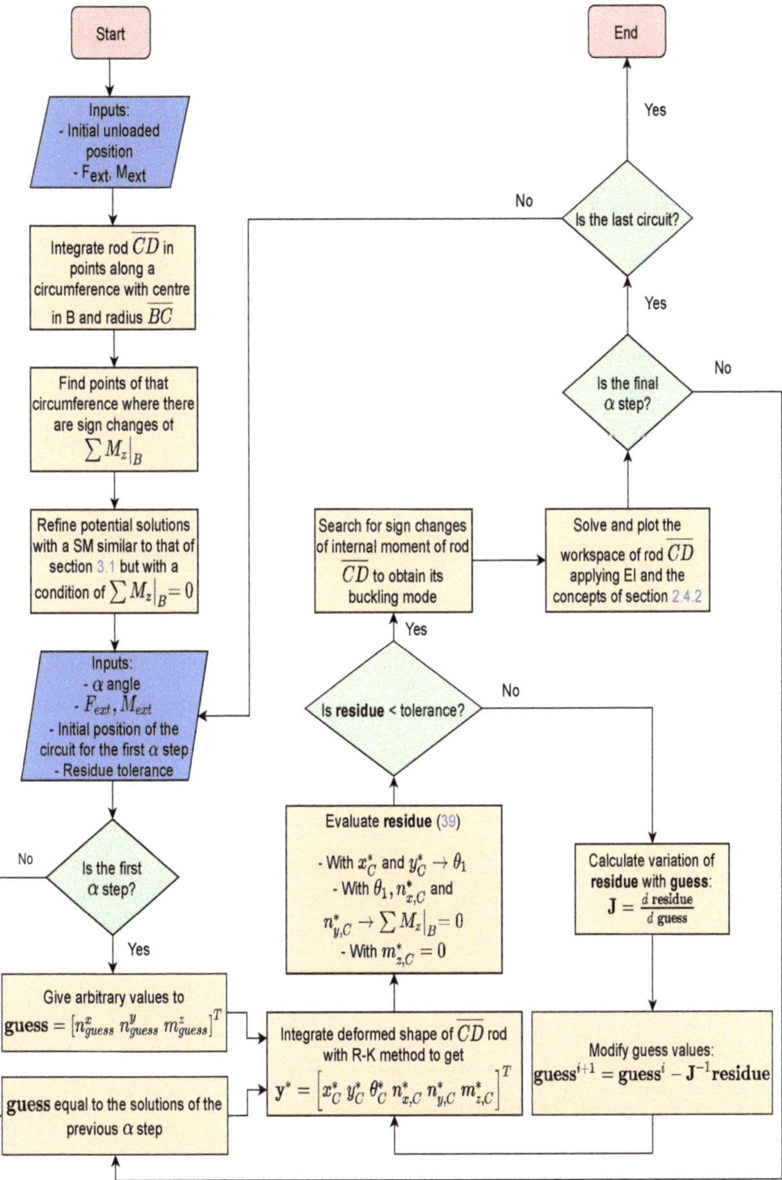

Figure 23. Flowchart of the entire algorithm.

4. Discussion

In this paper, the performance of the SM that combines the 4th order R-K method with a numerical minimization method, i.e., N-R scheme, to solve the position of hybrid compliant mechanisms is shown. This method not only allows solving the FKP, but also allows predicting the locus of all possible paths of the mechanism's coupler under different load cases. In this manner, this method can be used in the synthesis process of mechanisms to know whether the concerned mechanism can describe the desired trajectory supporting the external loading.

Furthermore, the SM is combined with the use of EI to obtain the workspace of the flexible rod, allowing detecting changes of buckling modes.

Future work could look for an algorithm able to detect abrupt changes or loops on the deformed shape of the compliant element. If this information is known, a highly detailed knowledge of the mechanism motion can be acquired. All this would allow designing mechanisms able to foresee the possible problems that could appear in the prototyping phase.

Further research can be conducted in the multiplicity of circuits of this kind of mechanism, applying the method here proposed. An ambitious goal could be an algorithm to obtain the loci of all trajectories for different load cases and all the mechanism circuits since, as can be seen in Figures 17–22, diverse circuits lead to different trajectories' loci. Detecting the influence of external loading in these circuits also should be studied.

5. Conclusions

The contribution of this paper is an algorithm that combines two numerical methods (a shooting method and elliptic integrals) to solve the movement of a hybrid rigid–flexible mechanism, taking into account the influence of the external load in that movement. The external load can affect the coupler trajectory by changing the deformed shape of the slender rod since, depending on the load, the flexible rod can work in different buckling modes, or even the whole mechanism can operate in different circuits. For this reason, the proposed algorithm is also developed to detect the circuit and buckling mode in which the mechanism works.

With this work, the potential of this method to study the behavior of compliant mechanisms is demonstrated. Its potential is such that it can be used, for instance, in the modeling process of humanoid robots, such as the one that appears in [23].

Supplementary Materials: The following are available online at https://www.mdpi.com/2227-7390/9/16/1869/s1: Video S1: Buckling mode change; Video S2: First mechanism circuit; Video S3: Second mechanism circuit; Video S4: Third mechanism circuit.

Author Contributions: Conceptualization, O.A., D.M.S., E.A. and V.P.; methodology, O.A., D.M.S., E.A. and V.P.; software, D.M.S.; validation, O.A., D.M.S., E.A. and V.P.; formal analysis, O.A. and D.M.S.; investigation, O.A., D.M.S., E.A. and V.P.; resources, O.A., E.A. and V.P.; data curation, O.A. and D.M.S.; writing—original draft preparation, O.A. and D.M.S.; writing—review and editing, O.A. and D.M.S.; visualization, O.A.; supervision, O.A., E.A. and V.P.; project administration, O.A. and V.P.; funding acquisition, O.A. and V.P. All authors have read and agreed to the published version of the manuscript.

Funding: This work was possible thanks to the funding of Project DPI2015-64450-R and PID2020-116176GB-100 (MINECO/FEDER, UE), and Project Ref. IT949-16 (Departamento de Educación, Política Lingüística y Cultura from the regional Basque Government).

Institutional Review Board Statement: Not applicable.

Informed Consent Statement: Not applicable.

Data Availability Statement: Not applicable.

Acknowledgments: The authors wish to acknowledge the financial support received from the Spanish government through the Ministerio de Economía y Competitividad (Project DPI2015-64450-R and PID2020-116176GB-100 (MINECO/FEDER, UE)), and the support for the research group through Project Ref. IT949-16, provided by the Departamento de Educación, Política Lingüística y Cultura from the regional Basque Government.

Conflicts of Interest: The authors declare no conflict of interest.

Abbreviations

Abbreviations

The following abbreviations are used in this manuscript:

BVP	Boundary Value Problem
EI	Elliptic Integrals
FK	Forward Kinematics
FKP	Forward Kinematic Problem
IKP	Inverse Kinematic Problem
N-R	Newton–Raphson
R-K	Runge–Kutta
SM	Shooting Method

References

1. IFToMM Terminology—Compliant Mechanisms. Available online: http://www.iftomm-terminology.antonkb.nl/2057/frames.html (accessed on 24 June 2021).
2. Görgülü, İ.; Can Dede, M.İ.; Carbone, G. Experimental Structural Stiffness Analysis of a Surgical Haptic Master Device Manipulator. *J. Med. Devices* **2021**, *15*, 011110. [CrossRef]
3. Hamida, I.B.; Laribi, M.A.; Mlika, A.; Romdhane, L.; Zeghloul, S.; Carbone G. On the Optimal Design of LAWEX for a Safe Upper Arm Rehabilitation Exercising. In *Mechanism Design for Robotics*; MEDER 2021; Zeghloul, S., Laribi, M.A., Arsicault, M., Mechanisms and Machine Science, Eds.; Springer: Cham, Switzerland, 2021; Volume 103, pp. 313–321. [CrossRef]
4. Velázquez, R.; Garzón-Castro, C.L.; Acevedo, M.; Orvañanos-Guerrero, M.T.; Ghavifekr, A.A. Design and Characterization of a Miniature Bio-Inspired Mobile Robot. In Proceedings of the 2021 12th International Symposium on Advanced Topics in Electrical Engineering (ATEE), Bucharest, Romania, 25–27 March 2021; pp. 1–5. [CrossRef]
5. Orvañanos-Guerrero, M.T.; Acevedo, M.; Sánchez, C.N.; Giannoccaro, N.I.; Visconti, P.; Velázquez, R. Efficient Balancing Optimization of a Simplified Slider-Crank Mechanism. In Proceedings of the 2020 IEEE ANDESCON, Quito, Ecuador, 13–16 October 2020; pp. 1–6. [CrossRef]
6. Myszka, D.H.; Murray, A.P.; Wampler, C.W. Mechanism branches, turning curves and critical points. In Proceedings of the IDETC/CIE 2012 ASME 2012 International Design Engineering Technical Conferences & Computers and Information in Engineering Conference, Chicago, IL, USA, 12–15 August 2012; pp. 1513–1525. [CrossRef]
7. Chase, T.; Mirth, J. Circuits and Branches of Single-Degree-of-Freedom Planar Linkages. *ASME J. Mech. Des.* **1993**, *115*, 223–230. [CrossRef]
8. Mirth, J.; Chase, T. Circuit Analysis of Watt Chain Six-Bar Mechanisms. *ASME J. Mech. Des.* **1993**, *115*, 214–222. [CrossRef]
9. Wantanabe, K.; Katoh, H. Identification of Motion Domains of Planar Six-Link Mechanisms of the Stephenson-Type. *Mech. Mach. Theory* **2004**, *39*, 1081–1099. [CrossRef]
10. Ma, F.; Chen, G. Modeling Large Planar Deflections of Flexible Beams in Compliant Mechanisms Using Chained Beam-Constraint-Model. *J. Mech. Robot.* **2016**, *8*, 021018. [CrossRef]
11. Kuo, C.H.; Chen, Y.C.; Pan, T.Y. Continuum Kinematics of a Planar Dual-Backbone Robot Based on Pseudo-Rigid-Body Model: Formulation, Accuracy, and Efficiency. In Proceedings of the ASME 2017 International Design Engineering Technical Conferences and Computers and Information in Engineering Conference, Cleveland, OH, USA, 6–9 August 2017; ASME: New York, NY, USA, 2017. [CrossRef]
12. Midha, A.; Bapat, S.G.; Mavanthoor, A.; Chinta, V. Analysis of a Fixed-Guided Compliant Beam With an Inflection Point Using the Pseudo-Rigid-Body Model Concept. *J. Mech. Robot.* **2015**, *7*, 031007. [CrossRef]
13. Antman, S.S. *Nonlinear Problems of Elasticity*, 2nd ed.; Springer: New York, NY, USA, 2005; Volume 107. [CrossRef]
14. Rucker, D.C.; Webster, R.J., III. Statics and dynamics of continuum robots with general tendon routing and external loading. *IEEE Trans. Robot.* **2011**, *27*, 1033–1044. [CrossRef]
15. Black, C.B.; Till, J.; Rucker, D.C. Parallel Continuum Robots: Modeling, Analysis, and Actuation-Based Force Sensing. *IEEE Trans. Robot.* **2018**, *34*, 29–47. [CrossRef]
16. Altuzarra, O.; Caballero, D.; Campa, F.J.; Pinto, C. Position analysis in planar parallel continuum mechanisms. *Mech. Mach. Theory* **2019**, *132*, 13–29. [CrossRef]
17. Gere, J.M.; Timoshenko, S.P. *Mechanics of Materials*, 4th ed.; PWS Publishing Company: Boston, MA, USA, 1997.
18. Kimball, C.; Tsai, L.W. Modeling of Flexural Beams Subjected to Arbitrary End Loads. *J. Mech. Des.* **2002**, *124*, 223–235. [CrossRef]
19. Holst, G.L.; Teicher, G.H.; Jesen, B.D. Modeling and Experiments of Buckling Modes and Deflection of Fixed-Guided Beams in Compliant Mechanisms. *J. Mech. Des.* **2011**, *133*, 051002. [CrossRef]
20. Zhang, A.; Chen, G. A Comprehensive Elliptic Integral Solution to the Large Deflection Problems of Thin Beams in Compliant Mechanisms. *J. Mech. Robot.* **2013**, *5*, 021006. [CrossRef]
21. Howell, L.L.; Magleby, S.P.; Olsen, B.M. *Handbook of Compliant Mechanisms*; John Wiley & Sons, Ltd.: Hoboken, NJ, USA, 2013. [CrossRef]

22. Rucker, D.C.; Jones, B.A.; Webster, R.J. A geometrically exact model for externally loaded concentric-tube continuum robots. *IEEE Trans. Robot. Publ. IEEE Robot. Autom. Soc.* **2010**, *26*, 769–780. [CrossRef] [PubMed]
23. Cafolla, D.; Ceccarelli, M. Design and FEM analysis of a novel humanoid torso. *Multibody Mechatron Syst. Mech. Mach. Sci.* **2015**, *25*, 477–488. [CrossRef]

 *mathematics*

Article

Adaptive Levenberg–Marquardt Algorithm: A New Optimization Strategy for Levenberg–Marquardt Neural Networks

Zhiqi Yan, Shisheng Zhong *, Lin Lin and Zhiquan Cui *

Department of Mechanical Engineering, Harbin Institute of Technology, Harbin 150000, China; cumtmaris@163.com (Z.Y.); waiwaiyl@163.com (L.L.)
* Correspondence: zhongss@hit.edu.cn (S.Z.); xiaocui2002yan@163.com (Z.C.)

Abstract: Engineering data are often highly nonlinear and contain high-frequency noise, so the Levenberg–Marquardt (LM) algorithm may not converge when a neural network optimized by the algorithm is trained with engineering data. In this work, we analyzed the reasons for the LM neural network's poor convergence commonly associated with the LM algorithm. Specifically, the effects of different activation functions such as Sigmoid, Tanh, Rectified Linear Unit (RELU) and Parametric Rectified Linear Unit (PRLU) were evaluated on the general performance of LM neural networks, and special values of LM neural network parameters were found that could make the LM algorithm converge poorly. We proposed an adaptive LM (AdaLM) algorithm to solve the problem of the LM algorithm. The algorithm coordinates the descent direction and the descent step by the iteration number, which can prevent falling into the local minimum value and avoid the influence of the parameter state of LM neural networks. We compared the AdaLM algorithm with the traditional LM algorithm and its variants in terms of accuracy and speed in the context of testing common datasets and aero-engine data, and the results verified the effectiveness of the AdaLM algorithm.

Keywords: Levenberg–Marquardt algorithm; convergence; neural networks; local minima; optimization

Citation: Yan, Z.; Zhong, S.; Lin, L.; Cui, Z. Adaptive Levenberg–Marquardt Algorithm: A New Optimization Strategy for Levenberg–Marquardt Neural Networks. *Mathematics* **2021**, *9*, 2176. https://doi.org/10.3390/math9172176

Academic Editors: Higinio Rubio Alonso, Alejandro Bustos Caballero, Jesus Meneses Alonso and Enrique Soriano-Heras

Received: 16 August 2021
Accepted: 1 September 2021
Published: 6 September 2021

Publisher's Note: MDPI stays neutral with regard to jurisdictional claims in published maps and institutional affiliations.

1. Introduction

When applied to real-world data interspersed with high nonlinearity and high-frequency noise, LM neural networks have irreplaceable advantages. They reduce the requirement of computing resources and guarantee high convergence speed, making their performance superior to other existing regression methods. The LM neural network is an artificial neural network with an LM optimizer. The LM algorithm is a second-order method for solving general least squares problems. As an optimizer, the LM algorithm has wide engineering applications in LM neural networks because of its fast convergence and small memory occupation [1–3].

However, the disadvantage of the LM algorithm makes LM neural networks have potential symptoms of divergence or convergence to a bad local minimum. The paper [4] reported that the LM algorithm may become stuck and fail to degrade the cost function. For a nonconvex optimization problem, such as neural network optimization, there can be many local minima. The cost function may end up in the "bad" local minima when it is difficult to guarantee the nonsingularity of the Jacobian matrix and Lipschitz continuity [5]. LM neural networks show higher errors than do regular neural networks in practical engineering application [6].

Most of the recent work have tried to help the LM algorithm find one of the "good" local minima by combining with other algorithms for better engineering applications, because it is generally acceptable in the modern neural network community that searching for a global minimum is often an unnecessary endeavor, e.g., the genetic algorithm (GA) can be used to optimize the initial weight of LM neural networks [7,8], and Bayesian estimation can be used to calibrate model variables. In Refs. [9,10], the wavelet method was used to preprocess data and the LM neural networks method was used to analyze them. In Ref. [11], the least squares support vector machine (LSSVM) and LM were combined into a hybrid model. The above hybrid methods have been proved to be effective by experiments.

Several other research groups have looked into improving the LM algorithm. Yang [12] presented a high-order LM method that has biquadratic convergence. Chen [13] presented a fourth-order method called the accelerated modified LM method. Derakhshandeh [14] presented a new three-step LM (TSLM) algorithm based on fuzzy logic theory (FLT). In Ref. [15], an adaptive Levenberg–Marquardt (LM) algorithm-based echo state network was proposed by adding a new adaptive damping term to the LM algorithm. These works proved the validity of the modified method algorithm by a trust region technique.

However, hybrid methods indicate a lot of computation, which weakens the advantage of the LM algorithm. The problem with improved algorithms is that their performance improvement is conditionally limited. These improved algorithms prove to be effective only under specific prerequisites:

(1). $F(w)$ and $f(w)$ are nonlinearly continuous and differentiable. The $f(w)$ is a neural network model, w is a parameter of the neural network model, and $F(w)$ is the cost function of the neural network. In Refs. [12–15], the global optimization ability of the proposed algorithm based on nonlinearly continuous and differential $F(w)$ and $f(w)$ was verified. In detail, the activation functions in the neural networks are not necessarily continuous and differentiable, so the above references are only valid in the current setting without proving that the global optimization can be achieved in the neural network model.

(2). $J = \partial F(w)/\partial w$. J is a Jacobian matrix. The Jacobian matrix in Refs. [16–18] was obtained by directly solving the system of nonlinear equations. The calculation results were relatively accurate, but the calculation process was very complex. In the neural networks, the Jacobian matrix participating in the proof process was obtained by back-propagation rather than by the derivation of the objective function. Therefore, the theory of the above references may be not suitable for the field of neural networks.

Rather, a new adaptive LM algorithm is proposed for neural networks in this paper. We explain in detail the specific factors that cause the cost function to fall into bad local minima by the original LM algorithm, by analyzing the output performance of several activation functions. In view of these factors, the new algorithm makes up for the deficiency of the LM algorithm to train a network efficiently.

The remaining parts are organized as follows. Section 2 introduces the LM algorithm. Section 3 proves that the original LM algorithm has the possibility of falling into local minima in neural networks. Section 4 gives the optimization of the LM algorithm. Section 5 verifies the correctness of this theory about experiments, and the conclusions are drawn in last section.

2. Explanation and Problem of LM Algorithms

2.1. LM Algorithm

The LM algorithm was first proposed for nonlinear least squares problems by scholars Levenberg and Marquardt [19]. Before introducing the LM and carrying out research, some parameters need to be defined, which are the same as those of [20].

Definition 1. *Given a neural network model f(w), the cost function is the least squares problem:*

$$F(w) = \Sigma(y_{\text{label}} - f(w))^2 \tag{1}$$

where w represents neural network parameters and y_{label} represents label data.

Definition 2. *A model L is assumed to describe the behavior of F in the current iteration*

$$L(h) \approx F(w + h) \equiv F(w) + h^{\mathrm{T}} J^{\mathrm{T}} f(w) + {}^{1}/_{2} h^{\mathrm{T}} J^{\mathrm{T}} Jh \tag{2}$$

Definition 3. *There is a method to evaluate the accuracy of model L, called gain ratio q:*

$$q = \frac{F(w) - F(w + h)}{L(0) - L(h)} \tag{3}$$

LM method minimizes $F(w)$ by iteration. h is introduced to describe the change direction of $F(w)$.

Definition 4. *h is a descent direction for F(w) at w, then F(w + h) < F(w).*

$$h = - (J^{\mathrm{T}} J + \mu I)^{-1} Jr \tag{4}$$

where J is the Jacoby matrix, r is the cost function error, and μ is the damping parameter.

Given a variable v, the damping parameter μ is adjusted by v in every iteration. The μ and v updating formulas are illustrated below [20]:

(1). If $q < 0$, $\mu = \mu \times v$, $v = v \times 2$;
(2). If $q > 0$, $\mu = \mu \times \max\{{}^{1}/_{3}, 1 - (2q - 1)^{3}\}$.

2.2. Problem of LM Algorithm

The problem of the LM algorithm is from the gain ratio q. A positive q means $L(h)$ is a good approximation to $F(w + h)$, and vice versa. The $L(0) - L(h)$ has proved to be positive [20], and its proof is:

$$
\begin{aligned}
L(0) - L(h) &= -h^{\mathrm{T}} J^{\mathrm{T}} f - \tfrac{1}{2} h^{\mathrm{T}} J^{\mathrm{T}} Jh \\
&= -\tfrac{1}{2} h^{\mathrm{T}} (2g + (J^{\mathrm{T}} J + \mu I - \mu I)h) \\
&= \tfrac{1}{2} h^{\mathrm{T}} (\mu h - g) \\
\text{where, } g &= J^{\mathrm{T}} f \\
h^{\mathrm{T}} h > 0, \; -h^{\mathrm{T}} g &= -h^{\mathrm{T}} (J^{\mathrm{T}} f) = -\left(- (J^{\mathrm{T}} J + \mu I)^{-1} g\right)^{\mathrm{T}} g \\
&= g^{\mathrm{T}} \left((J^{\mathrm{T}} J + \mu I)^{-1}\right)^{\mathrm{T}} g > 0 \\
L(0) - L(h) &= \tfrac{1}{2} h^{\mathrm{T}} (\mu h - g) > 0
\end{aligned}
\tag{5}
$$

In Definition 3, we know $q = (F(w) - F(w + h))/(L(0) - L(h))$, which indicates whether $L(h)$ is a good approximation to $F(w + h)$ dependences on $F(w) - F(w + h)$. As just said, if $q > 0$, $L(h)$ is a good approximation. If $q < 0$, the damping parameter μ in Equation (4) is adjusted and the current iteration is recalculated.

There is a situation where q will always remain negative regardless of how the damping parameter μ in Equation (4) is adjusted, while the cost function $F(w)$ in Equation (1) is still very large.

That situation indicates that iteration cannot continue, and that LM neural networks have fallen into the "bad" local minima or had divergence.

3. Analysis of Falling into "Bad" Local Minima

In the previous section, the possibility of the LM algorithm was analyzed. In this section, we focus on the case of where q would always remain negative and analyze the possible situation. If we find the condition where gain ratio $q < 0$ regardless of how the parameters are adjusted, we can prove that LM neural networks may diverge or converge to the poor local minima.

The proof process includes three steps: First, given the inequality $f(w) + J(w) < r(w^2)$, the conditions satisfying the inequality are analyzed. Secondly, we prove that "$\exists f(w)$, when $f(w) + J(w) < r(w^2)$, in Equation (3), the gain ratio $q < 0$". The conditions leading to $q < 0$ can be obtained. Thirdly, based on the above analysis, we discuss the conditions under which LM neural networks may diverge or converge to a worse local minimum.

Definition 5. *According to the structure of the neural network, the residual function is as follows:*

$$f(w) = \sigma(wx + b) - y_{\text{label}} \tag{6}$$

where σ is an activation function, x is the input of the neural network, the weight of the neural network is w, and the bias is b.

Definition 6. *The first-order Taylor expansion for the output function $f(w + h)$ of the neural network at h is:*

$$f(w + h) = f(w) + J(w)h + r(w^2) \tag{7}$$

where $r(w^2)$ is the remainder.

Lemma 1. *$\exists\{w,b,h\}$, when the absolute values of training data x and label data y of the neural network are large enough, $f(w) + J(w) < r(w^2)$.*

Proof. According to Equation (7):

$$r(w^2) = f(w + h) - f(w) - J(w)h \tag{8}$$

Substituting Equation (8) into $f(w) + J(w) < r(w^2)$ gives:

$$f(w + h) - 2f(w) - 2J(w)h > 0 \tag{9}$$

Thus, to prove $f(w) + J(w) < r(w^2)$, you only need to prove Equation (9). By Equation (6), we obtain:

$$f(w + h) = \sigma(wx + hx + b) - y_{label}$$
$$J(w) = \sigma'(wx + b) \tag{10}$$

Substituting Equation (6) and Equation (10) into Equation (9) gives:

$$\sigma(wx + hx + b) - 2\sigma(wx + b) - 2\sigma'(wx + b)h + y_{label} > 0 \tag{11}$$

Let:

$$G(w, h) = \sigma(wx + hx + b) - 2\sigma(wx + b) - 2\sigma'(wx + b)h + y_{label} \tag{12}$$

Thus, to prove $f(w) + J(w) < r(w^2)$, you only need to prove:

$$G(w, h) = \sigma(wx + hx + b) - 2\sigma(wx + b) - 2\sigma'(wx + b)h + y_{label} > 0 \tag{13}$$

In order to simplify the calculation, define:

$$w' = wx + b,\ h' = hx \tag{14}$$

Then:

$$\sigma(w') = \sigma(wx + b)$$
$$\sigma(w' + h') = \sigma(wx + hx + b)$$
$$\sigma'(w')h' = \sigma'(wx + b)\frac{\partial w}{\partial w'}h' = \sigma'(wx + b)\frac{h'}{x} = \sigma'(wx + b)h$$
$$G(w', h') = \sigma(w' + h') - 2\sigma(w') - 2\sigma'(w')h' + y_{label} =$$
$$\sigma(wx + hx + b) - 2\sigma(wx + b) - 2\sigma'(wx + b)h + y_{label}$$
$$= G(w, h)$$

(15)

The frequently used activation functions include the sigmoid function, tanh function, ReLU function, and PReLU function. Discussion:

(1). When σ is sigmoid:

$$\sigma(w' + h') = \frac{1}{1 + e^{-w' - h'}}$$
$$\sigma(w') = \frac{1}{1 + e^{-w'}}$$
$$\sigma'(w') = \frac{e^{-w'}}{1 + e^{-w'}}$$

(16)

Then, the function $G(w',h')$ is:

$$G(w', h') = \frac{1}{1 + e^{-w' - h'}} - \frac{2}{1 + e^{-w'}} - \frac{2he^{-w'}}{1 + e^{-w'}} + y_{label}$$

(17)

Due to the arbitrariness of the value of y_{label}, the first three terms of the function $G(w',h')$ are considered and expressed by $g(w',h')$:

$$g(w', h') = G(w', h') - y_{label} = \frac{1}{1 + e^{-w' - h'}} - \frac{2}{1 + e^{-w'}} - \frac{2he^{-w'}}{1 + e^{-w'}}$$

(18)

The image of $g(w',h')$ in Equation (18) drawn by OCTAVE is shown in Figure 1:

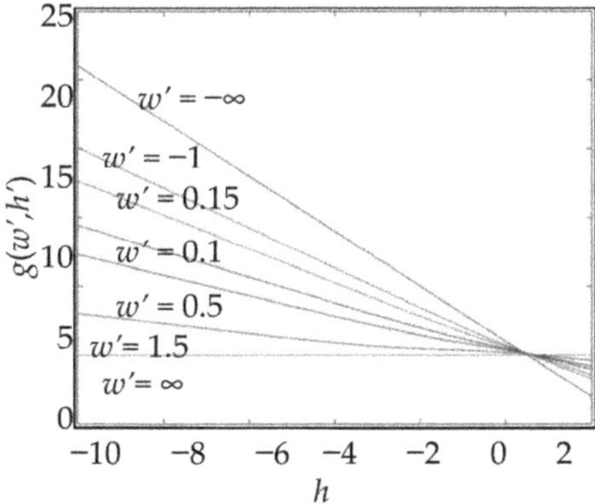

Figure 1. Regardless of the value of w', the $g(w',h')$ curve monotonically decreases and intersects the x-axis at the point (0.5, 0). This means that when the conditions (σ = sigmoid, $h' \leq 0.5$, and $y_{label} > 0$) are met, the algorithm has the possibility of divergence.

From Figure 1, it can be seen that the $g(w',h')$ curve is monotonously decreasing and converges at point ($h' = 0.5$, $g(w',h') = 0$); when $h' \leq 0.5$, $g(w',h') \geq 0$. That is:

$$g(w', h')|_{h' \leq 0.5} = G(w', h')|_{h' \leq 0.5} - y_{label} \geq 0$$

$$G(w', h')|_{h' \leq 0.5} \geq y_{label} \tag{19}$$

$$G(w, h) = G(w', h')|_{h' \leq 0.5} \geq y_{label}$$

Thus, when $h' \leq 0.5$ and $y_{label} > 0$, $G(w,h) > 0$. As $h' = hx$, for the arbitrariness of h, when the absolute values of x and y_{label} are large enough, we have $G(w,h) > 0$, that is, $f(w) + J(w) < r(w^2)$.

(2). When σ is tanh:

Similarly, let: $w' = wx + b$, $h' = hx$, then:

$$\sigma(w' + h') = \frac{e^{w'+h'} - e^{-w'-h'}}{e^{w'+h'} + e^{-w'-h'}}$$

$$\sigma(w') = \frac{e^{w'} - e^{-w'}}{e^{w'} + e^{-w'}} \tag{20}$$

$$\sigma'(w') = \frac{4}{2 + e^{-2w'} + e^{2w'}}$$

Then, the function $G(w',h')$ is:

$$G(w', h') = \frac{2}{1 + e^{2(w'+h')}} - \frac{4}{1 + e^{2w'}} - \frac{4h}{2 + e^{2w'} + e^{-2w'}} - 1 + y_{label} \tag{21}$$

For the arbitrariness of the value of y_{label}, the first three terms of the function $G(w',h')$ are considered and expressed by $g(w',h')$:

$$g(w', h') = G(w', h') - y_{label}$$
$$= \frac{2}{1 + e^{2(w'+h')}} - \frac{4}{1 + e^{2w'}} - \frac{4h}{2 + e^{2w'} + e^{-2w'}} - 1 \tag{22}$$

The image of $g(w',h')$ in Equation (21) drawn by OCTAVE is shown in Figure 2:

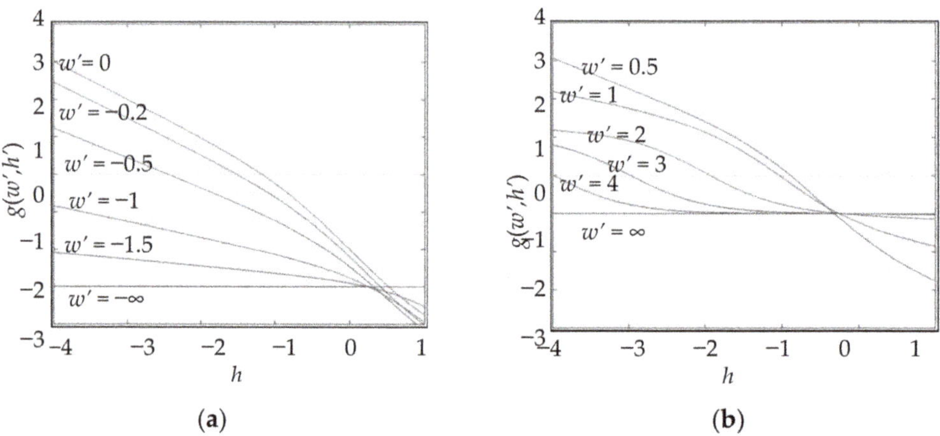

Figure 2. When σ = tanh, the curve $g(w',h')$ monotonically decreases: (a) when $h' < 0.3$ and $w < 0$, $\lim_{w \to \infty} g(w',h') = -3$; (b) when $h' < -0.3$ and $w > 0.5$, $\lim_{w \to \infty} g(w',h') = -1$. When the conditions ($\sigma$ = tanh, $h' < 0.3$, $w' < 0$, and $y_{label} > 3$) are met, the algorithm has the possibility of divergence.

In Figure 2, the curve $g(w',h')$ is also monotonically decreasing. When $h' < 0.3$, $g(w',h') \geq -3$, and so:

$$g(w', h')|_{h' \geq -3} = G(w', h')|_{h' \geq -3} - y_{label} \geq -3$$
$$G(w', h')|_{h' \geq -3} \geq y_{label} - 3 \tag{23}$$
$$G(w, h) = G(w', h')|_{h' \geq -3} \geq y_{label} - 3$$

Thus, when $h' < 0.3$ and $y_{label} > 3$, $G(w,h) > 0$. As $h' = hx$, for the arbitrariness of h, when the absolute values of x and y_{label} are large enough, we have $G(w,h) > 0$, that is, $f(w) + J(w) < r(w^2)$.

(3). When σ is ReLU:

Similarly, let: $w' = wx + b$, $h' = hx$, so:

$$\sigma(w' + h') = \max(0, w' + h')$$
$$\sigma(w') = \max(0, w') \tag{24}$$
$$\sigma'(w') = \max(0, 1)$$

then, $G(w', h')$ is:

$$G(w', h') = \max(0, w' + h') - \max(0, 2w') - \max(0, 2h) + y_{label} \tag{25}$$

Due to the arbitrariness of the value of y_{label}, the first three terms of the function $G(w', h')$ are considered and expressed by $g(w', h')$. The image of $g(w', h')$ is shown in Figure 3:

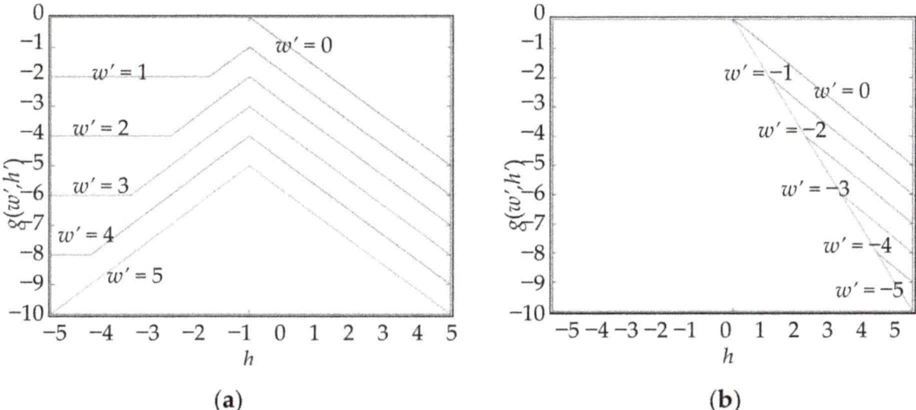

Figure 3. The curve $g(w', h')$ at σ = ReLU: (**a**) $w' > 0$; (**b**) $w' < 0$. When the conditions (σ = ReLU, $h' < 0.3$, $w' < 0$, and $y_{label} > 0$) are met, the algorithm has the possibility of divergence.

In Figure 3, when $w' < 0$ and $h' < 0$, $g(w', h') \geq 0$, and so:

$$g(w', h')|_{w' < 0, h' < 0} = G(w', h')|_{w' < 0, h' < 0} - y_{label} \geq 0$$
$$G(w', h')|_{w' < 0, h' < 0} \geq y_{label} \tag{26}$$
$$G(w, h) = G(w', h')|_{w' < 0, h' < 0} \geq y_{label}$$

Thus, when $w' < 0$, $h' < 0$, and $y_{label} > 0$, $G(w,h) > 0$. As $h' = hx$ and $w' = wx + b$, for the arbitrariness of h, w, b, when the absolute values of x and y_{label} are large enough, we have $G(w,h) > 0$, that is, $f(w) + J(w) < r(w^2)$.

(4). When σ is PReLU:

Similarly, let: $w' = wx + b$ and $h' = hx$, and draw the image of $g(w', h')$ as shown in Figure 4:

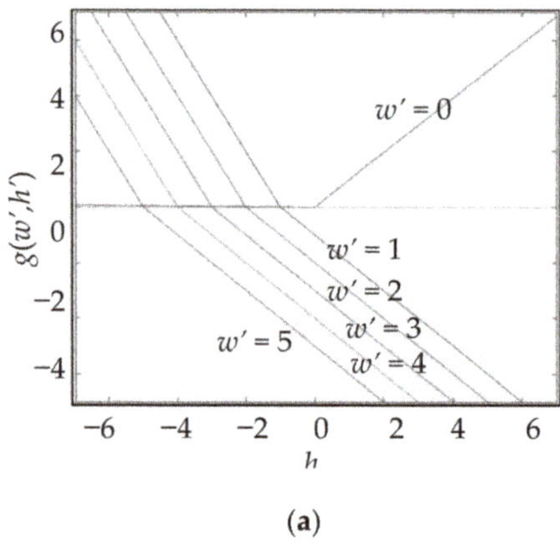

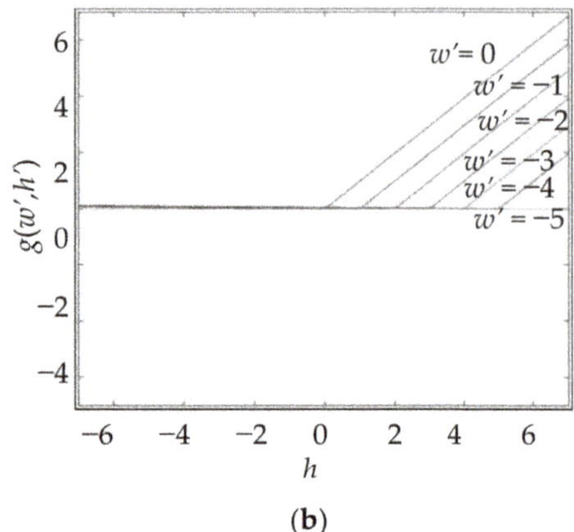

Figure 4. The curve $g(w',h')$ at σ = PReLU. (**a**) $w' > 0$; (**b**) $w' < 0$. When the conditions (σ = PReLU, $h' < 0$, $w' < 0$, and $y_{label} > 0$) are met, the algorithm has the possibility of divergence.

In Figure 4, when $w' < 0$, $g(w',h') > 0$, and so:

$$g(w',h')\big|_{w'<0,h'<0} = G(w',h')\big|_{w'<0,h'<0} - y_{label} > 0$$
$$G(w',h')\big|_{w'<0,h'<0} > y_{label} \tag{27}$$
$$G(w,h) = G(w',h')\big|_{w'<0,h'<0} > y_{label}$$

Thus, when $w' < 0$ and $y_{label} > 0$, $G(w,h) > 0$. As $h' = hx$ and $w' = wx + b$, for the arbitrariness of h,w,b, when the absolute values of x and y_{label} are large enough, we have $G(w,h) > 0$, that is, $f(w) + J(w) < r(w^2)$.

To sum up, when the absolute values of x and y_{label} are large enough, there are $f(w) + J(w) < r(w^2)$. $\square$

Lemma 2. $\exists f(w)$, *when* $f(w) + J(w) < r(w^2)$, *in Equation* (3), *the gain ratio* $q < 0$.

Proof. The cost function is: $F(w + h) = \frac{1}{2}||f(w + h)||^2$. The first-order Taylor expansion for $F(w + h)$ at w is:

$$F(w + h) = F(w) + h^\mathrm{T} J^\mathrm{T} f + \frac{1}{2}h^\mathrm{T} J^\mathrm{T} J h + R(w) = L(h) + R(w) \tag{28}$$

As $f(w) + J(w) < r(w^2)$, the $r(w^2)$ term of the residual function $f(w + h)$ after Taylor's first-order expansion cannot be omitted. Therefore, there exists $f(w)$ that makes it impossible to omit the $R(w)$ term of Taylor's first-order expansion for $F(w + h)$. This means: $L(w) < R(w)$.

$$\because L(h) = L(0) + \Delta L \tag{29}$$

In [19], it has been proven that:

$$\Delta L > 0, L(0) \simeq F(w) = \frac{1}{2}||f(w)||^2 > 0$$
$$\therefore \Delta L < L(h) < R(w)$$
$$\therefore q = \frac{F(w) - F(w+h)}{L(0) - L(h)} \tag{30}$$
$$= \frac{F(w) - F(w) - h^\mathrm{T} J^\mathrm{T} f - \frac{1}{2}h^\mathrm{T} J^\mathrm{T} J h - R(w)}{\Delta L}$$
$$= \frac{\Delta L - R(w)}{\Delta L} < 0$$

$\square$

Lemma 3. *In the case of Lemma 1 and Lemma 2, the LM algorithm causes cost functions to fall into "bad" local minima.*

Proof. In the kth iteration, we have $q_k < 0$; in the case of Lemma 2, according to the LM algorithm in the original, when $q_k < 0$, the trust region will be reduced, that is:

$$
\begin{aligned}
v &= v \times 2 \\
\mu &= \mu \times v \\
h &= -\left(J^{\mathrm{T}}J + \mu I\right)^{-1} Jr
\end{aligned}
\tag{31}
$$

At this point, the absolute value of step h decreases, and its sign remains unchanged. There is a situation where q is always negative when h is at any point in the positive or negative half-axis, and h will approach zero infinitely and the cost function cannot decrease. At this time, there is:

$$
q < 0, v \to +\infty, h \to 0
\tag{32}
$$

According to Equation (14), $h' = hx$, we have:

$$
q < 0, v \to +\infty, \frac{h'}{x} \to 0
\tag{33}
$$

We analyze the common activation functions one by one to find out whether they have the possibility of this situation:

For the activation function $\sigma = $ sigmoid, when $y_{\text{label}} > 1$, the area covered by the function is shown in Figure 5a. It can be seen that $g(w',h')$ will remain positive when h' is at any point on the negative half of the x-axis. In the case of Lemma 1 and Lemma 2, it is known that "$g(w',h')$ remains positive" means "q remains negative." According to Equation (14), $h' = hx$, so when $y_{\text{label}} > 1$ and $hx < 0$, q remains negative, and the LM algorithm will stop running, and the LM neural networks will also diverge or converge poorly.

Similarly, for the activation function $\sigma = $ tanh, when $w \le 0$ and $y_{\text{label}} > 3$, the area covered by the function is shown in Figure 5b. When $w > 0$ and $y_{\text{label}} > 1$, the area covered by the function is shown in Figure 5c. Thus, when $y_{\text{label}} > 3$, $x > 0$, and $hx < 0$, the LM neural network will diverge or converge poorly.

Similarly, for the activation function $\sigma = $ ReLU, when $w < 0$ and $y_{\text{label}} > 0$, the area covered by the function is shown in Figure 5d. When $w < 0$, $y_{\text{label}} > 0$, $x > 0$, and $hx < 0$, the LM neural networks will diverge or converge poorly.

Finally, for the activation function $\sigma = $ PReLU, when $w < 0$ and $y_{\text{label}} > 0$, the area covered by the function is shown in Figure 5d. When $y_{\text{label}} > 0$ and $w < 0$, the LM neural network will diverge or converge poorly. $\square$

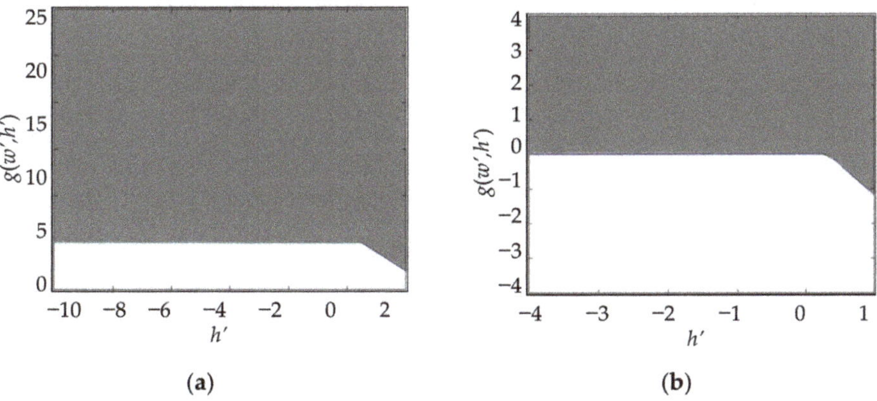

Figure 5. *Cont.*

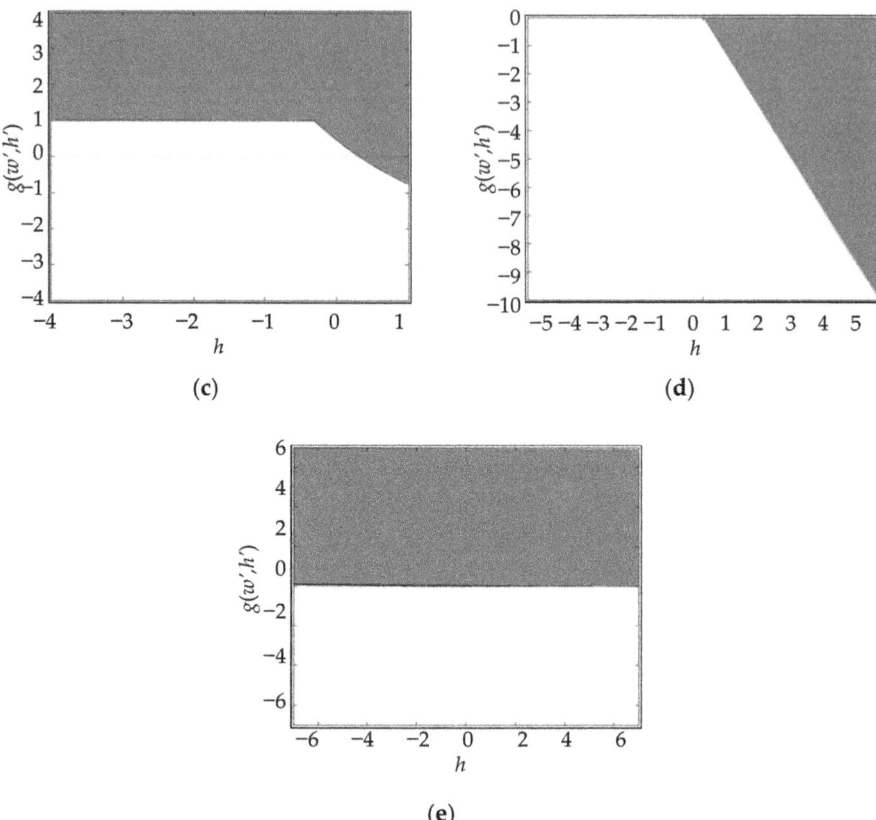

Figure 5. The gray part is the coverage area of $g(w',h')$: (**a**) The grey area covered by the $g(w',h')$ function when σ = sigmoid and $y_{\text{label}} > 1$. If $h' \leq 0.5$, $g(w',h')$ is always positive; (**b**) The grey area covered by the $g(w',h')$ function when σ = tanh, $y_{\text{label}} > 3$, and $w \leq 0$. If $h' \leq 0.3$, $g(w',h')$ is always positive; (**c**) The grey area covered by the $g(w',h')$ function when σ = tanh, $y_{\text{label}} > 1$, and $w > 0$. If $h' \leq 0$, $g(w',h')$ is always positive; (**d**) The grey area covered by the $g(w',h')$ function when σ = ReLU, $y_{\text{label}} > 0$, and $w < 0$. If $h' \leq 0$, $g(w',h')$ is always positive; (**e**) The grey area covered by the $g(w',h')$ function when σ = PReLU, $y_{\text{label}} > 0$, and $w < 0$. $g(w',h')$ is always positive.

4. The Proposed Algorithm: AdaLM

It can be seen from Section 3 that the parameter state of the model will affect the stability of the LM algorithm. In order to reduce the influence of the state of the model on the algorithm, the improved scheme starts from other algorithms to participate in the iterative operation, and then slowly transitions to the LM algorithm. Therefore, a more stable algorithm is needed, and the steepest descent method meets this requirement.

With this idea, this paper proposed the AdaLM algorithm that is close to the steepest descent method at the beginning and close to the LM algorithm after several iterations.

The core step calculation of the AdaLM algorithm can be expressed by Equation (34):

$$
\begin{aligned}
A &= J^{\mathrm{T}} J \\
g &= J^{\mathrm{T}} e \\
h_k &= -\frac{1}{1+e^{k-10}}\left(A + \mu I\right)^{-1} g - \frac{1}{1+e^{10-k}}\frac{g}{\|g\|}
\end{aligned}
\tag{34}
$$

where $A = J^T J$, α is the inertia parameter, V is the accumulated value of h, and k is the current number of iterations. The damping coefficient μ is calculated from q and v:

(1). If $q < 0$, $\mu = (\mu + 1) \times v$, $v = v \times 2$;
(2). If $q > 0$, $\mu = 0$, $v = 2$.

If AdaLM reaches the global minimum: $k \to \infty$ and $F(x_\infty) = g = 0$. Given $\varepsilon_1 > 0$, the first stop criteria can be set to:

$$\|g\|_\infty \leq \varepsilon_1 \tag{35}$$

Consistent with Reference [20], there is a second stop condition if the step size is very small:

$$\| h_k \| \leq \varepsilon_2(\|x\| + \varepsilon_2) \tag{36}$$

Finally, in all iterations, we need a protection against infinite loops, so there is a third stop condition:

$$k \geq k_{max} \tag{37}$$

Thus, the AdaLM algorithm can be described in Algorithm 1:

Algorithm 1 AdaLM algorithm.

Begin
  $k = 0; v = 2; x = x_0; \mu = 0;$
  $A = J^T J; g = J^T f(x);$
  While $(\| g \|_\infty \geq \varepsilon_1)$ **and** $(k < k_{max})$
  {
    $k = k + 1;$
    $h_k = - (A + \mu I)^{-1} g/(1 + \exp(k - 10)) - g/\| g \|(1 + \exp(k - 10));$
    if $(\| h_k \| \leq \varepsilon_2(\| x \| + \varepsilon_2))$
    break;
    else
    {
      $x_{new} = x + h_k;$
      $q = (F(x) - F(x_{new}))/(L(0) - L(h_k));$
      if $(q > 0)$
      {
        $x = x_{new};$
        $A = J^T J; g = J^T f(x);$
        if $(\| g \|_\infty \leq \varepsilon_1)$
        break;
        $\mu = 0; v = 2;$
      }
      else
      {
        $\}\mu = (\mu + 1) \times v; v = v \times 2;$
      }
  }
  }
End

---

5. Case Studies

5.1. Experimental Preparation

In order to evaluate the performance of AdaLM, it was arranged to compare with the original LM algorithm, as well as other popular LM variants, including High order Levenberg–Marquardt (HLM) and Three Step Levenberg–Marquardt (TSLM). Each algorithm is described in Table 1. In this paper, each algorithm and neural network were combined to build a prediction model, and the performance level of the algorithm was reflected by testing the prediction ability of the model.

Table 1. Description table of other LM family algorithms participating in the comparative test.

Algorithm	Full Name	Description
LM	LM	$\mathrm{LM}(\Delta x_k) = [J^{\mathrm{T}}(x_k)J(x_k) + \mu I]^{-1}\, J(x_k)e(x_k)$
HLM	High order LM	$\mathrm{HLM}(\Delta x_k) = \mathrm{LM}(\Delta x_k) + \mathrm{LM}(\Delta x_{k+1}) + \mathrm{LM}(\Delta x_{k+2})$
TSLM	Three step LM	$\mathrm{TSLM}(\Delta x_k) = \mathrm{LM}(\Delta x_k) + \alpha\mathrm{LM}(\Delta x_{k+1}) + \mathrm{LM}(\Delta x_{k+2})$

As for the neural network model, it has been proven that the three-layer neural network can simulate any complex nonlinear mapping if there are enough neurons in the hidden layer [21–23]. Therefore, the neural network models involved in this paper were all three-layer neural networks. We found that all the models involved in this paper will converge before 50 steps of iteration. Thus, we set the maximum number of iterations of the model to 50. In this study, the mean absolute error (MAE) was used to give the accuracy of the experimental results. MAE is the average value of absolute error, which can reflect well the actual situation of the predicted value error. Its formula is as follows:

$$\mathrm{MAE}(f(x), y) = \frac{1}{n}\sum_{i=1}^{n}|f(x_i) - y_i| \tag{38}$$

The number of nodes in the model affects the quality of the neural network. The input and output nodes of the model need to be set according to the requirements of the task. The number of hidden nodes of the model can be calculated according to the empirical formula given in [24]:

$$m = \sqrt{n \times l} \tag{39}$$

where m is the number of hidden layer nodes, n is the number of input layer nodes, and l is the number of output layer nodes.

The neural network model is shown in Figure 6. The model uses the square difference formula to calculate the error between the prediction data and the label data.

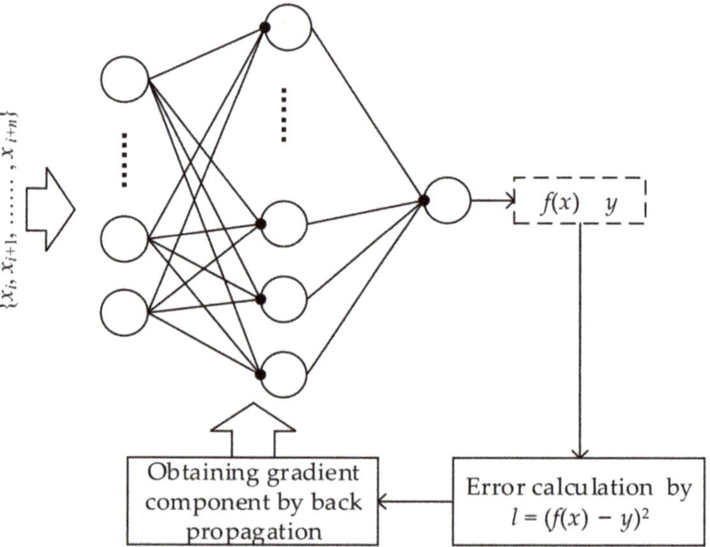

Figure 6. Neural network model.

In order to test the engineering application ability of each algorithm, the real prediction task of aero-engine performance parameters was arranged in this paper. In this paper, fuel flow data of an aero-engine were selected as training data. There is a large noise in the fuel flow data of the aero-engine due to environmental conditions, operation conditions, flight tasks, maintenance measures, etc. [25]. In Figure 7, the fuel flow data collected from the CFM56-5B engine records the fuel consumption rate of the A320 aircraft in pounds per hour.

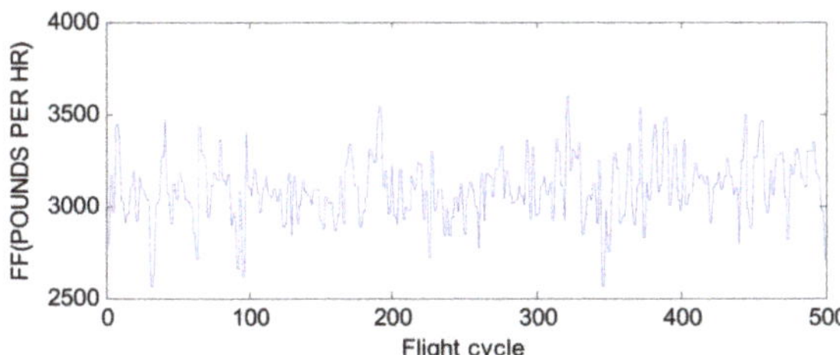

Figure 7. Aeroengine fuel flow data.

It is shown from Figure 7 that the data were highly nonlinear with many outliers. The data sample size was small. Therefore, the selection of the data can reflect the robustness and convergence of the algorithm. The specific implementation steps of the data prediction task for the fuel flow of the aeroengine are as follows:

Step 1: Obtain fuel flow data of the aeroengine. Set the total amount of data to N. Use normalization and data smoothing to preprocess the fuel flow data of the aeroengine, and then obtain the data $X = (x_1, x_2, x_3, \ldots \ldots, x_N)$. The purpose of normalization is to avoid numerical problems such as overflow and underflow, reducing the impact of model initialization, and improving the prediction accuracy. The purpose of data smoothing is to reduce the influence of noise and outliers on the model. The data smoothing method is given by Equation (40):

$$\mathrm{MAE}(f(x), y) = \frac{1}{n} \sum_{i=1}^{n} |f(x_i) - y_i| \tag{40}$$

Step 2: Organize data into datasets. In this step, the dataset is generated according to the number of input layer nodes n of neural network, and the determination method of n is given in step 5. The fuel flow data are divided into subsequences with length of n as the input data of the model and the adjacent element of the subsequence as the label data. Because this task is a one-step forecast, the label data dimension is 1. Input data X and label data Y are, respectively:

$$\begin{aligned}
\{x_1, x_2, \ldots \ldots, x_n\} &\to \{x_{n+1}\} \\
\{x_2, x_3, \ldots \ldots, x_{n+1}\} &\to \{x_{n+2}\} \\
&\cdots \\
\{x_i, x_{i+1}, \ldots \ldots, x_{i+n-1}\} &\to \{x_{i+n}\} \\
&\cdots \\
\{x_{N-n}, x_{N-n+1}, \ldots \ldots, x_{N-1}\} &\to \{x_N\}
\end{aligned} \tag{41}$$

Step 3: Split the dataset $\{X,Y\}$ into training set $\{X_{\mathrm{train}}, Y_{\mathrm{train}}\}$ and test set $\{X_{\mathrm{test}}, Y_{\mathrm{test}}\}$. According to experience, the first 95% of the data are used as training sets. The rest are the test set.

Step 4: Build the neural network model and use the training set $\{X_{\mathrm{train}}, Y_{\mathrm{train}}\}$ to train the model. The model uses the square difference formula to calculate the error between prediction data and the label data.

Step 5: Determine the number of nodes in the neural network model to make the model optimal. The enumeration method is used to determine the input node n of the neural network, with the range of 5−12. In detail, steps 2 to 4 are looped 8 times, and the input node n of the neural network is set to 5 to 12 in step 2 in each loop. The number of hidden layer nodes is determined by Equation (39), and the number of output nodes is 1. The MAE error in Equation (38) is used to describe the prediction accuracy of the experimental results. The number of input nodes and the corresponding model MAE are shown in Figure 8.

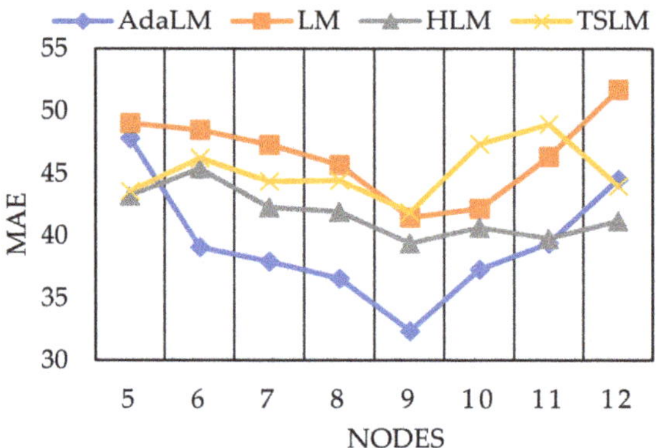

Figure 8. MAE of models with different neural network nodes.

Figure 8 shows that when the number of input nodes n is 9, the output errors of the algorithm LM reach the minimum, as well as HLM, TSLM, and AdaLM.

Step 6: Test the model. Input the test set X_{test} into the input model to obtain $f(X_{\text{test}})$.

5.2. The Influence of Each Algorithm on Activation Function

The purpose of this section is to evaluate the impact of the four algorithms on the activation function to evaluate the performance of each algorithm. In this paper, the AdaLM algorithm was compared with LM, HLM, and TSLM. As a contrast, two frameworks were set-up in this study to investigate the performance of these algorithms under different conditions.

The test results are listed in Table 2. "$\sqrt{}$" means convergence, "good" means local minima, "$\bigcirc$" means "bad" local minima, and "$\times$" means divergence or the term does not exist.

Table 2. Performance comparison of different activation functions.

Algorithm	Year	Sigmoid	Tanh	ReLU	PReLU
LM	1978	$\sqrt{}$	$\times$	$\times$	$\times$
HLM	2013	$\sqrt{}$	$\bigcirc$	$\sqrt{}$	$\bigcirc$
TSLM	2018	$\sqrt{}$	$\sqrt{}$	$\sqrt{}$	$\sqrt{}$
AdaLM	2020	$\sqrt{}$	$\sqrt{}$	$\sqrt{}$	$\sqrt{}$

Table 2 shows that the traditional LM model diverges when Tanh, ReLU, and PReLU are used as activation functions, which proves that the LM problem exists. HLM is a higher-order descent method. The HLM model could not converge when the activation functions were Tanh and PReLU, which indicates that over-high-order descent may produce negative effects. The TSLM model is an improved version of HLM. In the test, TSLM converged with four activation functions. The AdaLM algorithm proposed in this paper could change the descent direction in time and has good stability, so it achieved convergence.

5.3. Evaluation of Prediction Effect

From Figure 9a, the generated curve amplitude was very small with a large error. From the perspective of the prediction effect, the curve predicted by LM could not describe the future trend of real data. From Figure 9b,c, it could hardly reproduce the original data. After training, HLM and TSLM only achieved a line with slight fluctuation. Obviously, they could not predict the trend of future data in this test. From Figure 9d, the curves generated by AdaLM could fit the real data well. From the perspective of the prediction effect, the curve predicted by AdaLM could describe the future trend of real data.

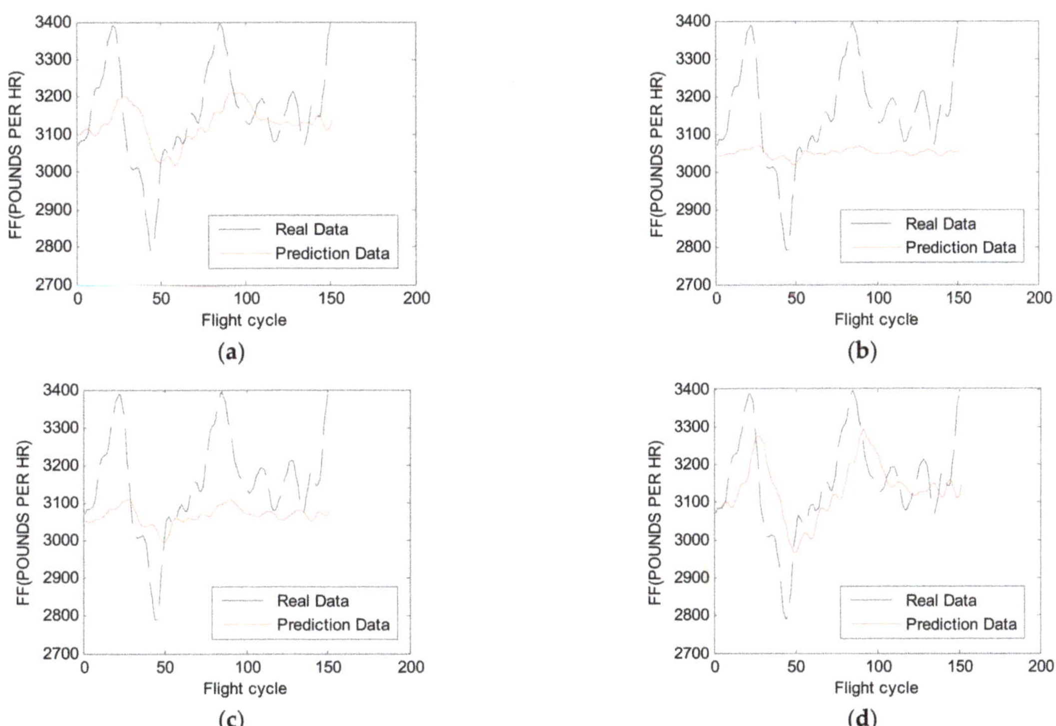

Figure 9. The prediction effect of each algorithm: (**a**) LM; (**b**) HLM; (**c**) TSLM; (**d**) AdaLM.

In order to quantify the prediction effect of each algorithm, the time consumption in 50 iterations and the prediction MAE of algorithms are shown in Table 3.

Table 3. MAE and time consumption of each algorithm.

Algorithm	LM	HLM	TSLM	AdaLM
Time/s	140	181	173	62
MAE	41.4847	38.2276	42.6715	32.9175

From Table 3, the neural network model with the original LM algorithm was trained in 140 s. The data predicted by the model after training were basically consistent with the original data, with an error of 41.4847. This shows that the original LM algorithm could basically make the neural network predict the fuel consumption data of the aeroengine, but the training took a lot of time because of the falling into a "bad" local minimum.

The neural network model with HLM algorithm completed the training in 181 s. The data predicted by the model after training fit the original data with an error of 38.2276. The HLM algorithm took the longest time to train in the prediction of fuel consumption data of an aeroengine. The HLM algorithm increased the unnecessary complexity and did not improve the performance.

The neural network model of the TSLM algorithm was trained in 173 s. The data predicted by the trained model fit the original data with an error of 42.6715. TSLM algorithm has the highest error in this task, and the complexity of the algorithm affects the prediction performance.

The neural network model with the AdaLM algorithm proposed in this paper was trained in 62 s, and the data predicted by the trained model fit the original data with an error of only 32.9175, which indicates that the AdaLM algorithm took into account high performance and high speed in predicting the fuel consumption data of aeroengine, with the least time consumption and the highest accuracy.

6. Conclusions

This paper pointed out that the LM neural networks may converge poorly or diverge because the training data are large and the weight is not selected properly. This work can guide researchers to carry out some appropriate strategies such as training data preprocessing and weights intervention at the beginning of training to avoid problems in the use of the LM neural networks when they still insist on using this model. Furthermore, this work can expand the application scope of LM neural networks and make LM neural networks play its role better in more situations.

This study proposed a new solution to the problem of LM neural networks: the AdaLM algorithm. The algorithm adds the weight term and error direction on the original algorithm, which makes the algorithm close to the steepest descent method at the beginning to preliminarily adjust the weight of the neural network. Then, the AdaLM algorithm gradually approaches the LM algorithm, which improves the stability based on inheriting the efficient optimization ability of LM. This paper compared the performance of traditional LM, HLM, TSLM, and AdaLM algorithms. We found that the performance improvement of the higher-order algorithm was very limited. Among many high-order algorithms, the third-order algorithm had the better performance. The AdaLM algorithm could not only make the neural network converge to a "good" local minimum, but also greatly shorten the operation time without losing the prediction accuracy.

The application of the AdaLM algorithm focused on a small sample and highly non-linear engineering data. It was especially effective for the analysis and prediction of engine performance parameters in the aviation field. In future research, we will collect engine data from more dimensions to comprehensively evaluate AdaLM's ability of analysis. The application scenarios of AdaLM will also be further explored.

Author Contributions: Writing—original draft preparation, Z.Y., S.Z., Z.C.; writing—review and editing, L.L., Z.C; funding acquisition, Z.C. All authors have read and agreed to the published version of the manuscript.

Funding: This research was funded by the National Natural Science Foundation of China, grant number U1733201.

Institutional Review Board Statement: Not applicable.

Informed Consent Statement: Not applicable.

Data Availability Statement: Not applicable.

Conflicts of Interest: The authors declare no conflict of interest.

References

1. Luo, G.; Zou, L.; Wang, Z.; Lv, C.; Ou, J.; Huang, Y. A novel kinematic parameters calibration method for industrial robot based on Levenberg-Marquardt and Differential Evolution hybrid algorithm. *Robot. Comput. Integr. Manuf.* **2021**, *71*, 102165. [CrossRef]
2. Kumar, S.S.; Sowmya, R.; Shankar, B.M.; Lingaraj, N.; Sivakumar, S. Analysis of Connected Word Recognition systems using Levenberg Mar-quardt Algorithm for cockpit control in unmanned aircrafts. *Mater. Today Proc.* **2021**, *37*, 1813–1819. [CrossRef]
3. Mahmoudabadi, Z.S.; Rashidi, A.; Yousefi, M. Synthesis of 2D-Porous MoS$_2$ as a Nanocatalyst for Oxidative Desulfuriza-tion of Sour Gas Condensate: Process Parameters Optimization Based on the Levenberg–Marquardt Algorithm. *J. Environ. Chem. Eng.* **2021**, *9*, 105200. [CrossRef]
4. Transtrum, M.K.; Machta, B.B.; Sethna, J.P. Why are nonlinear fits to data so challenging? *Phys. Rev. Lett.* **2010**, *104*, 060201. [CrossRef]
5. Amini, K.; Rostami, F. A modified two steps Levenberg–Marquardt method for nonlinear equations. *J. Comput. Appl. Math.* **2015**, *288*, 341–350. [CrossRef]
6. Kim, M.; Cha, J.; Lee, E.; Pham, V.H.; Lee, S.; Theera-Umpon, N. Simplified Neural Network Model Design with Sensitivity Analysis and Electricity Consumption Prediction in a Commercial Building. *Energies* **2019**, *12*, 1201. [CrossRef]
7. Zhao, L.; Otoo, C.O.A. Stability and Complexity of a Novel Three-Dimensional Envi-ronmental Quality Dynamic Evolution System. *Complexity* **2019**, *2019*, 3941920. [CrossRef]
8. Zhou, W.; Liu, D.; Hong, T. Application of GA-LM-BP Neural Network in Fault Prediction of Drying Furnace Equipment. *Matec. Web Conf.* **2018**, *232*, 01041. [CrossRef]
9. Jia, P.; Zhang, P. Type Identification of Coal Mining Face Based on Wavelet Packet Decomposition and LM-BP. In Proceedings of the 2018 IEEE 9th International Conference on Software Engineering and Service Science (ICSESS), Beijing, China, 23–25 November 2018.
10. Hua, L.; Bo, L.; Tong, L.; Wang, M.; Fu, H.; Guo, R. Angular Acceleration Sensor Fault Diagnosis Based on LM-BP Neural Network. In Proceedings of the 37th Chinese Control Conference, Wuhan, China, 25–27 July 2018; pp. 6028–6032.
11. Hossein, A.M.; Nazari, M.A.; Madah, R.G.H.; BehshadShafii, M.; Ahmadi, M.A. Thermal conductivity ratio prediction of Al2O3/water nanofluid by applying connectionist methods. *Colloids Surf. A Physicochem. Eng. Asp.* **2018**, *541*, 154–164.
12. Yang, X. A higher-order Levenberg–Marquardt method for nonlinear equations. *Appl. Math. Comput.* **2013**, *219*, 10682–10694. [CrossRef]
13. Chen, L. A high-order modified Levenberg–Marquardt method for systems of nonlinear equations with fourth-order convergence. *Appl. Math. Comput.* **2016**, *285*, 79–93. [CrossRef]
14. Derakhshandeh, S.Y.; Pourbagher, R.; Kargar, A. A novel fuzzy logic Leven-berg-Marquardt method to solve the ill-conditioned power flow problem. *Int. J. Electr. Power Energy Syst.* **2018**, *99*, 299–308. [CrossRef]
15. Qiao, J.; Wang, L.; Yang, C.; Gu, K. Adaptive levenberg-marquardt algorithm based echo state network for chaotic time series prediction. *IEEE Access* **2018**, *6*, 10720–10732. [CrossRef]
16. Ma, C.; Tang, J. The quadratic convergence of a smoothing Levenberg–Marquardt method for nonlinear complementarity problem. *Appl. Math. Comput.* **2008**, *197*, 566–581. [CrossRef]
17. Du, S.Q.; Gao, Y. Global convergence property of modified Levenberg-Marquardt meth-ods for nonsmooth equations. *Appl. Math.* **2011**, *56*, 481. [CrossRef]
18. Zhou, W. On the convergence of the modified Levenberg–Marquardt method with a non-monotone second order Armijo type line search. *J. Comput. Appl. Math.* **2013**, *239*, 152–161. [CrossRef]
19. Moré, J.J. The Levenberg-Marquardt algorithm: Implementation and theory. In *Numerical Analysis*; Springer: Berlin/Heidelberg, Germany, 1978; pp. 105–116.
20. Madsen, K.; Nielsen, H.B.; Tingleff, O. *Methods for Non-Linear Least Squares Problems*, 2nd ed.; Technical University of Denmark: Lyngby, Denmark, 2004.
21. Zhang, Z.; Ma, X.; Yang, Y. Bounds on the number of hidden neurons in three-layer binary neural networks. *Neural Netw.* **2003**, *16*, 995–1002. [CrossRef]
22. Liang, X.; Chen, R.C. A unified mathematical form for removing neurons based on or-thogonal projection and crosswise propagation. *Neural Comput. Appl.* **2010**, *19*, 445–457. [CrossRef]
23. Chua, C.G.; Goh, A.T.C. A hybrid Bayesian back-propagation neural network approach to multivariate modelling. *Int. J. Numer. Anal. Methods Geomech.* **2003**, *27*, 651–667. [CrossRef]
24. Sequin, C.H.; Clay, R.D. Fault tolerance in artificial neural networks. In Proceedings of the 1990 IJCNN International Joint Conference on Neural Networks, San Diego, CA, USA, 17–21 June 1990; pp. 703–708.
25. Li, Y.G.; Nilkitsaranont, P. Gas turbine performance prognostic for condition-based maintenance. *Appl. Energy* **2009**, *86*, 2152–2161. [CrossRef]

Article

Three-Legged Compliant Parallel Mechanisms: Fundamental Design Criteria to Achieve Fully Decoupled Motion Characteristics and a State-of-the-Art Review

Minh Tuan Pham [1,*], Song Huat Yeo [2] and Tat Joo Teo [3]

[1] Faculty of Mechanical Engineering, Ho Chi Minh City University of Technology, VNU-HCM, Ho Chi Minh City 700000, Vietnam
[2] School of Mechanical and Aerospace Engineering, Nanyang Technological University, Singapore 639798, Singapore; myeosh@ntu.edu.sg
[3] Robotics, Automation and Unmanned Systems of Home Team Science and Technology Agency, Singapore 138507, Singapore; daniel_teo@htx.gov.sg
* Correspondence: pminhtuan@hcmut.edu.vn

Abstract: A three-legged compliant parallel mechanism (3L-CPM) achieves fully decoupled motions when its theoretical 6×6 stiffness/compliance matrix is a diagonal matrix, which only contains diagonal components, while all non-diagonal components are zeros. Because the motion decoupling capability of 3L-CPMs is essential in the precision engineering field, this paper presents the fundamental criteria for designing 3L-CPMs with fully decoupled motions, regardless of degrees-of-freedom and the types of flexure element. The 6×6 stiffness matrix of a general 3L-CPM is derived based on the orientation of each flexure element, e.g., thin/slender beam and notch hinge, etc., and its relative position to the moving platform. Based on an analytical solution, several requirements for the flexure elements were identified and needed to be satisfied in order to design a 3L-CPM with a diagonal stiffness/compliance matrix. In addition, the developed design criteria were used to analyze the decoupled-motion capability of some existing 3L-CPM designs and shown to provide insight into the motion characteristics of any 3L-CPM.

Keywords: three-legged parallel mechanism; compliant mechanism; flexure-based mechanism; flexure; compliant joint; decoupled motion; coupled motion; stiffness; compliance

MSC: 70-10

Citation: Pham, M.T.; Yeo, S.H.; Teo, T.J. Three-Legged Compliant Parallel Mechanisms: Fundamental Design Criteria to Achieve Fully Decoupled Motion Characteristics and a State-of-the-Art Review. *Mathematics* **2022**, *10*, 1414. https://doi.org/10.3390/math10091414

Academic Editors: Higinio Rubio Alonso, Alejandro Bustos Caballero, Jesus Meneses Alonso and Enrique Soriano-Heras

Received: 24 February 2022
Accepted: 18 April 2022
Published: 22 April 2022

Publisher's Note: MDPI stays neutral with regard to jurisdictional claims in published maps and institutional affiliations.

1. Introduction

Compliant (or flexure-based) mechanisms have been widely used to develop precise devices, such as micro-scale manipulation grippers [1,2], nano-scale positioning actuators [3,4], alignment systems [5–7], etc., due to their frictionless elastic deformation behavior. Precise positioning is considered to be one of the most popular applications for a compliant mechanism, because it has the ability to deliver repeatable motion with high positioning resolution, which its traditional counterparts failed to achieve [8]. In the past few years, compliant mechanisms have been preferred in biological/surgical applications [9,10], micro-machining systems [11–13], and bulk lithography and microlithography three-dimensional fabrication processes [14,15]. It is seen that the abilities of producing precise and repeatable motions of compliant mechanisms are essential in these advanced applications. Thus, the motion property of compliant mechanisms needs to be thoroughly investigated to further improve their performance.

Compliant mechanisms in these applications can be designed through various techniques, such as the pseudo-rigid-body modeling approach [5,16–41], the exact constraint design approach [42–52], and topological/structural optimization methods [6,7,53–56]. In addition, parallel kinematic configuration is commonly adopted to design compliant positioning stages, due to its closed-loop and compact architecture. As a result, numerous three-legged compliant parallel mechanisms (3L-CPMs) with different numbers of degrees-of-freedom (DOF) have been developed over the past two decades, e.g., the 3-DOF in-plane motions (X-Y-θ_Z) 3L-CPMs [18–20,22,27,29,35,53,55], the 3-DOF out-of-plane motions (X-Y-Z) [24,25] and (θ_X-θ_Y-Z) [5–7] 3L-CPMs, and the 6-DOF motions 3L-CPMs [28,31,42,43,56]. These 3L-CPMs all adopted a three-legged parallel kinematic configuration as a base architecture. Due to their popularity, this paper focuses on the 3L-CPM with detailed studies on its motion decoupling capability.

In general, a 3L-CPM consists of three legs, wherein each leg is formed by either one or a series of flexure elements, e.g., thin/slender beams or notch hinges, etc., connected together. Hence, each leg can be partially compliant if a rigid-link is used to connect two flexure elements, or is fully compliant; or if there is no rigid-link between two flexure elements. Depending on the structure of the leg, 3L-CPMs can be classified into two types, i.e., single flexure serial chain and double reflecting flexure serial chains in a leg. For a 3L-CPM or any compliant mechanism, DOF represents the number of possible output motions that the moving platform can deliver, i.e., three translation motions along and three rotation motions about the respective X, Y, and Z axes. In an ideal case, the output motions have to be fully decoupled, i.e., delivering the desired DOF in the actuating directions and without any parasitic motion in the non-actuating directions. Based on Hooke's Law, the motion property (coupled or decoupled) of a compliant mechanism is governed by a 6×6 stiffness matrix where the diagonal components represent the stiffness characteristics of all six possible actuation directions, while the non-diagonal components are responsible for the off-axes (or non-actuating) stiffness characteristics.

Past works in the literature have shown that the motion properties of many existing 3L-CPMs were generally neglected [18–22,24,25,27–29,31,33,35,41–43,54]. Based on the derived 6×6 stiffness/compliance matrices, only few recent 3L-CPMs demonstrated decoupled motions [7,52,55,56], while most 3L-CPMs could only deliver coupled motions [5,6]. The main reason is that the existing 3L-CPMs were synthesized with the aim of achieving the desired DOF. As a result, they were able to deliver the motions in the desired actuating directions, but they also produced undesired parasitic motions in the non-actuating directions. More recent efforts have mainly focused on synthesizing 3L-CPMs with a high ratio between the non-actuating stiffness and actuating stiffness [6,7,53,55], so as to keep the undesired parasitic motions to a very small percentage as compared to the actuating motions. In addition, several design criteria for achieving 3L-CPMs with decoupled motion capability were recently presented in Reference [7]. Such criteria were obtained by substituting a number of discrete parameters (orientation and position) of flexure elements into the mathematical model of a 3L-CPM used in a specific structural optimization method [7,56]. Thus, these criteria are not general and cannot be applied in different design methods. Because motion decoupling is an important performance indicator for any positioning system, the criteria that can be used to design decoupled-motion 3L-CPMs regardless of design method and DOF are essential. This paper presents the fundamentals for designing any 3L-CPMs with fully decoupled motion characteristic. This includes several design criteria that need to be fulfilled in order to completely eliminate parasitic motions. The findings that arise from this work suggest that parametric features of flexure elements, such as the orientation and relative position to the end effector, will have a direct impact on the performance of any 3L-CPM in terms of the DOF, the constrained motions, and the parasitic motions.

The remainder of this paper is organized as follows: Section 2 describes the stiffness modeling of a typical 3L-CPM and the stiffness property of each leg. The criteria of flexure elements to design a 3L-CPM with fully decoupled motion are presented in Section 3, and a special case of 3L-CPMs having two reflecting flexure chains in a leg is discussed in Section 4. Section 5 presents a review on the decoupled-motion capability of an existing 3L-CPM, and Section 6 provides discussions about the findings in this work. Lastly, some conclusions are offered in Section 7.

2. Stiffness Modeling of a Decoupled-Motion 3L-CPM

In this work, a CPM is represented by a mechanism having three compliant legs that are distributed symmetrically about the center of the end effector. The legs are fixed at one end, while the free ends are connected with the end effector. Each legs contains a serial chain of flexure elements and rigid links, as illustrated in Figure 1. Here, the global frame, XYZ, is attached to the center of the end effector, and the local frame of each leg, $X'Y'Z'$, is attached at the free end of each leg. Note that the $X'Y'$ plane of the local frame of each leg lies on the same plane as the XY plane of the global frame.

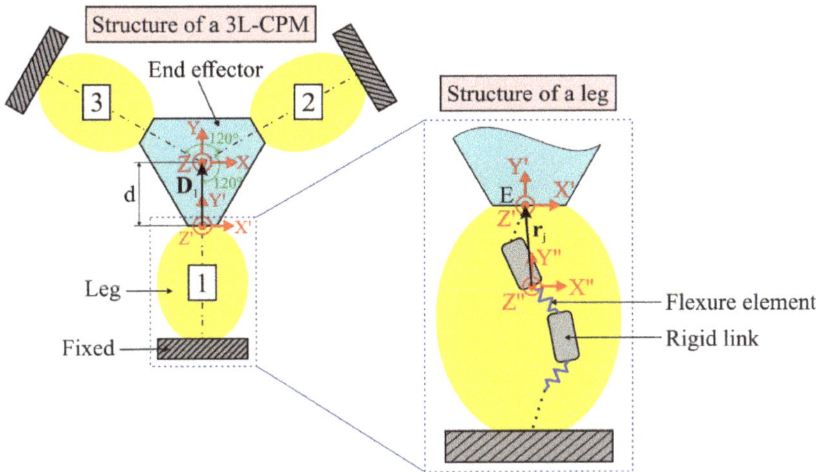

Figure 1. Construction of a general 3L-CPM.

The stiffness property of a 3L-CPM is governed by the stiffness of the legs and the moving stage (end-effector). The stiffness matrix of the leg along the Y axis is represented by $\mathbf{K}^l$ with respect to (w.r.t.) the local frame, i.e., at point E, as shown in Figure 1. With $\mathbf{D}$ as the vector that represents the distance between the local frame of the leg and the global frame of the 3L-CPM, the stiffness matrix of the entire 3L-CPM, $\mathbf{K}^m$, is expressed as follows:

$$\mathbf{K}^m = \sum_{i=1}^{3} \mathbf{J}_i^l \mathbf{R}_i^l \mathbf{K}^l \left(\mathbf{R}_i^l \right)^{-1} \left(\mathbf{J}_i^l \right)^T \tag{1}$$

where $i = 1, 2,$ and 3 denotes the three legs in the CPM, as illustrated in Figure 1; $\mathbf{J}_i^l$ is the translation matrix from the local frame of the ith leg to the global frame; and $\mathbf{R}_i^l$ is the rotation matrix about the Z axis of the ith leg. As three legs are symmetrical and $120°$ apart, $\mathbf{R}_i^l$ and $\mathbf{J}_i^l$ are written as follows:

$$\mathbf{R}_i^l = \begin{bmatrix} \mathbf{R}_z(\theta_i) & 0 \\ 0 & \mathbf{R}_z(\theta_i) \end{bmatrix} \text{ where } \mathbf{R}_z(\theta_i) = \begin{bmatrix} \cos\theta_i & -\sin\theta_i & 0 \\ \sin\theta_i & \cos\theta_i & 0 \\ 0 & 0 & 1 \end{bmatrix} \tag{2}$$

$$\mathbf{J}_i^l = \begin{bmatrix} \mathbf{I} & 0 \\ \mathbf{T}_i^l & \mathbf{I} \end{bmatrix} \text{ where } \mathbf{T}_i^l = \begin{bmatrix} 0 & D_{z_i} & -D_{y_i} \\ -D_{z_i} & 0 & D_{x_i} \\ D_{y_i} & -D_{x_i} & 0 \end{bmatrix} \tag{3}$$

In Equation (2), the values of θ_1, θ_2, and θ_3 are $0°$, $120°$, and $240°$, respectively. In Equation (3), D_{x_i}, D_{y_i}, and D_{z_i} are three components of $\mathbf{D}_i$ and represent the projections of the distance from each local frame to the global frame onto the X, Y, and Z axes, respectively.

Note that $D_{z_i} = 0$, since the $X'Y'$ plane of the local frames lies on the same plane as the XY plane of the global frame.

Here, the stiffness of a general leg, $\mathbf{K}^l$, is represented as follows:

$$
\mathbf{K}^l = \begin{bmatrix}
k_{11}^l & & & & & \\
k_{21}^l & k_{22}^l & & & \text{SYM} & \\
k_{31}^l & k_{32}^l & k_{33}^l & & & \\
k_{41}^l & k_{42}^l & k_{43}^l & k_{44}^l & & \\
k_{51}^l & k_{52}^l & k_{53}^l & k_{54}^l & k_{55}^l & \\
k_{61}^l & k_{62}^l & k_{63}^l & k_{64}^l & k_{65}^l & k_{66}^l
\end{bmatrix}
\tag{4}
$$

where the non-diagonal components are symmetrical. By substituting Equations (2)–(4) into Equation (1), the general form of the stiffness matrix of a 3L-CPM is as follows:

$$
\mathbf{K}^m = \begin{bmatrix}
k_{11}^m & & & & & \\
0 & k_{22}^m & & & \text{SYM} & \\
0 & 0 & k_{33}^m & & & \\
k_{41}^m & k_{42}^m & 0 & k_{44}^m & & \\
k_{51}^m & k_{52}^m & 0 & 0 & k_{55}^m & \\
0 & 0 & k_{63}^m & 0 & 0 & k_{66}^m
\end{bmatrix}
\tag{5}
$$

with d denoting the size of the end effector, as illustrated in Figure 1, the expressions of the non-zero components within the stiffness matrix of Equation (5) are expressed as follows:

$$
\begin{aligned}
k_{11}^m &= k_{22}^m = \tfrac{3}{2}\left(k_{11}^l + k_{22}^l\right) \\
k_{33}^m &= 3k_{33}^l \\
k_{44}^m &= k_{55}^m = \tfrac{3}{2}\left(d^2 k_{33}^l - 2d k_{43}^l + k_{44}^l + k_{55}^l\right) \\
k_{66}^m &= 3\left(d^2 k_{11}^l + 2d k_{61}^l + k_{66}^l\right) \\
k_{41}^m &= k_{52}^m = -\tfrac{3}{2}\left(d k_{31}^l - k_{41}^l - k_{52}^l\right) \\
k_{51}^m &= -k_{42}^m = \tfrac{3}{2}\left(k_{51}^l + d k_{32}^l - k_{42}^l\right) \\
k_{63}^m &= 3\left(d k_{31}^l + k_{63}^l\right)
\end{aligned}
\tag{6}
$$

By referring to Equation (6), the five non-diagonal components within $\mathbf{K}^m$ (k_{41}^m, k_{42}^m, k_{51}^m, k_{52}^m, and k_{63}^m) are represented by seven components within $\mathbf{K}^l$ (k_{31}^l, k_{32}^l, k_{41}^l, k_{42}^l, k_{51}^l, k_{52}^l, and k_{63}^l). To fulfill the requirements of a fully decoupled motion, all non-diagonal components in $\mathbf{K}^m$ must be zeros. In this work, the 6×6 stiffness matrix of a 3L-CPM with fully decoupled motion characteristic is termed as a diagonal stiffness matrix, as shown in Equation (7).

$$
\mathbf{K}^m = \begin{bmatrix}
k_{11}^m & & & & & \\
0 & k_{22}^m & & & \text{SYM} & \\
0 & 0 & k_{33}^m & & & \\
0 & 0 & 0 & k_{44}^m & & \\
0 & 0 & 0 & 0 & k_{55}^m & \\
0 & 0 & 0 & 0 & 0 & k_{66}^m
\end{bmatrix}
\tag{7}
$$

In order to obtain a diagonal stiffness matrix as shown in Equation (7), the five non-diagonal components in Equation (5), for which the expressions are written in Equation (6), must be zeros. With $k_{41}^m = k_{52}^m$ and $k_{51}^m = -k_{42}^m$, the relationship between seven variables (k_{31}^l, k_{32}^l, k_{41}^l, k_{42}^l, k_{51}^l, k_{52}^l, and k_{63}^l) in k_{41}^m, k_{42}^m, k_{51}^m, k_{52}^m, and k_{63}^m can only be represented by three equations, and this condition led to a multiple-solutions problem. Hence, it is important to note that this work only uses a special case (or solution) to demonstrate how the presented mathematical models can be used to synthesize a 3L-CPM with the aim of achieving fully decoupled motion capability. This special case is to make those seven components within $\mathbf{K}^l$ be zeros, as shown in Equation (8).

$$\mathbf{K}^l = \begin{bmatrix} k_{11}^l & & & & & \\ k_{21}^l & k_{22}^l & & & \text{SYM} & \\ 0 & 0 & k_{33}^l & & & \\ 0 & 0 & k_{43}^l & k_{44}^l & & \\ 0 & 0 & k_{53}^l & k_{54}^l & k_{55}^l & \\ k_{61}^l & k_{62}^l & 0 & k_{64}^l & k_{65}^l & k_{66}^l \end{bmatrix} \tag{8}$$

In this work, the two popular-choice flexure elements that are used to synthesize the compliant mechanisms are the beam type and the notch type, as shown in Figure 2. Both elements have thin features which permit elastic bending in a specific direction.

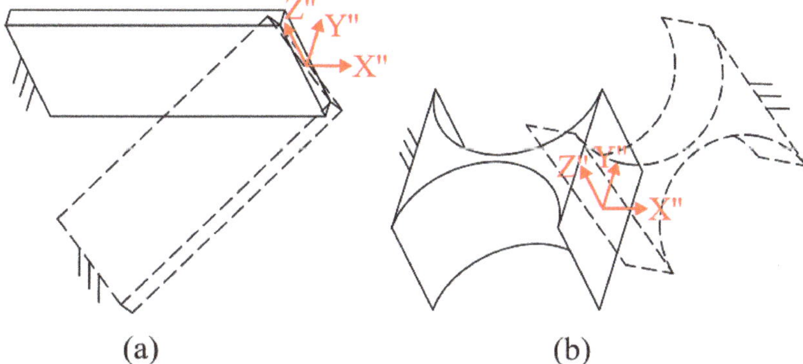

Figure 2. Original orientation of the flexure elements (solid lines): (**a**) beam type and (**b**) notch type, with the local frames, $X''Y''Z''$, attached at the free end and the arbitrary orientation of the flexure elements (dotted-lines) about these local frames.

Assuming that each leg being formed by a serial chain of flexure elements and rigid links where a rigid link has infinite stiffness (non-compliance) property, the compliance of each leg ($\mathbf{C}^l$) is governed by the compliance of each flexure element, $\mathbf{C}_j^e$, expressed as follows:

$$\mathbf{C}^l = \sum_{j=1}^{n} \mathbf{J}_j^e \mathbf{R}_j^e \mathbf{C}_j^e \left(\mathbf{R}_j^e \right)^{-1} \left(\mathbf{J}_j^e \right)^{T} \tag{9}$$

where n denotes the number of flexure elements, and $\mathbf{R}_j^e$ and $\mathbf{J}_j^e$ are the rotation matrix, and translation matrix of the jth flexure element respectively. Referring to Reference [57], the compliance matrix of each original flexure element, $\mathbf{C}_j^e$, with respect to the local frame, as illustrated in Figure 2, is defined as follows:

$$\mathbf{C}_j^e = \begin{bmatrix} c_{11}^e & & & & & \\ 0 & c_{22}^e & & & \text{SYM} & \\ 0 & 0 & c_{33}^e & & & \\ 0 & 0 & 0 & c_{44}^e & & \\ 0 & 0 & c_{53}^e & 0 & c_{55}^e & \\ 0 & c_{62}^e & 0 & 0 & 0 & c_{66}^e \end{bmatrix} \tag{10}$$

Equation (10) is applicable for both the beam type and notch type flexure elements, as illustrated in Figure 2 [57]. In addition, the geometry of each flexure element type can vary without changing the form of the compliant matrix expressed in Equation (10). Several kinds of flexure elements which have a similar form of compliance matrix are presented in

Appendix A. Note that the $X''Y''$ plane of the local frame of each flexure element (Figure 2) lies on the parallel plane with the XY plane of the global frame, as illustrated in Figure 1. By referring to Equation (9), we see that the rotation matrix, $\mathbf{R}_j^e$, is a 6×6 matrix and is defined as the multiplication of the rotation matrices about the X'', Y'', and Z'' axes ($\mathbf{R}_{x_j}^e$, $\mathbf{R}_{y_j}^e$ and $\mathbf{R}_{z_j}^e$). Hence, it is written as follows:

$$\mathbf{R}_j^e = \begin{bmatrix} \mathbf{R}_{z_j}^e \mathbf{R}_{y_j}^e \mathbf{R}_{x_j}^e & 0 \\ 0 & \mathbf{R}_{z_j}^e \mathbf{R}_{y_j}^e \mathbf{R}_{x_j}^e \end{bmatrix} \quad \text{where} \quad \mathbf{R}_{z_j}^e = \begin{bmatrix} \cos \gamma_j & -\sin \gamma_j & 0 \\ \sin \gamma_j & \cos \gamma_j & 0 \\ 0 & 0 & 1 \end{bmatrix},$$

$$\mathbf{R}_{y_j}^e = \begin{bmatrix} \cos \beta_j & 0 & \sin \beta_j \\ 0 & 1 & 0 \\ -\sin \beta_j & 0 & \cos \beta_j \end{bmatrix}, \tag{11}$$

$$\mathbf{R}_{x_j}^e = \begin{bmatrix} 1 & 0 & 0 \\ 0 & \cos \alpha_j & -\sin \alpha_j \\ 0 & \sin \alpha_j & \cos \alpha_j \end{bmatrix}$$

Here, α_j, β_j, and γ_j represent the rotation angles about the X'', Y'', and Z'' axes, respectively. The geometries of a flexure element before and after orientation are also illustrated in Figure 2. As for the 6×6 translation matrix, $\mathbf{J}_j^e$, it represents the projected distances onto the three axes (r_{x_j}, r_{y_j}, and r_{z_j}) from the jth flexure element to the local frame of the leg that are indicated by vector $\mathbf{r}_j$, as shown in Figure 1, written as follows:

$$\mathbf{J}_j^e = \begin{bmatrix} \mathbf{I} & \mathbf{r}_j^e \\ 0 & \mathbf{I} \end{bmatrix} \quad \text{where } \mathbf{r}_j^e = \begin{bmatrix} 0 & r_{z_j} & -r_{y_j} \\ -r_{z_j} & 0 & r_{x_j} \\ r_{y_j} & -r_{x_j} & 0 \end{bmatrix} \tag{12}$$

Using Equations (9)–(12), the compliance matrix of a leg, $\mathbf{C}^l$, can be obtained, and the stiffness matrix of each leg is given as $\mathbf{K}^l = \left(\mathbf{C}^l \right)^{-1}$.

As mentioned earlier, the stiffness matrix of each leg must follow the exact form shown in Equation (8), and this requirement applies to its corresponding compliance matrix too. The detailed derivation of the compliance matrix of each leg with the aim of achieving that requirement is presented in Appendix B. To summarize the results obtained from Appendix B, one condition which allows for the compliance matrix of a leg to become the exact same form as Equation (8) is for seven components within the compliance matrix of a leg to be zeros, i.e., $c_{31}^l = c_{32}^l = c_{41}^l = c_{42}^l = c_{51}^l = c_{52}^l = c_{63}^l = 0$. This condition offers simplicity during the design stage and can be used as the standard approach to synthesize 3L-CPMs with the aim of achieving fully decoupled motion capability. However, it also introduces a multiple-solutions problem to solve the corresponding components within the stiffness matrix, $\mathbf{K}^l$. Among a number of possible solutions, $k_{64}^l = k_{65}^l = 0$ is a unique solution that is used to fulfill the condition in this work. By adopting this unique solution, both the stiffness matrix and the compliance matrix of a leg will have the same form as expressed in Equation (13), and the expression of each component within $\mathbf{K}^l$ is given in Appendix C.

$$\underbrace{\begin{bmatrix} k_{11}^l & & & & & \\ k_{21}^l & k_{22}^l & & & \text{SYM} & \\ 0 & 0 & k_{33}^l & & & \\ 0 & 0 & k_{43}^l & k_{44}^l & & \\ 0 & 0 & k_{53}^l & k_{54}^l & k_{55}^l & \\ k_{61}^l & k_{62}^l & 0 & 0 & 0 & k_{66}^l \end{bmatrix}}_{\mathbf{K}^l} = \underbrace{\begin{bmatrix} c_{11}^l & & & & & \\ c_{21}^l & c_{22}^l & & & \text{SYM} & \\ 0 & 0 & c_{33}^l & & & \\ 0 & 0 & c_{43}^l & c_{44}^l & & \\ 0 & 0 & c_{53}^l & c_{54}^l & c_{55}^l & \\ c_{61}^l & c_{62}^l & 0 & 0 & 0 & c_{66}^l \end{bmatrix}^{-1}}_{\left(\mathbf{C}^l\right)^{-1}} \tag{13}$$

3. Characteristics of Flexure Elements in Decoupled-Motion 3L-CPMs

In this section, the desired rotation angles (α, β, and γ) and distances (r_x, r_y, and r_z) of flexure elements in a leg to achieve decoupled-motion capability are analyzed. First, Equation (9) is re-expressed as follows:

$$\mathbf{K}^l = \left(\mathbf{C}^l\right)^{-1} = \left(\sum_{j=1}^{n} \mathbf{J}_j^e \mathbf{R}_j^e \mathbf{C}_j^e \left(\mathbf{R}_j^e\right)^{-1} \left(\mathbf{J}_j^e\right)^T\right)^{-1} = \left(\sum_{j=1}^{n} \overline{\mathbf{C}}_j^e\right)^{-1} \tag{14}$$

where $\overline{\mathbf{C}}_j^e$ indicates the compliance matrix of the jth oriented flexure element referring to the local frame $X'Y'Z'$ attached to the free end of the leg, as illustrated in Figure 1. In order to achieve fully decoupled motion, the nine non-diagonal components (c_{31}^l, c_{41}^l, c_{51}^l, c_{32}^l, c_{42}^l, c_{52}^l, c_{63}^l, c_{64}^l, and c_{65}^l) of the compliance matrix, $\mathbf{C}^l$, must be zeros, as shown in Equation (13). As $\mathbf{C}^l$ is the sum of n sub-components, $\overline{\mathbf{C}}_j^e$, there could be numerous solutions, because all components within $\overline{\mathbf{C}}_j^e$ can have arbitrary values in general cases. In this work, a special case where all $\overline{\mathbf{C}}_j^e$ have the same form ($\overline{\mathbf{C}}^e$) is considered and yields the following:

$$\overline{\mathbf{C}}^e = \begin{bmatrix}
\overline{c}_{11}^e & & & & & \\
\overline{c}_{21}^e & \overline{c}_{22}^e & & & \text{SYM} & \\
\overline{c}_{31}^e = 0 & \overline{c}_{32}^e = 0 & \overline{c}_{33}^e & & & \\
\overline{c}_{41}^e = 0 & \overline{c}_{42}^e = 0 & \overline{c}_{43}^e & \overline{c}_{44}^e & & \\
\overline{c}_{51}^e = 0 & \overline{c}_{52}^e = 0 & \overline{c}_{53}^e & \overline{c}_{54}^e & \overline{c}_{55}^e & \\
\overline{c}_{61}^e & \overline{c}_{62}^e & \overline{c}_{63}^e = 0 & \overline{c}_{64}^e = 0 & \overline{c}_{65}^e = 0 & \overline{c}_{66}^e
\end{bmatrix} \tag{15}$$

Note that $\overline{\mathbf{C}}^e$ in Equation (15) is different from $\mathbf{C}^e$ in Equation (10), since $\overline{\mathbf{C}}^e = \mathbf{J}^e \mathbf{R}^e \mathbf{C}^e (\mathbf{R}^e)^{-1} (\mathbf{J}^e)^T$, as expressed in Equation (14). $\mathbf{R}^e$ and $\mathbf{J}^e$ are similar to Equations (11) and (12) with the subscript, j, being removed.

Equation (15) describes the condition to obtain a 3L-CPM with fully decoupled motion capability, expressed by nine equations that can be obtained based on those nine zero components. There are six unknowns in these equations, i.e., the rotation angles (α, β, and γ) and the distances (r_x, r_y, and r_z) measured from the moving end of the flexure element to the free end of the leg.

$$\begin{aligned}
\overline{c}_{31}^e =\ & -c_{11}^e \sin\beta \cos\beta \cos\gamma + c_{44}^e \cos\beta (r_y \cos\gamma - r_x \sin\gamma)(r_y \sin\beta + r_z \cos\beta \sin\gamma) - \\
& \cos\beta \sin\alpha \{\cos\gamma \sin\alpha (c_{62}^e r_z - c_{22}^e \sin\beta) + \cos\alpha [c_{62}^e r_y \cos\beta + (c_{22}^e - c_{62}^e r_z \sin\beta) \sin\gamma]\} + \\
& \cos\alpha \cos\beta \{\cos\alpha \cos\gamma (c_{53}^e r_z + c_{33}^e \sin\beta) + \sin\alpha [-c_{53}^e r_y \cos\beta + (c_{33}^e + c_{53}^e r_z \sin\beta) \sin\gamma]\} + \\
& [\sin\alpha \sin\beta (r_y \cos\gamma - r_x \sin\gamma) - \cos\alpha (r_x \cos\gamma + r_y \sin\gamma)] \cdot \\
& \{\cos\alpha \cos\gamma (c_{55}^e r_z + c_{53}^e \sin\beta) + \sin\alpha [-c_{55}^e r_y \cos\beta + (c_{53}^e + c_{55}^e r_z \sin\beta) \sin\gamma]\} + \\
& [\cos\gamma (r_x \sin\alpha + r_y \cos\alpha \sin\beta) + (r_y \sin\alpha - r_x \cos\alpha \sin\beta) \sin\gamma] \cdot \\
& \{\cos\gamma \sin\alpha (-c_{66}^e r_z + c_{62}^e \sin\beta) - \cos\alpha [c_{66}^e r_y \cos\beta + (c_{62}^e - c_{66}^e r_z \sin\beta) \sin\gamma]\}
\end{aligned} \tag{16}$$

$$\begin{aligned}
\overline{c}_{41}^e =\ & c_{44}^e \cos\beta \cos\gamma (r_y \sin\beta + r_z \cos\beta \sin\gamma) + (\cos\gamma \sin\alpha \sin\beta - \cos\alpha \sin\gamma) \cdot \\
& \{\cos\alpha \cos\gamma (c_{55}^e r_z + c_{53}^e \sin\beta) + \sin\alpha [-c_{55}^e r_y \cos\beta + (c_{53}^e + c_{55}^e r_z \sin\beta) \sin\gamma]\} + \\
& (\cos\alpha \cos\gamma \sin\beta + \sin\alpha \sin\gamma) \cdot \\
& \{\cos\gamma \sin\alpha (-c_{66}^e r_z + c_{62}^e \sin\beta) - \cos\alpha [c_{66}^e r_y \cos\beta + (c_{62}^e - c_{66}^e r_z \sin\beta) \sin\gamma]\}
\end{aligned} \tag{17}$$

$$\begin{aligned}
\overline{c}_{51}^e =\ & c_{44}^e \cos\beta \sin\gamma (r_y \sin\beta + r_z \cos\beta \sin\gamma) + (\cos\alpha \cos\gamma + \sin\alpha \sin\beta \sin\gamma) \cdot \\
& \{\cos\alpha \cos\gamma (c_{55}^e r_z + c_{53}^e \sin\beta) + \sin\alpha [-c_{55}^e r_y \cos\beta + (c_{53}^e + c_{55}^e r_z \sin\beta) \sin\gamma]\} + \\
& (-\cos\gamma \sin\alpha + \cos\alpha \sin\beta \sin\gamma) \cdot \\
& \{\cos\gamma \sin\alpha (-c_{66}^e r_z + c_{62}^e \sin\beta) - \cos\alpha [c_{66}^e r_y \cos\beta + (c_{62}^e - c_{66}^e r_z \sin\beta) \sin\gamma]\}
\end{aligned} \tag{18}$$

$$\begin{aligned}
\overline{c}^{\,e}_{32} &= -c^e_{11}\cos\beta\sin\beta\sin\gamma - c^e_{44}\cos\beta(r_z\cos\beta\cos\gamma + r_x\sin\beta)(r_y\cos\gamma - r_x\sin\gamma) + \\
&\quad \cos\beta\sin\alpha\{\cos\alpha[c^e_{62}r_x\cos\beta + \cos\gamma(c^e_{22} - c^e_{62}r_z\sin\beta)] + \sin\alpha(-c^e_{62}r_z + c^e_{22}\sin\beta)\sin\gamma\} + \\
&\quad \cos\alpha\cos\beta[c^e_{53}r_x\cos\beta\sin\alpha - \cos\gamma\sin\alpha(c^e_{33} + c^e_{53}r_z\sin\beta) + \cos\alpha(c^e_{53}r_z + c^e_{33}\sin\beta)\sin\gamma] + \\
&\quad \{\cos\alpha[c^e_{66}r_x\cos\beta + \cos\gamma(c^e_{62} - c^e_{66}r_z\sin\beta)] + \sin\alpha(-c^e_{66}r_z + c^e_{62}\sin\beta)\sin\gamma\}\cdot \\
&\quad [\cos\gamma(r_x\sin\alpha + r_y\cos\alpha\sin\beta) + (r_y\sin\alpha - r_x\cos\alpha\sin\beta)\sin\gamma] + \\
&\quad [c^e_{55}r_x\cos\beta\sin\alpha - \cos\gamma\sin\alpha(c^e_{53} + c^e_{55}r_z\sin\beta) + \cos\alpha(c^e_{55}r_z + c^e_{53}\sin\beta)\sin\gamma]\cdot \\
&\quad [\sin\alpha\sin\beta(r_y\cos\gamma - r_x\sin\gamma) - \cos\alpha(r_x\cos\gamma + r_y\sin\gamma)]
\end{aligned}\tag{19}$$

$$\begin{aligned}
\overline{c}^{\,e}_{42} &= -c^e_{44}\cos\beta\cos\gamma(r_z\cos\beta\cos\gamma + r_x\sin\beta) + (\cos\alpha\cos\gamma\sin\beta + \sin\alpha\sin\gamma)\cdot \\
&\quad \{\cos\alpha[c^e_{66}r_x\cos\beta + \cos\gamma(c^e_{62} - c^e_{66}r_z\sin\beta)] + \sin\alpha(-c^e_{66}r_z + c^e_{62}\sin\beta)\sin\gamma\} + \\
&\quad (\cos\gamma\sin\alpha\sin\beta - \cos\alpha\sin\gamma)\cdot \\
&\quad [c^e_{55}r_x\cos\beta\sin\alpha - \cos\gamma\sin\alpha(c^e_{53} + c^e_{55}r_z\sin\beta) + \cos\alpha(c^e_{55}r_z + c^e_{53}\sin\beta)\sin\gamma]
\end{aligned}\tag{20}$$

$$\begin{aligned}
\overline{c}^{\,e}_{52} &= -c^e_{44}\cos\beta(r_z\cos\beta\cos\gamma + r_x\sin\beta)\sin\gamma + (-\cos\gamma\sin\alpha + \cos\alpha\sin\beta\sin\gamma)\cdot \\
&\quad \{\cos\alpha[c^e_{66}r_x\cos\beta + \cos\gamma(c^e_{62} - c^e_{66}r_z\sin\beta)] + \sin\alpha(-c^e_{66}r_z + c^e_{62}\sin\beta)\sin\gamma\} + \\
&\quad (\cos\alpha\cos\gamma + \sin\alpha\sin\beta\sin\gamma)\cdot \\
&\quad [c^e_{55}r_x\cos\beta\sin\alpha - \cos\gamma\sin\alpha(c^e_{53} + c^e_{55}r_z\sin\beta) + \cos\alpha(c^e_{55}r_z + c^e_{53}\sin\beta)\sin\gamma]
\end{aligned}\tag{21}$$

$$\begin{aligned}
\overline{c}^{\,e}_{63} &= \tfrac{1}{2}\cos\beta\{(c^e_{62} + c^e_{53})\cos\beta\sin2\alpha + [2c^e_{44} - c^e_{55} - c^e_{66} + (c^e_{55} - c^e_{66})\cos2\alpha]\cdot \\
&\quad \sin\beta(-r_y\cos\gamma + r_x\sin\gamma) - (c^e_{55} - c^e_{66})\sin2\alpha(r_x\cos\gamma + r_y\sin\gamma)\}
\end{aligned}\tag{22}$$

$$\overline{c}^{\,e}_{64} = \cos\beta\left[\cos\gamma\left(-c^e_{44} + c^e_{66}\cos^2\alpha + c^e_{55}\sin^2\alpha\right)\sin\beta + (-c^e_{55} + c^e_{66})\cos\alpha\sin\alpha\sin\gamma\right]\tag{23}$$

$$\overline{c}^{\,e}_{65} = \cos\beta\left[(c^e_{55} - c^e_{66})\cos\alpha\cos\gamma\sin\alpha + c^e_{66}\cos^2\alpha\sin\beta\sin\gamma + \left(-c^e_{44} + c^e_{55}\sin^2\alpha\right)\sin\beta\sin\gamma\right]\tag{24}$$

This set of equations can be solved by considering Equation (22) first, because β is the only dominant angular variable, and the results are given as follows:

$$\overline{c}^{\,e}_{63} = 0 \Leftrightarrow \left[\begin{array}{l} \beta = 90° \\ \alpha = 0° \text{ and } \beta = 0°,\ 180° \\ \alpha = 90° \text{ and } \beta = 0°,\ 180° \end{array}\right. \quad ;\forall\gamma, r_x, r_y, r_z \tag{25}$$

Here, the rotation angle about the X'' axis, α, varies from $0°$ to $90°$, because of the symmetrical structure of the flexure elements, as illustrated in Figure 2. Equation (25) shows that there are three possible cases for, $\overline{c}^{\,e}_{63} = 0$ with the four remaining variables (γ, r_x, r_y, and r_z) being arbitrary values.

First, the case with $\beta = 90°$ is considered. With every component within the compliance matrix of the flexure element having a specific value, as shown in Equation (10), six Equations, from (16) to (21), are always different from zero with any value of γ, r_x, r_y, and r_z. Hence, $\beta = 90°$ is not a feasible solution.

Next, consider the second case with $\alpha = 0°$ and $\beta = 0°$, $180°$; the following results can be obtained:

$$\bar{c}^e_{31} = \pm r_z\left\{\mp c^e_{55} r_x \cos^2 \gamma \mp c^e_{44} r_x \sin^2 \gamma + \cos \gamma\left[c^e_{53} + (\pm c^e_{44} \mp c^e_{55}) r_y \sin \gamma\right]\right\}$$

$$\bar{c}^e_{41} = (c^e_{44} - c^e_{55}) r_z \cos \gamma \sin \gamma$$

$$\bar{c}^e_{51} = r_z\left(c^e_{55} \cos^2 \gamma + c^e_{44} \sin^2 \gamma\right)$$

$$\bar{c}^e_{32} = \pm r_z\left[\mp c^e_{44} r_y \cos^2 \gamma + (\pm c^e_{44} \mp c^e_{55}) r_x \cos \gamma \sin \gamma + \sin \gamma (c^e_{53} \mp c^e_{55} r_y \sin \gamma)\right] \tag{26}$$

$$\bar{c}^e_{42} = -r_z\left(c^e_{44} \cos^2 \gamma + c^e_{55} \sin^2 \gamma\right)$$

$$\bar{c}^e_{52} = (-c^e_{44} + c^e_{55}) r_z \cos \gamma \sin \gamma$$

$$\bar{c}^e_{64} = \bar{c}^e_{65} = 0$$

Here, the upper signs of "$\pm$" and "$\mp$" in Equation (26) represent the case of $\alpha = 0°$ and $\beta = 0°$, while the lower signs represent the case of $\alpha = 0°$ and $\beta = 180°$. To solve Equation (26), $\bar{c}^e_{51}$ is first considered to be equal to zero, a unique solution can be obtained $r_z = 0$ by, and γ can be any value. With $r_z = 0$, all the remaining equations in Equation (26) will also be equal to zeros. Hence, $r_z = 0$, $\alpha = 0°$, $\beta = 0°$ or $180°$ are solutions used to obtain a 3L-CPM with any DOF and decoupled motions.

Similarly, for the case with $\alpha = 90°$ and $\beta = 0°$ or $180°$, it can be shown that this is also a possible solution. In summary, the two feasible solutions found from Equation (25) with $r_z = 0$ are as follows:

$$\begin{bmatrix} \alpha = 0° \text{ and } \beta = 0°, 180° \\ \alpha = 90° \text{ and } \beta = 0°, 180° \end{bmatrix} ; \forall \gamma, r_x, r_y \tag{27}$$

Equation (27) provides the design criteria for the orientations and positions of the flexure elements that need to be satisfied to design a 3L-CPM with fully decoupled motion capability. Figure 3 illustrates the desired orientation of the flexure elements about the X'' and Y'' axes, with various orientations about the Z'' axis. With flexure elements in a leg having these orientations and distributing in the $X'Y'$ plane ($r_z = 0$), a 3L-CPM is able to achieve fully decoupled motions.

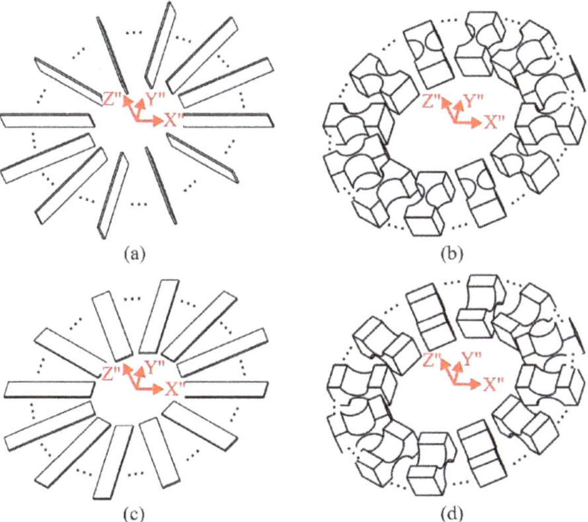

Figure 3. Orientations of the flexure elements in fully decoupled motion 3L-CPMs: (**a**) beam-type and (**b**) notch-type with $\alpha = 0°$, $\beta = 0°$ or $180°$, and $r_z = 0$, respectively. (**c**) Beam-type and (**d**) notch-type with $\alpha = 90°$, $\beta = 0°$, $180°$, and $r_z = 0$ respectively.

4. Stiffness Analysis of 3L-CPMs Containing Two Serial Flexure Chains in a Leg

In Section 2, the stiffness modeling of a general 3L-CPM, with each leg consisting of a single serial flexure chain, is presented. However, there are many existing 3L-CPMs synthesized by constraint-based and optimization methods that have two reflecting (or symmetrical) serial flexure chains, as shown in Figure 4 [6,7,28,31,42,43,53,55,56,58]. In this section, the analysis of such a leg configuration is presented.

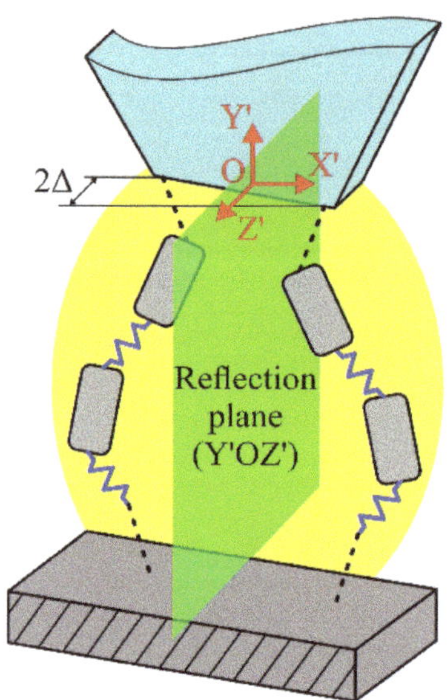

Figure 4. Construction of a 3L-CPM leg containing two reflecting serial flexure chains.

From the literature [6,7,28,31,42,43,53,55,56,58], we can see that the double flexure chains are either on the same plane or have an offset distance of 2Δ along the Z' axis, as shown in Figure 4. The stiffness matrix of each leg is expressed as follows:

$$\mathbf{K}^l = \left(\sum_{j=1}^{n} \mathbf{J}_{(+\Delta)} \mathbf{C}^{sc} \mathbf{J}_{(+\Delta)}^{T} \right)^{-1} + \left(\sum_{j=1}^{n} \mathbf{J}_{(-\Delta)} \mathbf{M} \mathbf{C}^{sc} \mathbf{M}^{T} \mathbf{J}_{(-\Delta)}^{T} \right)^{-1} \tag{28}$$

where $\mathbf{M}$ is the reflection matrix about the $Y'Z'$ plane given in Equation (29); and $\mathbf{J}_{(+\Delta)}$ and $\mathbf{J}_{(-\Delta)}$ represent the offset matrices used to shift the original flexure chain and the reflecting flexure chain along the Z'-axis distances of $+\Delta$ and $-\Delta$, respectively, as given in Equation (30).

$$\mathbf{M} = \begin{bmatrix} -1 & 0 & 0 & 0 & 0 & 0 \\ 0 & 1 & 0 & 0 & 0 & 0 \\ 0 & 0 & 1 & 0 & 0 & 0 \\ 0 & 0 & 0 & -1 & 0 & 0 \\ 0 & 0 & 0 & 0 & 1 & 0 \\ 0 & 0 & 0 & 0 & 0 & 1 \end{bmatrix} \tag{29}$$

$$\mathbf{J}_{(\pm\Delta)} = \begin{bmatrix} 1 & 0 & 0 & 0 & 0 & 0 \\ 0 & 1 & 0 & 0 & 0 & 0 \\ 0 & 0 & 1 & 0 & 0 & 0 \\ 0 & \pm\Delta & 0 & 1 & 0 & 0 \\ \mp\Delta & 0 & 0 & 0 & 1 & 0 \\ 0 & 0 & 0 & 0 & 0 & 1 \end{bmatrix} \tag{30}$$

Here, $\mathbf{C}^{sc}$ is the compliance matrix of a serial flexure chain that can be calculated by Equation (9). The results from Section 3 are used to analyze the stiffness property of 3L-CPMs with two reflecting flexure chains. After substituting Equations (9), (14), (29) and (30) into Equation (28), the results show that the decoupled motion capability can only be achieved when the offset distance $2\Delta = 0$ (or $\Delta = 0$). The offset distance can be considered as the translation component along the Z axis of each flexure elements (r_{z_j}) that can lead coupled motions, as mentioned before. Most important, to achieve fully decoupled output motion, $\mathbf{C}^{sc}$ must be in the following form:

$$
\mathbf{C}^{sc} = \begin{bmatrix}
c_{11}^{sc} & & & & & \\
c_{21}^{sc} & c_{22}^{sc} & & & \text{SYM} & \\
0 & 0 & c_{33}^{sc} & & & \\
0 & 0 & c_{43}^{sc} & c_{44}^{sc} & & \\
0 & 0 & c_{53}^{sc} & c_{54}^{sc} & c_{55}^{sc} & \\
c_{61}^{sc} & c_{62}^{sc} & 0 & 0 & 0 & c_{66}^{sc}
\end{bmatrix}
\tag{31}
$$

5. Review and Analysis of the Motion Characteristics of Existing 3L-CPMs

In Sections 2–4, 3L-CPMs constructed with single or double serial flexure chains in a leg together with their corresponding stiffness modeling and decoupled design criteria were presented. In this section, the proposed criteria are used to analyze the decoupled-motion capability of some existing 3L-CPM designs. It can be seen that the analysis provides insight into the motion characteristics of 3L-CPMs. With the design criteria for synthesizing 3L-CPMs with fully decoupled motions being provided and the motion-decoupling capability of popular 3L-CPM designs being clearly defined, designers can select or synthesize any 3L-CPM with suitable motion property (coupled or decoupled) for their specific applications.

This section presents a short review on the motion characteristics from some of the existing 3L-CPMs. For each CPM, either the compliance matrix of a leg/serial flexure chain or the stiffness/compliance matrix of the entire mechanism was derived and used to determine the theoretical motion characteristics. For 3L-CPMs having a single serial flexure chain in each leg, fully decoupled motion can be achieved if the compliance matrix of the leg is in the form of Equation (13). The full stiffness matrix of the entire 3L-CPMs can then be calculated by using Equations (5), (6) and (14). For 3L-CPMs containing two reflecting serial flexure chains in each leg, fully decoupled motion can be obtained if the compliance matrix of a serial flexure chain is similar to Equation (31), and the full stiffness matrix of entire CPM can be calculated based on Equations (5), (6) and (28). Note that the analyses were conducted for general cases where most parameters, such as rotations angles and distances, are symbolic. Therefore, the obtained results are in general forms, while the results for some specific designs were obtained by substituting values into the parameters.

5.1. Three-Legged Revolute–Revolute–Revolute and Three-Legged Prismatic–Revolute–Revolute CPMs

The Three-legged Revolute–Revolute–Revolute (3RRR) configuration has been the most popular design for developing 3-DOF 3L-CPMs with (X-Y-θ_Z) planar motions, which are widely used in positioning/alignment systems [18–20,22,27,29,35,41]. The schematic diagram of a typical 3RRR-CPM is illustrated in Figure 5a, while Figure 6a illustrates the three-legged Prismatic–Revolute–Revolute (3PRR) configuration, which is a variant of 3RRR, where a revolute joint is replaced by a prismatic joint. The 3PRR-CPMs are preferred to create 3-DOF planar motions in MEMS devices, due to their advantage in actuation [59,60]. The prismatic joint can be designed as an active joint and easily driven by a linear actuator. Note that all flexure elements must be in the XY plane to create three desired motions. The physical prototypes of positioning systems developed based on 3RRR- and 3PRR-CPMs are shown in Figures 5b and 6b, respectively.

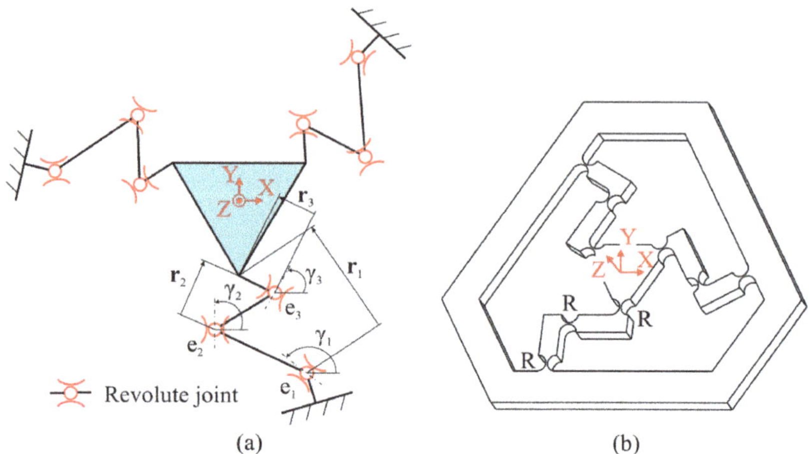

Figure 5. (**a**) Schematic diagram of 3RRR-CPM and (**b**) micro-motion stage developed based on 3RRR-CPM.

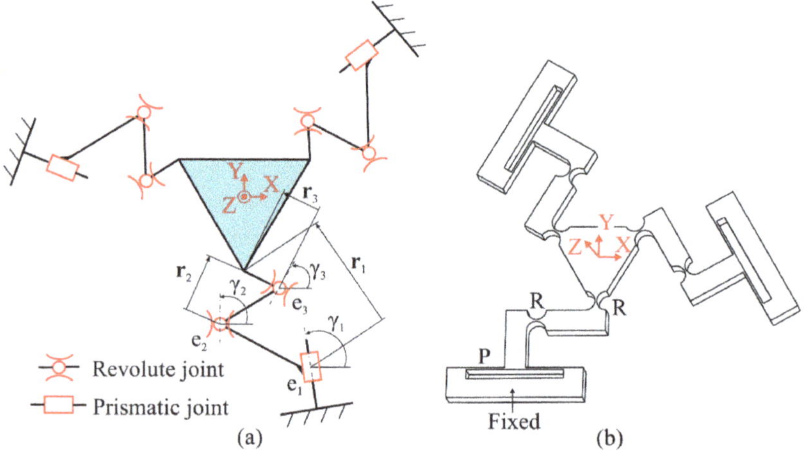

Figure 6. (**a**) Schematic diagram of 3PRR-CPM and (**b**) a positioning stage developed based on 3PRR-CPM.

To determine the theoretical motion characteristics of 3RRR-CPM or 3PRR-CPM, the compliance matrix of each flexure element is represented by $\mathbf{C}_1^e$, $\mathbf{C}_2^e$, and $\mathbf{C}_3^e$, respectively, and each compliance matrix has a similar form to the one shown in Equation (10). Here, $\mathbf{C}_1^e$ is the compliance matrix of either the first revolute joint within a 3RRR-CPM or the first prismatic joint within a 3PRR-CPM. $\mathbf{C}_2^e$ and $\mathbf{C}_3^e$ are the compliance matrices of the second and third revolute joints, respectively. As illustrated in Figures 5a and 6a, each joint has a different orientation about the Z'' axis, γ_j, and a different distance to the moving end point, $\mathbf{r}_j$, with $j = 1, 2$, and 3. The projections of $\mathbf{r}_j$ onto the X and Y axes are r_{x_j} and r_{y_j}, respectively. Based on Equation (9), the formula to calculate the compliance matrix of a leg is written as follows:

$$\mathbf{C}^l = \sum_{j=1}^{3} \mathbf{J}_j^e \mathbf{R}_j^e \mathbf{C}_j^e \left(\mathbf{R}_j^e\right)^{-1} \left(\mathbf{J}_j^e\right)^T \tag{32}$$

where $\mathbf{J}_j^e$ and $\mathbf{R}_j^e$ are obtained by using Equations (12) and (11), respectively. The result of Equation (32) is shown in Equation (33), and the detailed expression of each component is given in Appendix D. It is observed that the compliance matrix of each leg within the 3RRR-CPMs and 3PRR-CPMs is similar to the form expressed in Equation (13). Hence, this observation suggests that the 3RRR and 3PRR configurations are able to deliver fully decoupled motion. This performance indicator also highlighted why both configurations are popular designs for developing state-of-the-art 3L-CPMs.

$$
\mathbf{C}^l = \begin{bmatrix}
c^l_{11} & & & & & \\
c^l_{21} & c^l_{22} & & & \text{SYM} & \\
0 & 0 & c^l_{33} & & & \\
0 & 0 & c^l_{43} & c^l_{44} & & \\
0 & 0 & c^l_{53} & c^l_{54} & c^l_{55} & \\
c^l_{61} & c^l_{62} & 0 & 0 & 0 & c^l_{66}
\end{bmatrix}
\tag{33}
$$

5.2. Three-Legged Prismatic–Prismatic–Spherical and Three-Legged Revolute–Prismatic–Spherical CPMs

Based on past works in the literature, the three-legged Prismatic–Prismatic–Spherical (3PPS) [5,32] and the three-legged Revolute–Prismatic–Spherical (3RPS) [21,61] configurations were adopted to develop 3-DOF 3L-CPMs with (θ_X-θ_Y-Z) out-of-plane motions, as shown in Figure 7a,b respectively. For both 3L-CPMs, each leg employed three serially connected flexure elements along the Z axis to create the desired motions. As a result, the component r_{z_j} in Equation (12) exists and generates the off-axis components in the compliance matrix of a flexure element after transformation ($\overline{\mathbf{C}}^e_j$). Here, the components within $\mathbf{C}^l$ can be derived from the following:

$$
c^l_{ab} = \sum_{j=1}^{3} \overline{c}^{ej}_{ab}
\tag{34}
$$

where $\overline{c}^{ej}_{ab}$ is a component on row a and column b in $\overline{\mathbf{C}}^e_j$ calculated by Equation (14).

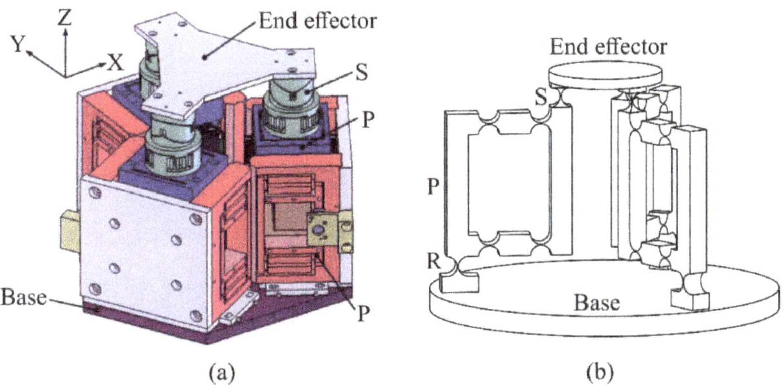

(a)   (b)

Figure 7. Structure of (**a**) 3PPS CPM [62] and (**b**) 3RPS CPM modeled based on the design presented in [21].

Based on the results obtained from Equation (16) to Equation (24), the components, i.e., $\overline{c}^{ej}_{31}$, $\overline{c}^{ej}_{41}$, $\overline{c}^{ej}_{51}$, $\overline{c}^{ej}_{32}$, $\overline{c}^{ej}_{42}$, and $\overline{c}^{ej}_{52}$, are non-zeros because of the existence of r_{z_j}. Thus, the corresponding components in $\mathbf{C}^l$ are also non-zeros. Therefore, the legs' compliance matrices within the 3PPS- and 3RPS-CPMs do not satisfy Equation (13). As a result, both the 3-DOF 3PPS- and 3RPS-CPMs will generate coupled motions. This coupled motion property can be observed from Equation (35), which is the 6×6 stiffness matrix of the 3PPS-CPM (Figure 7a) taken from Reference [5]. Having those five non-diagonal components indicates that the developed 3PPS-CPM cannot deliver fully decoupled motion.

$$\mathbf{K} = \begin{bmatrix} 2.57 \times 10^5 & & & & & \\ 0 & 2.57 \times 10^5 & & & \text{SYM} & \\ 0 & 0 & 3.99 \times 10^5 & & & \\ 5.47 \times 10^1 & 5.18 \times 10^3 & 0 & 9.51 \times 10^2 & & \\ -5.18 \times 10^3 & 5.47 \times 10^1 & 0 & 0 & 9.51 \times 10^2 & \\ 0 & 0 & -5.63 \times 10^1 & 0 & 0 & 1.72 \times 10^3 \end{bmatrix} \tag{35}$$

5.3. Three-Legged Prismatic–Revolution–Prismatic–Revolution CPMs

Based on past works in the literature, several 3-DOF 3L-CPMs with spatial motions capability have been developed by using the three-legged Prismatic–Resolution–Prismatic–Revolution (3PRPR) configuration. Depending on the orientations of the prismatic and revolute joints, a 3PRPR-CPM can deliver either $(X\text{-}Y\text{-}Z)$ [24,25] or $(\theta_X\text{-}\theta_Y\text{-}Z)$ [16,17] spatial motions, as shown in Figure 8a,b respectively. For the 3PRPR-CPM with $(X\text{-}Y\text{-}Z)$ spatial motion, as shown in Figure 8a, all the P-joints and R-joints within each leg have their respective $X''Y''$ planes. The P-joints operate in their respective $X''Y''$ planes, while the R-joints rotate out of these planes. To analyze its compliance behavior, the leg parallel to the Y axis of the global frame was selected. Note that detailed modeling follows the procedures presented in Section 2. Here, each leg consists of four flexure elements (e_1, e_2, e_3, and e_4), representing the P-, R-, P-, and R-joints, respectively. The local frame $X'Y'Z'$ is attached at the free end of the leg. The compliance matrix of each flexure element, $\mathbf{C}_j^e$ (with $j = 1, 2, 3,$ 4), is given as follows:

$$\mathbf{C}_1^e = \mathbf{C}_3^e = \begin{bmatrix} c_{11}^P & & & & & \\ 0 & c_{22}^P & & & \text{SYM} & \\ 0 & 0 & c_{33}^P & & & \\ 0 & 0 & 0 & c_{44}^P & & \\ 0 & 0 & c_{53}^P & 0 & c_{55}^P & \\ 0 & c_{62}^P & 0 & 0 & 0 & c_{66}^P \end{bmatrix}, \text{ and } \mathbf{C}_2^e = \mathbf{C}_4^e = \begin{bmatrix} c_{11}^R & & & & & \\ 0 & c_{22}^R & & & \text{SYM} & \\ 0 & 0 & c_{33}^R & & & \\ 0 & 0 & 0 & c_{44}^R & & \\ 0 & 0 & c_{53}^R & 0 & c_{55}^R & \\ 0 & c_{62}^R & 0 & 0 & 0 & c_{66}^R \end{bmatrix} \tag{36}$$

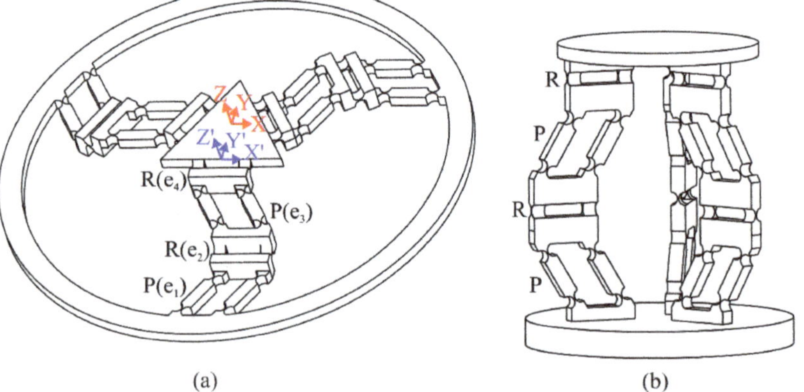

Figure 8. Prototypes of 3PRPR-CPMs with (**a**) $X\text{-}Y\text{-}Z$ motions modeled based on the design presented in [24] and (**b**) $\theta_X\text{-}\theta_Y\text{-}Z$ motions modeled based on the design presented in [17], respectively.

Based on Figures 2b, 8a, and A1b, the two P-joints can have arbitrary rotation angles (γ_1 and γ_3) about their Z'' axes, while the two R-joints have two rotations, i.e., about the X'' axes with an angle of 90° ($\alpha_2 = \alpha_4 = 90°$) and about the Z'' axes with an angle of 90° ($\gamma_2 = \gamma_4 = 90°$). All flexure elements have no rotation about their Y'' axes ($\beta_1 = \beta_2 = \beta_3 = \beta_4 = 0$). In addition, as the first three flexure elements (e_1, e_2, and e_3) are located at specific distances from the local frame of the leg $X'Y'Z'$, the distance vectors are $\mathbf{r}_1 = \{r_{x_1}, r_{y_1}, r_{z_1}\}$,

$\mathbf{r}_2 = \{r_{x_2}, r_{y_2}, r_{z_2}\}$, and $\mathbf{r}_3 = \{r_{x_3}, r_{y_3}, r_{z_3}\}$, respectively, as illustrated in Figure 1 and Equation (12). Moreover, $\mathbf{r}_4$ is a zero vector, because the free end of the last R-joint (e_4) coincides with the frame $X'Y'Z'$. By substituting these parameters into Equations (9), (11) and (12), the compliance matrix of the proposed leg can be expressed as follows:

$$
\mathbf{C}^l =
\begin{bmatrix}
c_{11}^l & & & & & \\
c_{21}^l & c_{22}^l & & & \text{SYM} & \\
c_{31}^l & c_{32}^l & c_{33}^l & & & \\
c_{41}^l & c_{42}^l & c_{43}^l & c_{44}^l & & \\
c_{51}^l & c_{52}^l & c_{53}^l & c_{54}^l & c_{55}^l & \\
c_{61}^l & c_{62}^l & 0 & 0 & 0 & c_{66}^l
\end{bmatrix}
\tag{37}
$$

The expression of each component in Equation (37) is given in Appendix E. It is observed that the compliance matrix expressed in Equation (37) does not match the form given in Equation (13) with six non-diagonal components (c_{31}^l, c_{41}^l, c_{51}^l, c_{32}^l, c_{42}^l, and c_{52}^l) not equal to zero. This is due to the non-zero translation components along the Z axis of the first three flexure elements (r_{z_1}, r_{z_2}, $r_{z_3} \neq 0$) making the oriented compliance matrix of each flexure element different from the condition in Equation (15). Thus, the output motions of the entire 3PRPR 3L-CPM are coupled.

As for the 3PRPR-CPM with θ_X-θ_Y-Z spatial motion shown in Figure 8b, each leg has its respective $X''Y''$ plane. Based on the design of each leg, it is observed that the P-joints (also illustrated in Figure A1b) would operate in their respective $X''Y''$ plane, while the R-joints were designed to rotate out of that working plane. In addition, the translation of each flexure element consists of three projection components onto three axes. Similar to the 3-DOF 3PRPR 3L-CPM presented above, this design also generates coupled motions, since the flexure elements are not distributed in the XY plane ($r_{z_j} \neq 0$) and the orientations of the flexure elements also include the rotation out of the $X''Y''$ plane ($\beta_j \neq 0$), making the non-diagonal components within the compliance matrix of each leg become non-zeros, as proven in Section 3. Lastly, a similar analysis can be adopted to demonstrate that the 3L-CPM using three serial Prismatic–Universal–Prismatic–Universal (3-PUPU) flexure chains presented in Reference [33] will also generate coupled motions.

5.4. Three-Legged CPM with Paired Prismatic–Spherical–Spherical Configuration

A 6-DOF 3L-CPM [28] was developed using a pair of Prismatic–Spherical–Spherical (PSS) serial chain configuration to create each leg as illustrated in Figure 9. By considering the leg in the front of the 3L-CPM, it is observed that two PSS serial chains located in the $X'Z'$ plane and reflect about the $Y'Z'$ plane. Based on Figure A1, the P-joint is oriented about the X'' axis ($\alpha_1 = 90°$) and Z'' axis ($\gamma_1 = 90°$) respectively to obtain the desired orientation. In addition, the two S-joints are oriented 45° about their Y'' axes ($\beta_2 = \beta_3 = 45°$) and the distance vector from each joint to the local frame $X'Y'Z'$ at the moving end of the leg only consists of two components, i.e., $\mathbf{r}_1 = \{r_{x_1}, 0, r_{z_1}\}$, $\mathbf{r}_2 = \{r_{x_2}, 0, r_{z_2}\}$ and $\mathbf{r}_3 = \{r_{x_3}, 0, r_{z_3}\}$. Subsequently, the compliance matrix of each flexure element is given as follows:

$$
\mathbf{C}_1^e =
\begin{bmatrix}
c_{11}^P & & & & & \\
0 & c_{22}^P & & & \text{SYM} & \\
0 & 0 & c_{33}^P & & & \\
0 & 0 & 0 & c_{44}^P & & \\
0 & 0 & c_{53}^P & 0 & c_{55}^P & \\
0 & c_{62}^P & 0 & 0 & 0 & c_{66}^P
\end{bmatrix}
, \text{ and } \mathbf{C}_2^e = \mathbf{C}_3^e =
\begin{bmatrix}
c_{11}^S & & & & & \\
0 & c_{22}^S & & & \text{SYM} & \\
0 & 0 & c_{33}^S & & & \\
0 & 0 & 0 & c_{44}^S & & \\
0 & 0 & c_{53}^S & 0 & c_{55}^S & \\
0 & c_{62}^S & 0 & 0 & 0 & c_{66}^S
\end{bmatrix}
\tag{38}
$$

where c_{uv}^P and c_{uv}^S (with $u, v = 1, 2, \ldots, 6$) are the components within the compliance matrices of the P-joints and S-joints, respectively.

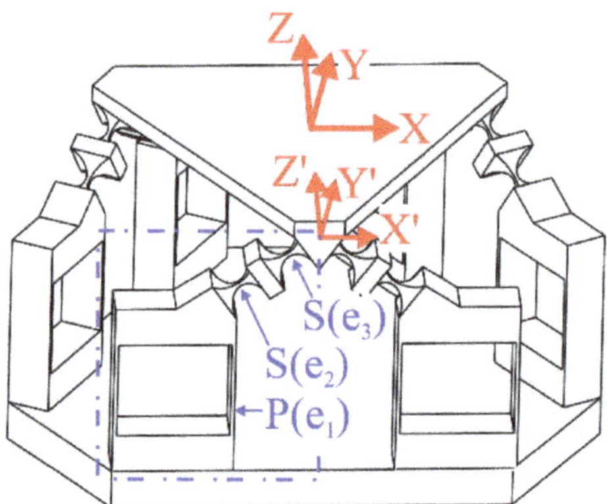

Figure 9. Six-DOF 3L-CPM with two reflecting PSS chain in a leg modeled based on the design presented in [28].

In this design, the two symmetrical PSS serial chains have no offset distance. Hence, the offset matrices, $\mathbf{J}_{(\pm\Delta)}$, as shown in Equation (30), become identity matrices. By using Equations (9), (11) and (12) with the defined parameters, the compliance matrix of a serial flexure chain in the proposed leg, $\mathbf{C}^{sc}$, is written in Equation (39), and the result of each component is given in Appendix F.

$$
\mathbf{C}^{sc} = \begin{bmatrix}
c_{11}^{sc} & & & & & \\
c_{21}^{sc} & c_{22}^{sc} & & & \text{SYM} & \\
c_{31}^{sc} & c_{32}^{sc} & c_{33}^{sc} & & & \\
c_{41}^{sc} & c_{42}^{sc} & c_{43}^{sc} & c_{44}^{sc} & & \\
c_{51}^{sc} & c_{52}^{sc} & c_{53}^{sc} & c_{54}^{sc} & c_{55}^{sc} & \\
c_{61}^{sc} & c_{62}^{sc} & 0 & 0 & 0 & c_{66}^{sc}
\end{bmatrix}
\tag{39}
$$

Based on the obtained $\mathbf{C}^{sc}$, the stiffness matrix of each leg, $\mathbf{K}^l$, can be calculated by Equation (28). It is realized that the form of $\mathbf{C}^{sc}$ in this design is different from the condition shown in Equation (31), as the flexure elements are not located in the XY plane ($r_{z_j} \neq 0$) and the P-joint has a rotation about the Y'' axis ($\beta_1 \neq 0$). Therefore, even the offset distance between two serial flexure chains in each leg is zero ($\Delta = 0$), and the output motions of this 6-DOF 3L-CPM are coupled based on the results obtained from Section 4. Using a similar analysis, the 6-DOF 3L-CPM presented in Reference [31] also generates coupled motions.

5.5. Six-DOF 3L-CPM Synthesized by Constrained-Based Method

A 6-DOF 3L-CPM using double flexure chains configuration, as shown in Figure 10, was presented in Reference [42]. The construction of each leg contains two horizontal beam-type flexure elements at both sides and one vertical beam flexure at the center. This structure can be considered as a reflecting double chain about the $Y'Z'$ plane, where each serial flexure chain consists of one horizontal beam and a half of the vertical beam, as illustrated in Figure 10. Referring to Figure 2a, it is observed that the vertical beam (e_2) has a rotation of 90° about its Z'' axis ($\gamma_2 = 90°$) while the horizontal beam (e_1) remains in its original orientation. In addition, the translations of e_1 and e_2 are represented by $\mathbf{r}_1 = \{r_{x_1}, r_{y_1}, 0\}$ and $\mathbf{r}_2 = \{0, 0, 0\}$, respectively, since the free end of e_2 coincides with the frame $X'Y'Z'$ of the leg. The compliance matrices of two flexure elements in a serial flexure chain are given as follows:

$$\mathbf{C}_1^e = \begin{bmatrix} c_{11}^{e_1} & & & & & \\ 0 & c_{22}^{e_1} & & \text{SYM} & & \\ 0 & 0 & c_{33}^{e_1} & & & \\ 0 & 0 & 0 & c_{44}^{e_1} & & \\ 0 & 0 & c_{53}^{e_1} & 0 & c_{55}^{e_1} & \\ 0 & c_{62}^{e_1} & 0 & 0 & 0 & c_{66}^{e_1} \end{bmatrix}, \text{ and } \mathbf{C}_2^e = \begin{bmatrix} c_{11}^{e_2} & & & & & \\ 0 & c_{22}^{e_2} & & \text{SYM} & & \\ 0 & 0 & c_{33}^{e_2} & & & \\ 0 & 0 & 0 & c_{44}^{e_2} & & \\ 0 & 0 & c_{53}^{e_2} & 0 & c_{55}^{e_2} & \\ 0 & c_{62}^{e_2} & 0 & 0 & 0 & c_{66}^{e_2} \end{bmatrix} \tag{40}$$

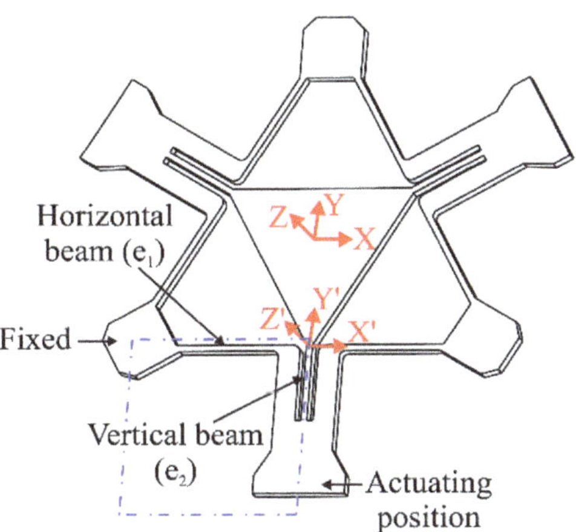

Figure 10. Six-DOF 3L-CPM modeled based on the design presented in [42].

Based on Equations (9), (11) and (12) and the defined parameters, the obtained compliance matrix of the serial flexure chain shown in Figure 10, $\mathbf{C}^{sc}$, is written as follows:

$$\mathbf{C}^{sc} = \begin{bmatrix} \begin{matrix} c_{11}^{e_1} + c_{22}^{e_2} + c_{66}^{e_1} r_{y_1}^2 \\ -(c_{62}^{e_1} + c_{66}^{e_1} r_{x_1}) r_{y_1} \end{matrix} & & & & & \\ 0 & c_{22}^{e_1} + c_{11}^{e_2} + r_{x_1}(2c_{62}^{e_1} + c_{66}^{e_1} r_{x_1}) & & & \text{SYM} & \\ 0 & 0 & \begin{matrix} c_{33}^{e_1} + c_{33}^{e_2} - 2c_{53}^{e_1} r_{x_1} + c_{55}^{e_1} r_{x_1}^2 + c_{44}^{e_1} r_{y_1}^2 \end{matrix} & & & \\ 0 & 0 & -c_{53}^{e_2} + c_{44}^{e_1} r_{y_1} & c_{44}^{e_1} + c_{55}^{e_2} & & \\ 0 & 0 & c_{53}^{e_1} - c_{55}^{e_1} r_{x_1} & 0 & c_{55}^{e_1} + c_{44}^{e_2} & \\ -c_{62}^{e_2} - c_{66}^{e_1} r_{y_1} & c_{62}^{e_1} + c_{66}^{e_1} r_{x_1} & 0 & 0 & 0 & c_{66}^{e_1} + c_{66}^{e_2} \end{bmatrix} \tag{41}$$

It is observed that the compliance matrix presented in Equation (41) matches the form expressed in Equation (31), and all flexure elements are distributed in the XY plane ($\Delta = 0$ and $\mathbf{J}_{(\pm\Delta)}$ become identity matrices). Note that, while c_{54}^{sc} is zero, those nine essential non-diagonal components (c_{31}^{sc}, c_{41}^{sc}, c_{51}^{sc}, c_{32}^{sc}, c_{42}^{sc}, c_{52}^{sc}, c_{63}^{sc}, c_{64}^{sc}, and c_{65}^{sc}) are zero and, thus, satisfy the special case proposed in this work. As a result, it shows that this design is able to produce fully decoupled motion. This can be demonstrated by the resulting diagonal matrix obtained by applying Equations (1) and (28) to analyze the stiffness property of the entire 3L-CPM. Several micro-scale 6-DOF 3L-CPMs which developed based on the same concept were proposed in Reference [43], and a variant of this design was presented in Reference [58].

5.6. Six-DOF 3L-CPM Synthesized by Optimization Method

A recent 6-DOF 3L-CPM synthesized by the structural optimization method [56] is shown in Figure 11. Its leg consists of two reflecting flexures about the $Y'Z'$ plane, and each flexure is constructed by a curved beam. Here, this curved beam can be considered as a series of straight beam-type flexure elements ($e_1, e_2, \ldots, e_n$) with various orientations about their respective Z'' axes ($\gamma_1, \gamma_2, \ldots, \gamma_n$), as illustrated in Figure 3a. Note that each beam-type flexure element has the compliance matrix in the form shown in Equation (10).

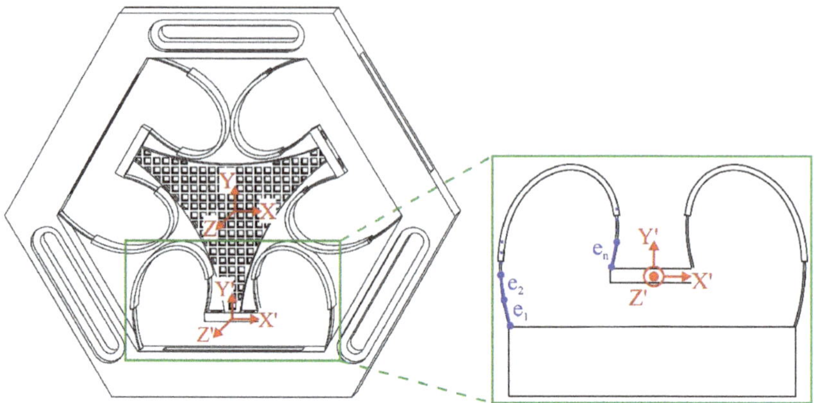

Figure 11. Six-DOF 3L-CPM synthesized by the structural optimization method.

In Figure 11, the translation of each flexure element can be represented by $\mathbf{r}_{e_j} = \left\{ r_{x_j}, r_{y_j}, 0 \right\}$ (where $j = 1, 2, \ldots, n$) because all flexure elements are located in the XY plane. It is observed that the structure of this CPM matches the case shown in Figure 4, with zero offset distance between two reflecting serial flexure chains ($\Delta = 0$). Thus, this CPM is able to achieve full decoupling. Referring to the results in Reference [56], we can see that the stiffness matrix of this CPM has the form of a diagonal matrix, as shown in Equation (42), and its motion-decoupling capability was also demonstrated.

$$
\mathbf{C} =
\begin{bmatrix}
3.67 \times 10^{-5} & & & & & \\
0 & 3.67 \times 10^{-5} & & & \text{SYM} & \\
0 & 0 & 9.70 \times 10^{-5} & & & \\
0 & 0 & 0 & 3.06 \times 10^{-2} & & \\
0 & 0 & 0 & 0 & 3.06 \times 10^{-2} & \\
0 & 0 & 0 & 0 & 0 & 3.47 \times 10^{-2}
\end{bmatrix}
\tag{42}
$$

5.7. Three-DOF Planar-Motion 3L-CPM Synthesized by Optimization Method

The three-legged configuration with double serial flexure chains within a leg is preferred in 3L-CPMs that were synthesized by the topological optimization method. Here, an optimized 3-DOF (X-Y-θ_Z) planar motion 3L-CPM [53] shown in Figure 12a was analyzed to study its motion property. It is observed that each leg contains two reflecting flexure chains about the $Y'Z'$ plane, and each flexure chain has four beam-type flexure elements (e_1, e_2, e_3, and e_4) connected together in a series. The thicker segments with each flexure chain are considered as rigid-body links in the stiffness analysis. With this 3L-CPM being a planar structure, the translation of each flexure element to the end effector is located in the XY plane, and there is no projection component (r_{z_j}) in the Z axis. By referring to Reference [53], the 6×6 stiffness matrix of the entire 3L-CPM is as follows:

$$
\mathbf{K} =
\begin{bmatrix}
2.0 \times 10^4 & & & & & \\
0 & 2.0 \times 10^4 & & & \text{SYM} & \\
0 & 0 & 2.6 \times 10^6 & & & \\
0 & -545 & 0 & 1.3 \times 10^3 & & \\
545 & 0 & 0 & 0 & 1.3 \times 10^3 & \\
0 & 0 & 0 & 0 & 0 & 12
\end{bmatrix}
\tag{43}
$$

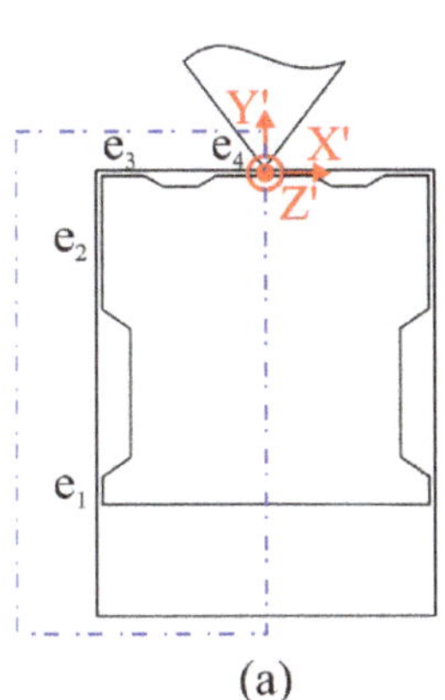

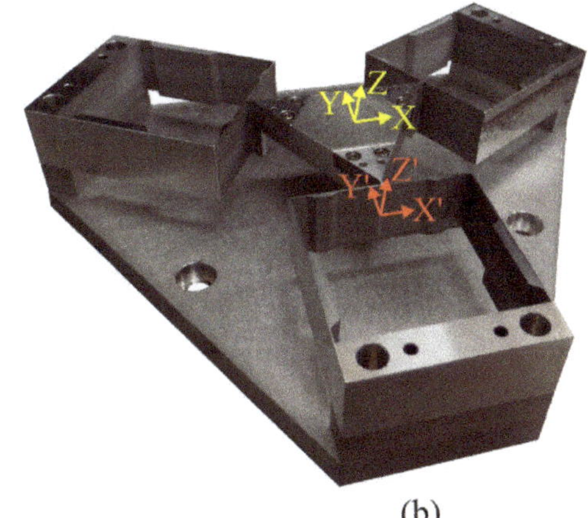

(a)   (b)

Figure 12. Three-DOF (X-Y-θ_Z) 3L-CPM designed by using the optimization method (**a**) structure of a leg and (**b**) prototype built based on the design presented in [53].

From Figure 12a, we can see that the 3L-CPM has two reflecting flexure chains with no offset distance, so that, depending on the results presented in Section 4, its output motions are decoupled. However, Equation (43) suggests that this 3L-CPM will generate coupled motions, since there are two non-diagonal components. This result can be explained by the stiffness modeling used in Reference [53]; the compliance matrices of the leg were calculated at the local frame $X'Y'Z'$ in the middle plane, while the stiffness matrix of the entire 3L-CPM was derived at the global frame XYZ (located at the center of the end effector) in the top plane, as illustrated in Figure 12b. That generates the translation components D_{z_j} along the Z axis in the translation matrices of the legs, as demonstrated in Equation (3), and creates the off-axis stiffness components within the final stiffness matrix. Similar designs presented in Reference [55] will also generate decoupled motions. If the global frame is attached at the middle plane of the end effector instead, the non-diagonal stiffness will be eliminated, and the 3L-CPM will have fully decoupled motion characteristic.

5.8. Three-DOF Spatial-Motion 3L-CPMs Synthesized by Optimization Method

Two designs of 3-DOF (θ_X-θ_Y-Z) spatial-motion 3L-CPMs, as shown in Figure 13, were synthesized by the structural optimization method [6,7]. Here, the desired motions were generated by two reflecting beam-type flexure elements about the $Y'Z'$ plane in each leg. The design shown in Figure 13a has a small offset distance along the Z axis ($\Delta \neq 0$) between two flexure chains, as compared to the other design, which is shown in Figure 13b, where all flexure elements are located in the same XY plane ($\Delta = 0$).

$$\mathbf{C}_a = \begin{bmatrix} 3.21 \times 10^{-8} & & & & & & \\ 0 & 3.21 \times 10^{-8} & & & \text{SYM} & & \\ 0 & 0 & 8.98 \times 10^{-5} & & & & \\ -3.31 \times 10^{-6} & 7.84 \times 10^{-8} & 0 & 3.05 \times 10^{-2} & & & \\ -7.84 \times 10^{-8} & -3.31 \times 10^{-6} & 0 & 0 & 3.05 \times 10^{-2} & & \\ 0 & 0 & 4.28 \times 10^{-6} & 0 & 0 & 1.50 \times 10^{-4} \end{bmatrix} \tag{44}$$

$$\mathbf{C}_b = \begin{bmatrix} 5.53 \times 10^{-8} & & & & & & \\ 0 & 5.53 \times 10^{-8} & & & \text{SYM} & & \\ 0 & 0 & 1.99 \times 10^{-4} & & & & \\ 0 & 0 & 0 & 1.32 \times 10^{-1} & & & \\ 0 & 0 & 0 & 0 & 1.32 \times 10^{-1} & & \\ 0 & 0 & 0 & 0 & 0 & 2.33 \times 10^{-4} \end{bmatrix} \tag{45}$$

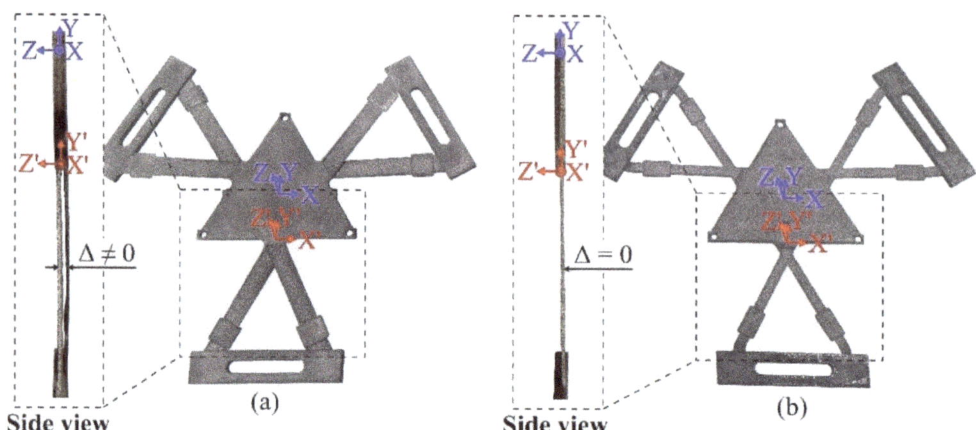

Figure 13. Three-DOF (θ_X-θ_Y-Z) 3L-CPMs designed by structural optimization method (**a**) with offset distance and (**b**) without offset distance between two flexure chains in a leg [63].

For the 3L-CPM with $\Delta \neq 0$, as shown in Figure 13a, the results presented in Reference [6] agree with the findings described in Section 4, where its compliance matrix, $\mathbf{C}_a$, has five non-diagonal components, as derived theoretically in Equation (44). For the other 3L-CPM with $\Delta = 0$, as shown in Figure 13b [7], the compliance matrix, $\mathbf{C}_b$, only has diagonal components, as demonstrated theoretically in Equation (45). In summary, the motion property of 3L-CPMs having two reflecting flexure chains in a leg was demonstrated. The 3L-CPM having an offset distance in the Z axis between two flexure chains will generate coupled motions, while the other having both flexure chains located in the same plane will generate decoupled motions.

6. Discussion

In order to deliver fully decoupled motion, a 3L-CPM must have a 6×6 diagonal stiffness/compliance matrix whereby all non-diagonal components are zero. Due to the property of the parallel architecture, the stiffness matrix of a 3L-CPM can be calculated based on the stiffness matrices of its legs. However, the stiffness matrix of each leg cannot be derived directly, since it can be constructed by one or two serial chains; each is formed by a number of flexure elements and rigid links, and, thus, its characteristic is defined by the compliance. Due to the challenges in converting between the stiffness and compliance matrices, existing 3L-CPMs failed to analyze their motion property analytically. To overcome this limitation and, most importantly, for designing a 3L-CPM to obtain a full-decoupled motion characteristic, the conditions for the compliance matrix of the single serial flexure chain, i.e., Equation (13), and for the compliance matrix of the double serial flexure chains, i.e., Equation (31), are provided in this work. A short review of various state-of-the-art 3L-CPMs presented in Section 5 showed that the conditions of the compliance matrix can be used to identify the motion property of these 3L-CPMs. In order to satisfy these conditions, analytical analyses show that every flexure element within each leg must be located in the global XY plane with only two orientations ($0°$ and $90°$) about its local X'' axis, as illustrated in Figure 3, and the offset distance along the Z axis between two serial flexure chains, as shown in Figure 4, must be zero ($\Delta = 0$). In other words, these design criteria can be used to synthesize a 3L-CPM that aims to achieve fully decoupled motion capability. Moreover, the findings in this work are applicable to any synthesis method, e.g., traditional pseudo-rigid-body model, constraint-based and topology/structural optimization methods, etc. Consequently, these design criteria and conditions for the compliance matrices of the flexure chains can be used as the fundamental design guidelines for the syntheses of 3L-CPMs to achieve desired motion property that can be either decoupled or coupled.

As the 3L-CPM plays an important role in precise motion systems, the motion decoupling capability needs to be clearly defined in the design process to make the control simpler and the output motions more accurate as well. Based on the literature, several compliant systems which are capable of producing precise motions with a simple control method, due to the defined motion characteristics of their compliant structures, have been developed. In particular, the 3-DOF spatial-motion (θ_X-θ_Y-Z) manipulator [6,7] and the flexure-based electromagnetic nano-positioning actuator [3] are able to produce a large workspace with high resolutions, using simple open-loop control systems. Moreover, 3L-CPMs with decoupled motions can also be applied to design micro-fabrication systems, e.g., the flexural spindle head in a micro drilling machine tool [11,12], the motion stage in a micro milling system [13], and the flexural stage to adjust the angles of mirror in advanced three-dimensional fabrication methods [14,15]. In addition, the benefits offered by decoupled 3L-CPMs have been recently exploited in biomedical applications, such as the flexural micro-dissection device [10]. Since the application range of decoupled-motion 3L-CPMs has been increasing, it can be said that the fundamental criteria for synthesizing 3L-CPMs with fully decoupled motions and the motion property of some existing designs presented in this paper are an important background for developing advanced flexure-based systems.

7. Conclusions

This paper presented the fundamental design criteria for synthesizing any 3L-CPM with fully decoupled motion capability regardless of the targeted DOF. The stiffness characteristics of a 3L-CPM were analytically modeled. The derived criteria suggested that the flexure elements in each leg must be distributed in the same plane with the end effector of the 3L-CPM in order to fulfill the decoupled motions requirements. In the case where each leg contains two parallel reflecting flexure chains, such requirements are valid if both flexure chains are located in the same plane with no offset distance. To demonstrate the effectiveness of the design criteria and conditions obtained from this work, several state-of-the-art 3L-CPMs were analyzed for their stiffness characteristics and compared with these criteria. The presented cases show that the proposed design criteria can be applied to accurately determine the motion characteristics of any 3L-CPM through the analysis of its stiffness/compliance matrix; only 3L-CPMs having diagonal stiffness/compliance matrices are able to achieve fully decoupled motions. Findings from this work can be used to define the motion property of any form of 3L-CPM during the design process.

In this paper, only some special solutions were considered to make the non-diagonal components within the stiffness/compliance matrix of a 3L-CPM equal to zero; there could be other solutions that need to be explored. Future work will focus on investigating more general design criteria for synthesizing any CPMs with desired DOF and motion property.

Author Contributions: Conceptualization, methodology, software, validation, formal analysis, visualization, and writing—original draft preparation, M.T.P.; writing—review and editing, and supervision, T.J.T. and S.H.Y. All authors have read and agreed to the published version of the manuscript.

Funding: This research is funded by Vietnam National University HoChiMinh City (VNU-HCM), under grant number C2020-20-01.

Institutional Review Board Statement: Not applicable.

Informed Consent Statement: Not applicable.

Data Availability Statement: Not applicable.

Conflicts of Interest: The authors declare no conflict of interest.

Appendix A. Stiffness Characteristics of Some Common Types of Flexure Element

Referring to References [22,57], some popular flexure elements have the same compliance matrix form as expressed in Equation (10) and are shown in Figure A1. They can be revolute hinge; thin beam, as illustrated in Figure 2; and also can be some other forms, such as spherical joint (Figure A1a) or prismatic joint (linear spring), as shown in Figure A1b. The notch of spherical joint can have a circular or square cross-sectional area, while the linear spring can be constructed by four notch hinges or a pair of cantilever beams.

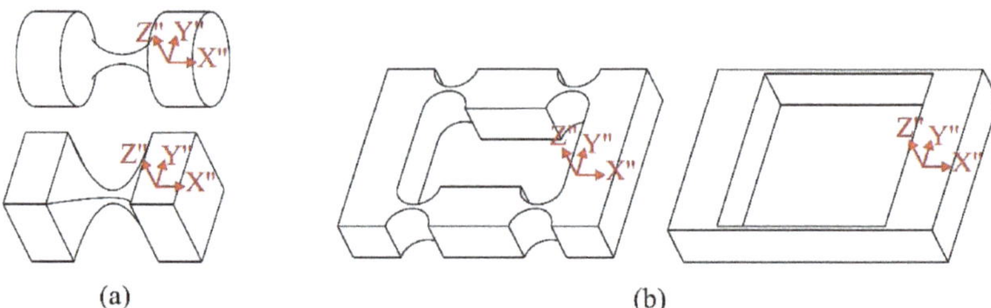

Figure A1. Flexure elements: (**a**) spherical joint and (**b**) linear spring.

Appendix B. Conditions of the Leg's Compliance Matrix for Decoupled Motions

The results of the inversion of Equation (8) are written as follows:

$$
\underbrace{\begin{bmatrix}
k_{11}^l & & & & & \\
k_{21}^l & k_{22}^l & & & \text{SYM} & \\
0 & 0 & k_{33}^l & & & \\
0 & 0 & k_{43}^l & k_{44}^l & & \\
0 & 0 & k_{53}^l & k_{54}^l & k_{55}^l & \\
k_{61}^l & k_{62}^l & 0 & k_{64}^l & k_{65}^l & k_{66}^l
\end{bmatrix}}_{(\mathbf{K}^l)^{-1}}^{-1}
=
\underbrace{\begin{bmatrix}
c_{11}^l & & & & & \\
c_{21}^l & c_{22}^l & & & \text{SYM} & \\
c_{31}^l & c_{32}^l & c_{33}^l & & & \\
c_{41}^l & c_{42}^l & c_{43}^l & c_{44}^l & & \\
c_{51}^l & c_{52}^l & c_{53}^l & c_{54}^l & c_{55}^l & \\
c_{61}^l & c_{62}^l & c_{63}^l & c_{64}^l & c_{65}^l & c_{66}^l
\end{bmatrix}}_{\mathbf{C}^l}
\tag{A1}
$$

The expression of each component in $\mathbf{C}^l$ is given as follows:

$$
c_{31}^l = \frac{1}{\varsigma}\left(k_{22}^l k_{61}^l - k_{21}^l k_{62}^l\right)\left[k_{64}^l\left(k_{53}^l k_{54}^l - k_{43}^l k_{55}^l\right) + k_{65}^l\left(-k_{44}^l k_{53}^l + k_{43}^l k_{54}^l\right)\right]
\tag{A2}
$$

$$
c_{41}^l = -\frac{1}{\varsigma}\left(k_{22}^l k_{61}^l - k_{21}^l k_{62}^l\right)\left[k_{64}^l\left(\left(k_{53}^l\right)^2 - k_{33}^l k_{55}^l\right) + k_{65}^l\left(-k_{43}^l k_{53}^l + k_{33}^l k_{54}^l\right)\right]
\tag{A3}
$$

$$
c_{51}^l = -\frac{1}{\varsigma}\left(k_{22}^l k_{61}^l - k_{21}^l k_{62}^l\right)\left[k_{64}^l\left(-k_{43}^l k_{53}^l + k_{33}^l k_{54}^l\right) + k_{65}^l\left(\left(k_{43}^l\right)^2 - k_{33}^l k_{44}^l\right)\right]
\tag{A4}
$$

$$
c_{32}^l = -\frac{1}{\varsigma}\left(k_{21}^l k_{61}^l - k_{11}^l k_{62}^l\right)\left[k_{64}^l\left(k_{53}^l k_{54}^l - k_{43}^l k_{55}^l\right) + k_{65}^l\left(-k_{44}^l k_{53}^l + k_{43}^l k_{54}^l\right)\right]
\tag{A5}
$$

$$
c_{42}^l = \frac{1}{\varsigma}\left(k_{21}^l k_{61}^l - k_{11}^l k_{62}^l\right)\left[k_{64}^l\left(\left(k_{53}^l\right)^2 - k_{33}^l k_{55}^l\right) + k_{65}^l\left(-k_{43}^l k_{53}^l + k_{33}^l k_{54}^l\right)\right]
\tag{A6}
$$

$$
c_{52}^l = \frac{1}{\varsigma}\left(k_{21}^l k_{61}^l - k_{11}^l k_{62}^l\right)\left[k_{64}^l\left(-k_{43}^l k_{53}^l + k_{33}^l k_{54}^l\right) + k_{65}^l\left(\left(k_{43}^l\right)^2 - k_{33}^l k_{44}^l\right)\right]
\tag{A7}
$$

$$
c_{63}^l = \frac{1}{\varsigma}\left(\left(k_{21}^l\right)^2 - k_{11}^l k_{22}^l\right)\left[k_{64}^l\left(k_{53}^l k_{54}^l - k_{43}^l k_{55}^l\right) + k_{65}^l\left(-k_{44}^l k_{53}^l + k_{43}^l k_{54}^l\right)\right]
\tag{A8}
$$

$$
c_{64}^l = -\frac{1}{\varsigma}\left(\left(k_{21}^l\right)^2 - k_{11}^l k_{22}^l\right)\left[k_{64}^l\left(\left(k_{53}^l\right)^2 - k_{33}^l k_{55}^l\right) + k_{65}^l\left(-k_{43}^l k_{53}^l + k_{33}^l k_{54}^l\right)\right]
\tag{A9}
$$

$$
c_{65}^l = -\frac{1}{\varsigma}\left(\left(k_{21}^l\right)^2 - k_{11}^l k_{22}^l\right)\left[k_{64}^l\left(-k_{43}^l k_{53}^l + k_{33}^l k_{54}^l\right) + k_{65}^l\left(\left(k_{43}^l\right)^2 - k_{33}^l k_{44}^l\right)\right]
\tag{A10}
$$

$$c_{21}^l = \tfrac{1}{\zeta}\Big\{ -k_{21}^l \left(k_{53}^l\right)^2 \left(k_{64}^l\right)^2 + 2k_{43}^l k_{53}^l \left(k_{54}^l k_{61}^l k_{62}^l + k_{21}^l k_{64}^l k_{65}^l - k_{21}^l k_{54}^l k_{66}^l\right) -$$
$$\left(k_{43}^l\right)^2 \left(k_{55}^l k_{61}^l k_{62}^l + k_{21}^l \left(k_{65}^l\right)^2 - k_{21}^l k_{55}^l k_{66}^l\right) +$$
$$k_{33}^l \left[k_{21}^l k_{55}^l \left(k_{64}^l\right)^2 + k_{54}^l \left(-k_{54}^l k_{61}^l k_{62}^l - k_{21}^l k_{64}^l k_{65}^l + k_{21}^l k_{54}^l k_{66}^l\right)\right] +$$
$$k_{44}^l \left[\left(k_{53}^l\right)^2 \left(-k_{61}^l k_{62}^l + k_{21}^l k_{66}^l\right) + k_{33}^l \left(k_{55}^l k_{61}^l k_{62}^l + k_{21}^l \left(k_{65}^l\right)^2 - k_{21}^l k_{55}^l k_{66}^l\right)\right] \Big\} \tag{A11}$$

$$c_{61}^l = \frac{1}{\zeta}\Big\{ \left[-2k_{43}^l k_{53}^l k_{54}^l + k_{33}^l \left(k_{54}^l\right)^2 + \left(k_{43}^l\right)^2 k_{55}^l + k_{44}^l \left(\left(k_{53}^l\right)^2 - k_{33}^l k_{55}^l\right)\right] \left(k_{22}^l k_{61}^l - k_{21}^l k_{62}^l\right) \Big\} \tag{A12}$$

$$c_{62}^l = -\frac{1}{\zeta}\Big\{ \left[-2k_{43}^l k_{53}^l k_{54}^l + k_{33}^l \left(k_{54}^l\right)^2 + \left(k_{43}^l\right)^2 k_{55}^l + k_{44}^l \left(\left(k_{53}^l\right)^2 - k_{33}^l k_{55}^l\right)\right] \left(k_{21}^l k_{61}^l - k_{11}^l k_{62}^l\right) \Big\} \tag{A13}$$

$$c_{43}^l = \tfrac{1}{\zeta}\Big\{ 2k_{21}^l \left(k_{53}^l k_{54}^l - k_{43}^l k_{55}^l\right)k_{61}^l k_{62}^l + k_{11}^l \left(-k_{53}^l k_{54}^l - k_{43}^l k_{55}^l\right)\left(k_{62}^l\right)^2 +$$
$$\left(k_{21}^l\right)^2 \left(k_{53}^l k_{64}^l k_{65}^l - k_{43}^l \left(k_{65}^l\right)^2 - k_{53}^l k_{54}^l k_{66}^l + k_{43}^l k_{55}^l k_{66}^l\right) +$$
$$k_{22}^l \left[-k_{53}^l \left(k_{54}^l \left(k_{61}^l\right)^2 + k_{11}^l k_{64}^l k_{65}^l - k_{11}^l k_{54}^l k_{66}^l\right) + k_{43}^l \left(k_{55}^l \left(k_{61}^l\right)^2 + k_{11}^l \left(k_{65}^l\right)^2 - k_{11}^l k_{55}^l k_{66}^l\right)\right] \Big\} \tag{A14}$$

$$c_{53}^l = \tfrac{1}{\zeta}\Big\{ 2k_{21}^l \left(k_{43}^l k_{53}^l - k_{33}^l k_{54}^l\right)k_{61}^l k_{62}^l + k_{11}^l \left(-k_{43}^l k_{53}^l + k_{33}^l k_{54}^l\right)\left(k_{62}^l\right)^2 -$$
$$\left(k_{21}^l\right)^2 \left(k_{33}^l k_{64}^l k_{65}^l + k_{43}^l k_{53}^l k_{66}^l - k_{33}^l k_{54}^l k_{66}^l\right) +$$
$$k_{22}^l \left[k_{43}^l k_{53}^l \left(-\left(k_{61}^l\right)^2 + k_{11}^l k_{66}^l\right) + k_{33}^l \left(k_{54}^l \left(k_{61}^l\right)^2 + k_{11}^l k_{64}^l k_{65}^l - k_{11}^l k_{54}^l k_{66}^l\right)\right] \Big\} \tag{A15}$$

where we have the following:

$$\zeta = -2k_{21}^l \Big\{ -2k_{43}^l k_{53}^l k_{54}^l + k_{33}^l \left(k_{54}^l\right)^2 + \left(k_{43}^l\right)^2 k_{55}^l + k_{44}^l \left[\left(k_{53}^l\right)^2 - k_{33}^l k_{55}^l\right]\Big\} k_{61}^l k_{62}^l +$$
$$k_{11}^l \Big\{ -2k_{43}^l k_{53}^l k_{54}^l + k_{33}^l \left(k_{54}^l\right)^2 + \left(k_{43}^l\right)^2 k_{55}^l + k_{44}^l \left[\left(k_{53}^l\right)^2 - k_{33}^l k_{55}^l\right]\Big\} \left(k_{62}^l\right)^2 +$$
$$\left(k_{21}^l\right)^2 \Big\{ \left(k_{53}^l\right)^2 \left[-\left(k_{64}^l\right)^2 + k_{44}^l k_{66}^l\right] + 2k_{43}^l k_{53}^l \left(k_{64}^l k_{65}^l - k_{54}^l k_{66}^l\right) + \left(k_{43}^l\right)^2 \left[-\left(k_{65}^l\right)^2 + k_{55}^l k_{66}^l\right] +$$
$$k_{33}^l \left[k_{55}^l \left(k_{64}^l\right)^2 - 2k_{54}^l k_{64}^l k_{65}^l + k_{44}^l \left(k_{65}^l\right)^2 + \left(k_{54}^l\right)^2 k_{66}^l - k_{44}^l k_{55}^l k_{66}^l\right]\Big\} +$$
$$k_{22}^l \Big\{ k_{11}^l \left(k_{53}^l\right)^2 \left(k_{64}^l\right)^2 - 2k_{43}^l k_{53}^l \left[k_{54}^l \left(k_{61}^l\right)^2 + k_{11}^l k_{64}^l k_{65}^l - k_{11}^l k_{54}^l k_{66}^l\right] +$$
$$\left(k_{43}^l\right)^2 \left[k_{55}^l \left(k_{61}^l\right)^2 + k_{11}^l \left(k_{65}^l\right)^2 - k_{11}^l k_{55}^l k_{66}^l\right] +$$
$$k_{33}^l \left[-k_{11}^l k_{55}^l \left(k_{64}^l\right)^2 + k_{54}^l \left(k_{54}^l \left(k_{61}^l\right)^2 + 2k_{11}^l k_{64}^l k_{65}^l - k_{11}^l k_{54}^l k_{66}^l\right)\right] +$$
$$k_{44}^l \left[\left(k_{53}^l\right)^2 \left(\left(k_{61}^l\right)^2 - k_{11}^l k_{66}^l\right) - k_{33}^l \left(k_{55}^l \left(k_{61}^l\right)^2 + k_{11}^l \left(k_{65}^l\right)^2 - k_{11}^l k_{55}^l k_{66}^l\right)\right]\Big\} \tag{A16}$$

The form of $\mathbf{C}^l$ needs to be specified as a standard for the design process of decoupled-motion 3L-CPMs. It is observed that the expressions of seven compliance components (c_{31}^l, c_{32}^l, c_{41}^l, c_{42}^l, c_{51}^l, c_{52}^l, and c_{63}^l) corresponding to the seven zero-components in the stiffness matrix have similar forms as shown in Equations (A2) to (A8). In this paper, these seven compliance components are required to be zeros, so that the form of the leg's compliance matrix will be the same with its stiffness matrix. This special form offers simplicity during the design process and can be used as the standard to define the decoupled-motion capability of various 3L-CPMs. The requirements to make the seven compliance components equal to zeros are written in Equation (A17).

$$
\begin{cases}
k_{22}^l k_{61}^l - k_{21}^l k_{62}^l = 0 \\
k_{21}^l k_{61}^l - k_{11}^l k_{62}^l = 0 \\
\left(k_{21}^l\right)^2 - k_{11}^l k_{22}^l = 0
\end{cases}
\text{ or }
\begin{cases}
k_{64}^l\left(k_{53}^l k_{54}^l - k_{43}^l k_{55}^l\right) + k_{65}^l\left(-k_{44}^l k_{53}^l + k_{43}^l k_{54}^l\right) = 0 \\
k_{64}^l\left(\left(k_{53}^l\right)^2 - k_{33}^l k_{55}^l\right) + k_{65}^l\left(-k_{43}^l k_{53}^l + k_{33}^l k_{54}^l\right) = 0 \\
k_{64}^l\left(-k_{43}^l k_{53}^l + k_{33}^l k_{44}^l\right) + k_{65}^l\left(\left(k_{43}^l\right)^2 - k_{33}^l k_{44}^l\right) = 0
\end{cases}
\tag{A17}
$$

As the diagonal components in the stiffness matrix are always non-zeros, while the non-diagonal components can be zeros or non-zeros, the non-diagonal components are considered as unknowns, and the diagonal ones are parameters. In the first set of equations, one of the first two equations can be redundant. The answers to the first set of equations are as follows:

$$
\begin{cases}
k_{21}^l = \pm\sqrt{k_{11}^l k_{22}^l} \\
k_{61}^l = \pm\sqrt{\dfrac{k_{11}^l}{k_{22}^l}}\, k_{62}^l
\end{cases}
\tag{A18}
$$

The second set of equations contain three equations with five unknowns, so that there could be many solutions. Here, two simple solutions are proposed, and their answers are given as follows:

$$
\begin{cases}
k_{64}^l = 0 \\
k_{65}^l = 0
\end{cases}
\text{ or }
\begin{cases}
k_{53}^l k_{54}^l - k_{43}^l k_{55}^l = 0 \\
-k_{44}^l k_{53}^l + k_{43}^l k_{54}^l = 0 \\
\left(k_{53}^l\right)^2 - k_{33}^l k_{55}^l = 0 \\
-k_{43}^l k_{53}^l + k_{33}^l k_{54}^l = 0 \\
\left(k_{43}^l\right)^2 - k_{33}^l k_{44}^l = 0
\end{cases}
\Leftrightarrow
\begin{cases}
k_{64}^l = 0 \\
k_{65}^l = 0
\end{cases}
\text{ or }
\begin{cases}
k_{43}^l = \pm\sqrt{k_{33}^l k_{44}^l} \\
k_{53}^l = \pm\sqrt{k_{33}^l k_{55}^l} \\
k_{54}^l = \pm\sqrt{k_{44}^l k_{55}^l}
\end{cases}
\tag{A19}
$$

Appendix C. Inversion of the Compliance Matrix of a Leg in a Typical Decoupled-Motion 3L-CPM

$$
k_{11}^l = \frac{\left(c_{62}^l\right)^2 - c_{22}^l c_{66}^l}{c_{22}^l\left(c_{61}^l\right)^2 - 2c_{21}^l c_{61}^l c_{62}^l + c_{11}^l\left(c_{62}^l\right)^2 + \left(c_{21}^l\right)^2 c_{66}^l - c_{11}^l c_{22}^l c_{66}^l}
\tag{A20}
$$

$$
k_{21}^l = \frac{-c_{61}^l c_{62}^l + c_{21}^l c_{66}^l}{-2c_{21}^l c_{61}^l c_{62}^l + c_{11}^l\left(c_{62}^l\right)^2 + \left(c_{21}^l\right)^2 c_{66}^l + c_{22}^l\left[\left(c_{61}^l\right)^2 - c_{11}^l c_{66}^l\right]}
\tag{A21}
$$

$$
k_{61}^l = \frac{c_{22}^l c_{61}^l - c_{21}^l c_{62}^l}{c_{22}^l\left(c_{61}^l\right)^2 - 2c_{21}^l c_{61}^l c_{62}^l + c_{11}^l\left(c_{62}^l\right)^2 + \left(c_{21}^l\right)^2 c_{66}^l - c_{11}^l c_{22}^l c_{66}^l}
\tag{A22}
$$

$$
k_{22}^l = \frac{\left(c_{61}^l\right)^2 - c_{11}^l c_{66}^l}{c_{22}^l\left(c_{61}^l\right)^2 - 2c_{21}^l c_{61}^l c_{62}^l + c_{11}^l\left(c_{62}^l\right)^2 + \left(c_{21}^l\right)^2 c_{66}^l - c_{11}^l c_{22}^l c_{66}^l}
\tag{A23}
$$

$$
k_{62}^l = \frac{-c_{21}^l c_{61}^l + c_{11}^l c_{62}^l}{-2c_{21}^l c_{61}^l c_{62}^l + c_{11}^l\left(c_{62}^l\right)^2 + \left(c_{21}^l\right)^2 c_{66}^l + c_{22}^l\left[\left(c_{61}^l\right)^2 - c_{11}^l c_{66}^l\right]}
\tag{A24}
$$

$$k_{33}^l = \frac{\left(c_{54}^l\right)^2 - c_{44}^l c_{55}^l}{c_{44}^l \left(c_{53}^l\right)^2 - 2c_{43}^l c_{53}^l c_{54}^l + c_{33}^l \left(c_{54}^l\right)^2 + \left(c_{43}^l\right)^2 c_{55}^l - c_{33}^l c_{44}^l c_{55}^l} \tag{A25}$$

$$k_{43}^l = \frac{-c_{53}^l c_{54}^l + c_{43}^l c_{55}^l}{-2c_{43}^l c_{53}^l c_{54}^l + c_{33}^l \left(c_{54}^l\right)^2 + \left(c_{43}^l\right)^2 c_{55}^l + c_{44}^l \left[\left(c_{53}^l\right)^2 - c_{33}^l c_{55}^l\right]} \tag{A26}$$

$$k_{53}^l = \frac{c_{44}^l c_{53}^l - c_{43}^l c_{54}^l}{c_{44}^l \left(c_{53}^l\right)^2 - 2c_{43}^l c_{53}^l c_{54}^l + c_{33}^l \left(c_{54}^l\right)^2 + \left(c_{43}^l\right)^2 c_{55}^l - c_{33}^l c_{44}^l c_{55}^l} \tag{A27}$$

$$k_{44}^l = \frac{\left(c_{53}^l\right)^2 - c_{33}^l c_{55}^l}{c_{44}^l \left(c_{53}^l\right)^2 - 2c_{43}^l c_{53}^l c_{54}^l + c_{33}^l \left(c_{54}^l\right)^2 + \left(c_{43}^l\right)^2 c_{55}^l - c_{33}^l c_{44}^l c_{55}^l} \tag{A28}$$

$$k_{54}^l = \frac{-c_{43}^l c_{53}^l + c_{33}^l c_{54}^l}{-2c_{43}^l c_{53}^l c_{54}^l + c_{33}^l \left(c_{54}^l\right)^2 + \left(c_{43}^l\right)^2 c_{55}^l + c_{44}^l \left[\left(c_{53}^l\right)^2 - c_{33}^l c_{55}^l\right]} \tag{A29}$$

$$k_{55}^l = \frac{\left(c_{43}^l\right)^2 - c_{33}^l c_{44}^l}{c_{44}^l \left(c_{53}^l\right)^2 - 2c_{43}^l c_{53}^l c_{54}^l + c_{33}^l \left(c_{54}^l\right)^2 + \left(c_{43}^l\right)^2 c_{55}^l - c_{33}^l c_{44}^l c_{55}^l} \tag{A30}$$

$$k_{66}^l = \frac{\left(c_{21}^l\right)^2 - c_{11}^l c_{22}^l}{c_{22}^l \left(c_{61}^l\right)^2 - 2c_{21}^l c_{61}^l c_{62}^l + c_{11}^l \left(c_{62}^l\right)^2 + \left(c_{21}^l\right)^2 c_{66}^l - c_{11}^l c_{22}^l c_{66}^l} \tag{A31}$$

Appendix D. Compliance Matrix of a Leg in 3RRR- and 3PRR-3L-CPMs

$$c_{11}^l = \frac{1}{2} \sum_{j=1}^{3} \left[c_{11}^{e_j} + c_{22}^{e_j} + 2c_{66}^{e_j} r_{y_j}^2 + \left(c_{11}^{e_j} - c_{22}^{e_j} \right) \cos 2\gamma_j + 4c_{62}^{e_j} r_{y_j} \sin \gamma_j \right] \tag{A32}$$

$$c_{21}^l = \sum_{j=1}^{3} \left\{ -c_{66}^{e_j} r_{x_j} r_{y_j} - c_{62}^{e_j} r_{x_j} \sin \gamma_j + \cos \gamma_j \left[-c_{62}^{e_j} r_{y_j} + \left(c_{11}^{e_j} - c_{22}^{e_j} \right) \sin \gamma_j \right] \right\} \tag{A33}$$

$$c_{61}^l = \sum_{j=1}^{3} \left(-c_{66}^{e_j} r_{y_j} - c_{62}^{e_j} \sin \gamma_j \right) \tag{A34}$$

$$c_{22}^l = \frac{1}{2} \sum_{j=1}^{3} \left[c_{11}^{e_j} + c_{22}^{e_j} + 2c_{66}^{e_j} r_{x_j}^2 + 4c_{62}^{e_j} r_{x_j} \cos \gamma_j + \left(-c_{11}^{e_j} + c_{22}^{e_j} \right) \cos 2\gamma_j \right] \tag{A35}$$

$$c_{62}^l = \sum_{j=1}^{3} \left(c_{66}^{e_j} r_{x_j} + c_{62}^{e_j} \cos \gamma_j \right) \tag{A36}$$

$$c_{33}^l = \frac{1}{2} \sum_{j=1}^{3} \left\{ 2c_{33}^{e_j} + \left(c_{44}^{e_j} + c_{55}^{e_j} \right) \left(r_{x_j}^2 + r_{y_j}^2 \right) - \left(c_{44}^{e_j} - c_{55}^{e_j} \right) \left(r_{x_j} - r_{y_j} \right) \left(r_{x_j} + r_{y_j} \right) \cos 2\gamma_j - \right.$$
$$\left. 4c_{53}^{e_j} r_{y_j} \sin \gamma_j - 4r_{x_j} \cos \gamma_j \left[c_{53}^{e_j} + \left(c_{44}^{e_j} - c_{55}^{e_j} \right) r_{y_j} \sin \gamma_j \right] \right\} \tag{A37}$$

$$c_{43}^l = \frac{1}{2} \sum_{j=1}^{3} \left[\left(c_{44}^{e_j} + c_{55}^{e_j} \right) r_{y_j} - 2c_{53}^{e_j} \sin \gamma_j + \left(c_{44}^{e_j} - c_{55}^{e_j} \right) \left(r_{y_j} \cos 2\gamma_j - r_{x_j} \sin 2\gamma_j \right) \right] \tag{A38}$$

$$c_{53}^l = \frac{1}{2}\sum_{j=1}^{3}\left[-\left(c_{44}^{e_j}+c_{55}^{e_j}\right)r_{x_j}+2c_{53}^{e_j}\cos\gamma_j+\left(c_{44}^{e_j}-c_{55}^{e_j}\right)\left(r_{x_j}\cos 2\gamma_j+r_{y_j}\sin 2\gamma_j\right)\right] \tag{A39}$$

$$c_{44}^l = \frac{1}{2}\sum_{j=1}^{3}\left[c_{44}^{e_j}+c_{55}^{e_j}+\left(c_{44}^{e_j}-c_{55}^{e_j}\right)\cos 2\gamma_j\right] \tag{A40}$$

$$c_{54}^l = \sum_{j=1}^{3}\left[\left(c_{44}^{e_j}-c_{55}^{e_j}\right)\cos\gamma_j\sin\gamma_j\right] \tag{A41}$$

$$c_{55}^l = \frac{1}{2}\sum_{j=1}^{3}\left[c_{44}^{e_j}+c_{55}^{e_j}+\left(-c_{44}^{e_j}+c_{55}^{e_j}\right)\cos 2\gamma_j\right] \tag{A42}$$

$$c_{66}^l = \sum_{j=1}^{3}\left(c_{66}^{e_j}\right) \tag{A43}$$

Appendix E. Components in the Compliance Matrix of a Leg in the 3-DOF (*X-Y-Z*) 3PRPR-3L-CPM

$$c_{11}^l = 2c_{33}^R - 2c_{53}^R r_{y_2} + c_{55}^R r_{y_2}^2 + c_{66}^P\left(r_{y_1}^2+r_{y_3}^2\right) + c_{44}^R r_{z_2}^2 + \left(c_{11}^P+c_{55}^P r_{z_1}^2\right)\cos^2\gamma_1 + \left(c_{11}^P+c_{55}^P r_{z_3}^2\right)\cos^2\gamma_3 + \\ \sin\gamma_1\left[2c_{62}^P r_{y_1}+\left(c_{22}^P+c_{44}^P r_{z_1}^2\right)\sin\gamma_1\right]+2c_{62}^P r_{y_3}\sin\gamma_3+\left(c_{22}^P+c_{44}^P r_{z_3}^2\right)\sin^2\gamma_3 \tag{A44}$$

$$c_{21}^l = r_{x_2}\left(c_{53}^R-c_{55}^R r_{y_2}\right)-c_{66}^P\left(r_{x_1}r_{y_1}+r_{x_3}r_{y_3}\right)+\cos\gamma_1\left\{-c_{62}^P r_{y_1}+\left[c_{11}^P-c_{22}^P+\left(-c_{44}^P+c_{55}^P\right)r_{z_1}^2\right]\sin\gamma_1\right\}- \\ c_{62}^P\left(r_{x_1}\sin\gamma_1+r_{x_3}\sin\gamma_3\right)+\cos\gamma_3\left\{-c_{62}^P r_{y_3}+\left[c_{11}^P-c_{22}^P+\left(-c_{44}^P+c_{55}^P\right)r_{z_3}^2\right]\sin\gamma_3\right\} \tag{A45}$$

$$c_{31}^l = -c_{44}^R r_{x_2}r_{z_2}-c_{55}^P r_{x_1}r_{z_1}\cos^2\gamma_1-c_{55}^P r_{x_3}r_{z_3}\cos^2\gamma_3+r_{z_1}\cos\gamma_1\left[c_{53}^P+\left(c_{44}^P-c_{55}^P\right)r_{y_1}\sin\gamma_1\right]+ \\ r_{z_3}\cos\gamma_3\left[c_{53}^P+\left(c_{44}^P-c_{55}^P\right)r_{y_3}\sin\gamma_3\right]-c_{44}^P\left(r_{x_1}r_{z_1}\sin^2\gamma_1+r_{x_3}r_{z_3}\sin^2\gamma_3\right) \tag{A46}$$

$$c_{41}^l = \left(c_{44}^P-c_{55}^P\right)\left(r_{z_1}\cos\gamma_1\sin\gamma_1+r_{z_3}\cos\gamma_3\sin\gamma_3\right) \tag{A47}$$

$$c_{51}^l = c_{44}^R r_{z_2}+c_{55}^P r_{z_1}\cos^2\gamma_1+c_{55}^P r_{z_3}\cos^2\gamma_3+c_{44}^P r_{z_1}\sin^2\gamma_1+c_{44}^P r_{z_3}\sin^2\gamma_3 \tag{A48}$$

$$c_{61}^l = 2c_{53}^R-c_{55}^R r_{y_2}-c_{66}^P\left(r_{y_1}+r_{y_3}\right)-c_{62}^P\left(\sin\gamma_1+\sin\gamma_3\right) \tag{A49}$$

$$c_{22}^l = 2c_{11}^R+c_{55}^R r_{x_2}^2+c_{66}^P\left(r_{x_1}^2+r_{x_3}^2\right)+c_{66}^R r_{z_2}^2+2c_{62}^P r_{x_1}\cos\gamma_1+\left(c_{22}^P+c_{44}^P r_{z_1}^2\right)\cos^2\gamma_1+2c_{62}^P r_{x_3}\cos\gamma_3+ \\ \left(c_{22}^P+c_{44}^P r_{z_3}^2\right)\cos^2\gamma_3+\left(c_{11}^P+c_{55}^P r_{z_1}^2\right)\sin^2\gamma_1+\left(c_{11}^P+c_{55}^P r_{z_3}^2\right)\sin^2\gamma_3 \tag{A50}$$

$$c_{32}^l = -\left(c_{62}^R+c_{66}^R r_{y_2}\right)r_{z_2}-c_{44}^P r_{y_1}r_{z_1}\cos^2\gamma_1+\left(c_{44}^P-c_{55}^P\right)r_{x_1}r_{z_1}\cos\gamma_1\sin\gamma_1+r_{z_1}\sin\gamma_1\left(c_{53}^P-c_{55}^P r_{y_1}\sin\gamma_1\right)+ \\ r_{z_3}\left[-c_{44}^P r_{y_3}\cos^2\gamma_3+\left(c_{44}^P-c_{55}^P\right)r_{x_3}\cos\gamma_3\sin\gamma_3+\sin\gamma_3\left(c_{53}^P-c_{55}^P r_{y_3}\sin\gamma_3\right)\right] \tag{A51}$$

$$c_{42}^l = -c_{66}^R r_{z_2}-c_{44}^P r_{z_1}\cos^2\gamma_1-c_{44}^P r_{z_3}\cos^2\gamma_3-c_{55}^P\left(r_{z_1}\sin^2\gamma_1+r_{z_3}\sin^2\gamma_3\right) \tag{A52}$$

$$c_{52}^l = -\left(c_{44}^P-c_{55}^P\right)\left(r_{z_1}\cos\gamma_1\sin\gamma_1+r_{z_3}\cos\gamma_3\sin\gamma_3\right) \tag{A53}$$

$$c_{62}^l = c_{55}^R r_{x_2}+c_{66}^P\left(r_{x_1}+r_{x_3}\right)+c_{62}^P\left(\cos\gamma_1+\cos\gamma_3\right) \tag{A54}$$

$$c_{33}^l = 2c_{33}^P + 2c_{22}^R + c_{44}^R r_{x_2}^2 + r_{y_2}\left(c_{62}^R + c_{66}^R r_{y_2}\right) + c_{44}^P\left(r_{y_1}\cos\gamma_1 - r_{x_1}\sin\gamma_1\right)^2 - c_{53}^P\left(r_{x_1}\cos\gamma_1 + r_{y_1}\sin\gamma_1\right) +$$
$$\left(r_{x_1}\cos\gamma_1 + r_{y_1}\sin\gamma_1\right)\left(-c_{53}^P + c_{55}^P r_{x_1}\cos\gamma_1 + c_{55}^P r_{y_1}\sin\gamma_1\right) + c_{44}^P\left(r_{y_3}\cos\gamma_3 - r_{x_3}\sin\gamma_3\right)^2 -$$
$$c_{53}^P\left(r_{x_3}\cos\gamma_3 + r_{y_3}\sin\gamma_3\right) + \left(r_{x_3}\cos\gamma_3 + r_{y_3}\sin\gamma_3\right)\left(-c_{53}^P + c_{55}^P r_{x_3}\cos\gamma_3 + c_{55}^P r_{y_3}\sin\gamma_3\right) \tag{A55}$$

$$c_{43}^l = 2c_{62}^R + c_{66}^R r_{y_2} + c_{44}^P r_{y_1}\cos^2\gamma_1 + c_{44}^P r_{y_3}\cos^2\gamma_3 + \left(-c_{44}^P + c_{55}^P\right)r_{x_1}\cos\gamma_1\sin\gamma_1 + c_{55}^P r_{y_1}\sin^2\gamma_1 +$$
$$\left(-c_{44}^P + c_{55}^P\right)r_{x_3}\cos\gamma_3\sin\gamma_3 + c_{55}^P r_{y_3}\sin^2\gamma_3 - c_{53}^P(\sin\gamma_1 + \sin\gamma_3) \tag{A56}$$

$$c_{53}^l = -c_{44}^R r_{x_2} - c_{55}^P r_{x_1}\cos^2\gamma_1 - c_{55}^P r_{x_3}\cos^2\gamma_3 + \cos\gamma_1\left[c_{53}^P + \left(c_{44}^P - c_{55}^P\right)r_{y_1}\sin\gamma_1\right] +$$
$$\cos\gamma_3\left[c_{53}^P + \left(c_{44}^P - c_{55}^P\right)r_{y_3}\sin\gamma_3\right] - c_{44}^P\left(r_{x_1}\sin^2\gamma_1 + r_{x_3}\sin^2\gamma_3\right) \tag{A57}$$

$$c_{44}^l = 2c_{66}^R + c_{44}^P\left(\cos^2\gamma_1 + \cos^2\gamma_3\right) + c_{55}^P\left(\sin^2\gamma_1 + \sin^2\gamma_3\right) \tag{A58}$$

$$c_{54}^l = \left(c_{44}^P - c_{55}^P\right)\left(\cos\gamma_1\sin\gamma_1 + \cos\gamma_3\sin\gamma_3\right) \tag{A59}$$

$$c_{55}^l = 2c_{44}^R + c_{55}^P\left(\cos^2\gamma_1 + \cos^2\gamma_3\right) + c_{44}^P\left(\sin^2\gamma_1 + \sin^2\gamma_3\right) \tag{A60}$$

$$c_{66}^l = 2\left(c_{66}^P + c_{55}^R\right) \tag{A61}$$

Appendix F. Components in the Compliance Matrix of a Leg in the 6-DOF 6PSS-3L-CPM

$$c_{11}^{sc} = \frac{1}{2}\left[2c_{33}^P + 2c_{11}^S + 2c_{22}^S + 2c_{44}^P r_{z_1}^2 + \left(c_{44}^S + c_{55}^S\right)\left(r_{z_2}^2 + r_{z_3}^2\right)\right] \tag{A62}$$

$$c_{21}^{sc} = \frac{1}{2}\left[2c_{11}^S - 2c_{22}^S + 2c_{53}^P r_{x_1} - \sqrt{2}c_{62}^S(r_{x_2} + r_{x_3}) - \left(c_{44}^S - c_{55}^S\right)\left(r_{z_2}^2 + r_{z_3}^2\right)\right] \tag{A63}$$

$$c_{31}^{sc} = \frac{1}{2}\left[-2c_{44}^P r_{x_1}r_{z_1} + \sqrt{2}c_{53}^S(r_{z_2} + r_{z_3}) - \left(c_{44}^S + c_{55}^S\right)(r_{x_2}r_{z_2} + r_{x_3}r_{z_3})\right] \tag{A64}$$

$$c_{41}^{sc} = \frac{1}{2}\left(c_{44}^S - c_{55}^S\right)(r_{z_2} + r_{z_3}) \tag{A65}$$

$$c_{51}^{sc} = \frac{1}{2}\left[2c_{44}^P r_{z_1} + \left(c_{44}^S + c_{55}^S\right)(r_{z_2} + r_{z_3})\right] \tag{A66}$$

$$c_{61}^{sc} = c_{53}^P - \sqrt{2}c_{62}^S \tag{A67}$$

$$c_{22}^{sc} = c_{11}^P + c_{11}^S + \frac{1}{2}\left[2c_{22}^S + 2c_{55}^P r_{x_1}^2 + 2\sqrt{2}c_{62}^S(r_{x_2} + r_{x_3}) + 2c_{66}^S\left(r_{x_2}^2 + r_{x_3}^2\right) + 2c_{66}^P r_{z_1}^2 + \left(c_{44}^S + c_{55}^S\right)\left(r_{z_2}^2 + r_{z_3}^2\right)\right] \tag{A68}$$

$$c_{32}^{sc} = \frac{1}{2}\left[-c_{62}^P r_{z_1} + \sqrt{2}c_{53}^S(r_{z_2} + r_{z_3}) + \left(c_{44}^S - c_{55}^S\right)(r_{x_2}r_{z_2} + r_{x_3}r_{z_3})\right] \tag{A69}$$

$$c_{42}^{sc} = \frac{1}{2}\left[-2c_{66}^P r_{z_1} - \left(c_{44}^S + c_{55}^S\right)(r_{z_2} + r_{z_3})\right] \tag{A70}$$

$$c_{52}^{sc} = -\frac{1}{2}\left(c_{44}^S - c_{55}^S\right)(r_{z_2} + r_{z_3}) \tag{A71}$$

$$c_{62}^{sc} = \sqrt{2}c_{62}^S + c_{55}^P r_{x_1} + c_{66}^S(r_{x_2} + r_{x_3}) \tag{A72}$$

$$c_{33}^{sc} = \frac{1}{2}\left[2c_{22}^{P} + 4c_{33}^{S} + 2c_{44}^{P}r_{x_1}^2 - 2\sqrt{2}c_{53}^{S}(r_{x_2} + r_{x_3}) + \left(c_{44}^{S} + c_{55}^{S}\right)\left(r_{x_2}^2 + r_{x_3}^2\right)\right] \tag{A73}$$

$$c_{43}^{sc} = \frac{1}{2}\left[2c_{62}^{P} - 2\sqrt{2}c_{53}^{S} - \left(c_{44}^{S} - c_{55}^{S}\right)(r_{x_2} + r_{x_3})\right] \tag{A74}$$

$$c_{53}^{sc} = \frac{1}{2}\left[2\sqrt{2}c_{53}^{S} - 2c_{44}^{P}r_{x_1} - \left(c_{44}^{S} + c_{55}^{S}\right)(r_{x_2} + r_{x_3})\right] \tag{A75}$$

$$c_{44}^{sc} = c_{66}^{P} + c_{44}^{S} + c_{55}^{S} \tag{A76}$$

$$c_{54}^{sc} = c_{44}^{S} - c_{55}^{S} \tag{A77}$$

$$c_{55}^{sc} = c_{44}^{P} + c_{44}^{S} + c_{55}^{S} \tag{A78}$$

$$c_{66}^{sc} = c_{55}^{P} + 2c_{66}^{S} \tag{A79}$$

References

1. Wang, D.H.; Yang, Q.; Dong, H.M. A Monolithic Compliant Piezoelectric-Driven Microgripper: Design, Modeling, and Testing. *IEEE/ASME Trans. Mechatron.* **2013**, *18*, 138–147. [CrossRef]
2. Wang, F.; Liang, C.; Tian, Y.; Zhao, X.; Zhang, D. Design and Control of a Compliant Microgripper with a Large Amplification Ratio for High-Speed Micro Manipulation. *IEEE/ASME Trans. Mechatron.* **2016**, *21*, 1262–1271. [CrossRef]
3. Teo, T.J.; Yang, G.; Chen, I.M. A flexure-based electromagnetic nanopositioning actuator with predictable and re-configurable open-loop positioning resolution. *Precis. Eng.* **2015**, *40*, 249–260. [CrossRef]
4. Teo, T.J.; Bui, V.P.; Yang, G.; Chen, I.M. Millimeters-Stroke Nanopositioning Actuator with High Positioning and Thermal Stability. *IEEE/ASME Trans. Mechatron.* **2015**, *20*, 2813–2823. [CrossRef]
5. Teo, T.J.; Chen, I.M.; Yang, G. A large deflection and high payload flexure-based parallel manipulator for UV nanoimprint lithography: Part II. Stiffness modeling and performance evaluation. *Precis. Eng.* **2014**, *38*, 872–884. [CrossRef]
6. Pham, M.T.; Teo, T.J.; Yeo, S.H.; Wang, P.; Nai, M.L.S. A 3D-printed Ti-6Al-4V 3-DOF compliant parallel mechanism for high precision manipulation. *IEEE/ASME Trans. Mechatron.* **2017**, *22*, 2359–2368. [CrossRef]
7. Pham, M.T.; Yeo, S.H.; Teo, T.J.; Wang, P.; Nai, M.L.S. Design and Optimization of a Three Degrees-of-Freedom Spatial Motion Compliant Parallel Mechanism with Fully Decoupled Motion Characteristics. *J. Mech. Robot.* **2019**, *11*, 051010. [CrossRef]
8. Teo, T.J.; Yang, G.; Chen, I.M. Compliant Manipulators. In *Handbook of Manufacturing Engineering and Technology*; Springer: London, UK, 2014; pp. 2229–2300.
9. Thomas, T.L.; Kalpathy Venkiteswaran, V.; Ananthasuresh, G.K.; Misra, S. Surgical Applications of Compliant Mechanisms: A Review. *J. Mech. Robot.* **2021**, *13*, 020801. [CrossRef]
10. Huang, H.; Pan, Y.; Pang, Y.; Shen, H.; Gao, X.; Zhu, Y.; Chen, L.; Sun, L. Piezoelectric Ultrasonic Biological Microdissection Device Based on a Novel Flexure Mechanism for Suppressing Vibration. *Micromachines* **2021**, *12*, 196. [CrossRef]
11. Shinde, S.M.; Lekurwale, R.R. Synthesising of flexural spindle head micro drilling machine tool in PLM environment. *Int. J. Virtual Technol. Multimed.* **2021**, *1*, 246–264. [CrossRef]
12. Shinde, S.M.; Lekurwale, R.R. Radial stiffness computation of single Archimedes spiral plane supporting spring loaded in flexural mechanism mounted in spindle head of micro drilling machine tool. *Mech. Based Des. Struct. Mach.* **2022**, 1–21. [CrossRef]
13. Lv, B.; Lin, B.; Cao, Z.; Li, B.; Wang, G. A parallel 3-DOF micro-nano motion stage for vibration-assisted milling. *Mech. Mach. Theory* **2022**, *173*, 104854. [CrossRef]
14. Gandhi, P.; Bhole, K. 3D Microfabrication Using Bulk Lithography. In Proceedings of the ASME 2011 International Mechanical Engineering Congress and Exposition, Denver, CO, USA, 11–17 November 2011; pp. 393–399.
15. Gandhi, P.; Deshmukh, S.; Ramtekkar, R.; Bhole, K.; Baraki, A. "On-Axis" Linear Focused Spot Scanning Microstereolithography System: Optomechatronic Design, Analysis and Development. *J. Adv. Manuf. Syst.* **2013**, *12*, 43–68. [CrossRef]
16. Tanikawa, T.; Arai, T.; Koyachi, N. Development of small-sized 3 DOF finger module in micro hand for micro manipulation. In Proceedings of the 1999 IEEE/RSJ International Conference on Intelligent Robots and Systems, Kyongju, Korea, 17–21 October 1999; pp. 876–881.
17. Tanikawa, T.; Ukiana, M.; Morita, K.; Koseki, Y.; Ohba, K.; Fujii, K.; Arai, T. Design of 3-DOF parallel mechanism with thin plate for micro finger module in micro manipulation. In Proceedings of the IEEE/RSJ International Conference on Intelligent Robots and Systems, Lausanne, Switzerland, 30 September–4 October 2002; pp. 1778–1783.
18. Yi, B.-J.; Chung, G.B.; Na, H.Y.; Kim, W.K.; Suh, I.H. Design and experiment of a 3-DOF parallel micromechanism utilizing flexure hinges. *IEEE Trans. Robot. Autom.* **2003**, *19*, 604–612. [CrossRef]

19. Pham, H.-H.; Chen, I.-M.; Yeh, H.-C. Micro-motion selective-actuation XYZ flexure parallel mechanism: Design and modeling. *J. Micromechatronics* **2003**, *3*, 51–73. [CrossRef]

20. Lu, T.-F.; Handley, D.C.; Yong, Y.K.; Eales, C. A three-DOF compliant micromotion stage with flexure hinges. *Ind. Robot. Int. J.* **2004**, *31*, 355–361. [CrossRef]

21. Chao, D.; Zong, G.; Liu, R. Design of a 6-DOF compliant manipulator based on serial-parallel architecture. In Proceedings of the 2005 IEEE/ASME International Conference on Advanced Intelligent Mechatronics, Monterey, CA, USA, 24–28 July 2005; pp. 765–770.

22. Pham, H.-H.; Chen, I.M. Stiffness modeling of flexure parallel mechanism. *Precis. Eng.* **2005**, *29*, 467–478. [CrossRef]

23. Choi, Y.-J.; Sreenivasan, S.V.; Choi, B.J. Kinematic design of large displacement precision XY positioning stage by using cross strip flexure joints and over-constrained mechanism. *Mech. Mach. Theory* **2008**, *43*, 724–737. [CrossRef]

24. Yao, Q.; Dong, J.; Ferreira, P.M. A novel parallel-kinematics mechanisms for integrated, multi-axis nanopositioning: Part 1. Kinematics and design for fabrication. *Precis. Eng.* **2008**, *32*, 7–19. [CrossRef]

25. Dong, J.; Yao, Q.; Ferreira, P.M. A novel parallel-kinematics mechanism for integrated, multi-axis nanopositioning: Part 2: Dynamics, control and performance analysis. *Precis. Eng.* **2008**, *32*, 20–33. [CrossRef]

26. Dong, W.; Sun, L.; Du, Z. Stiffness research on a high-precision, large-workspace parallel mechanism with compliant joints. *Precis. Eng.* **2008**, *32*, 222–231. [CrossRef]

27. Wang, H.; Zhang, X. Input coupling analysis and optimal design of a 3-DOF compliant micro-positioning stage. *Mech. Mach. Theory* **2008**, *43*, 400–410. [CrossRef]

28. Wu, T.L.; Chen, J.H.; Chang, S.H. A six-DOF prismatic-spherical-spherical parallel compliant nanopositioner. *IEEE Trans. Ultrason. Ferroelectr. Freq. Control.* **2008**, *55*, 2544–2551. [CrossRef] [PubMed]

29. Yong, Y.K.; Lu, T.-F. Kinetostatic modeling of 3-RRR compliant micro-motion stages with flexure hinges. *Mech. Mach. Theory* **2009**, *44*, 1156–1175. [CrossRef]

30. Li, Y.; Xu, Q. A Totally Decoupled Piezo-Driven XYZ Flexure Parallel Micropositioning Stage for Micro/Nanomanipulation. *IEEE Trans. Autom. Sci. Eng.* **2011**, *8*, 265–279. [CrossRef]

31. Liang, Q.; Zhang, D.; Chi, Z.; Song, Q.; Ge, Y.; Ge, Y. Six-DOF micro-manipulator based on compliant parallel mechanism with integrated force sensor. *Robot. Comput.-Integr. Manuf.* **2011**, *27*, 124–134. [CrossRef]

32. Yang, G.; Teo, T.J.; Chen, I.M.; Lin, W. Analysis and design of a 3-DOF flexure-based zero-torsion parallel manipulator for nano-alignment applications. In Proceedings of the 2011 IEEE International Conference on Robotics and Automation (ICRA), Shanghai, China, 9–13 May 2011; pp. 2751–2756.

33. Yun, Y.; Li, Y. Optimal design of a 3-PUPU parallel robot with compliant hinges for micromanipulation in a cubic workspace. *Robot. Comput.-Integr. Manuf.* **2011**, *27*, 977–985. [CrossRef]

34. Li, Y.; Huang, J.; Tang, H. A Compliant Parallel XY Micromotion Stage with Complete Kinematic Decoupling. *IEEE Trans. Autom. Sci. Eng.* **2012**, *9*, 538–553. [CrossRef]

35. Kim, H.-Y.; Ahn, D.-H.; Gweon, D.-G. Development of a novel 3-degrees of freedom flexure based positioning system. *Rev. Sci. Instrum.* **2012**, *83*, 055114. [CrossRef]

36. Xiao, S.; Li, Y. Optimal Design, Fabrication, and Control of an XY Micropositioning Stage Driven by Electromagnetic Actuators. *IEEE Trans. Ind. Electron.* **2013**, *60*, 4613–4626. [CrossRef]

37. Bhagat, U.; Shirinzadeh, B.; Clark, L.; Chea, P.; Qin, Y.; Tian, Y.; Zhang, D. Design and analysis of a novel flexure-based 3-DOF mechanism. *Mech. Mach. Theory* **2014**, *74*, 173–187. [CrossRef]

38. Clark, L.; Shirinzadeh, B.; Zhong, Y.; Tian, Y.; Zhang, D. Design and analysis of a compact flexure-based precision pure rotation stage without actuator redundancy. *Mech. Mach. Theory* **2016**, *105*, 129–144. [CrossRef]

39. Hao, G.; Yu, J. Design, modelling and analysis of a completely-decoupled XY compliant parallel manipulator. *Mech. Mach. Theory* **2016**, *102*, 179–195. [CrossRef]

40. Li, Y.; Wu, Z. Design, analysis and simulation of a novel 3-DOF translational micromanipulator based on the PRB model. *Mech. Mach. Theory* **2016**, *100*, 235–258. [CrossRef]

41. Wang, R.; Zhang, X. A planar 3-DOF nanopositioning platform with large magnification. *Precis. Eng.* **2016**, *46*, 221–231. [CrossRef]

42. Culpepper, M.L.; Anderson, G. Design of a low-cost nano-manipulator which utilizes a monolithic, spatial compliant mechanism. *Precis. Eng.* **2004**, *28*, 469–482. [CrossRef]

43. Chen, S.-C.; Culpepper, M.L. Design of a six-axis micro-scale nanopositioner—μHexFlex. *Precis. Eng.* **2006**, *30*, 314–324. [CrossRef]

44. Awtar, S.; Slocum, A.H. Constraint-Based Design of Parallel Kinematic XY Flexure Mechanisms. *J. Mech. Des.* **2006**, *129*, 816–830. [CrossRef]

45. Hopkins, J.B.; Culpepper, M.L. Synthesis of multi-degree of freedom, parallel flexure system concepts via Freedom and Constraint Topology (FACT)—Part I: Principles. *Precis. Eng.* **2010**, *34*, 259–270. [CrossRef]

46. Hopkins, J.B.; Culpepper, M.L. Synthesis of multi-degree of freedom, parallel flexure system concepts via freedom and constraint topology (FACT). Part II: Practice. *Precis. Eng.* **2010**, *34*, 271–278. [CrossRef]

47. Hopkins, J.B.; Culpepper, M.L. Synthesis of precision serial flexure systems using freedom and constraint topologies (FACT). *Precis. Eng.* **2011**, *35*, 638–649. [CrossRef]

48. Awtar, S.; Ustick, J.; Sen, S. An XYZ Parallel-Kinematic Flexure Mechanism with Geometrically Decoupled Degrees of Freedom. *J. Mech. Robot.* **2012**, *5*, 015001. [CrossRef]

49. Hopkins, J.B.; Panas, R.M. Design of flexure-based precision transmission mechanisms using screw theory. *Precis. Eng.* **2013**, *37*, 299–307. [CrossRef]
50. Li, H.; Hao, G. A constraint and position identification (CPI) approach for the synthesis of decoupled spatial translational compliant parallel manipulators. *Mech. Mach. Theory* **2015**, *90*, 59–83. [CrossRef]
51. Li, H.; Hao, G.; Kavanagh, R. A New XYZ Compliant Parallel Mechanism for Micro-/Nano-Manipulation: Design and Analysis. *Micromachines* **2016**, *7*, 23. [CrossRef]
52. Hao, G. Design and analysis of symmetric and compact 2R1T (in-plane 3-DOC) flexure parallel mechanisms. *Mech. Sci.* **2017**, *8*, 1–9. [CrossRef]
53. Lum, G.Z.; Teo, T.J.; Yeo, S.H.; Yang, G.; Sitti, M. Structural optimization for flexure-based parallel mechanisms—Towards achieving optimal dynamic and stiffness properties. *Precis. Eng.* **2015**, *42*, 195–207. [CrossRef]
54. Jin, M.; Zhang, X. A new topology optimization method for planar compliant parallel mechanisms. *Mech. Mach. Theory* **2016**, *95*, 42–58. [CrossRef]
55. Lum, G.Z.; Pham, M.T.; Teo, T.J.; Yang, G.; Yeo, S.H.; Sitti, M. An XY & thetaZ flexure mechanism with optimal stiffness properties. In Proceedings of the 2017 IEEE International Conference on Advanced Intelligent Mechatronics (AIM), Munich, Germany, 3–7 July 2017; pp. 1103–1110.
56. Pham, M.T.; Yeo, S.H.; Teo, T.J.; Wang, P.; Nai, M.L.S. A Decoupled 6-DOF Compliant Parallel Mechanism with Optimized Dynamic Characteristics Using Cellular Structure. *Machines* **2021**, *9*, 5. [CrossRef]
57. Lobontiu, N. *Compliant Mechanisms: Design of Flexure Hinges*; CRC Press: Boca Raton, FL, USA, 2010.
58. Akbari, S.; Pirbodaghi, T. Precision positioning using a novel six axes compliant nano-manipulator. *Microsyst. Technol.* **2017**, *23*, 2499–2507. [CrossRef]
59. de Jong, B.R.; Brouwer, D.M.; de Boer, M.J.; Jansen, H.V.; Soemers, H.M.; Krijnen, G.J. Design and Fabrication of a Planar Three-DOFs MEMS-Based Manipulator. *J. Microelectromech. Syst.* **2010**, *19*, 1116–1130. [CrossRef]
60. Mukhopadhyay, D.; Dong, J.; Pengwang, E.; Ferreira, P. A SOI-MEMS-based 3-DOF planar parallel-kinematics nanopositioning stage. *Sens. Actuators A Phys.* **2008**, *147*, 340–351. [CrossRef]
61. Lee, K.M.; Arjunan, S. A three-degrees-of-freedom micromotion in-parallel actuated manipulator. *IEEE Trans. Robot. Autom.* **1991**, *7*, 634–641. [CrossRef]
62. Teo, T.J. Flexure-Based Electromagnetic Parallel-Kinematics Manipulator System. Ph.D. Thesis, Nanyang Technological University, Singapore, 2009.
63. Pham, M.T. Design and 3D Printing of Compliant Mechanisms. Ph.D. Thesis, Nanyang Technological University, Singapore, 2019.

Article

Predictive Suspension Algorithm for Land Vehicles over Deterministic Topography

Alejandro Bustos *, Jesus Meneses, Higinio Rubio and Enrique Soriano-Heras

Department of Mechanical Engineering, Universidad Carlos III de Madrid, 28911 Leganes, Spain; meneses@ing.uc3m.es (J.M.); hrubio@ing.uc3m.es (H.R.); esoriano@ing.uc3m.es (E.S.-H.)
* Correspondence: albustos@ing.uc3m.es

Abstract: A good suspension system is mandatory for ensuring stability, comfort and safety in land vehicles; therefore, advanced semi and fully active suspension systems have been developed along with their associated management strategies to overcome the limitations of passive suspensions. This paper presents a suspension algorithm for land vehicles traveling through a deterministic topography. The kinematics of a half-vehicle model and the algorithm are implemented in Simulink. The algorithm's inputs are the measurements provided by a position scanner located on the front wheel of the vehicle. Based on this input, the algorithm reconstructs the topography in real-time and sends the corresponding command to an actuator located on the rear wheel to compensate for the irregularities of the terrain. The actuation is governed by the parameter "D", which represents the distance over which the algorithm averages the height of the terrain. Two ground profiles were tested and sensitivity analysis of the parameter "D" was performed. Results show that larger values of "D" usually yield less vibration on the actuated mass, but this value also depends on the irregularities of the terrain.

Keywords: suspension algorithm; predictive suspension; land vehicle; sprung mass

MSC: 70B15

Citation: Bustos, A.; Meneses, J.; Rubio, H.; Soriano-Heras, E. Predictive Suspension Algorithm for Land Vehicles over Deterministic Topography. *Mathematics* **2022**, *10*, 1467. https://doi.org/10.3390/math10091467

Academic Editor: Andrey Jivkov

Received: 28 March 2022
Accepted: 25 April 2022
Published: 27 April 2022

Publisher's Note: MDPI stays neutral with regard to jurisdictional claims in published maps and institutional affiliations.

1. Introduction

Suspension is an essential system for land vehicles to ensure their stability, comfort and safety; however, increasing the vehicle stability usually results in a worsening of comfort and vice versa [1,2]. In an attempt to improve both stability and comfort, semi-active and active suspensions and their corresponding management strategies have been developed in recent years.

Suspension systems can be classified into three types according to their operation: passive, semi-active and active.

The fundamental elements of passive suspensions are springs, dampers and the required rigid links to connect the sprung and unsprung masses. In this kind of suspension, the parameters of springs and dampers are fixed; therefore, the dynamic response of the system is fixed too. The design of these suspensions must achieve a trade-off solution for different conditions due to the different surfaces the vehicle travels through; however, non-linear springs [3,4], non-linear dampers [5] and hydro–gas rotary suspension systems [6] have been used in land vehicles to obtain better response over the widest possible range of conditions.

Semi-active suspensions can adjust the parameters of passive suspensions, but the generated forces will always be dependent on the speed of the damper [7]. Hence, semi-active suspensions are not able to control ride height, roll or pitch angles. Nonetheless, their simplicity requires less power consumption than fully active suspensions and can be implemented easily:

- The stiffness of semi-active suspensions is modified by using magnetorheological dampers within a specific mechanism [8] or by governing smart air springs [9].
- Variable damping is the most common strategy in the development of semi-active suspensions, having different solutions:
  - Hydraulic dampers with electrically controlled valves for adjusting the fluid flow rate from one chamber to the other [10].
  - Magnetorheological [11] and electrorheological [12,13] dampers are similar to hydraulic dampers but they contain a magnetorheological or electrorheological fluid. This fluid, usually non-Newtonian, reacts to the application of magnetic or electrical fields and changes the properties of the damper. The use of these devices is widely spread in road and rail vehicles [14–16].
  - Electromagnetic dampers take advantage of the known interaction between a moving coil and the magnetic field generated by a permanent magnet or electromagnet to generate a damping effect [17]. Karnopp studies the use of permanent magnet linear motors as variable mechanical dampers for vehicle suspensions [18].
- The third component of suspension systems, inertia, is modified through a variable inerter. The inerter is a mechanical device in which the force applied is proportional to the relative accelerations between the two terminals of the device [19]. These systems have been proposed to improve the comfort of rail vehicles [20].

Active suspensions allow full management of the suspension motion, as the actuator force is independent of the speed direction. On the other hand, the main disadvantage of these suspension types is the high energy consumption that they require. The devices used in these suspension systems are, typically:

- Electromechanical actuators can use an electric motor attached to a mechanism that converts rotary motion into linear motion [21] or can use linear motors [22].
- Hydraulic actuators require the installation of pumps, valves and all the devices needed for the proper operation of the system. These actuators have been applied to improve the quality of the ride on both road [23] and railway vehicles [24].

Regarding railway vehicles, their own nature allows the development of tilting technology, which rolls the car body in curves with the aim of reducing the lateral acceleration perceived by the passengers. Usually, the movement of the car body is achieved by using hydraulic or electromechanical actuators [25]; however, Talgo introduced in the 1980s a passive natural tilting system in its trains [26].

The management of active and semi-active suspension systems is performed through different strategies. Several concepts are found in the literature: classic control [27,28], Model Predictive Control [29], fuzzy logic [30], skyhook control [31] and H2 and H∞ control [32]. All of them have proved an ability to achieve a reasonable trade-off between stability and comfort with the appropriate tuning conditions.

Another approach for managing the suspension system of light vehicles is proposed. This new approach is based on mechanical principles and relies on the measurements acquired by a position scanner installed at the front end of the vehicle to actuate on the following suspension systems; therefore, the unevenness of the terrain can be counteracted. This paper presents a predictive suspension algorithm for land vehicles traveling through a deterministic topography. The kinematics of a half vehicle model is posed and then implemented in Simulink, including the suspension algorithm. Our algorithm "reads" the surface the vehicle is traveling through thanks to a sensor installed in the front wheel of the vehicle and acts over a device installed in the rear wheel to compensate for the irregularities of the terrain. The actuation is governed by the parameter "D", which represents the distance over which the algorithm averages the height of the terrain. Two ground profiles were tested and sensitivity analysis of the parameter "D" was performed.

The structure of the paper is as follows: Section 2 describes the equations that define the vehicle model and the suspension algorithm. Section 3 shows the results obtained after simulating the vehicle and the response of the algorithm in two types of terrain. The fourth and last section presents the conclusions of this work.

2. Materials and Methods

The predictive suspension algorithm is implemented on a two-axle vehicle, whose kinematics must be solved first. For simplicity, only 1/2 vehicle is modeled and symmetry action on both wheels of the same axle is assumed, so the problem becomes bi-dimensional. The vehicle will travel through a deterministic topography. Hence, the vertical position of any wheel can be established from the horizontal position by the function that relates both variables.

The vehicle model also assumes that the mechanical energy remains constant, only wheels have mass and pure rolling (no sliding) is considered as it is a condition for the good performance of the suspension algorithm. This model and the algorithm were implemented in Simulink.

2.1. Vehicle Model

The mechanical energy of the vehicle is given by Equation (1)

$$E = \frac{1}{2}m_1 v_1^2 + \frac{1}{2}I_1 \omega_1^2 + m_1 \cdot g \cdot y_1 + \frac{1}{2}m_2 v_2^2 + \frac{1}{2}I_2 \omega_2^2 + m_2 \cdot g \cdot y_2 \tag{1}$$

where E is the mechanical energy, m_1 and m_2 are the masses of both wheels, I_1 and I_2 are the inertias of both wheels, v_1 and v_2 are the linear velocities of the wheels, ω_1 and ω_2 are the angular speeds of the wheels, y_1 and y_2 are the vertical positions of the wheels and g is the gravity.

By assuming pure rolling motion and the wheels as solid disks, Equation (1) can be simplified; therefore, the initial energy of the system is obtained from Equation (2), where the subscript 0 means "initial conditions".

$$E_0 = \frac{1}{2}2 \cdot m \cdot v_0^2 + \frac{1}{4}2 \cdot m \cdot v_0^2 + 2 \cdot m \cdot g \cdot y_0 = \frac{3}{2} \cdot m \cdot v_0^2 + 2 \cdot m \cdot g \cdot y_0 \tag{2}$$

The distance between the front and rear wheels (that is, the wheelbase) L is known and can be described mathematically as a circumference centered on the front wheel and radius L. The position of the rear wheel is obtained by computing the intersection of this circumference with the trajectory described by the ground function $y = f(x)$. Hence, the solution of the equation system given by Equation (3) yields the position of the rear wheel.

$$\begin{cases} (x_2 - x_1)^2 + (y_2 - y_1)^2 = L^2 \\ y = f(x) \end{cases} \tag{3}$$

This system of equations is solved by iteration, as the ground function can be any type of function, even not analytical. The flowchart describing the iteration process is shown in Figure 1.

The velocity of both wheels is computed from the conservation of energy, the time derivative of the constraint equation and the ground function, obtaining the equation system built up by Equations (4)–(6).

$$E_0 - (m_1 \cdot g \cdot y_1 + m_2 \cdot g \cdot y_2) = \frac{3}{4}m_1 v_1^2 + \frac{3}{4}m_2 v_2^2 \tag{4}$$

$$2(\dot{x}_2 - \dot{x}_1)(x_2 - x_1) + 2(\dot{y}_2 - \dot{y}_1)(y_2 - y_1) = 0 \tag{5}$$

$$\dot{y} = \frac{d\,f(x)}{dt} \tag{6}$$

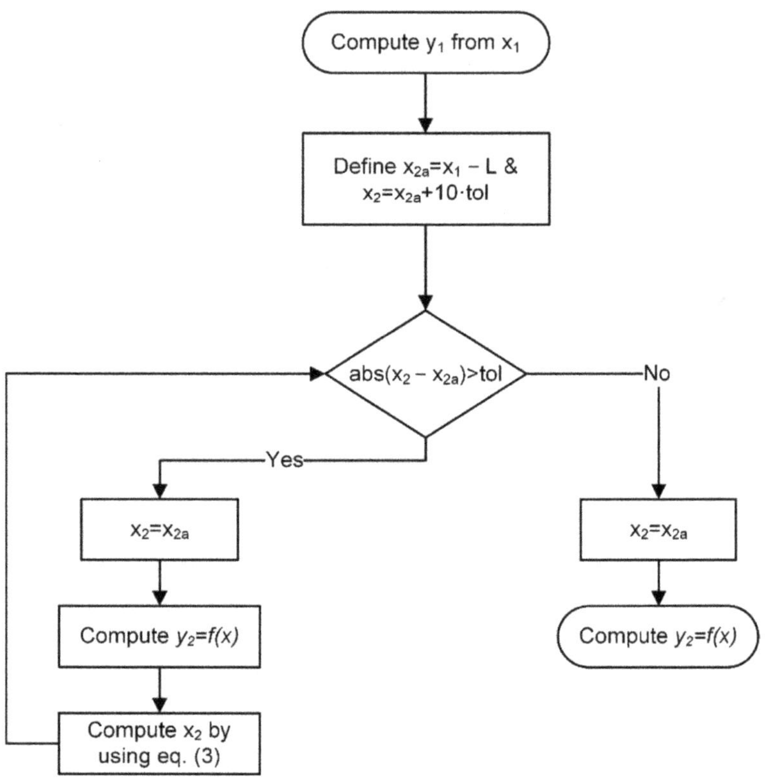

Figure 1. Flowchart for computing the position of the rear wheel.

In order to solve this system more efficiently, polar coordinates are used. This implies defining the pitch angle θ as Equation (7).

$$\theta = \pi + arctg\left(\frac{y_1 - y_2}{x_1 - x_2}\right) \tag{7}$$

Therefore, the constraint Equation (3) is rewritten as Equation (8).

$$\begin{cases} x_2 = x_1 + L\cos\theta \\ y_2 = y_1 + Lsin\theta \end{cases} \tag{8}$$

The combination of the conservation of energy with the time derivatives of the ground function and the constraint equation in polar coordinates yields Equations (9)–(12), which allows computing all the components of the speed of both wheels.

As previously, this system of equations is solved by iteration following the flowchart in Figure 2.

$$E_0 - (m_1 \cdot g \cdot y_1 + m_2 \cdot g \cdot y_2) = \frac{3}{4}m_1 v_1^2 + \frac{3}{4}m_2 v_2^2 = \frac{3}{4}m_1(\dot{x}_1^2 + \dot{y}_1^2) + \frac{3}{4}m_2(\dot{x}_2^2 + \dot{y}_2^2) \tag{9}$$

$$\dot{y}_1 = \frac{d\,f(x_1)}{dt} \tag{10}$$

$$\dot{x}_2 = \dot{x}_1 - L\cdot\dot{\theta}\cdot sen\theta \tag{11}$$

$$\dot{y}_2 = \dot{y}_1 + L\cdot\dot{\theta}\cdot\cos\theta \tag{12}$$

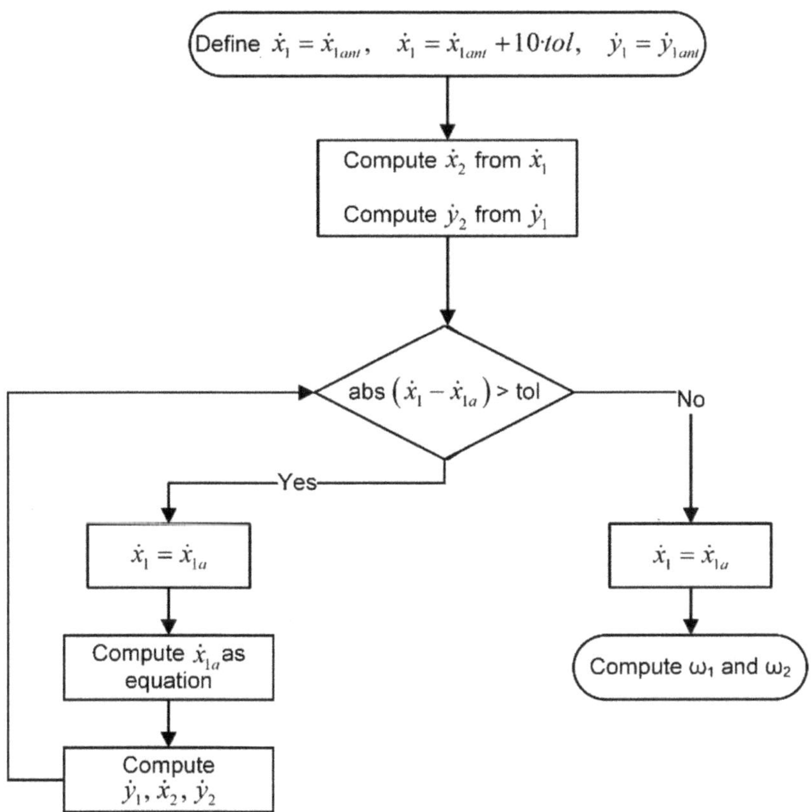

Figure 2. Flowchart for computing the angular speeds.

Two different ground topographies and several values of the parameter D were proposed for testing the algorithm. The first topography (profile 1) simulates a wiggly surface and is defined by Equation (13)—where A_1 is the amplitude of the wave and λ_1 is the wavelength, both in meters.

$$y = A_1 sin\left(\frac{2\pi}{\lambda_1}x\right) \tag{13}$$

The second profile is a combination of two sinusoidal irregularities according to Equation (14):

$$y = A_0 + A_1 sen\left(\frac{2\pi}{\lambda_1}(x-a)\right) + A_2 sen\left(\frac{2\pi}{\lambda_2}x\right) \tag{14}$$

A_1 and A_2 are the amplitudes of both sines, λ_1 and λ_2 are the wavelengths, A_0 is the reference height and a is a parameter to guarantee continuity in the first derivative; therefore, A_0 and a are such that:

$$a = \frac{1}{2}\lambda_1\left(1 - \frac{1}{\pi}\right)arccos\left(\frac{A_2\lambda_1}{A_1\lambda_2}\right) \tag{15}$$

$$A_0 = A_1 sen\left(\frac{2\pi a}{\lambda_1}\right) \tag{16}$$

The model parameters and initial conditions used in the simulations of both topography profiles are summarized in Table 1. The input values of both topography profiles are depicted in Table 2, as well as the computed values of variables A_0 and a.

Table 1. Model parameters.

Parameter	Value
Simulation time	10 s
Horizontal velocity at t = 0	3 m/s
Vertical velocity at t = 0	0 m/s
Wheel diameter	1 m
Wheel mass	60 kg
Wheelbase	3 m

Table 2. Ground function parameters.

Profile 1		Profile 2	
Parameter	Value	Parameter	Value
A_1	0.05 m	A_1	0.05 m
λ_1	4 m	λ_1	2 m
		A_2	−0.25 m
		λ_2	10 m
		A_0	0 m
		a	0 m

2.2. Suspension Algorithm

The suspension algorithm is applied to the 1/2 vehicle model described in the previous subsection. The frontwheel acts as a probe wheel, whereas the rear wheel is the actuated wheel. The probe wheel is instrumented with a vertical accelerometer and both wheels should be equipped with an accurate system for measuring their angular positions.

In Figure 3, two kinds of terrain defects (peak and corrugated) are sketched, as well as the trajectories described by the wheel centers and the trajectory that the sprung mass should follow (a straight line).

Let Δt be the sampling time interval of the accelerometer and $\ddot{y}_n = \ddot{y}(n\Delta t)$ the vertical acceleration registered at time $t = n\Delta t$. Assuming a constant vertical jerk along the interval $[(n\text{-}1)\Delta t, n\Delta t]$ yields Equation (17).

$$\dddot{y}_n = \frac{\ddot{y}_n - \ddot{y}_{n-1}}{\Delta t} = cte \tag{17}$$

Then the expressions for the vertical velocity and position of the wheel center are, respectively:

$$\dot{y}_n = \dot{y}_{n-1} + \ddot{y}_{n-1}\Delta t + \frac{1}{2}\dddot{y}_n\Delta t^2 = \dot{y}_{n-1} + \frac{1}{2}(\ddot{y}_{n-1} + \ddot{y}_n)\Delta t \tag{18}$$

$$y_n = y_{n-1} + \dot{y}_{n-1}\Delta t + \frac{1}{2}\ddot{y}_{n-1}\Delta t^2 + \frac{1}{6}\dddot{y}_n\Delta t^3 = y_{n-1} + \dot{y}_{n-1}\Delta t + \frac{1}{6}(2\ddot{y}_{n-1} + \ddot{y}_n)\Delta t^2 \tag{19}$$

Although the corresponding velocity and position errors grow with time, $n\Delta t$, and time squared, $(n\Delta t)^2$, respectively, (easy to show); this is not too much of a problem since the algorithm does not include the absolute height, but its variation over a certain distance, "D". In other words, the database of values obtained in Equations (18) and (19), will be dynamically refreshed, so that n, and therefore the temporal interval $n\Delta t$, does not increase indefinitely.

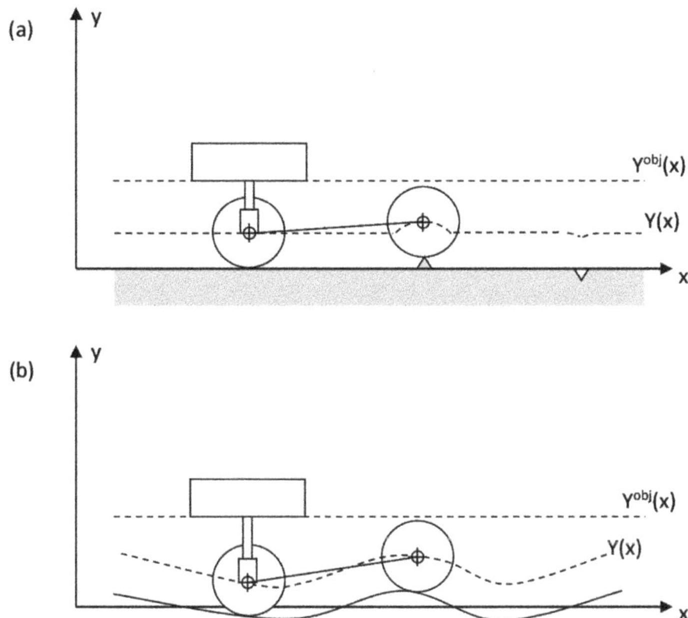

Figure 3. Two kinds of terrain defects: (**a**) peak defects and (**b**) smooth or corrugated terrain.

The vertical actuator (between the actuated or rear wheel and the sprung mass) has to maintain the latter at the objective height. In this approach, the objective height is the result of the moving average of the calculated height over the rolling distance $2D$. Actually, the larger this distance, the softer the objective function, but also the bigger error is accumulated in calculating the height from the accelerometer reading. Thus, parameter "D" should be adjusted according to the kind of roughness the topography has (with the sole requirement not to exceed the distance between wheels).

For the proposed algorithm, it is important to know exactly the position of both wheels (which are supposed to roll without slipping) on the ground. This is achieved by recording the corresponding angles, φ_n^{pro}, φ_n^{act} (angle turned by the probe wheel, and actuated wheel, respectively), having set both to zero at the initial position ($n = 0$). The model is sketched in Figure 4 at the initial position and at the time $t = n\Delta t$. The actual trajectories followed by the center of the wheels and the trajectory the end of the actuator should follow are also represented.

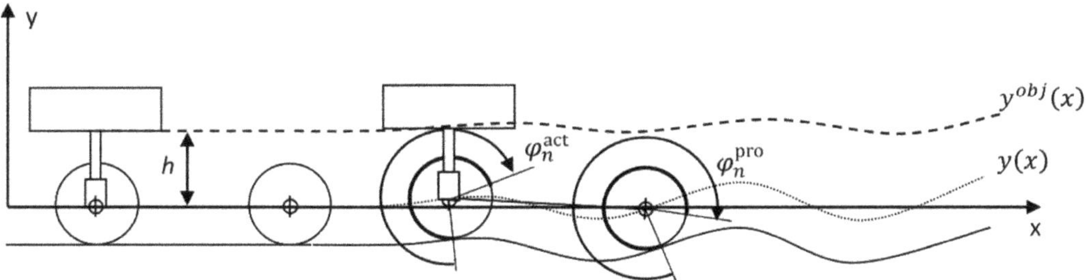

Figure 4. Model of the vehicle sketched at the initial position (t = 0) and at the time $t = n\Delta t$, where the angles are indicated. Dotted and dashed curves represent respectively the actual trajectory followed by the center of the wheels and that the actuated mass should follow.

At time $t = n\Delta t$, the vertical acceleration of the probe wheel, $\ddot{y}_n$, and the angles turned by the wheels, are measured. Then, the velocity, $\dot{y}_n$, and vertical position, y_n, are calculated by using Equations (18) and (19), respectively. All these values are recorded in a database in the way indicated in Table 3.

Table 3. Database of recorded and computed values.

# Record	Probe Wheel Angle Turned	Actuated Wheel Angle Turned	Vertical Acceleration	Vertical Velocity	Vertical Position	Time
1	φ_1^{pro}	φ_1^{act}	$\ddot{y}_1$	$\dot{y}_1$	y_1	t_1
$\vdots$	$\vdots$	$\vdots$	$\vdots$	$\vdots$	$\vdots$	$\vdots$
n	φ_n^{pro}	φ_n^{act}	$\ddot{y}_n$	$\dot{y}_n$	y_n	t_n

Once enough records have been acquired, the algorithm is ready to provide the current length to be taken by the actuator. The objective height is calculated as the moving average centered on the record corresponding to the actuated wheel and over a sufficient number of records; speaking in terms of rolling distances, the moving average is calculated over a rolling distance of 2D. This is achieved by identifying the records m, i and f that, respectively, define the current position for the actuated wheel and the first and last of the records to average over. In other words, at $t = m\Delta t$ the probe wheel was just on the same point of the trajectory that the actuated wheel is at $t = n\Delta t$ (i.e., at present). In Figure 5, the model at moments corresponding to records m, n, i and f is sketched.

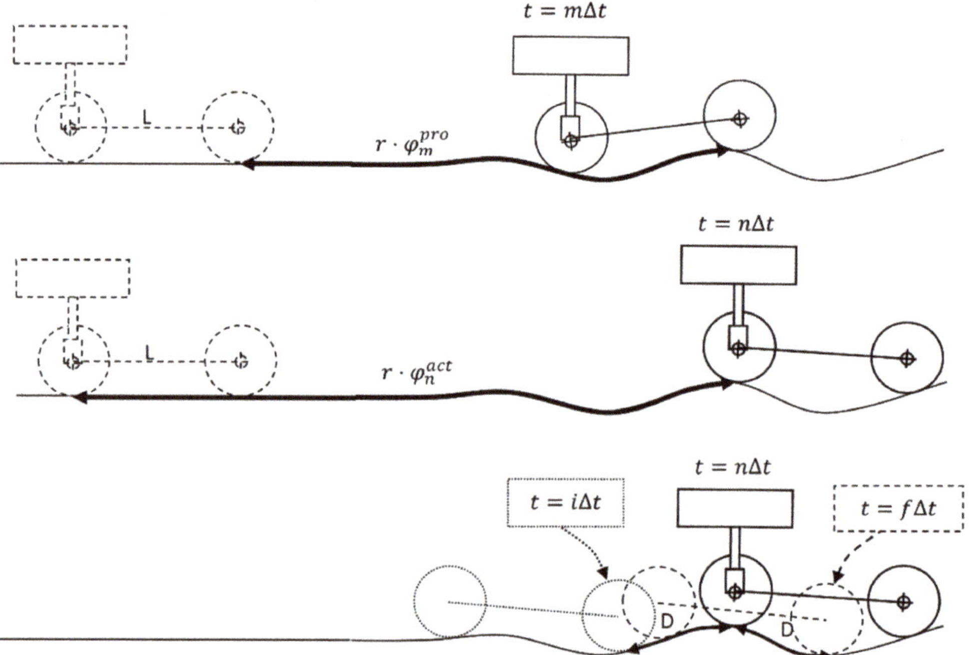

Figure 5. Sketch of the model at the key moments for the algorithm: at the initial position (dashed line in (**up**) and (**middle**) diagrams); when the record m is being registered (**up**); when the record n is being registered (**middle**); when the record i is being registered (dotted line, (**down**)); when the record f is being registered (dash line, (**down**)).

On seeing Figure 5, if we consider both wheels have the same radius, r, the following rules for identifying the numbers m, i and f are deduced:

m is such that:

$$\varphi_m^{pro} \approx \varphi_n^{act} - \frac{L}{r} \tag{20}$$

i is such that:

$$\varphi_i^{pro} \approx \varphi_m^{pro} - \frac{D}{r} \approx \varphi_n^{act} - \frac{L+D}{r} \tag{21}$$

f is such that:

$$\varphi_f^{pro} \approx \varphi_m^{pro} + \frac{D}{r} \approx \varphi_n^{act} + \frac{L+D}{r} \tag{22}$$

With "$\approx$" we mean "is the nearest value to" (note that $\left\{\varphi_j^{pro}\right\}$ is a discrete set of values, so in general, none of the last equalities could be exactly satisfied). We have also considered that the first section of the trajectory is flat and horizontal, with a minimum distance that is equal to the wheelbase, L.

Once the records i, m and f have been identified, the algorithm calculates the objective height (see Figure 6) as:

$$y^{obj} = \frac{1}{f-i+1}\sum_{j=i}^{f} y_j + h \tag{23}$$

and the actuated distance results:

$$y^{act} = y^{obj} - y_m = \frac{1}{f-i+1}\sum_{j=i}^{f} y_j + h - y_m \tag{24}$$

where h is the initial objective height (i.e., the actuator initial length, see Figures 4 and 6).

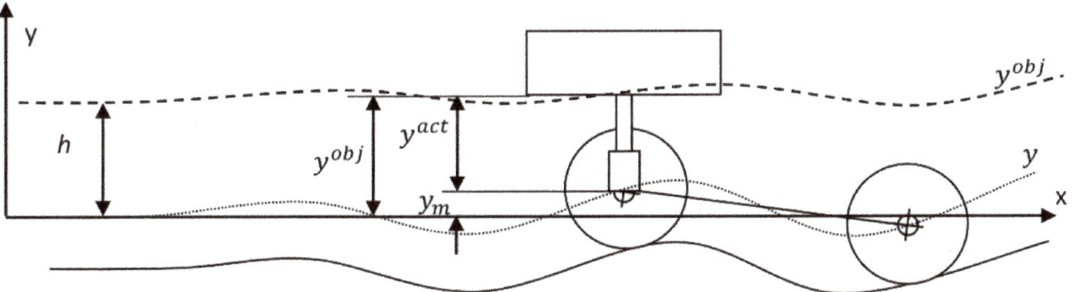

Figure 6. Heights that are involved in the calculus of the actuated distance.

Finally, the database is refreshed by deleting the records with subscripts less than i, and decreasing the remainder subscripts in i units:

$$\text{if } j < i\text{:, delete record}$$

$$\text{if } j \geq i, j = j - i$$

In this way, the database does not grow indefinitely, and only the records to be used by the algorithm at the current time of calculation remain.

In summary, the algorithm can be considered as consisting of three tasks:

1. Reading (angles and vertical acceleration) and computing vertical velocity and position.
2. Identifying the indexes i, m and f.
3. Obtaining the actuated distance and refreshing the database.

To understand how the algorithm performs the second task, suppose the database at time $n\Delta t$ is as shown in Table 4. For a given angle turned by the actuated wheel φ_n^{act}, the algorithm computes the theoretical angle turned by the probe wheel at the three key moments $i\Delta t$, $m\Delta t$ and $f\Delta t$ ($\varphi_n^{act} - \frac{L+D}{r}$, $\varphi_n^{act} - \frac{L}{r}$ and $\varphi_n^{act} + \frac{L+D}{r}$, respectively), using Equations (20)–(22). Then, it searches the closest actual angles recorded in the database to those computed and determines the indexes i, m and f. As the probe wheel angle turned does not decrease with n (hence $i \leq m \leq f$), the search can be performed sequentially.

Table 4. Time "$n\Delta t$" database.

# Record	Angle Turned by Probe Wheel	Angle Turned by Actuated Wheel
1	φ_1^{pro}	φ_1^{act}
$\vdots$	$\vdots$	$\vdots$
i	φ_i^{pro}	φ_i^{act}
$\vdots$	$\vdots$	$\vdots$
m	φ_m^{pro}	φ_m^{act}
$\vdots$	$\vdots$	$\vdots$
f	φ_f^{pro}	φ_f^{act}
$\vdots$	$\vdots$	$\vdots$
n	φ_n^{pro}	φ_n^{act}

The whole algorithm flowchart is represented in Figure 7.

2.3. Implementation in Simulink

The models described above to define the kinematics of the 1/2 vehicle and the suspension algorithm are implemented in Simulink, and more specifically, in Simscape. The main diagram is shown in Figure 8 and is composed of six big blocks. The input block is the "Ground function" block, which defines the parameters of the trajectory that the vehicle must follow. The "Energy" block performs all the calculations (Equations (1)–(13)) to compute the kinematics of the half vehicle model. Front and rear wheel blocks contain the relationships between the wheel bodies and the main reference frame of Simscape. The rear wheel block also has the actuator to move the sprung mass according to the commands given by the suspension algorithm. The computation of the rotation angles turned by both wheels is carried out within the rotation block. Finally, the block "Suspension algorithm" performs all the calculations (Equations (14)–(21)) to actuate the sprung mass.

The "Energy" block is composed of another block called "Calculations", which is the block that actually computes the kinematics of the model. The inputs of this block are the ground profile, the wheel properties and the position of the front wheel. The outputs of this block are the kinematics of both wheels. Within this block, an S-function is responsible for executing the algorithms that compute the kinematics. To perform this action, this function is feedbacked with the positions and velocities computed in the previous integration step.

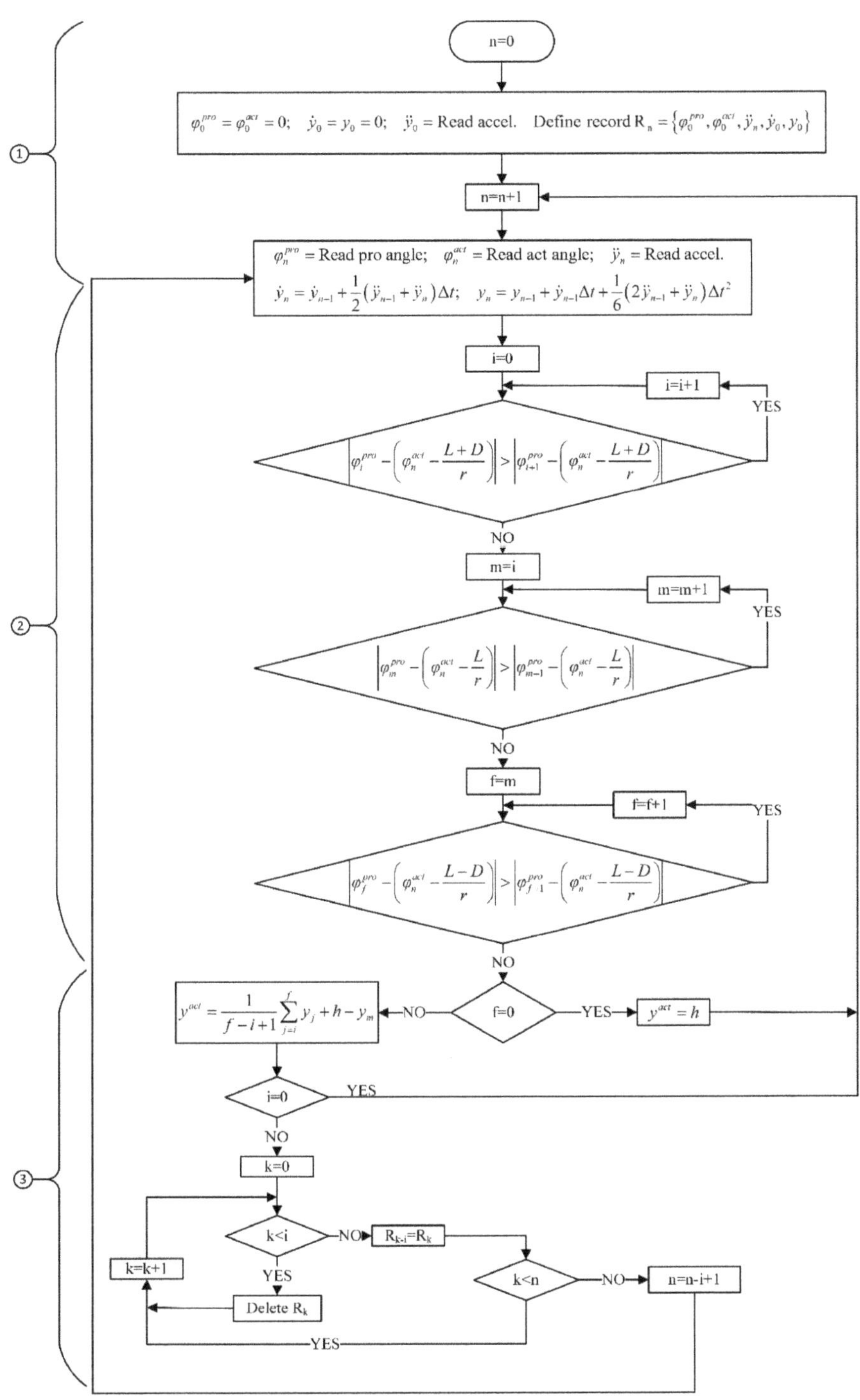

Figure 7. Suspension algorithm flowchart.

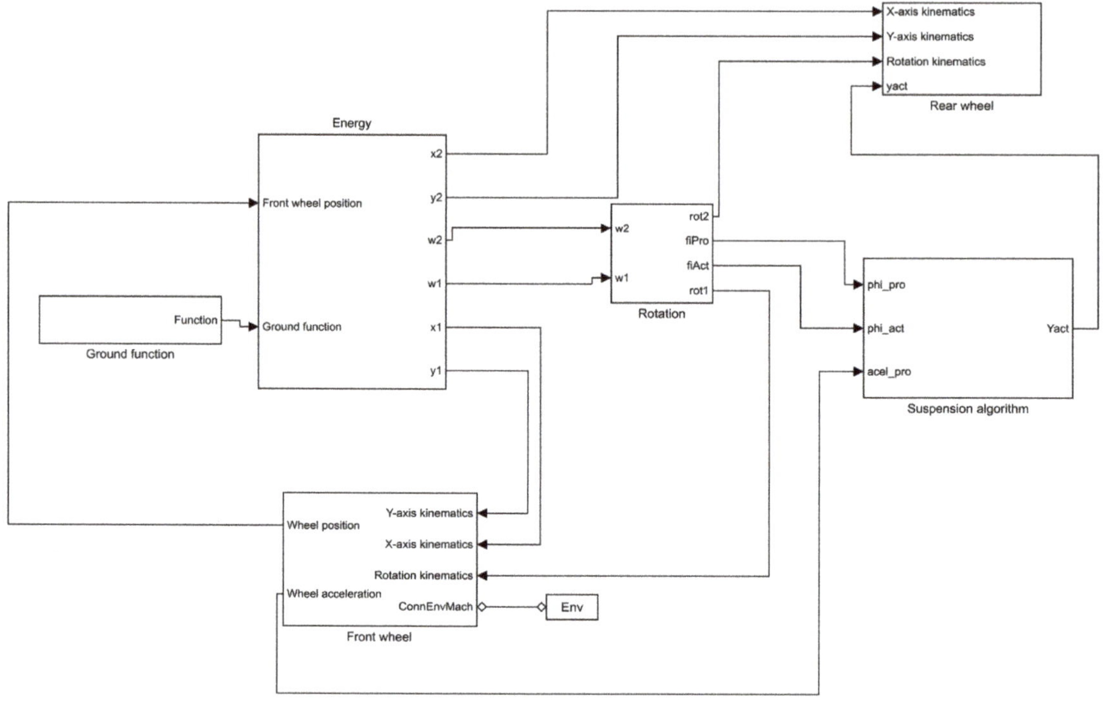

Figure 8. Diagram of the Simulink implementation of the whole model.

3. Results

This section presents the results obtained after testing the predictive suspension algorithm over the two ground topographies defined in Section 2.1. The first topography (profile 1) simulates a wiggly surface; the second topography is a combination of two sinusoidal irregularities. The parameters of both topographies are summarized in Table 2.

As is described in Section 2.2, the algorithm's key parameter to define the motion of the sprung or actuated mass is the distance "D". This parameter is set from $D = 0$ m (no actuation over the sprung mass) to $D = 3$ m (the vehicle's wheelbase) in steps of 0.5 m. Hence, seven values of "D" were tested.

Additionally, the parameter "D" is normalized to the wheelbase of the vehicle by defining a new variable $d = D/L$, where L is the wheelbase. As the value of the "D" is limited by the wheelbase (it is not possible to obtain information on positions not yet traveled by the front wheel), the value of "d" will always be between 0 and 1.

3.1. Profile Type 1

The results of the simulation of the first ground profile are shown in Figures 9–12. Figure 9 illustrates the kinematics of both wheels of the 1/2 vehicle. The trajectory of both wheels follows the input ground profile as expected. The velocity in both axes varies according to the topography of the ground profile and the energy conservation. Similar behavior is noticed in the accelerations. The significant acceleration peak at the horizontal position equal to 0 is due to the sudden transition from flat terrain to an irregular surface.

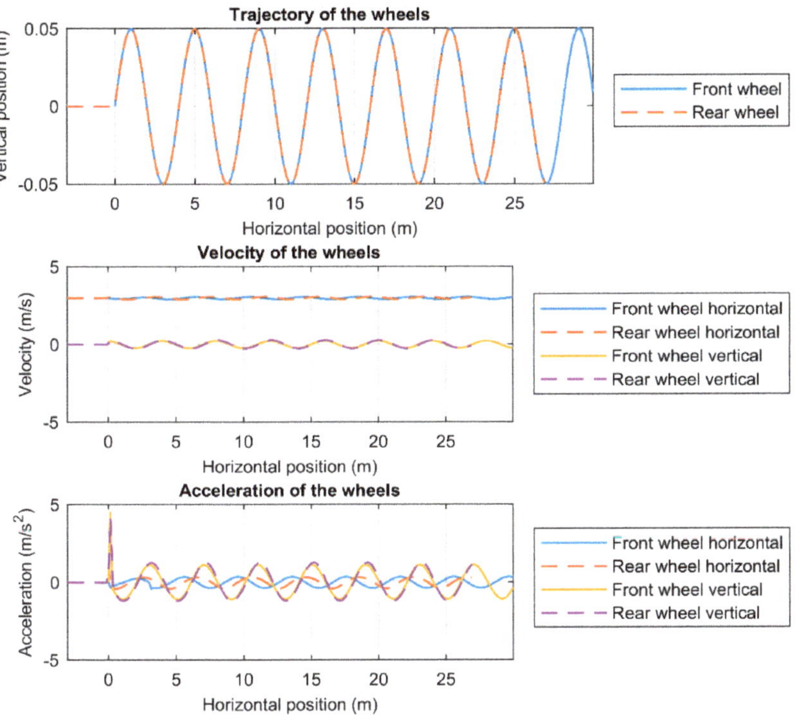

Figure 9. Kinematics of the vehicle when traveling through the ground profile 1.

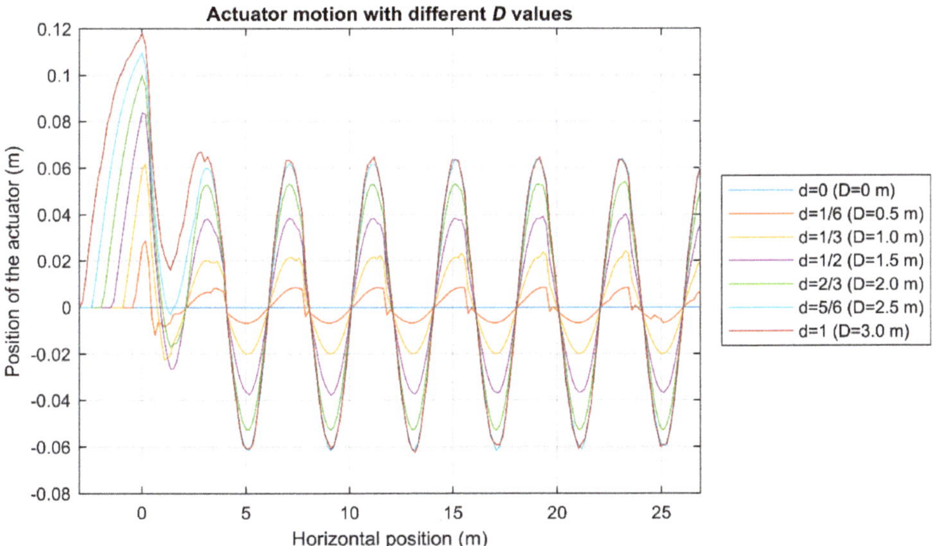

Figure 10. The motion of the actuator in the first ground profile.

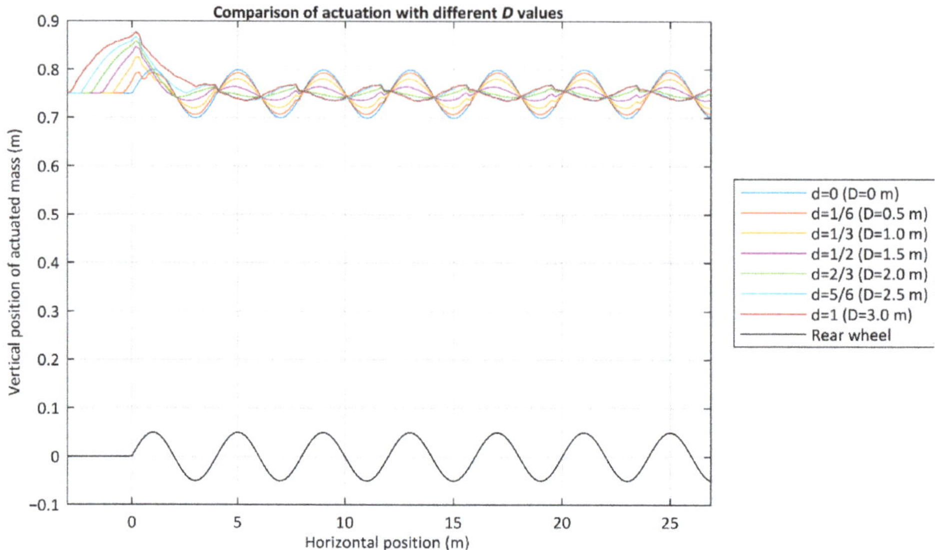

Figure 11. Displacement of the actuated mass in the first ground profile.

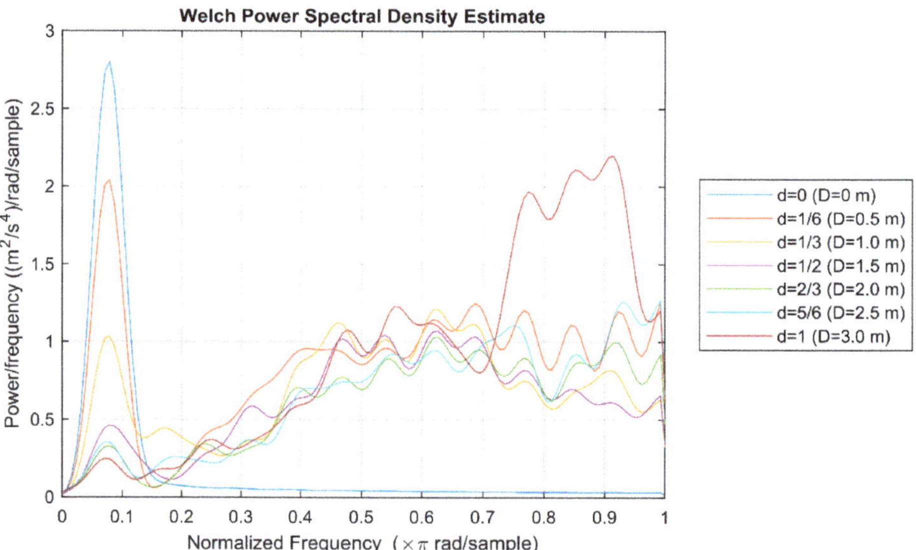

Figure 12. Acceleration spectra of the actuated mass in the first ground profile tested.

The displacement of the actuator placed between the rear wheel and the sprung mass is plotted in Figure 10. As it can be seen, the actuator does not move when the value of the normalized parameter "*d*" is 0. When increasing the value of this parameter, the actuator moves to compensate for the irregularities of the ground, which even leads to a wave out phased 180° to the ground profile. Lower values of parameter "*d*" yield smaller displacements of the actuator, though it also generates some discontinuities. Larger values of parameter "*d*" depict better an out phased trajectory from the terrain, but the amplitude of the movement is larger too. The actuator always works in extension (positive displacement) at the beginning of these simulations due to the initial elevation of the front wheel.

Figure 11 shows the path followed by the center of the rear wheel and the position of the actuated mass for a better comparison of the followed trajectory and the actuation of the suspension system. It is observed that the movement of the sprung mass is the opposite of the rear wheel when the parameter "*d*" is larger than 2/3, that is, when the distance "*D*" over which the moving average is computed is larger than 2 m. In this situation, when the rear wheel moves upwards, the sprung mass is moved downwards, and vice versa. For values of parameter "*D*" below 2 m ($d < 2/3$), the sprung mass moves vertically in the same direction as the rear wheel, but this movement is limited.

In order to establish what is the best value of the normalized parameter "*d*" among all the tested and for ground profile 1, the spectra of the vertical accelerations experienced by the sprung mass are computed. The results obtained are plotted in Figure 12. The analysis of the graph shows that the largest values of the parameter "*d*" yield the lowest accelerations at low frequencies; however, it appears to be an undesirable vibration at high frequencies; therefore, according to this criterion, the best value will be one that obtains the best arrangement between low and high-frequency vibrations.

To make easier the search for the optimum value of parameter "*d*", Table 5 summarizes the maximum amplitude and the RMS of the spectra shown in Figure 12. Combining the spectra and the data in the table, it can be concluded that the best value for parameter "*d*" in the conditions of this simulation is $d = 2/3$, which means the distance $D = 2\ m$.

Table 5. Characterization of the acceleration spectra in the first ground profile tested.

	$d = 0$	$d = 1/6$	$d = 1/3$	$d = 1/2$	$d = 2/3$	$d = 5/6$	$d = 1$
Maximum	2.8035	2.0397	1.2130	1.0728	1.0361	1.2695	2.1981
RMS	0.6057	0.9422	0.7551	0.6772	0.6728	0.7363	1.1141

3.2. Profile Type 2

The results obtained from the simulation of the second ground profile are shown in Figures 13–16. Figure 13 displays the kinematics of both wheels of the 1/2 vehicle. As in the previous case, the trajectory of both wheels follows the input ground profile as expected. The speed varies according to the topography of the ground profile and the energy conservation. Velocity variations are slower on the horizontal axis than on the vertical axis, where the velocity is more sensitive to profile variations. Similar behavior is noticed in the accelerations, being the vertical accelerations larger than the horizontal ones.

The displacement of the actuator is plotted in Figure 14. As it can be seen, the actuator does not move when the value of the parameter "*d*" is zero. Once the simulation begins, the actuator starts to move in compression (negative displacement) to compensate initial descent of the vehicle. As it happens in the first set of simulations, the larger the value of the parameter "*d*", the larger the displacement of the actuator. The use of the largest values results in an actuation that clearly highlights the two harmonics that compose the second ground profile.

Figure 15 shows the path followed by the center of the rear wheel and the position of the actuated mass for a better comparison of the followed trajectory and the actuation of the suspension system. The movement of the mass follows the movement of the rear wheel, but the larger the value of "*d*", the smoother the displacement of the actuated mass. In fact, the algorithm moves the mass in advance to the irregularity of the surface the vehicle will encounter to try to achieve a smoother ride.

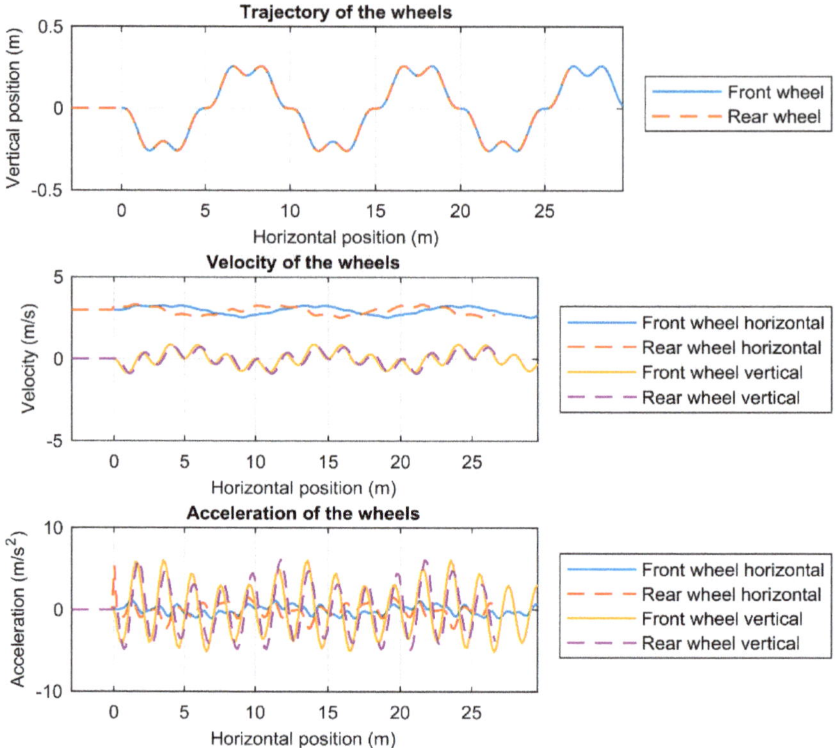

Figure 13. Kinematics of the vehicle when traveling through the second ground profile.

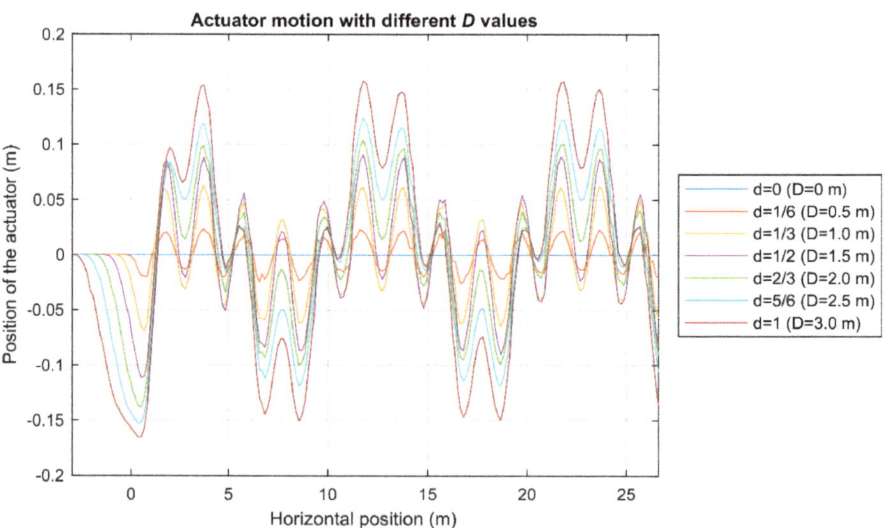

Figure 14. The motion of the actuator in the first ground profile.

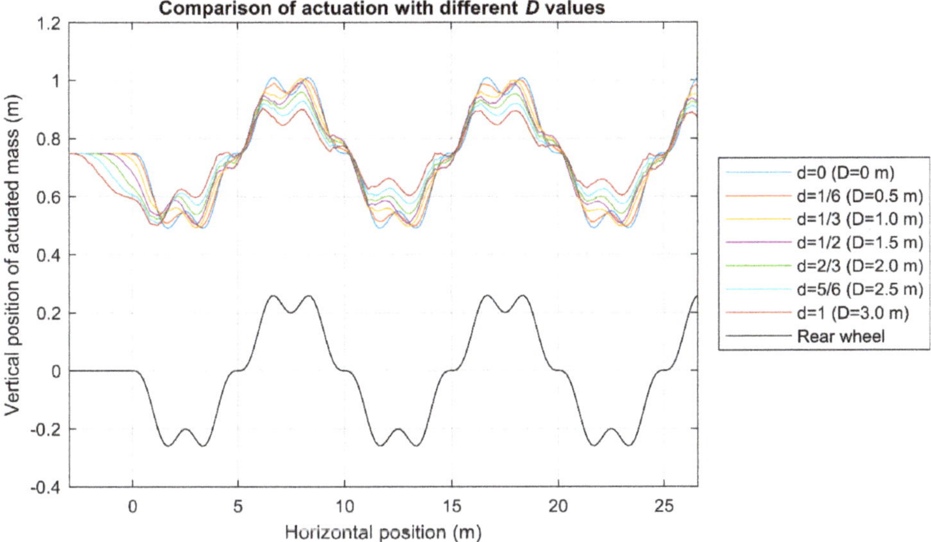

Figure 15. Displacement of the actuated mass in the second ground profile tested.

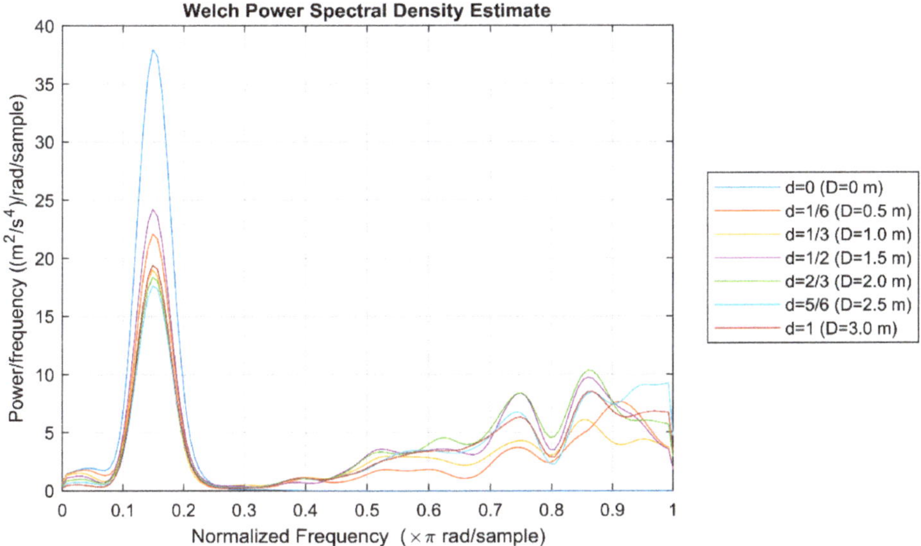

Figure 16. Acceleration spectra of the actuated mass in the second ground profile tested.

The PSD of the vertical accelerations experienced by the sprung mass through the second ground profile is computed to establish what is the best value of the parameter "*d*" among all the tested values. The results obtained are plotted in Figure 16. As was seen in the previous subsection, there is a big peak at low frequencies and also some undesirable accelerations at high frequencies; however, in this set of simulations, a larger value of the parameter "*d*" does not mean lower acceleration amplitudes at low frequencies. In fact, larger acceleration amplitudes are achieved with d = 1 than with d = 5/6, which indicates that moving the sprung mass in advance to the terrain elevation is not as good as it can be expected.

Once again, to make easier the search for the optimum value of parameter "*d*", the maximum amplitude and the RMS of the spectra are summarized in Table 6. Combining the spectra and the numbers in the table, it can be concluded that the best value for parameter "*d*" in the conditions of this simulation is $d = 5/6$, as it achieves the lowest maximum amplitude and RMS values. The normalized parameter d = 5/6 means that the distance "*D*" is equal to 2.5 m.

Table 6. Characterization of the acceleration spectra in the second ground profile tested.

	$d = 0$	$d = 1/6$	$d = 1/3$	$d = 1/2$	$d = 2/3$	$d = 5/6$	$d = 1$
Maximum	37.9703	22.1259	19.0021	24.2496	18.3643	17.6156	19.4286
RMS	8.4317	5.6769	5.0129	6.7450	5.9922	5.7087	5.7104

4. Conclusions

This paper presents a predictive suspension algorithm for land vehicles traveling through a deterministic topography. Prior to the virtual implementation of the algorithm, the kinematics of a two-wheel 2D (1/2) vehicle were proposed and solved. The algorithm inputs are the measurements given by a position scanner installed in the first wheel of the vehicle. From these inputs, the algorithm reconstructs the terrain and averages its height over a distance "*D*". Then, it generates the command for moving the actuator installed in the rear wheel and connected to the sprung mass. Both the algorithm and the kinematic model of the 1/2 vehicle were implemented in Simulink.

The algorithm was tested on two terrains with different profiles. Furthermore, sensitivity analyses were carried out in order to establish the best value of the parameter "*D*" in each profile. The performance of the algorithm is quantified for several values of the parameter "*D*", using the spectral analysis of the accelerations experienced by the sprung mass.

The results show that, in general, increasing the parameter "*D*" reduces the motion of the sprung/actuated mass and, hence, improves comfort; however, there is an optimum value of the parameter "*D*" for each ground profile, and this value does not coincide necessarily with the highest value of "*D*". In fact, the optimum "*D*" value will depend on the terrain irregularities and their characteristic length. For the first ground profile, the optimum value of the parameter "*D*" is 2 m ($d = 2/3$), which is half the wavelength of the unevenness of the terrain. The optimum value for the second terrain is $D = 2.5$ m ($d = 5/6$).

The value of the "*D*" parameter is limited by the wheelbase since it is not possible to obtain information on positions not yet traveled by the front wheel. In fact, the maximum value of the "*D*" parameter must be slightly less than the wheelbase, as the data processing and calculation time of the algorithm must be taken into account. This is not of major significance in the current development status of the research as the Simulink implementation computes the algorithm in real-time, but it should be taken into account when developing the physical device to install in a light vehicle for real testing.

Finally, it can be concluded that the algorithm works correctly and decreases the movement of the sprung/actuated mass when the vehicle travels through a deterministic topography. The next steps in the research would be the development of a physical device that executes the proposed predictive algorithm and its implementation in a light vehicle.

Author Contributions: Individual contributions are as follows: original concept: J.M.; mathematical development: J.M. and A.B.; implementation, simulation and draft preparation: A.B.; review, editing, validation, and formal analysis: E.S.-H., H.R. and J.M. All authors have read and agreed to the published version of the manuscript.

Funding: The research work described in this paper is part of the R&D and Innovation projects MC4.0 PID2020-116984RB-C21 and MC4.0 PID2020-116984RB-C22 supported by the MCIN/AEI/10.13039/501100011033.

Institutional Review Board Statement: Not applicable.

Informed Consent Statement: Not applicable.

Data Availability Statement: Not applicable.

Conflicts of Interest: The authors declare no conflict of interest.

References

1. Reimpell, J.; Stoll, H.; Betzler, J.W. *The Automotive Chassis: Engineering Principles*, 2nd ed.; Butterworth Heinemann: Oxford, UK, 2001; ISBN 978-0-7506-5054-0.
2. Theunissen, J.; Tota, A.; Gruber, P.; Dhaens, M.; Sorniotti, A. Preview-Based Techniques for Vehicle Suspension Control: A State-of-the-Art Review. *Annu. Rev. Control* **2021**, *51*, 206–235. [CrossRef]
3. Luo, R.; Shi, H.; Guo, J.; Huang, L.; Wang, J. A Nonlinear Rubber Spring Model for the Dynamics Simulation of a High-Speed Train. *Veh. Syst. Dyn.* **2020**, *58*, 1367–1384. [CrossRef]
4. Anubi, O.M.; Patel, D.R.; Crane, C.D., III. A New Variable Stiffness Suspension System: Passive Case. *Mech. Sci.* **2013**, *4*, 139–151. [CrossRef]
5. Kalyan Raj, A.H.; Padmanabhan, C. A New Passive Non-Linear Damper for Automobiles. *Proc. Inst. Mech. Eng. Part J. Automob. Eng.* **2009**, *223*, 1435–1443. [CrossRef]
6. Solomon, U.; Padmanabhan, C. Hydro-Gas Suspension System for a Tracked Vehicle: Modeling and Analysis. *J. Terramech.* **2011**, *48*, 125–137. [CrossRef]
7. Fu, B.; Giossi, R.L.; Persson, R.; Stichel, S.; Bruni, S.; Goodall, R. Active Suspension in Railway Vehicles: A Literature Survey. *Railw. Eng. Sci.* **2020**, *28*, 3–35. [CrossRef]
8. Anubi, O.M.; Crane, C.A. New Semiactive Variable Stiffness Suspension System Using Combined Skyhook and Nonlinear Energy Sink-Based Controllers. *IEEE Trans. Control Syst. Technol.* **2015**, *23*, 937–947. [CrossRef]
9. Sun, S.; Deng, H.; Li, W. Variable Stiffness and Damping Suspension System for Train. In Proceedings of the SPIE Volume 9057, Active and Passive Smart Structures and Integrated Systems 2014, San Diego, CA, USA, 9–13 March 2014; p. 90570P.
10. Faraj, R.; Graczykowski, C.; Holnicki-Szulc, J. Adaptable Pneumatic Shock Absorber. *J. Vib. Control* **2019**, *25*, 711–721. [CrossRef]
11. Spencer, B.F.; Dyke, S.J.; Sain, M.K.; Carlson, J.D. Phenomenological Model for Magnetorheological Dampers. *J. Eng. Mech.* **1997**, *123*, 230–238. [CrossRef]
12. Gavin, H.P.; Hanson, R.D.; Filisko, F.E. Electrorheological Dampers, Part I: Analysis and Design. *J. Appl. Mech.* **1996**, *63*, 669–675. [CrossRef]
13. Dixon, J. *The Shock Absorber Handbook*; John Wiley & Sons, Ltd.: Hoboken, NJ, USA, 2008; ISBN 978-0-470-51642-3.
14. Mei, T.X.; Zaeim, A.; Li, H. Control of Railway Wheelsets—A Semi-Active Approach. In *Advances in Dynamics of Vehicles on Roads and Tracks*; Klomp, M., Bruzelius, F., Nielsen, J., Hillemyr, A., Eds.; Lecture Notes in Mechanical Engineering; Springer International Publishing: Cham, Swizerland, 2020; pp. 16–23. ISBN 978-3-030-38076-2.
15. Yang, J.; Ning, D.; Sun, S.S.; Zheng, J.; Lu, H.; Nakano, M.; Zhang, S.; Du, H.; Li, W.H. A Semi-Active Suspension Using a Magnetorheological Damper with Nonlinear Negative-Stiffness Component. *Mech. Syst. Signal Process.* **2021**, *147*, 107071. [CrossRef]
16. Tudon-Martinez, J.C.; Hernandez-Alcantara, D.; Amezquita-Brooks, L.; Morales-Menendez, R.; Lozoya-Santos, J.D.J.; Aquines, O. Magneto-Rheological Dampers—Model Influence on the Semi-Active Suspension Performance. *Smart Mater. Struct.* **2019**, *28*, 105030. [CrossRef]
17. Soliman, A.; Kaldas, M. Semi-Active Suspension Systems from Research to Mass-Market—A Review. *J. Low Freq. Noise Vib. Act. Control* **2021**, *40*, 1005–1023. [CrossRef]
18. Karnopp, D. Permanent Magnet Linear Motors Used as Variable Mechanical Dampers for Vehicle Suspensions. *Veh. Syst. Dyn.* **1989**, *18*, 187–200. [CrossRef]
19. Smith, M.C. Synthesis of Mechanical Networks: The Inerter. *IEEE Trans. Autom. Control* **2002**, *47*, 1648–1662. [CrossRef]
20. Lewis, T.D.; Jiang, J.Z.; Neild, S.A.; Gong, C.; Iwnicki, S.D. Using an Inerter-Based Suspension to Improve Both Passenger Comfort and Track Wear in Railway Vehicles. *Veh. Syst. Dyn.* **2019**, *58*, 472–493. [CrossRef]
21. Kawamoto, Y.; Suda, Y.; Inoue, H.; Kondo, T. Electro-Mechanical Suspension System Considering Energy Consumption and Vehicle Manoeuvre. *Veh. Syst. Dyn.* **2008**, *46*, 1053–1063. [CrossRef]
22. Klimenko, Y.I.; Batishchev, D.V.; Pavlenko, A.; Grinchenkov, V.P. Design of a Linear Electromechanical Actuator with an Active Vehicle Suspension System. *Russ. Electr. Eng.* **2015**, *86*, 588–593. [CrossRef]
23. Kilicaslan, S. Control of Active Suspension System Considering Nonlinear Actuator Dynamics. *Nonlinear Dyn.* **2018**, *91*, 1383–1394. [CrossRef]
24. Goodall, R.; Freudenthaler, G.; Dixon, R. Hydraulic Actuation Technology for Full- and Semi-Active Railway Suspensions. *Veh. Syst. Dyn.* **2014**, *52*, 1642–1657. [CrossRef]
25. Goodall, R. Tilting Trains and Beyond-the Future for Active Railway Suspensions. 1. Improving Passenger Comfort. *Comput. Control Eng. J.* **1999**, *10*, 153–160. [CrossRef]

26. Carballeira, J.; Baeza, L.; Rovira, A.; García, E. Technical Characteristics and Dynamic Modelling of Talgo Trains. *Veh. Syst. Dyn.* **2008**, *46*, 301–316. [CrossRef]
27. Jamil, I.A.A.; Moghavvemi, M. Optimization of PID Controller Tuning Method Using Evolutionary Algorithms. In Proceedings of the 2021 Innovations in Power and Advanced Computing Technologies (i-PACT), Kuala Lumpur, Malaysia, 27–29 November 2021; pp. 1–7.
28. Di Gialleonardo, E.; Facchinetti, A.; Bruni, S. Control of an Integrated Lateral and Roll Suspension for a High-Speed Railway Vehicle. *Veh. Syst. Dyn.* **2022**, 1–27. [CrossRef]
29. Rodriguez-Guevara, D.; Favela-Contreras, A.; Beltran-Carbajal, F.; Sotelo, D.; Sotelo, C. Active Suspension Control Using an MPC-LQR-LPV Controller with Attraction Sets and Quadratic Stability Conditions. *Mathematics* **2021**, *9*, 2533. [CrossRef]
30. Palanisamy, S.; Karuppan, S. Fuzzy Control of Active Suspension System. *J. Vibroeng.* **2016**, *18*, 3197–3204. [CrossRef]
31. Díaz-Choque, C.S.; Félix-Herrán, L.C.; Ramírez-Mendoza, R.A. Optimal Skyhook and Groundhook Control for Semiactive Suspension: A Comprehensive Methodology. *Shock Vib.* **2021**, *2021*, 8084343. [CrossRef]
32. Boada, B.L.; Boada, M.J.L.; Vargas-Melendez, L.; Diaz, V. A Robust Observer Based on H ∞ Filtering with Parameter Uncertainties Combined with Neural Networks for Estimation of Vehicle Roll Angle. *Mech. Syst. Signal Process.* **2018**, *99*, 611–623. [CrossRef]

Article

A Novel Denoising Method for Retaining Data Characteristics Brought from Washing Aeroengines

Zhiqi Yan [1], Ming Zu [2], Zhiquan Cui [3,*] and Shisheng Zhong [4]

[1] School of Mechatronics Engineering, Harbin Institute of Technology, Harbin 150001, China; cumtmaris@163.com
[2] School of Electronic Information & Media, Dongying Vocational Institute, Dongying 257091, China; dyzm2006@126.com
[3] School of Automotive Engineering, Harbin Institute of Technology, Weihai 264209, China
[4] School of Ocean Engineering, Harbin Institute of Technology, Weihai 264209, China; zhongss@hit.edu.cn
* Correspondence: xiaocui2002yan@163.com

Abstract: Airlines evaluate the energy-saving and emission reduction effect of washing aeroengines by analyzing the exhaust gas temperature margin (EGTM) data of aeroengines so as to formulate a reasonable washing schedule. The noise in EGTM data must be reduced because they interfere with the analysis. EGTM data will show several step changes after cleaning the aeroengine. These step changes increase the difficulty of denoising because they will be smoothed in the denoising. A denoising method for aeroengine data based on a hybrid model is proposed to meet the needs of accurately evaluating the washing effect. Specifically, the aeroengine data is first decomposed into several components by time and frequency. The amplitude of the component containing the most noise is amplified, and Gaussian noise is added to generate noise-amplified data. Second, a Gated Recurrent Unit Autoencoder (GAE) model is proposed to capture engine data features. The GAE is trained to reconstruct the original data from the amplified noise data to develop its noise reduction ability. The experimental results show that, compared with the current popular algorithms, the proposed denoising method can achieve a better denoising effect, retaining the key characteristics of the aeroengine data.

Keywords: autoencoder; aeroengine; denoising; GRU

MSC: 68T09

Citation: Yan, Z.; Zu, M.; Cui, Z.; Zhong, S. A Novel Denoising Method for Retaining Data Characteristics Brought from Washing Aeroengines. *Mathematics* **2022**, *10*, 1485. https://doi.org/10.3390/math10091485

Academic Editors: Higinio Rubio Alonso, Alejandro Bustos Caballero, Jesus Meneses Alonso and Enrique Soriano-Heras

Received: 14 March 2022
Accepted: 26 April 2022
Published: 29 April 2022

1. Introduction

An aeroengine will be polluted after long-term operation. Pollution of aeroengines leads to mechanical failure, degraded engine performance, and pollutant gas emission. Washing is a main approach to controlling aeroengine pollution specified in aircraft maintenance manuals. Engine washing constitutes spraying cleaning fluid into the engine through multiple spray guns to remove fouling in the Engine. Washing an aeroengine is expensive. In order to minimize the washing cost while maintaining the normal performance of the engine, airlines must formulate a reasonable aircraft engine washing schedule.

A reasonable washing schedule can reduce 2735 tons of fuel consumption and 8626 tons of harmful gas emissions per 100 engines per year, saving USD 10.08 million [1]. However, the mechanical failure of the polluted engine causes high-frequency noise in the aeroengine data. The noise of aeroengine data will interfere with the reasonable formulation of an aircraft engine cleaning schedule, so it is necessary for airlines to denoise aeroengine data.

There are two difficulties in denoising aeroengine data: the noise of the engine itself is difficult to eliminate, and the step change of data generated by washing will be smoothed. Aeroengines are man-made systems. Different from noise in nature, the probability distribution of the engine's own noise does not obey the normal distribution, which increases the difficulty of denoising. After washing the aircraft engine, the data will experience a step change. Traditional denoising methods often treat the step changes in data as noise and smooth them out, making it difficult for the data to accurately reveal the washing effect, adding difficulties for subsequent evaluation. Therefore, a new denoising model is needed to eliminate the noise of the engine itself and retain the impact of cleaning on the data, as shown in Figure 1.

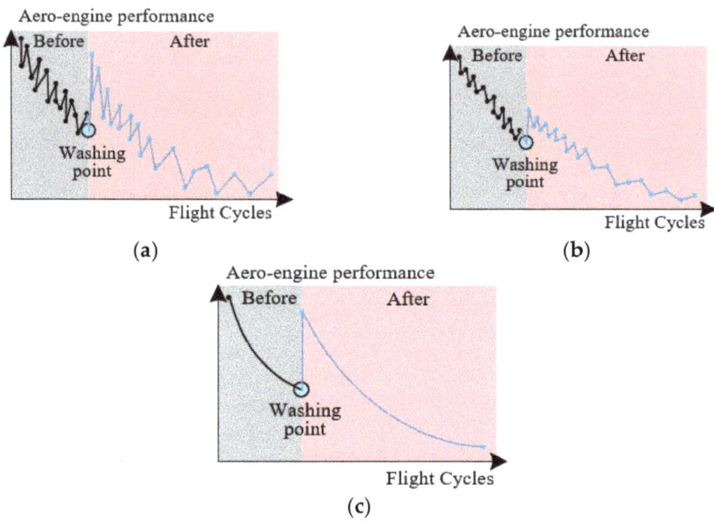

Figure 1. Demand for denoising aeroengine data. (**a**) Raw engine data; (**b**) Denoising effect of traditional methods; (**c**) Denoising required for aircraft engine

Many noise reduction methods have been proposed, including empirical mode decomposition (EMD), wavelet threshold denoising, and filtering. Xue Feng et al. [2] decomposed the data with EMD after adding Gaussian noise to the data and successfully separated the noise from the signal by searching the dominant component of noise through the continuous mean square error criterion. Lu et al. [3] randomly shuffled the high-frequency noise part of the data and then decomposed the data with EMD to achieve noise reduction. Kai et al. [4] denoised complex images with a non-sampling wavelet transform method. Mohammad Saleh Sadooghi et al. [5] denoised the compressor vibration signal of aeroengines with the wavelet threshold denoising method. Because the effect of wavelet threshold denoising is related to the threshold and threshold function, the author evaluates 84 matching effects of the threshold and threshold function to search for the most suitable method for denoising engine compressor vibration signals. P. Maragos et al. [6] proposed morphological filters, which are filters composed of basic operations of mathematical morphology. Morphological filters can selectively suppress image noise. M. H. Sedaaghi et al. [7] proposed mediated morphological filters by combining morphological filters with media filtering and classical gray scale morphological operators. Mediated morphological filters have a good effect on eliminating salt and pepper noise in images. Yang Chao et al. [8] denoised the temperature data with a two-way Kalman filter method. They improved the computational efficiency by simplifying the Kalman filter algorithm by precalculating the filter coefficients.

The above methods show some disadvantages. The EMD algorithm must calculate the dividing point between noise components and key characters; the Wavelet threshold denoising algorithm must find the threshold and threshold function manually. These two algorithms are inefficient, and the calculation results can only be approximate. The Kalman filter is suitable for linear systems rather than highly nonlinear systems such as aeroengines. The above three methods are not adaptive methods. Researchers must select or set model parameters according to data characteristics [9]. Therefore, these methods are not suitable for aeroengine data with non-uniform data distribution, hardly meeting the needs of accurately evaluating the washing effect.

Autoencoders are adaptive artificial intelligence algorithms that are frequently used as noise reduction models. Denoising Autoencoders (DAE) were first proposed by Bourlard H. [10] to extract features from raw data. The idea of the denoising autoencoder is to make the autoencoder reconstruct the original data from the noise-added data to obtain its noise reduction ability. Vincent et al. [11] enhanced the robustness of the model by setting the input of the autoencoder to zero at a specific scale. Song Hui et al. [12] used seismic data to verify the above algorithm and found that the algorithm can filter out random noise with strong intensity, but the algorithm's efficiency is low. According to the characteristics of underwater heterogeneous information data, Wang et al. [13] combined a three-layer sparse autoencoder with two-layer convolution to build a model with strong noise reduction ability. Peng et al. [14] added convolutional layers to the autoencoder, which made the autoencoder more robust with a smaller reconstruction error. However, the signal amplitude is reduced by this. Hui et al. [15] constructed a convolutional autoencoder to filter seismic data with a low signal-to-noise ratio and achieved desirable results. Alexander Kensert et al. [16] denoised the chromatogram to a completely or almost completely noise-free state with a deep convolutional autoencoder.

However, the DAE also smooths the edge features of the data while denoising. Although the literature [11,13] preserves the characteristics of the data to the greatest extent, the premise is that the noise of underwater heterogeneous information data and seismic data comes from nature, conforming to the Gaussian noise distribution. The noise of aeroengine data comes from the engine system rather than nature. There is no reference documenting that the engine noise conforms to the Gaussian noise distribution. Therefore, the above-mentioned algorithms are not very applicable to engine data. Therefore, the existing denoising methods based on autoencoders can only eliminate the conventional noise, but it is difficult to eliminate the specific noise of the aeroengine. These methods are not suitable for engine data and hardly meet the need to accurately evaluate the washing effect.

To meet the needs of accurately evaluating the washing effect, a denoising method for aeroengine data based on a hybrid model is proposed. This method can filter out the noise of the aeroengine data, retaining the edge features of the data. The method first splits the data by washing time, and then decomposes the data into several components by frequency. Second, the method finds the component that contains the most noise, amplifies its magnitude, and adds Gaussian noise to compose the noise-amplified data. Finally, the method inputs the noise-amplified data together with the original data into the proposed autoencoders, training the model's ability to recover the detailed information of the engine data from the noise-amplified data. Figure 2 shows the principle of the aeroengine data denoising method based on the hybrid model.

The rest of the paper is as follows: The second part presents the decomposition method of the aeroengine data and the identification method of the data frequency band containing the most noise. A GAE model is proposed to denoise aeroengine data. The third part first introduces the source of the engine data and secondly gives the identification results of the data frequency bands containing the most noise and the determination results of the hyperparameters of the GAE model. Finally, this part tests the noise reduction effect of the EMD model, the EAD model, and the GAE model on aeroengine data. The fourth part summarizes the superiority of the proposed aeroengine data denoising method through the analysis of the above noise reduction testing results.

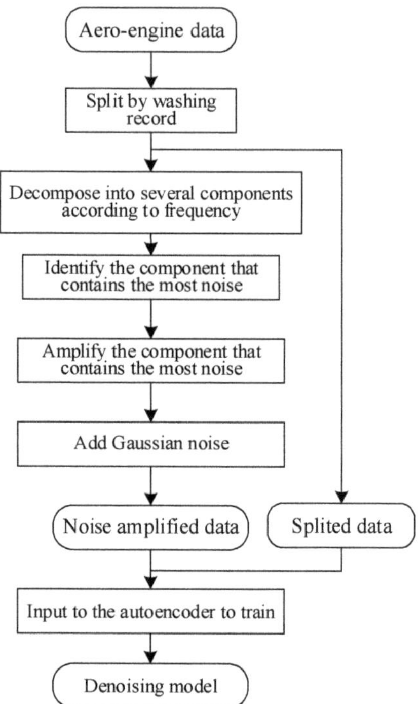

Figure 2. Denoising model principle.

2. Denoising Method for Aeroengine Data Based on a Hybrid Model

The paper involves more abbreviations, and their meanings are given in Table 1.

Table 1. Meaning of abbreviations.

Abbreviations	Full Title
DAE	Denoising Autoencoders
EMD	Empirical Mode Decomposition
EGTM	Exhaust Gas Temperature Margin
DTW	Dynamic Time Warping
IMF	Intrinsic Mode Functions
GRU	Gated Recurrent Unit Autoencoder
OEM	Original Equipment Manufacturer
MAE	Mean Absolute Error

2.1. Analysis of Noise Reduction Problems and Corresponding Solutions

The problems in eliminating engine noise and remaining data step sizes from washing are analyzed from the following three aspects, and methods are proposed.

Problem 1: Difficulty in retaining the data step size generated by washing the engine.

Aeroengines are complex nonlinear systems. Engine data is the output of a complex system that characterizes the performance and condition of the engine. Cleaning can be viewed as a stimulus applied to the engine system. The engine system has changed state due to being purged. As the system changes, the distribution of data output by the system changes. This change will be removed as noise, thus smoothing out the data's features.

The solution to this problem is to divide aeroengine data into several data segments according to the washing time. The data distribution within each data segment does not change, avoiding the smoothing of data features.

Problem 2: Difficulty in identifying engine-specific noise.

Each data segment contains an engine characteristic part and an engine-specific noise part. The process of denoising is performed to maximize the separation of engine-specific noise components from data segments. Traditional methods can identify noise with a Gaussian distribution, but it is difficult to identify engine-specific noise.

The solution to this problem is to decompose the data segment into several components with different frequencies. The similarity of each component to the noise data specific to the aeroengine is calculated to identify the component that contains the most noise.

Problem 3: Difficulty in reducing engine-specific noise.

Because the components containing the most noise still contain engine characteristic information, it is unscientific to directly delete the data components containing the most noise. Removing these components results in a loss of the feature components in the aeroengine data, smoothing the edge features of the data.

To reduce the loss of feature components, a new denoising method is proposed. First, the amplitudes of the data components that contain the most noise are amplified. Artificial noise is added to the data. Secondly, the autoencoder model is built. Finally, let the model reconstruct the original data from the noise-amplified data, training its ability to perform denoising.

Based on the above analysis, a denoising method for aeroengine data based on a mixed model is proposed. The noise reduction process of this method is shown in Figure 3. After the EGTM data were collected, they were split into several sets of data according to the water wash records. The data components containing the most noise as well as artificial Gaussian noise are superimposed on the original data. The original data are input into the autoencoder for training to generate several sets of denoised EGTM data. Finally, several sets of data are assembled into denoised complete EGTM data.

As shown in Figure 3, the denoising method for aeroengine data based on mixed models includes a proposed method for identifying data components that contain noise and a proposed denoising method.

Method 1: The proposed method for identifying data components that contain noise.

The paper first decomposes the aeroengine data according to the washing records; secondly, the paper decomposes the data into several components by frequency with the EMD algorithm; lastly, the noise content in the data components is evaluated with the distance calculation method based on the DTW algorithm to find the data components that contain the most noise.

Method 2: The proposed denoising method.

First, the amplitudes of the data components that contain the most noise are amplified. Artificial noise is added to the data. Secondly, the autoencoder model is built. Finally, let the model reconstruct the original data from the noise-amplified data, training its ability to perform denoising.

2.2. Identification Method of Aeroengine Data Components Containing Noise Based on EMD and DTW

This section is about splitting the data, decomposing the data, and identifying noise.

The washing time is used as the basis for splitting the engine data. The engine data in this study come from OEM factories, and they are recorded with the flight cycle as the observation time, so the aeroengine data is time series data. The engine washing records are tables that record the time, location, engine type, and aircraft type when the engine was washed. What can be used in this study is wash time. Data is denoted as $X = \{x_1, x_2, x_3, \ldots \}$. If the number of washings is $n - 1$, the washing records can be defined as: $T_{\text{washing}} = \{t_1, t_2, \ldots, t_{n-1}\}$. T_{washing} can split data X into n pieces of split data. Let $X(i)$ denote split data, which can be given by Equation (1).

$$X(i) = \left\{ x_{t_{(i-1)}+1}, x_{t_{(i-1)}+2}, \cdots, x_{t_i} \right\}, i = 1, 2, \cdots, n \tag{1}$$

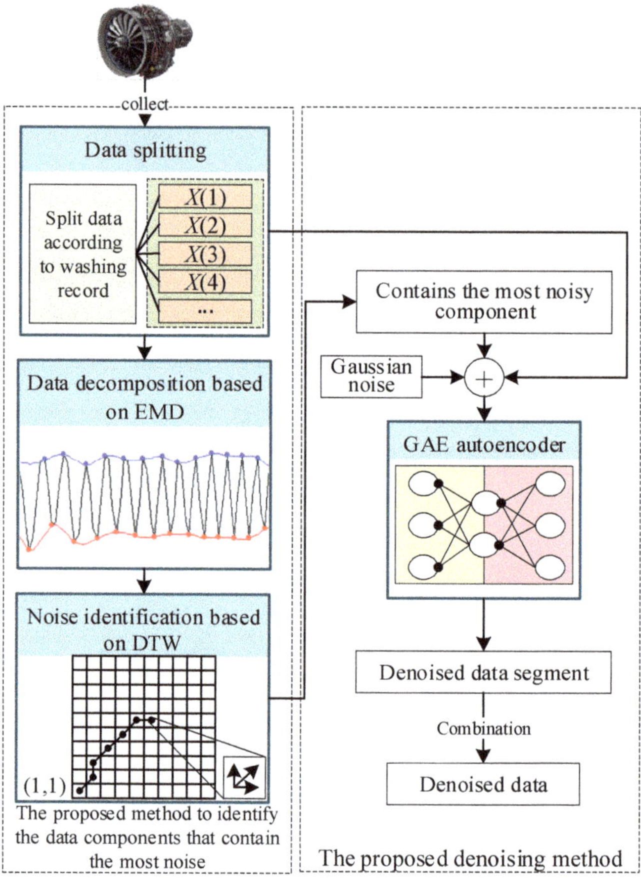

Figure 3. Denoising method for aeroengine data based on a hybrid model.

Then $X(i)$ is decomposed in frequency with the EMD algorithm. The EMD algorithm is a signal analysis algorithm, which decomposes the signal according to the timescale of the aeroengine data without setting any basis functions in advance. Aeroengine data can be divided into several "intrinsic mode functions" (IMF) by the EMD method without deviating from the time domain. The aeroengine data can be expressed as the trend component and the sum of several IMF functions.

The steps of EMD decomposing the data segment $X(i)$ obtained by dividing the aeroengine data are as follows:

Step 1: Find all the extreme points of the aeroengine data, use the spline curve to connect all the maximum points into the upper envelope, and connect all the minimum points to the lower envelope.

Step 2: Calculate the average value $m(i)$ of the upper and lower envelopes and the engine data IMF component $c(i)$ according to Equation (2).

$$c(i) = X(i) - m(i) \tag{2}$$

Nordne E. Hunag [17] put forward the concept of IMF and defined two conditions of IMF: 1. In the whole time range, the number of local extreme points and zero-crossing points of IMF differs by at most one; 2. The mean value of the upper and lower envelopes is zero. If $c(i)$ does not satisfy these two conditions, repeat steps 1 and 2 until they are satisfied, and then go to step 3.

Step 3: Separate $c(i)$ from the aeroengine data $X(i)$: $X(i) - c(i)$. Repeat step 1, step 2, and step 3 until $X(i) - c(i)$ becomes a monotonic sequence. Given $r(i)$, defined by Equation (3):

$$r(i) = X(i) - \sum c(i) \tag{3}$$

where $r(i)$ is the residual component of engine data $X(i)$.

The pseudocode of the EMD algorithm is given by Algorithm 1.

Algorithm 1. EMD algorithm

```
IMFs = []
While haspeaks(X(i)):
    maximum_points = search(X(i), maximum)
    minimum_points = search(X(i), minimum)
    max_envelope = fitting(maximum_points)
    min_envelope = fitting(minimum_points)
    c(i) = (max_envelope + min_envelope)/2
    r(i) = X(i) − c(i)
IMFs.append(data_imf)
```

Real engine data can be seen as a superposition of ideal data and noise data caused by minor faults. To identify the characteristic noise components of aeroengines, four kinds of fault data were prepared in this study, including compressor fault data, fan fault data, high-pressure turbine fault data, and low-pressure turbine fault data. The distance of the fault data from each data component is calculated to find the data component with the most noise.

However, each component of the engine data has unequal data lengths and non-existing linear correspondences with the four types of fault data. There are phenomena such as amplitude scaling and linear drift between data [18]. The Euclidean distance is directly used to represent the similarity between each component of the engine data and the four types of fault data, which may encounter the problem of information flooding to a certain extent, resulting in an unrealistic distance between each component of the engine data and the four types of fault data. Therefore, a distance calculation method that is compatible with various data lengths is required.

The DTW algorithm is proposed for measuring the distance between two time series of different lengths. DTW is widely used in the field of speech recognition, and it is also suitable for recognizing two similar aeroengine data sets. In this study, DTW is used to calculate the minimum distance between the four types of fault data and $c(i)$ or $r(i)$.

The process of the DTW algorithm is shown in Figure 4. First, a matrix grid is created, and the length and width are respectively the lengths of the engine data component and the length of the fault data. The elements of the matrix represent the Euclidean distance d of the data at the corresponding position of the engine data component and the fault data. Then take the lower-left corner (1,1) point of the matrix grid as the starting point and move to the adjacent grid corresponding to the smallest element of the right and upper elements until it reaches the upper-right corner of the matrix grid to form a path. Finally, calculate the cumulative value of the matrix elements that the path passes through. This accumulated value is the distance between the engine data component and the fault data.

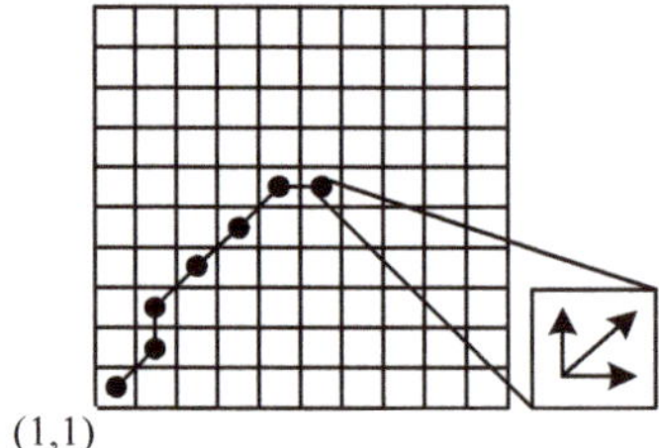

Figure 4. Principle of DTW algorithm.

The pseudocode of the DTW algorithm is given by Algorithm 2.

Algorithm 2. DTW algorithmM

---
n = length(*engine_data*)
m = length(*fault_data*)
dtw_matrix = $[\infty]_{(n+1)\times(m+1)}$
dtw_matrix(0,0) = 0
for i = 2:n + 1
  for j = 2:m + 1
    dtw_matrix(i,j) = $\sum[engine_data(i-1) - fault_data(j-1)]^2$ + min[*dtw_matrix*($i-1,j$),
    dtw_matrix($i-1,j-1$), *dtw_matrix*($i,j-1$)]
  end
end
return *dtw_matrix*(n + 1, m + 1)

Define the distance as *dis*. The distance of $c(i)$ or $r(i)$ and the four kinds of fault data are given in Table 2, as are symbols and explanations.

Table 2. Distance table.

dis	Interpretation of dis
$dis_{c(i)}$	Distance sum between $c(i)$ and all fault data
$dis_{r(i)}$	Distance sum between $r(i)$ and all fault data
$dis_{c(i),\ \text{FAN}}$	Distance between $c(i)$ and fan fault data
$dis_{c(i),\ \text{COMP}}$	Distance between $c(i)$ and compressor fault data
$dis_{c(i),\ \text{HPT}}$	Distance between $c(i)$ and high-pressure turbine fault data
$dis_{r(i),\ \text{FAN}}$	Distance between $r(i)$ and fan failure data
$dis_{r(i),\ \text{COMP}}$	Distance between $r(i)$ and compressor failure data
$dis_{r(i),\ \text{HPT}}$	Distance between $r(i)$ and high-pressure turbine failure data
$dis_{r(i),\ \text{LPT}}$	Distance between $r(i)$ and low-pressure turbine failure data

In Table 2, the distance between $c(i)$ and all fault data can be given by Equation (4).

$$dis_{c(i)} = \sum(dis_{c(i),\ \text{FAN}},\ dis_{c(i),\ \text{COMP}},\ dis_{c(i),\ \text{HPT}},\ dis_{c(i),\ \text{LPT}}) \tag{4}$$

The distance between $r(i)$ and all fault data can be given by Equation (5).

$$dis_{r(i)} = \sum(dis_{r(i),\ \text{FAN}},\ dis_{r(i),\ \text{COMP}},\ dis_{r(i),\ \text{HPT}},\ dis_{r(i),\ \text{LPT}}) \tag{5}$$

The smaller the distance, the more noise it contains. The engine data component corresponding to the smallest distance is selected as the noise component of the aeroengine.

2.3. Gated Recurrent Unit Autoencoder(GAE): A Proposed Denoising Autoencoder Model for Aeroengine Data

DAE is a promising method for data denoising, which can be used for aeroengine data denoising. It is a feature extractor with a denoising function whose purpose is to convert noisy aeroengine data into clean aeroengine data.

In addition, the aeroengine data is a time series, and each element in the sequence is affected by all the previous elements, which requires the model to be able to learn the influence from the historical accumulation of the aeroengine data. Real engine data is difficult to obtain, and the limited amount of data has difficulty supporting large-scale models. This requires the model to have a simple structure, and the GRU (Gated Recurrent Unit) model meets the requirements. The GRU was proposed by Cho et al. [19]. GRUs have fewer parameters, making training faster and requiring less data [20].

Therefore, in this paper, the Gated Recurrent Unit Autoencoder (GAE) model is proposed as a denoising module for aeroengine data by combining the autoencoder with the GRU. The structure of the GAE model is shown in Figure 5. The model includes encoders and decoders with special structures. The input end of the encoder and the output end of the decoder are both GRU modules, and a three-layer autoencoder is used as the connection in the middle. The three layers of the autoencoder are marked as h_1, h_2, and h_3. In the coding stage, the aeroengine data is continuously input to the GRU and the characteristic data is output through h_1 and h_2. In the decoding stage, the feature data is input into the h_3 layer, and the denoised data is output through the GRU module.

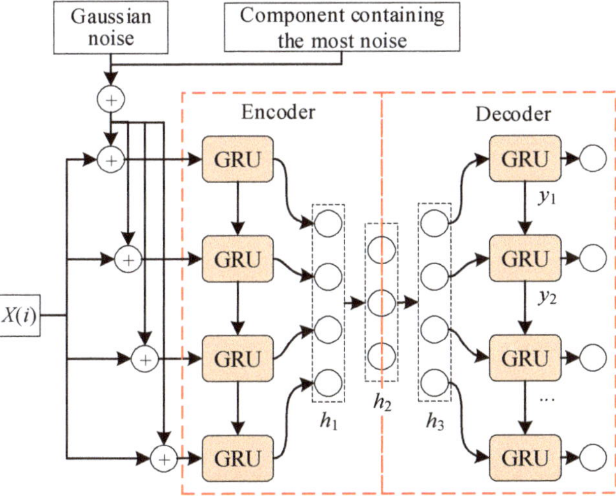

Figure 5. GAE Model.

In Figure 5, the input data is the aeroengine data that amplifies its own noise and adds Gaussian noise, which can be expressed as $X(i)_{\text{noise}}$; the label data is the raw aeroengine data, expressed as $X(i)$. The GRU involved in Figure 5 has two doors: update door and reset door. the basic structure of the GRU is shown in Figure 6.

In Figure 6, the definitions are as follows: z is the update gate, r is the reset gate, h is the current state, and h_{-1} is the previous state. The update gate z is used to control the amount of previous state information which is brought into the current state. The reset gate r controls how much information from the previous state is written to the current candidate set. The smaller the reset gate, the less information from the previous state is written. Define the current candidate set as $\widetilde{h}$. For the input data $X(i)_{\text{noise}}$, the forward propagation formula of the network can be constructed, which is given by Equation (6):

$$\begin{cases} r = \sigma(W_r \cdot [h_{-1}, X(i)_{\text{noise}}]) \\ z = \sigma(W_z \cdot [h_{-1}, X(i)_{\text{noise}}]) \\ \widetilde{h} = \tanh(W_h \cdot [r \cdot h_{-1}, X(i)_{\text{noise}}]) \\ h = (1 - z) \cdot h_{-1} + z \cdot \widetilde{h} \\ y = \sigma(W_y \cdot h) \end{cases} \tag{6}$$

where r is the weight from the input and the hidden layer at the previous moment to the reset gate r; W_z is the weight from the input and the hidden layer at the previous moment to the update gate z; W_h is the input and the hidden layer of the previous moment to $\widetilde{h}$; W_y is the weights from the hidden layer to the output layer.

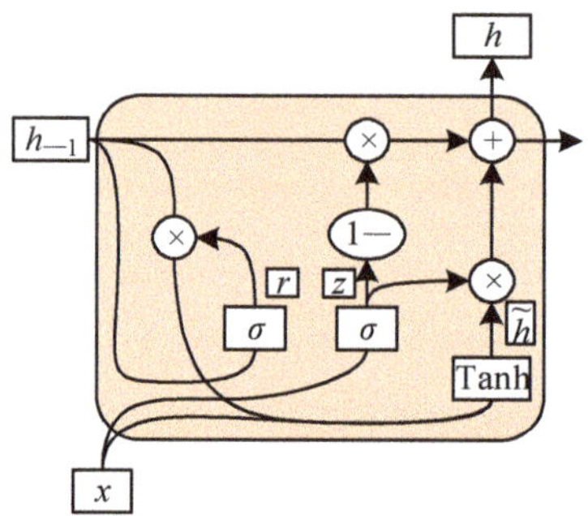

Figure 6. The structure of GRU.

The h_2 layer's output of the denoising autoencoder model is the feature code of the aeroengine data. The encoding process is that the encoding function maps high-dimensional aeroengine data vectors to low-dimensional feature vectors. The encoder function is the activation function of the h_2 layer of the autoencoder. The activation function is defined as the sigmoid function. The feature vector output by the h_2 layer is ***co***. Then the aeroengine data encoding process of the h_2 layer is given by Equation (7).

$$co = \frac{1}{1 + e^{-w_{\text{hid}} \cdot y - b_{\text{hid}}}} \tag{7}$$

where w_{hid} is the h_2 layer weight matrix; b_{hid} is the h_2 layer bias vector.

The decoding process maps the feature vector ***co*** to the reconstructed aeroengine data. The activation function of the h_3 layer in the decoder is defined as a linear function, and the feature vector output by the h_3 layer is ***deco***, which is given by Equation (8).

$$\boldsymbol{deco} = w_{\text{out}} \cdot \boldsymbol{h} + \boldsymbol{b}_{\text{out}} \tag{8}$$

where w_{out} is the h_3 layer weight matrix; b_{out} is the h_3 layer bias vector.

Finally, ***deco*** is output by the GRU module as denoised data, which is defined as $X(i)_{\text{denoise}}$. The loss function in this model is the mean absolute error (MAE). Define the number of output nodes of the GAE model as n_{out}, then MAE is given by Equation (9).

$$\text{MAE}(X(i)_{\text{denoise}}, X(i)_{\text{noise}}) = \frac{1}{n_{\text{out}}} \sum_{j=1}^{n_{\text{out}}} (X(i)_{\text{denoise}} - X(i)_{\text{noise}})^2 \tag{9}$$

3. Experiment

All functions are written on the python 3.5 platform of the Windows system with Tensorflow as the framework. Tensorflow is an open source machine learning platform. The hardware platforms involved in the calculation are CPU and GPU; the models are Intel(R) Core (TM) i3-9100F CPU_3.60GHz and NVIDIA GeForce GTX TITAN X, respectively.

The data involved in this paper is aeroengine EGTM data. Airlines take EGTM as a reference for aeroengine performance. EGTM refers to the difference between the red line value of engine exhaust temperature and the exhaust temperature when the engine takes off at full thrust. When the engine is washed, there are step changes in the EGTM data. During denoising, these step changes may be recognized as noise and smoothed. Therefore, selecting EGTM data as the noise reduction object can better verify the performance of the proposed method.

Since exhaust gas temperature margin (EGTM) data is an important indicator for evaluating the efficiency of engine washing, the data form of the paper is EGTM data that includes the number of cycles recorded in the water wash. The EGTM data is collected at the outlet of the low-pressure turbine of the engine, and Figure 7 is the measurement point of the exhaust temperature of the aeroengine.

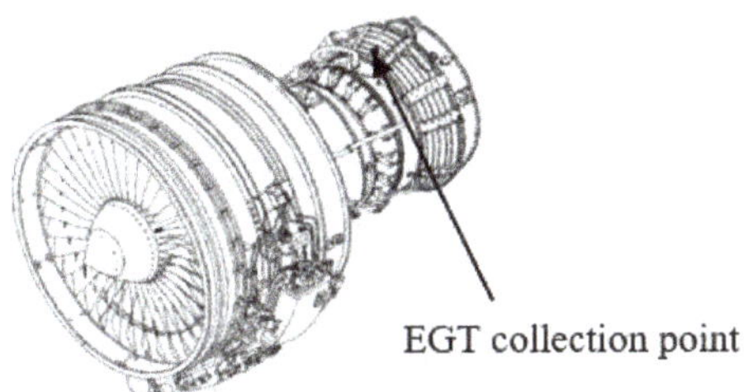

Figure 7. Aeroengine data sources.

EGTM data is real data provided by the OEM. In this study, the flight cycle was used as a time unit to record the aeroengine data from takeoff to landing. Therefore, aeroengine data is time series data. In practice, the airline provided three materials for an aeroengine, including OEM data in Table 3, fault records in Table 4, and washing records in Table 5.

Table 3. OEM data.

ID	ESN	Time	EGTM
B-5793	657208	16 September 2013 1:52	90.845
B-5793	657208	16 September 2013 5:09	84.52
B-5793	657208	16 September 2013 8:50	87.208
B-5793	657208	16 September 2013 12:31	89.306
B-5793	657208	17 September 2013 8:00	82.397
B-5793	657208	17 September 2013 11:45	85.755
B-5793	657208	18 September 2013 9:44	85.973
B-5793	657208	21 September 2013 0:06	66.281
B-5793	657208	21 September 2013 4:08	86.617
B-5793	657208	21 September 2013 8:36	75.281

Table 4. Fault records.

ID	Full Title	Noise Source	Noise Type
B2530	1 August 2009	Abnormal left engine	Compressor fault noise
B2588	2 January 2014	Fan seal broken	Fan fault noise
B6076	26 February 2002	Turbine oil leakage	Turbine fault noise
B6070	2 January 2014	Abnormal turbine blade	Turbine fault noise

Table 5. Washing records.

Date	ID	Base	CAMP
8 September 2014 0:00	B-1816	Beijing	A320 720000-CCA-C-02
8 September 2014 0:00	B-1816	Beijing	A320 720000-CCA-C-02
28 September 2014 0:00	B-2210	Hangzhou	A320 720000-CCA-C-02
16 January 2015 0:00	B-2210	Hangzhou	A320 720000-CCA-C-02
9 March 2014 0:00	B-2210	Hangzhou	A320 720000-CCA-C-03
19 April 2014 0:00	B-2210	Hangzhou	A320 720000-CCA-C-03
16 April 2014 0:00	B-2364	Chengdu	A320 720000-CCA-C-02
16 April 2014 0:00	B-2364	Chengdu	A320 720000-CCA-C-02

The data required for the study are spread out across three tables. The collection process of the data to be denoised is: According to the aircraft registration number (ID) and water washing date in Table 5, the IDs of the OEM data in Table 3 are located to collect all the corresponding EGTM data; the collection process of the fault data is: According to the ID in the fault record in Table 4, the ID of the OEM data in Table 3 is located to collect all the corresponding EGTM data. Figure 8 shows the aeroengine EGTM data. Figure 9 shows all types of data.

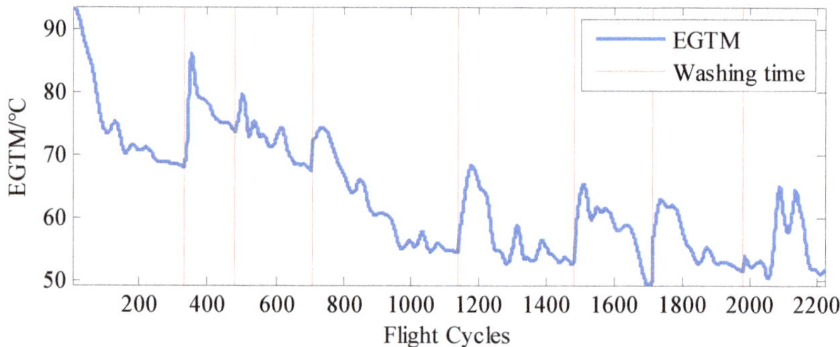

Figure 8. The engine data involved in this paper.

During the wing period, the engine was washed a total of nine times. Taking the washing time as the time point, the aeroengine data is split into eight segments. The first six segments are used as the training set, and the last two segments are used as the testing set. Number the eight segments sequentially to obtain Table 6.

Table 6. Split aeroengine data grouping table.

	Data Number	Flight Cycles
Training data	1	1~329
	2	330~481
	3	482~708
	4	709~1142
	5	1143~1484
	6	1485~1711
Testing Data	7	1712~1981
	8	1982~2222

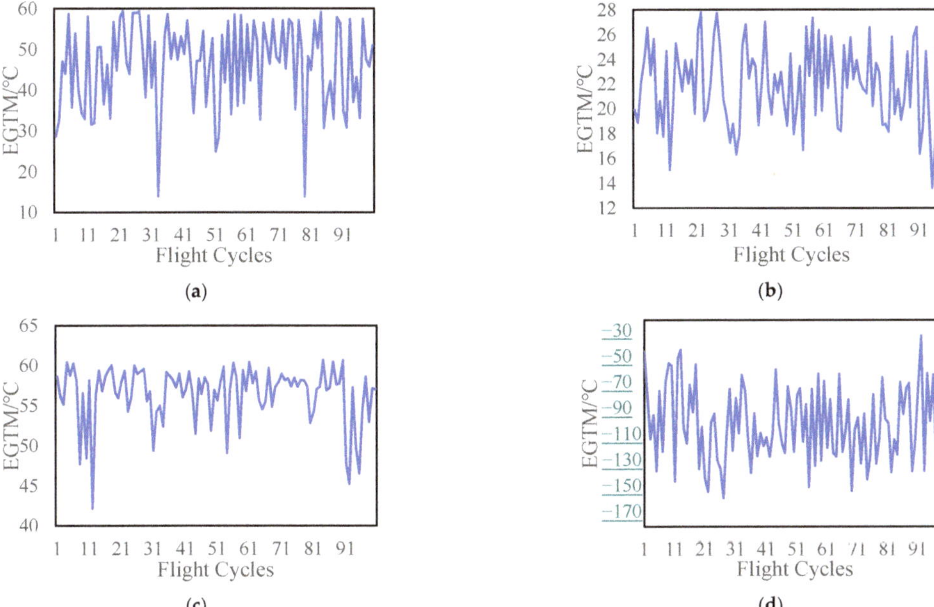

Figure 9. The fault data involved in this paper. (**a**) EGTM data for compressor failure; (**b**) EGTM data for fan failure; (**c**) EGTM data for high pressure turbine failure; (**d**) EGTM data for low pressure turbine failure.

3.1. Noise Identification Results for Data Components

To identify the data components that contain the most noise, the aeroengine data components are first decomposed based on EMD after being collected. Secondly, based on DTW technology, the noise identification of each component is carried out. The aeroengine EGTM data in Figure 8 are decomposed into residual components and IMF components. Figure 10 shows the two components derived from the decomposition of the six-segment aeroengine training data.

Based on the DTW algorithm, the distances were calculated between the two components of aeroengine EGTM data and four kinds of fault data, such as compressor fault data, fan fault data, high-pressure turbine fault data and low-pressure turbine fault data. In the calculation results, a very large number ($>1 \times 10^{308}$) appeared in the calculation of the distances, marked as "∞". Table 7 records the distances between the IMF component of the six-segment aeroengine training data and the four kinds of fault data. Table 8 records the distances between the residual component of the six-segment aeroengine training data and the four types of fault data.

Table 7. Distances of IMF components.

Data Number	Fan Fault Data	Compressor Fault Data	High Pressure Turbine Fault Data	Low Pressure Turbine Fault Data
1	5.236	5.227	6.493	6.199
2	4442.453	5582.838	∞	∞
3	10,463.488	12,162.886	∞	∞
4	2.907	4.205	2.521	5.467
5	3.167	3.981	3.0158	6.385
6	6209.860	7909.258	∞	∞

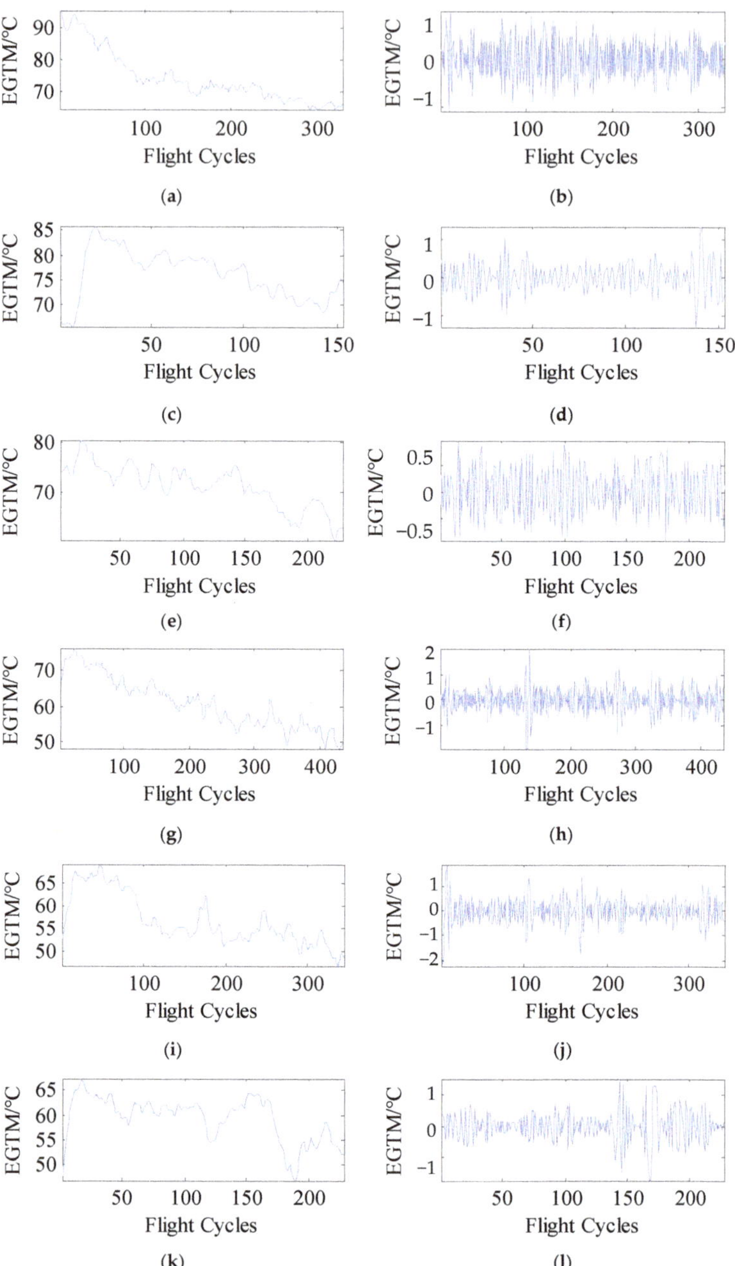

Figure 10. Components of the six-segment aeroengine training data. (**a**) No. 1 data residual component; (**b**) No. 1 data IMF component; (**c**) No. 2 data residual component; (**d**) No. 2 data IMF component; (**e**) No. 3 data residual component; (**f**) No. 3 data IMF component; (**g**) No. 4 data residual component; (**h**) No. 4 data IMF component; (**i**) No. 5 data residual component; (**j**) No. 5 data IMF component; (**k**) No. 6 data residual component; (**l**) No. 6 data IMF component.

Table 8. Distances of residual components.

Data Number	Fan Fault Data	Compressor Fault Data	High Pressure Turbine Fault Data	Low Pressure Turbine Fault Data
1	10.703	13.716	9.991	13.171
2	∞	∞	∞	∞
3	∞	∞	∞	∞
4	8.680	11.051	8.301	11.654
5	9.745	11.480	9.929	14.420
6	∞	∞	∞	∞

In Tables 7 and 8, the distances between the residual components of No. 2, 3, and 6 in the training data and the four types of fault data are all "∞", indicating that these three sets of residual component data hardly contain noise components. The distances between the residual components of No. 1, 4, and 5 and the four kinds of fault data are all within 15, while the distances between the IMF components of these three groups and the four kinds of fault data are smaller than those of the residual components, which shows that more noise is included in the IMF component of the aeroengine EGTM data. Therefore, the IMF component is selected as the data component containing the most noise. Separate the data components that contain the most noise in the testing data (No. 7 and No. 8), as shown in Figure 11.

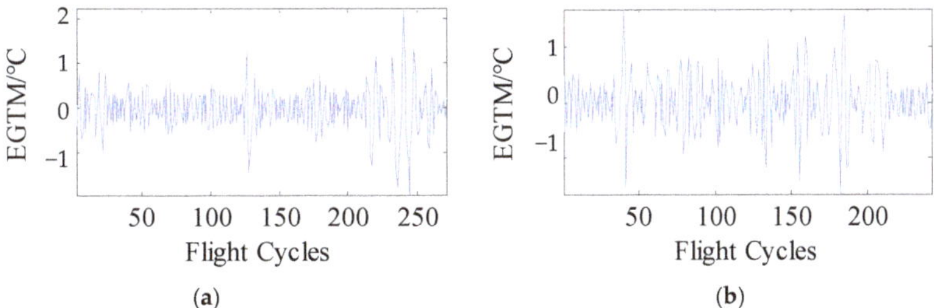

Figure 11. Components that contain the most noise in the testing data. (**a**) Data No. 7 contains the most noise data component; (**b**) Data No. 8 contains the most noise data component.

3.2. Hyperparameter Settings for GAE

In this part, the number of nodes in the h_1 layer and the number of nodes in the h_2 layer of the GAE model are determined. Define the number of nodes in the h_1 layer as n_{in}, the number of nodes in the h_2 layer as n_{hid}, and the number of nodes in the h_3 layer as n_{out}. Since the number of nodes in the h_1 layer is same as that in the h_3 layer, $n_{in} = n_{out}$.

There is currently no formula defining the n_{in}. To obtain reliable results, the enumeration method is used to determine the most suitable n_{in}. The enumeration range is 15–25. The mean absolute error (MAE) is used to describe the reconstruction error of the denoising autoencoder for the null engine data.

The model uses the Adam algorithm as the descent algorithm. The number of iterations is 3000 and the training batch is 100. The learning rate is 0.001, and the component containing the most noise is amplified by a factor of 1.1. The model was run 10 times, and the n_{in} and the reconstruction error were plotted as shown in Figure 12.

The vertical line in Figure 12 marks the n_{in} when the reconstruction error is the smallest. In order to analyze the results in more detail, the mean, standard deviation, and minimum and maximum values of 10 experiments are given in Table 9. The data in the Table 9 are reserved for three significant figures.

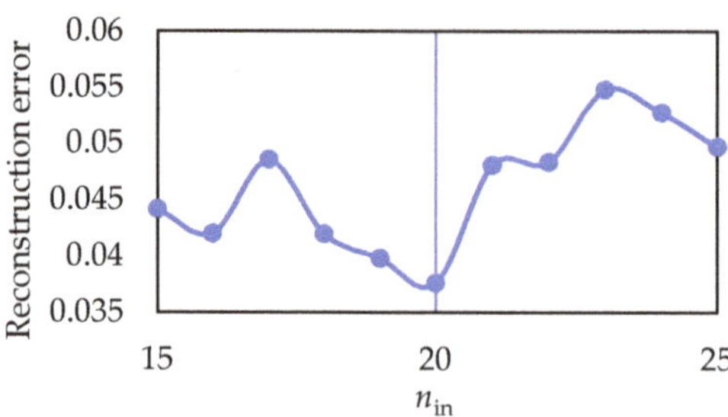

Figure 12. Relationship between the number of nodes in the h_1 layer and reconstruction error.

Table 9. Table of the nodes number in the h_1 layer and reconstruction error data.

n_{in}	Mean Reconstruction Error	Standard Deviation	Minimum	Maximum
15	0.0442	0.0476	0.0174	0.152
16	0.0420	0.0237	0.0208	0.0863
17	0.0486	0.0317	0.0236	0.111
18	0.0420	0.0286	0.0190	0.114
19	0.0398	0.0272	0.0197	0.117
20	0.0376	0.0397	0.0170	0.140
21	0.0481	0.0349	0.0261	0.150
22	0.0484	0.0441	0.0173	0.168
23	0.0548	0.0356	0.0254	0.114
24	0.0528	0.0496	0.0194	0.117
25	0.0497	0.0507	0.0223	0.186

From the data given in Table 9, the average reconstruction error is the lowest when n_{in} is 20. The average reconstruction error is slightly higher when n_{in} is 19. However, the standard deviation when n_{in} is 20 is larger than that when n_{in} is 19, being 0.0397 and 0.0272, respectively. The minimum error value when n_{in} is 20 is smaller than that when n_{in} is 19. The maximum error when n_{in} is 20 is greater than that when n_{in} is 19. Table 5 shows that the output is unstable when n_{in} is 20, although the average reconstruction error is small. Although the average reconstruction error is slightly larger, the output is stable when n_{in} is 19. Therefore, n_{in} is determined to be 19 in this paper.

Similarly, to minimize reconstruction error, n_{hid} must be determined. At present, researchers have developed empirical formulas [21] for n_{hid}, which help narrow the search for optimal nodes. The empirical formula is given by Equation (10).

$$n_{\text{hid}} = \sqrt{n_{\text{in}} \times n_{\text{out}}} \tag{10}$$

According to Equation (10), it can be roughly determined that n_{hid} is around 19. The optimal n_{hid} is searched by the enumeration method in the interval of 15–25. After the experiment is carried out 10 times, n_{hid} and the reconstruction error are plotted in Figure 13.

As can be seen from Figure 13, when n_{hid} is 21, the reconstruction error is the lowest. To analyze the results in more detail, in Table 10, the mean reconstruction error, standard deviation, minimum error, and maximum error of 10 experiments are given. The above four types of data in Table 10 all retain three significant figures.

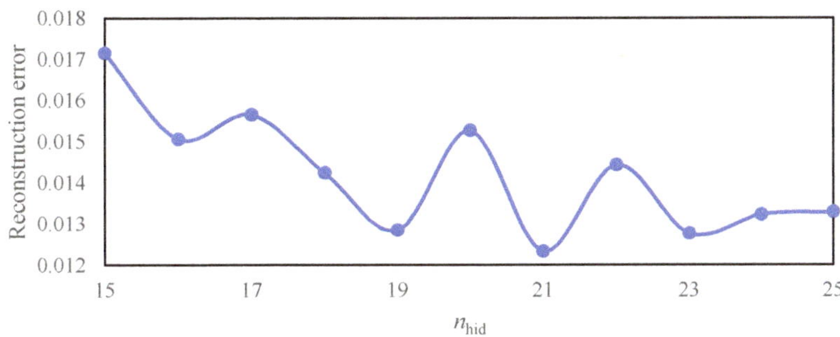

Figure 13. Relationship between n_{hid} and reconstruction error.

Table 10. Table of n_{hid} and reconstruction error.

n_{hid}	Mean Reconstruction Error	Standard Deviation	Minimum	Maximum
15	0.0172	0.00522	0.00900	0.0260
16	0.0151	0.00460	0.00900	0.0220
17	0.0157	0.00748	0.00900	0.0350
18	0.0143	0.00487	0.0100	0.0270
19	0.0129	0.00375	0.00900	0.0190
20	0.0153	0.00503	0.00800	0.0260
21	0.0123	0.00343	0.00800	0.0180
22	0.0144	0.00469	0.00900	0.0220
23	0.0128	0.00418	0.00700	0.0190
24	0.0132	0.00486	0.00900	0.0230
25	0.0133	0.00421	0.00800	0.0230

From the data given in Table 10, the mean reconstruction error is the lowest when n_{hid} is 21. The standard deviation is the smallest when n_{hid} is 21. This means that the model's output is currently stable. Therefore, the most suitable n_{hid} is determined to be 21.

3.3. Validation of Aeroengine Data Denoising Method

3.3.1. Reconstruction Accuracy Verification of GAE

First, the reconstruction accuracy of the GAE model is verified. Figure 14 visually shows the reconstruction errors of the three denoising models. Data No. 1–6 are the reconstruction error of the training data in the figure, and data No. 7–8 are the reconstruction error of the Testing data. The figure shows that the reconstruction error of the denoised data output by the EMD model is much larger than that of the GAE model and the EAD model. Except for data No. 4 and No. 8, the reconstruction error of the GAE model is smaller than that of the EAD model.

The reconstruction errors of the DAE, EMD, and GAE models on the training data are given in Table 11.

Table 11. Reconstruction errors of training data.

Model	Reconstruction Error of Training Data						Average
GAE	0.0810	0.0909	0.0637	0.0681	0.0585	0.0861	0.0747
DAE	0.0853	0.0995	0.0991	0.0628	0.0901	0.0817	0.0864
EMD	0.185	0.3578	0.320	0.288	0.248	0.705	0.351

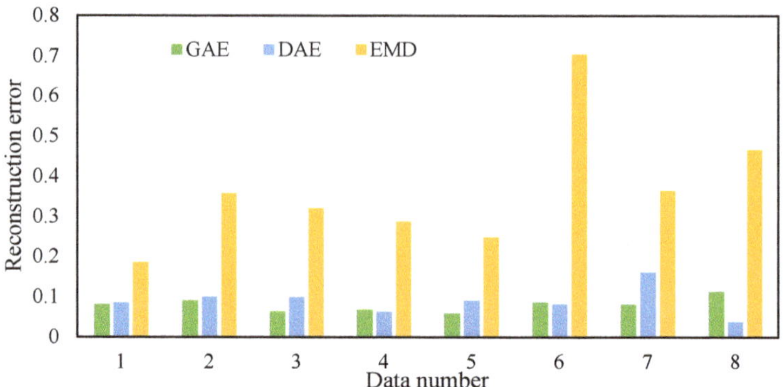

Figure 14. Comparison of reconstruction accuracy of three noise reduction models.

Tables 11 and 12 show that the MAE error of GAE training data is $(0.0864 - 0.0747)/0.0864 \times 100\% = 13.54\%$ lower than that of DAE and 78.72% lower than that of EMD; the MAE error of GAE test data is 3.11% lower than that of DAE and 76.80% lower than that of EMD.

Table 12. Reconstruction errors of testing data.

Model	Reconstruction Error of Testing Data		Average
GAE	0.0807	0.0807	0.0965
DAE	0.161	0.161	0.0996
EMD	0.365	0.467	0.416

The MAE error of GAE is relatively smaller than that of DAE, which reflects that the GRU in GAE can learn the time series relationship of aeroengine data so as to better reconstruct pure data. The combination of the GRU module and the DAE model effectively improves the noise reduction effect, making the MAE error smaller than that of the DAE model.

The MAE error of GAE is much smaller than that of EMD. The reason for this is that the algorithm of EMD is susceptible to noise interference. The algorithm of EMD separates the noise signal by calculating the envelopes of the extrema of the data. Noise makes the extremum of the data change, and the envelopes follow. The changes of the envelope also make the EMD algorithm unstable, resulting in a large error.

To summarize, Tables 8 and 11 reflect that the noise reduction effect of the DAE model is slightly better than that of the EMD model. The mean reconstruction error of the GAE model is smaller than that of the EAD model. The data reconstruction ability of the GAE model is stronger than that of the DAE model and the EMD model. It is proved that the GAE model can preserve the characteristics of the aeroengine better.

3.3.2. The Effectiveness of the Proposed Noise Reduction Method Based on Hybrid Model Is Verified

Figure 15 shows the reconstruction curves of DAE, EMD, and the proposed model on the EGTM Testing data of an aeroengine. In the figure, the original data are marked with black curves, the denoised data of the proposed model are marked with green dotted lines, the denoised data of the EAD model are marked with light blue, and the noise reduction of the EMD model are marked with dark blue.

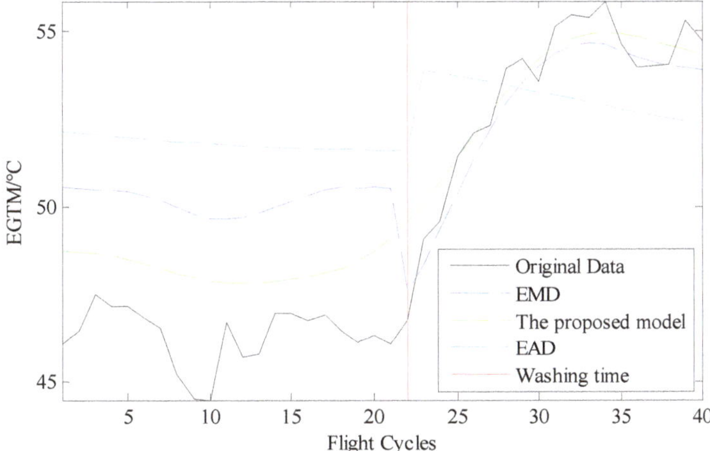

Figure 15. Comparison of noise reduction effects of three noise reduction models.

The red vertical line in Figure 15 represents the time of washing the aerocngine. Due to the washing, there is a step change in the original EGTM data. EAD, EMD, and the proposed method show different performances under the influence of cleaning. The EAD method significantly smoothed the step change of EGTM. It shows that the EAD model has an underfitting problem in training. The EMD method separates the step change as noise from the EGTM, thus smoothing the step change of the EGTM.

In Figure 15, the proposed model can reconstruct the EGTM sequence and express data mutation well. The denoised data of the proposed model in the figure is closer to the original data than DAE and EMD, which shows that the denoised data of the proposed model is more suitable for the subsequent analysis of washing effects. In order to more accurately express the noise reduction effect of DAE, EMD, and the proposed model on aeroengine data, the EGTM step value after aeroengine washing is investigated.

The influence of the denoising model on the subsequent evaluation of the washing effect can be known by calculating step size after washing. The step size after washing is a direct expression of the washing effect. Therefore, calculating the post-washing step size of the denoised data is an important indicator to investigate whether the denoising model will affect the evaluation of the washing effect. Since the data after washing has a gradual increase rather than a sudden increase, the step size calculated in the paper is the difference between the average values of 10 data before and after washing. The step size is calculated from the No. 2 data.

Figure 16 visually shows the steps sizes of No. 2–8 raw data and the step sizes of denoised data of DAE, EMD and the proposed model. Table 13 gives the step sizes of the training data and testing data. The data in Table 13 are the steps sizes of each piece of data. Since No. 1 is unwashed engine data, the step size is recorded from No. 2 data.

Table 13. Step of each segment of data.

Model	Step Size of the EGTM/°C						
	Training Data					Testing Data	
Original data	11.326	5.212	7.199	12.206	12.078	5.613	6.971
The proposed model	10.604	4.841	6.832	11.292	11.555	5.584	5.108
DAE	9.649	4.525	5.435	9.979	10.958	5.871	2.709
EMD	10.220	2.813	5.067	8.137	9.469	10.288	0.738

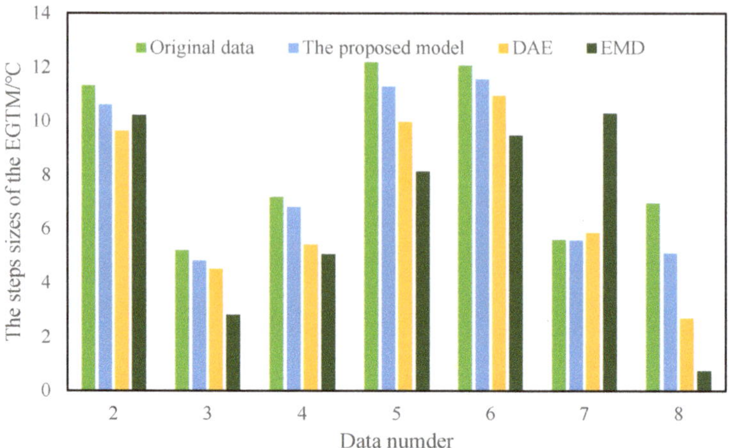

Figure 16. The step sizes from data of all models.

Table 13 is the variance of the steps of the original data and the steps of the denoised data of DAE, EMD, and the proposed model.

It can be seen from Table 14 that EMD has a large variance in both the testing set and the training set. The variances of the DAE in the training set and testing set are much smaller than those of the EMD, but they are still large. The variances of the proposed model are extremely small in both the testing set and the training set. The step sizes calculated from the denoised data of the proposed model are the closest to the original data. Therefore, the proposed model has the least impact on the analysis of subsequent washing effects. It is more suitable for aeroengine data noise reduction than other models.

Table 14. Step value variance between noise reduction data and original data.

Model	Training Data	Testing Data
The proposed model	0.381	1.736
DAE	2.522	9.116
EMD	6.978	30.358

4. Conclusions

To improve prediction accuracy, a denoising method for aeroengine data based on a hybrid model is proposed in this paper. The method first amplifies the noise part of the data, and then adds Gaussian noise to the data as the input of the autoencoder. Let the autoencoder reconstruct the original data from the amplified noise data so that the autoencoder can perform targeted noise reduction. In the paper, the proposed model is compared with EMD and DAE, which reflects that the proposed model can effectively denoise the data and retain mutation characteristics after aeroengine washing.

The autoencoder involved in the hybrid-model-based aeroengine data denoising method is the GAE model proposed in this paper. The GAE model is composed of three fully connected layers connecting two GRU modules. The model is good at working with time series data. After testing with real aeroengine data, compared with EMD and DAE, the reconstruction error of the GAE model is the smallest, preserving the data features to the greatest extent.

The model proposed in this paper has an ideal effect on the denoising of EGTM data after aeroengine washing. This model is applicable for denoising the various data with sudden changes, such as the gas path data of aeroengines or gas turbines after maintenance. In the future, we will plan to collect more real data to improve our methods.

Author Contributions: Methodology, M.Z.; writing—original draft preparation, Z.Y., Z.C., and S.Z.; writing—review and editing, Z.C.; funding acquisition, M.Z. and Z.C. All authors have read and agreed to the published version of the manuscript.

Funding: This research was funded by the National Natural Science Foundation of China, grant number U2133202; and the National Natural Science Foundation of China, grant number 51975157.

Institutional Review Board Statement: Not applicable.

Informed Consent Statement: Not applicable.

Data Availability Statement: Not applicable.

Conflicts of Interest: The authors declare no conflict of interest.

References

1. Chen, D.; Sun, J. Fuel and emission reduction assessment for civil aircraft engine fleet on-wing washing. *Transp. Res. Part D Transp. Environ.* **2018**, *65*, 324–331. [CrossRef]
2. Xue, F.; Sun, X.; Dong, Z.; Yang, H.; Wang, H. Research on Data Denoising Algorithm Based on EEMD. *Mech. Eng. Autom.* **2021**, *5*, 9–11.
3. Lu, T.; Qian, W.; He, X.; Le, Y.; Huang, J. An improved EMD noise reduction method based on noise statistical characteristics, Bull. *Surv. Mapp.* **2020**, *11*, 71–75.
4. Hu, K.; Cheng, Q.; Li, B.; Gao, X. The complex data denoising in MR images based on the directional extension for the undecimated wavelet transform. *Biomed. Signal Process. Control* **2018**, *39*, 336–350. [CrossRef]
5. Sadooghi, M.S.; Khadem, S.E. A new performance evaluation scheme for jet engine vibration signal denoising. *Mech. Syst. Signal Process.* **2016**, *76*, 201–212. [CrossRef]
6. Maragos, P.; Schafer, R.W. Morphological Filters. Part 1. Their Set-Theoretic Analysis and Relations to Linear Shift-Invariant Filters. *IEEE Trans. Acoust. Speech Signal Process.* **1987**, *35*, 1153–1169.
7. Sedaaghi, M.H.; Daj, R.; Khosravi, M. Mediated morphological filters. In Proceedings of the 2001 International Conference on Image Processing (Cat. No.01CH37205), Thessaloniki, Greece, 7–10 October 2001; pp. 692–695.
8. Yang, C.; Li, J.; Yang, W.; Yang, W. Denoising Method for Temperature Log Data Based on A Kalman Filter. *Well Logging Technol.* **2020**, *2*, 168–171.
9. Li, Y.; Wang, C.; Tian, Y.; Wang, S. Parameter-shared variational auto-encoding adversarial network for desert seismic data denoising in Northwest China. *J. Appl. Geophys.* **2021**, *11*, 104428. [CrossRef]
10. Bourlard, H.; Kamp, Y. Auto-association by multilayer perceptrons and singular value decomposition. *Biol. Cybern.* **1988**, *4*, 291–294. [CrossRef] [PubMed]
11. Vincent, P.; Larochelle, H.; Bengio, Y.; Manzagol, P.A. Extracting and Composing Robust Features with Denoising Autoencoders. In Proceedings of the 25th international conference on Machine Learning, Helsinki, Finland, 5–9 June 2008.
12. Song, H.; Gao, Y.; Chen, W.; Zhang, X. Seismic noise suppression based on convolutional denoising autoencoder. *Oil Geophys. Prospect.* **2020**, *6*, 1210–1219.
13. Wang, X.; Zhao, Y.; Teng, X.; Sun, W. A stacked convolutional sparse denoising autoencoder model for underwater heterogeneous information data. *Appl. Acoust.* **2020**, *167*, 107391. [CrossRef]
14. Peng, F.; Gao, Y. BPSK Signal Denoise Based on Convolution Auto-Encoder Network. *Inf. Commun.* **2020**, *8*, 41–44.
15. Song, H.; Gao, Y.; Chen, W.; Xue, Y.J.; Zhang, H.; Zhang, X. Seismic random noise suppression using deep convolutional autoencoder neural network. *J. Appl. Geophys.* **2020**, *178*, 104071. [CrossRef]
16. Kensert, A.; Collaerts, G.; Efthymiadis, K.; Van Broeck, P.; Desmet, G.; Cabooter, D. Deep convolutional autoencoder for the simultaneous removal of baseline noise and baseline drift in chromatograms. *J. Chromatogr. A* **2021**, *1*, 462093. [CrossRef] [PubMed]
17. Wu, Z.; Huang, N.E. A study of the characteristics of white noise using the empirical mode decomposition method. *Proc. R. Soc. London. Ser. A Math. Phys. Eng. Sci.* **2004**, *460*, 1597–1611. [CrossRef]
18. Park, S.; Chu, W.W.; Yoon, J.; Won, J. Similarity Search of Time-Warped Subsequences via a Suffix Tree. *Inf. Syst.* **2003**, *7*, 867–883. [CrossRef]
19. Chung, J.; Gulcehre, C.; Cho, K.H.; Bengio, Y. Empirical Evaluation of Gated Recurrent Neural Networks on Sequence Modeling. *arXiv* **2014**, arXiv:1412.3555.
20. Olah, C. Understanding Lstm Networks. 2015. Available online: http://colah.github.io/posts/2015-08-Understanding-LSTMs (accessed on 9 March 2021).
21. Sequin, C.H.; Clay, R.D. Fault tolerance in artificial neural networks. In Proceedings of the 1990 IJCNN International Joint Conference on Neural Networks, San Diego, CA, USA, 17–21 June 1990; pp. 703–708.

 *mathematics*

MDPI

Article

Adaptive Rejection of a Sinusoidal Disturbance with Unknown Frequency in a Flexible Rotor with Lubricated Journal Bearings

Gerardo Amato [1], Roberto D'Amato [2,3,*] and Alessandro Ruggiero [4]

[1] Nonlinear and Adaptive Controls Laboratory, Department of Electronic Engineering, University of Rome "Tor Vergata", Via del Politecnico, 00133 Rome, Italy; amato@ing.uniroma2.it

[2] Escuela Técnica Superior de Ingeniería y Diseño Industrial, Universidad Politécnica de Madrid, Ronda de Valencia, 28012 Madrid, Spain

[3] Structural Materials Research Center (CIME-UPM), Departamento de Ciencia de Materiales ETS de Ingenieros de Caminos, Canales y Puertos, C/Profesor Aranguren S/N, 28040 Madrid, Spain

[4] Department of Industrial Engineering, University of Salerno, Via Giovanni Paolo II, 84084 Fisciano, Italy; ruggiero@unisa.it

* Correspondence: r.damato@upm.es

Abstract: Frequency-dependent adaptive noise cancellation-based tracking controller (ANC-TC) is a known technique for the stabilization of several nonlinear dynamical systems. In recent years, this control strategy has been introduced and applied for the stabilization of a flexible rotor supported on full-lubricated journal bearings. This paper proposes a theoretical investigation, based on robust immersion and invariance (I&I) theory, of a novel ANC-frequency estimation (FE) technique designed to stabilize a flexible rotor shaft affected by self-generated sinusoidal disturbances, generalized to the case of unknown frequency. A structural proof, under assumptions on closed-loop output signals, shows that the sinusoidal disturbance rejection is exponential. Numerical simulations are presented to validate the mathematical results in silico. The iterative Inexact Newton method is applied to the disturbance frequency and phase estimation error point series. The data fitting confirms that the phase estimation succession has an exponential convergence behavior and that the asymptotical frequency estimation is a warm-up phase of the overall close-loop disturbance estimation process. In two different operating conditions, the orders of convergence obtained by phase and frequency estimate timeseries are $p_\varphi = 1$, $p_{\omega,unc} = 0.9983$ and $p_{\omega,cav} = 1.005$. Rejection of the rotor dynamic disturbance occurs approximately 76% before in the cavitated than in the uncavitated condition, 2 (s) and 8.5 (s), respectively.

Keywords: rotordynamic; adaptive rejection control; sinusoidal disturbance; flexible rotor; hydrodynamic journal bearing

MSC: 37M05; 37N35

Citation: Amato, G.; D'Amato, R.; Ruggiero, A. Adaptive Rejection of a Sinusoidal Disturbance with Unknown Frequency in a Flexible Rotor with Lubricated Journal Bearings. *Mathematics* **2022**, *10*, 1703. https://doi.org/10.3390/math10101703

Academic Editors: Higinio Rubio Alonso, Alejandro Bustos Caballero, Jesus Meneses Alonso and Enrique Soriano-Heras

Received: 15 April 2022
Accepted: 10 May 2022
Published: 16 May 2022

Publisher's Note: MDPI stays neutral with regard to jurisdictional claims in published maps and institutional affiliations.

1. Introduction

1.1. Problem Statement: Flexible Rotor with Lubricated Journal Bearings

In rotating machines, bearings are used to transfer radial and axial forces in a supporting structure ensuring a low value of the coefficient of friction and good system stability [1,2]. For cases with high loads and high rotational speeds, fluid film lubricated bearings are often preferred. The latter can be of two main categories: hydrodynamic bearings and hydrostatic bearings [3]. In the case of hydrodynamic bearings, the relative movement between the two coupled surfaces (journal and bearing) generates a pressurized fluid film that reduces the friction forces, ensuring the necessary support for the external load [3]. In the case of hydrostatic bearings, the pressure of the lubricating fluid is guaranteed by the action of an external pump, which injects pressurized lubricant into the bearing [4]. The most common version of hydrodynamic bearings is the full journal bearing, which consists of a rotor (journal) that rotates inside a bearing with a diameter slightly larger than that of the rotor, and the fluid film exists in the small space between the two (meatus) [3]. In the unsteady dynamical behavior, fluid film bearings can, however, present oscillating behavior (typical elliptical orbits) due to destabilizing cross coupling forces caused by the nonlinear fluid dynamic phenomena in the oil film [2,5]. To obtain acceptable operating conditions, it is necessary to simultaneously analyze the rotor-journal bearing couple in the case of a flexible rotor. These types of dynamic systems exhibit a particular type of self-excited vibration due to fluid dynamics phenomena in the oil film [2] known as *oil whirl* and *oil whip*, and are characterized by subsynchronous processional motions [6,7]. These vibrations appear when the subsynchronous vortex frequency reaches the natural frequency of the system [8] and is typically characterized by high vibration amplitudes. In recent years, research has shown that oil whirl phenomena are generated also when the journal bearing runs on micropolar lubricant [9–11]. Many authors have studied the effect of lubricant contamination [12] and the nonlinear behavior of film-oil main bearings in rotating machines [13]. Furthermore, the performance of the bearing, in order to avoid instability phenomena, has been analyzed, studied and simulated considering factors such as misalignment [14], elasticity of the bearing liner [15], dynamic conditions and coupled surface roughness [16].

1.2. Literary Review: Active Noise Rejection Control

One of the objectives of recent research has been to propose control strategies for solving the classical problem of vibration rejection, in the frame of flexible mechanical structures [17,18], flexible rotor bearings control [19–23] and their active balancing [21]. The aim is to absorb, or reject, any vibrational phenomena on a rotating shaft connected to a motor, and stabilize it around an arbitrary equilibrium position [24], for example, using a closed-loop controller with notch filtering actions [24–26]. Good attenuation features are obtained if the disturbance frequency is well-known. Indeed, the stop bands of these filtering-based controllers exhibit very steep edges. Usually, in the presence of unmodelled dynamics, observer-based and sliding mode observer-based controls allow for the tracking control precision to be improved by adapting to the unmatched uncertainties [18]. Nevertheless, the use of an adaptive observer to solve this category of problems, the vibration suppression in flexible mechanical system, is not so widespread. For example [26], avoids the application of an adaptive observer due to the risk of control spillover. Conversely, the ANC control in D'Amato et al. [24] is implemented as an observer-based control. In [24], the equations of the elastic contributions in the journal bearings and the relative vibrations are modelled as a separated exosystem [27], which represents the disturbance acting on the control. Then, the ANC operation consists of the observation of such disturbance elastic contributions, so that the control injects, on purpose, a counterphase oscillation, aiming to delete them. This makes the closed-loop control independent of the operating conditions of the bearings: cavitated and uncavitated.

First results concerning controllability and observability problems in rotordynamic systems date back to the 1980s. The flexible rotor and the flexibly-mounted journal bearings were modelled by Stanway and Burrows, 1981 [22], as a spring-damping-mass system, with well-known natural vibration modes. They determined the system observability and controllability conditions in order to assign directly the closed-loop eigenvalues with a linear controller. A linear control approach was also addressed by Lei and Palazzolo, 2008 [25]. To this aim they used the Finite-Element-Method (FEM) to model the rotor dynamic system. In their approach, active magnetic bearings (AMB) were employed to handle the rotordynamic systems by acting coaxially to the inertial axis through a linear control law. In the framework of control techniques which uses AMB, Kumar and Tiwari, 2020 [28], investigated the rotordynamic system modeling unbalance and misalignment with the rigid body theory. The possibility of modeling the flexible system with a new set of nonlinear closed-form equations is a definite novelty introduced by D'Amato et al., 2022 [24], and retrieved in this presented study. By considering the lubricated bearing dynamic effects and with lubricant film cavitation conditions, an input–output feedback linearizing controller can be implemented. The proposed control law is adaptive with respect to the control and disturbance parameter (i.e., phase and frequency of the sinusoidal disturbance, corresponding to rotor angular speed of the rotor).

1.3. Proposed Control Strategy: ANC with Frequency Estimation

This paper proposes a theoretical investigation of a novel ANC-frequency estimation (FE) technique designed to stabilize a flexible rotor shaft affected by self-generated sinusoidal disturbances. This work extends the adaptive noise cancellation tracking control (ANC-TC) algorithm presented by D'Amato et al. in 2022 [24,29], generalized to the case of unknown frequency. The disturbance frequency corresponds to the rotor operating angular speed (ω) [20,24,30], which is driven by an external actuator, so that uncertainties may arise in the frequency actual value due to actuation operating point fluctuations. Other incoming nonlinear phenomena, such as gyroscopic moments acting on the disk—for example due to asymmetries of the rotor support [31]-can make the operating frequency vary. The ANC-TC in [24,29] uses such a frequency value as a known parameter, which is also a parameter of the dynamics of the system. With respect to the specific uncavitated case study [29], a theoretical formulation has been provided in [24]: structural proofs for noise suppression estimation, cancellation and system stabilization are given when the disturbance frequency is known but not the initial condition. In this paper, inspired by immersion and invariance (I&I) robust control techniques [18,32,33], the possibility of estimating online the constant unknown frequency of the disturbance is considered. Following I&I-based approaches, the studied controlled system is immersed into a target dynamical model so that its trajectories are an image of the target dynamic system into the immersion map. Then, the aim of this kind of approach is to make the target system image an attractive manifold, so that the controlled system is made globally, uniformly, asymptotically stable around its invariant trajectories [18,32,33].

The approach of [33] was convoluted in the hybrid logic (even if in a more general sense with respect to [32]). The general multi-harmonic disturbance case may be considered [17,33]. An I&I-based strategy, adaptive with respect to perturbation parameters, was recently presented and applied to a reversible cold strip rolling mill to control the speed and tension system [18], characterized by severe coupling effects, multiple state-variables, nonlinearities and model uncertainties. In [18], it was shown that the estimation errors follow an exponential convergence.

A recent research work [17] deals with the problem of active multi-harmonic disturbances cancellation in flexible mechanical structures, under the analysis of passivity properties of a closed-loop system. The controller operates an inner loop, which performs the position control of the system, while the frequency estimation is demanded to an outer control loop, elaborating the position measurements of the plant. A robust linear compensator scheme, non-accurate model-based, was employed and experimentally validated. An exponential cancellation of the disturbance was obtained.

The ANC-TC in [24,29] already showed an intrinsic robustness since the vibrational disturbances acting on the rotor were cancelled. Consequently, the introduction of a frequency identifier for vibrational modes improves the robustness of the method.

The closed-loop sinusoidal noise rejection is conceived in two phases: first, the operating frequency identification is performed as a combination of state-observer and asymptotical parameter estimation [32,33]; second, the asymptotical frequency estimate is fed back to the closed-loop adaptive noise cancellation control (ANC), which hooks the phase of the disturbance by injecting on rotor dynamics a counterphase sinusoid acceleration. The frequency estimation is in practice a warm-up phase of the overall closed-loop control [34,35]. This study proposed a unified theoretical structure including an I&I-based frequency estimator as a plug-in control block. The proposed control architecture design concerns the externalization of the FE process, which appears as an additive standalone block downstream to the ANC closed-loop, elaborating its output data and generating the asymptotical frequency estimate. Once such an asymptotical estimate converges at the steady-state closed-loop, the ANC control exponentially hooks the sinusoidal disturbance, following [24].

Table 1 itemizes and catalogues the scientific literary references that are investigated as background for the proposed study.

Table 1. Literary review.

Authors	Year	Source	Section Content	Authors	Year	Source	Section Content
Poritsky	1953	[8]	PS	Avramov and Borysiuk	2012	[31]	PCS
Stanway and Burrows	1981	[22]	LR	Harika et al.	2013	[12]	PS
Burrows and Sahinkaya	1983	[19]	LR	Marino and Tomei	2016	[27]	LR
Lund	1987	[2]	PS	Carnevale	2016	[33]	PCS
Vance	1988	[30]	PCS	Salazar and Santos	2017	[4]	PS
Muszynska	1988	[6]	PS	Zheng et al.	2017	[23]	LR
D'Agostno et al.	2001	[7]	PS	Ruggiero et al.	2018	[20]	LR, PCS
Zhou and Shi	2001	[21]	LR	Ballnus et al.	2018	[35]	PCS
Hamrock and Schimd	2004	[3]	PS	Das and Guha	2019	[10]	PS
Sukumaran Nair and Prabhakaran	2004	[15]	PS	Bhattacharjee et al.	2019	[11]	PS
Das et al.	2005	[14]	PS	D'Amato et al.	2019	[29]	PCS
Ypma	2006	[36]	PCS	Marko et al.	2020	[26]	LR
Prabhakaran Nair et al.	2007	[16]	PS	Kumar and Tiwari	2020	[28]	LR
Lei and Palazzolo	2008	[25]	LR	Chen et al.	2021	[5]	PS
Carnevale and Astolfi	2008	[32]	PCS	Tripathy and Bhattacharyya	2022	[9]	PS
Friswell et al.	2010	[1]	PS	Liu et al.	2022	[18]	LR, PCS
Hoad et al.	2010	[34]	PCS	Marko et al.	2022	[17]	LR, PCS
Vania et al.	2012	[13]	PS	D'Amato et al.	2022	[24]	PCS

PS: Problem Statement
- flexible rotor with lubricated journal bearings;
- fluid dynamics;
- self-excited vibrations due to fluid dynamics;

LR: Literary Review
- flexible rotor bearings control;
- active rejection of vibrational phenomena;
- notch filtering;
- observer-based control;

PCS: Proposed Control Strategy
- adaptive noise cancellation-rejection;
- immersion and invariance;
- method comparison;
- numerical analysis methods;

A mathematical proposition for the ANC-FE is formulated and a structural proof is given under assumptions on closed-loop output signals, showing that the sinusoidal disturbance rejection is exponentially performed. Numerical simulations are presented to validate the theoretical results in silico. The simulations are designed to illustrate the adaptive rejection performance in terms of time and residual vibration amplitude. Moreover, it is shown that the proposed I&I-based asymptotical frequency estimation may have a practical application since its duration is acceptable as an initialization process. This is a training/warm-up phase for the ANC controller. In addition, in order to evaluate the order of convergence of the disturbance frequency and phase estimation error point series the iterative Inexact Newton method [36] was applied.

The main novelties of this study are summarized:

i   generalization of ANC-TC control devised in [24] to the unknown/uncertain operating frequency;

ii   application of I&I robust technique for frequency estimation (FE), considering as input an analytical reconstruction of the closed-loop output noise signals;

iii   plug-in integration of the FE module with the ANC-TC control;

iv   formulation of a mathematical proposition for the novel ANC-FE control;

v   structural proof of the proposition under assumptions on closed-loop output signals.

This paper has the following outline: Section 2 introduces the closed formulation of the rotor dynamics model and ANC-FE control conceptualization and design; in Section 3 the theoretical proposition is formulated and its proof is given; Section 4 reports numerical simulation results with a detailed description of the data fitting analysis; in Section 5 the numerical simulation results are discussed and some conclusions are reported.

2. Materials and Methods

The dynamical system under investigation is composed by an unbalanced thin disk, symmetrically disposed on a flexible shaft, and supported by two hydrodynamic full (short) journal bearings, as represented in [24]. The mass of the rotor is 2 m and the flexible shaft is a thin bar of a circular section with a negligible mass with respect to the disk mass. According to Figure 1, during the motion, the disk remains in plane motion in x–y plane without gyroscopic effects. The lubricated dynamical system can be studied by applying the classical approach based on the Reynolds equation [20] in cylindrical coordinates and under the classical hypothesis of short bearing approximation [30]. The scalar nonlinear differential motion equations of the system are [20]

$$\begin{cases} m\ddot{x} + k(x_c - x_j) = mu\omega^2 \sin\omega t \\ m\ddot{y} + k(y_c - y_j) = -mu\omega^2 \cos\omega t - W \\ F_x\left(x_j, \dot{x}_J, y_j, \dot{y}_J\right) = -k(x_c - x_j) \\ F_y\left(x_j, \dot{x}_J, y_j, \dot{y}_J\right) = -k(y_c - y_j) \end{cases} \tag{1a}$$

where the forces F_x, F_y are defined in [24].

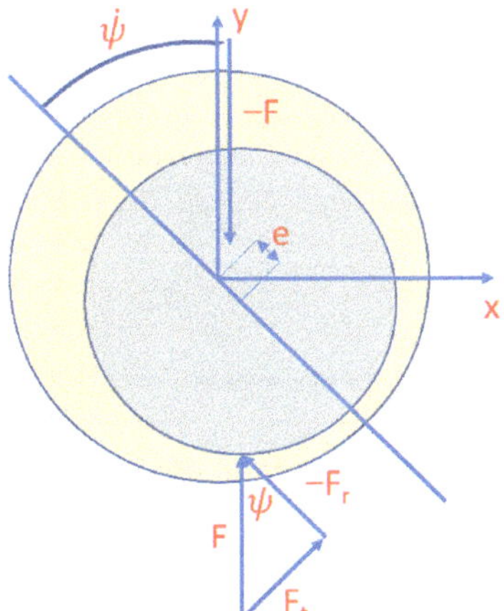

Figure 1. Journal bearing geometry.

2.1. Adaptive Noise Cancellation Tracking Control (ANC-TC)

In the following, for the lubricated dynamical system, the cavitated (π-film) and the uncavitated short bearing model are considered. A closed-form model formulation for the controlled model, which has been introduced in [24], is written as follows

$$\begin{cases} \ddot{x}(t) = \begin{bmatrix} \ddot{x}_c \\ \ddot{y}_c \end{bmatrix} \doteq -\frac{k}{m}(x - z_j) - d_w - \ddot{d}_e(t, \omega) + \vec{u}_x \\ \dot{z}_j(t, \omega, i) = \begin{bmatrix} \dot{x}_j \\ \dot{y}_j \end{bmatrix} \doteq ke^3 l^{-1}(t, z_j, e, i) \left\{ \Sigma(t, z_j, e, \omega, i) \begin{bmatrix} x_j \\ y_j \end{bmatrix} + \frac{1}{\mu_0} \begin{bmatrix} x_c \\ y_c \end{bmatrix} \right\} \end{cases} \tag{1b}$$

where: $x = (x_c, y_c)^T$ and $z_j = (x_j, y_j)^T$ denote the disc center and journal center coordinates, respectively; $d_w = [0, W/m]^T$ is a constant load term and $\vec{u}_x$ is the control input. The vector of sinusoidal disturbances acting on the two coordinate dynamics is denoted by $d_e(t, \omega) = u[\sin(\omega t + \varphi_{01}), -\cos(\omega t + \varphi_{02})]^T$ ($\varphi_{0_{1,2}}$ unknown initial phase), so that

$$\dot{d}_e(t, \omega) = \Omega \cdot d_e(t, \omega) \qquad \Rightarrow \qquad \dot{d}_e(t, \omega) = \begin{bmatrix} 0 & -\omega \\ \omega & 0 \end{bmatrix} \cdot \begin{bmatrix} u\,\sin(\omega t + \varphi_{01}) \\ -u\,\cos(\omega t + \varphi_{02}) \end{bmatrix} \tag{2}$$

The dynamics of acceleration noise in Equation (2) has the canonical parametrization (Ω) of a disturbance with ellipsoidal trajectories, whose dimensions depend on the values of the system parameters, as found by the numerical simulations of the analytical model Equation (1a) presented in [24,29]. The ω is the frequency of the disturbance and corresponds to the operating rotation speed of the flexible rotor. The two matrices l^{-1} and Σ are defined as follows

$$
l^{-1}(t,z_j,e,i)i \cdot \Lambda^{-1}(t,z_j,e,i) = i\left\{\left[\begin{pmatrix} zx_j & -2\zeta\pi^{(i-1)}ey_j \\ zy_j & +2\zeta\pi^{(i-1)}ex_j \end{pmatrix} + (1-i)\cdot 8\lambda\begin{pmatrix} -y_j & ex_j \\ x_j & ey_j \end{pmatrix}\right]\cdot\begin{pmatrix} ex_j & ey_j \\ -y_j & x_j \end{pmatrix}\right\}^{-1}
$$
$$
\Sigma(t,z_j,e,\omega,i)\begin{pmatrix} -\frac{1}{\mu_0} & -\omega\frac{\zeta}{k}\frac{\pi}{i}^{(i-1)} \\ \omega\frac{\zeta}{k}\frac{\pi}{i}^{(i-1)} & -\frac{1}{\mu_0} \end{pmatrix} + (1-i)\cdot\begin{pmatrix} \frac{2\lambda}{k}\omega & \frac{t_0}{ke} \\ -\frac{t_0}{ke} & \frac{2\lambda}{k}\omega \end{pmatrix}
$$

$$(3)$$

where: $e = \sqrt{x_j^2 + y_j^2}$ denotes the orbit eccentricity; the $i \in \{1,2\}$ is the summary index, which allows us to switch the model representation among uncavitated ($i = 1$) and cavitated ($i = 2$) operating conditions, respectively. In [1], it was shown-analytically and by numerical simulations- that the dynamical time-variant matrix Λ is always nonsingular with a high determinant.

To reduce Equation (1a) into Equations (1b)–(3), the following analytical positions are introduced in [24]

$$
z = \frac{2\pi\zeta(1+2\varepsilon^2)}{(1-\varepsilon^2)} \qquad \zeta = \frac{1}{2(1-\varepsilon^2)^{3/2}} \qquad \mu_0 = \mu R\left(\frac{L}{c}\right)^3
$$
$$
\sigma = \frac{e}{(c^2-e^2)^{1/2}} \qquad \lambda = \sigma\zeta
$$
$$
\rho_0 = 2RLp_0 \qquad \varepsilon = \frac{e}{c} \qquad t_0 = \frac{\rho_0}{\mu_0}
$$

$$(4)$$

where: $k, m, \mu, R, L, c, \pi, p_0, u$ are dynamical and geometrical parameters of the system collected in Table 1 [24].

In the case of known operating frequency ω, in [24] the closed-loop ANC-TC algorithm $\vec{u}_x$ in Equations (5)–(7) was proposed for the flexible rotor Equations (1a) and (2), and an exponential tracking of a desired trajectory for the disc center $[\vec{x}_r; \dot{\vec{x}}_r; \ddot{\vec{x}}_r] = [x_{cr}, y_{cr}\,; \dot{x}_{cr}, \dot{y}_{cr}\,; \ddot{x}_{cr}, \ddot{y}_{cr}]$ is obtained (the detailed formal proof is reported in [24]). The ANC-TC includes two main parts: it allows both the sinusoidal disturbance in closed-loop and the springback force contribution to be adaptively canceled due to the always nonzero displacement between disc and journal (eccentricity). The overall vectorial control law is reported in Equations (5)–(7) (the vector symbols are omitted for convenience)

$$
\begin{cases} u_x \triangleq \hat{u}_{x0} + \ddot{\hat{d}}_e(t,\omega) \\ \hat{u}_{x0} \triangleq \frac{k}{m}x + d_w + \hat{v}_x \\ \hat{v}_x \triangleq -\frac{k}{m}\hat{z}_j + \ddot{x}_r - k_1\dot{\tilde{x}} - k_2\tilde{x} \end{cases}
$$

$$(5)$$

$$
\begin{cases}
\ddot{\hat{d}}(t,\omega) \triangleq \omega^2\begin{bmatrix} -usin(\omega t + \hat{\varphi}(t)) \\ ucos(\omega t + \hat{\varphi}(t)) \end{bmatrix} \\
\hat{\varphi}(t,\omega) \triangleq \int_0^t\left(\phi(\tau,\omega)\begin{bmatrix} k_{\varphi_1} & 0 \\ 0 & k_{\varphi_2} \end{bmatrix}\cdot\left(\omega^2\tilde{d}_e(\tau,\omega)\right)\right)d\tau = \left(\phi(\tau,\omega)\begin{bmatrix} k_{\varphi_1} & 0 \\ 0 & k_{\varphi_2} \end{bmatrix}\cdot\left(\ddot{\tilde{x}} + k_1\dot{\tilde{x}} + k_2\tilde{x}\right)\right)d\tau \\
\phi(t,\omega) \triangleq u\begin{bmatrix} cos(\omega t + \phi_0) & 0 \\ 0 & sin(\omega t + \phi_0) \end{bmatrix}
\end{cases}
$$

$$(6)$$

$$
\begin{cases} \hat{z}_j \triangleq -\frac{m}{k}(\hat{\eta} + c_\eta\cdot\dot{x}) \\ \dot{\hat{\eta}} \triangleq -c_\eta\cdot\hat{v}_x + c_\eta\cdot\hat{\eta} + c_\eta^2\cdot\dot{x} \end{cases} \qquad\Rightarrow\qquad \hat{\eta}(0) \doteq 0\ ,\quad \hat{z}_j \triangleq -\frac{m}{k}\cdot c_\eta\cdot\dot{x}(0),
$$

$$(7)$$

where: $\tilde{x} = (x - x_r), \dot{\tilde{x}} = \dot{x} - \dot{x}_r, \ddot{\tilde{x}} = \ddot{x} - \ddot{x}_r$ are the tracking errors of position, velocity and acceleration of disc center, respectively; $\hat{d}_e(t,\omega)$ is the sinusoidal disturbance estimate;

$\hat{\varphi}(t) = [\hat{\varphi}_1, \hat{\varphi}_2]^T$ is the phase estimate vector, which is responsible for matching at steady-state the initial conditions of the noise ($\overline{\varphi_0} = [\varphi_0, \varphi_0]^T = [\varphi_{01}, \varphi_{02}]^T$) in (2), provided that the amplitude (u) and frequency (ω) are known; $\phi_0 = \hat{\varphi}_i(0) = \mathrm{rand}(-2\pi, 2\pi)$ is a random value chosen as the (arbitrary) initial value for the phase estimation. According to [24], $\omega^2 \tilde{d}_e = \ddot{\tilde{x}} + k_1 \dot{\tilde{x}} + k_2 \tilde{x}$ is a known term, which has been modeled in Equation (6) in terms of residual spurious dynamics, under the assumption that the asymptotic control $\hat{u}_{x0}$—linked to the estimation of the displacement Equation (7)—is already converged on target value. Equation (6) is written according to a linear parametrization of the estimation error $\tilde{d}_e(t, \omega) = d_e - \hat{d}_e$ [37]: it appears as product of regressor $\Phi(t, \omega)$ and the parameter estimation error, namely $\tilde{d}_e = \Phi^T(t, \omega) \cdot \tilde{\varphi}(t)$ (see [24]).

The control gains $k_1, k_2, k_{\varphi_1}, k_{\varphi_2}$ and c_η are found by trial-and-error procedures reported in Table 2.

Table 2. Simulation parameters.

Dynamical Parameters	Values	Initial Conditions	Values	Operating Parameters	Values
m (kg)	1.5	$x_c(0)$	0	f (Hz)	500
K (N/m)	4×10^6	$\dot{x}_c(0)$	0	ω (rad/s)	3141
u (m)	10^{-3}	$y_c(0)$	-10^{-5}	μ_0 (m kg/s)	7.06×10^4
R (m)	1.6×10^{-2}	$\dot{y}_c(0)$	0	ρ_0 (N)	1×10^{-3}
L (m)	1.6×10^{-2}	$x_j(0)$	0	t_0 (s^{-1})	1.45×10^{-8}
μ (kg/s)	3.4×10^{-2}	$y_j(0)$	-10^{-5}	p_0 (bar)	2
c (m)	3.16×10^{-5}	φ_0(rad)	[5.2, 5.2]		
g (m/s^2)	9.81				
π	180°				
	Step		5×10^{-6}		
	Time (s)		12		

The phase estimator part of Equation (6) is devised since the main problem in sinusoidal noise rejection is the knowledge of the initial disturbance phase [27,38], so it acts as a dynamic Phase-Locked-Loop (PLL [39]). The displacement estimation part of Equation (7) represents a reduced-order observer of the dynamical subsystem $[\dot{x}_c, \dot{y}_c, x_j, y_j]$ of the overall system Equation (1b). It is required due to technical limitation to install a sensor to measure the center position of the journal. For the same reasons, moreover, even if an analytical value of z_j was considered by Equation (1b), even in this case, the uncertainty on the initial condition $z_{j0} = z_j(0)$ may have a fundamental role in the vibration attenuation at steady-state.

Remark 1. *It is noteworthy that only the disc center position (x), speed ($\dot{x}$) and acceleration ($\ddot{x}$) measurements are required and fed back for control closed-loop adaptation Equation (7), as well as for frequency estimation (Figure 2).* $\lozenge$

2.2. Adaptive Noise Cancellation with Frequency Estimation (ANC-FE) Control

Now, even if the operating frequency was known, which is used as a parameter in the ANC-TC [24], it can be affected by a percentage uncertainty due to possible miscalibration of the actuator, which is keeping the rotor in rotation at ω speed. In practice, this makes the operating frequency, and consequently the disturbance frequency, unknown. Moreover, other incoming nonlinear phenomena, such as gyroscopic moments acting on the disk, for example, due to asymmetries of the rotor support [31], can make the operating frequency vary. For this scope, consider the problem of estimating the operating frequency of the system. The following control scheme is adopted (Figure 2). The scheme recovers the one adopted in [24], where the exosystem Equation (2) (hereafter EXO) acts as an external dynamics. Downstream of the control loop, a further stage is added: the frequency estimator (FE). This module elaborates the analytical output y_{an} of the closed-loop system Equations (1a)–(7) in loop with the frequency online estimate $\hat{\omega}(t)$. Eventually, the FE module feeds back to the control loop the value of estimated frequency ω_{est}, which is used in the ANC-TC module as an input.

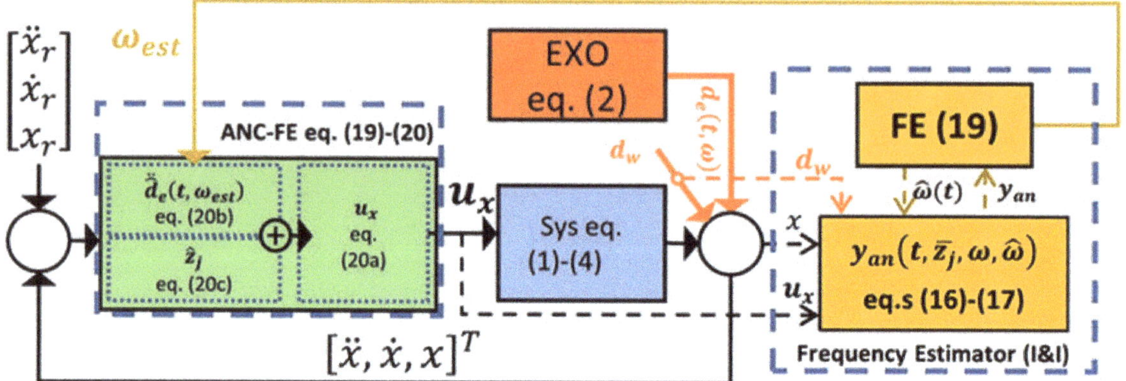

Figure 2. ANC-FE (adaptive noise cancellation with frequency estimation) [1].

2.2.1. ANC-FE Control Design with Measured Output Noise (y)

In the following, a mathematical formulation of the novel adaptive noise cancellation control with frequency estimation (ANC-FE) is given. It is inspired by immersion and invariance (I&I) approach (see [32,33]), which unifies the asymptotic parameter and state estimation problems as reduced-order-observer-based problems (see Remarks 3–5). Let some notations be introduced. From Equation (2)

$$\dot{d}_e(t, \omega) = \Omega \begin{bmatrix} u\sin(\omega t + \varphi_0) \\ -u\cos(\omega t + \varphi_0) \end{bmatrix}$$

$$\begin{bmatrix} \vdots \\ \dot{d}_{ei} \\ \vdots \end{bmatrix}(t, \omega) = \Omega \cdot \begin{bmatrix} \vdots \\ d_{ei} \\ \vdots \end{bmatrix} = \Omega \begin{bmatrix} d_{e1} \\ d_{e2} \end{bmatrix} \quad \Rightarrow \quad y(t, \omega) d_{ei}(t, \omega) \quad \Rightarrow \quad \begin{cases} \dot{y} = \dot{d}_{ei} = \varepsilon \\ \dot{\varepsilon} = -\varsigma y = -\omega^2 y \end{cases} \tag{8}$$

First, the signal y can be regarded as a (measured) scalar output of the closed-loop system Equations (1a)–(7) (Figure 1); $\varsigma = \omega^2$, u and φ_0 are related to the initial states $y(0) = y(t = 0)$ and $\varepsilon(0) = \varepsilon(t = 0)$.

Remark 2. *The following computation is performed for the only scalar variable d_{ei} first since the two components' analyses are analogous and since the two disturbance components are shifted of $\pi/2$, so that the estimation is shared among the two components [24]. For convention, let the first component $y = d_{e1} = u\,sin(\omega t + \varphi_0)$ be used.* ◊

Consistently with the notation used in Equations (5)–(7) (see [24]), by calling $\hat{\varepsilon}$ and $\hat{\varsigma}$ the online state and parameter estimates of the disturbance, respectively, the following injected scalar estimation errors [18] are defined as follows

$$z_\varepsilon(y) = k_\varepsilon y + \left\{ \hat{\varepsilon}\left(k_\varepsilon^2 + \varsigma\right) - \varepsilon \right\} \doteq \beta_\varepsilon(y) + (\alpha_\varepsilon)\,\hat{\varepsilon} - \varepsilon \; ; z_\varsigma(y) = y\,\hat{\varepsilon} + \left\{ \hat{\varsigma} - \varsigma \right\} \doteq \beta_\varsigma(y, \hat{\varepsilon}) + \tilde{\varsigma} \tag{9}$$

where $k_\varepsilon > 0$ is an arbitrary control parameter reported in Table 2. The two errors $[z_\varepsilon, z_\varsigma]$ in Equation (9) differ from the classical error definition $\tilde{\varepsilon} = \hat{\varepsilon} - \varepsilon$ and $\tilde{\varsigma} = \hat{\varsigma} - \varsigma$. They contain two injection terms $[\beta_\epsilon(y), \beta_\varsigma(y, \hat{e})]$ acting as a dynamical correction; $\alpha_\epsilon = (k_\epsilon^2 + \varsigma) > 0$ acts a kind of scaling. The Equation (9) is written as follows

$$\epsilon = -z_\epsilon(y) + \hat{e}\,(\alpha_\epsilon) + \beta_\epsilon(y); \qquad \varsigma = -z_\varsigma(y) + \hat{\varsigma} + \beta_\varsigma(y, \hat{e}). \tag{10}$$

By setting at zero the estimation errors $z_\epsilon(y), z_\varsigma(y)$ in Equations (9) and (10), it follows that

$$\epsilon_{est} = \left(k_\epsilon^2 + \varsigma_{est}\right)\hat{e} + k_\epsilon y; \varsigma_{est} = \hat{\varsigma} + y\,\hat{e}, \tag{11}$$

where, $\epsilon_{est}(0) = \left(k_\epsilon^2 + \varsigma_{est}(0)\right)\hat{e}(0) + k_\epsilon y(0)$ and $\varsigma_{est}(0) = \hat{\varsigma}(0) + y(0)\hat{e}(0)$.

The two asymptotic estimates $[\epsilon_{est}, \varsigma_{est}]$ are found as with a PI (proportional integral) law with respect to the measured output y. Indeed, both $(\alpha_\epsilon)\,\hat{e}$ and $\hat{\varsigma}$ turn out to be an integration process, α_ϵ denotes the gain of the integral part, while both the two injection terms represent the proportional part. To prove this point, derive in time the errors $[z_\epsilon, z_\varsigma](y)$ in Equation (9). From Equation (8) it follows that

$$\dot{z}_\epsilon(y) = (\alpha_\epsilon)\dot{\hat{e}} + \varsigma y + k_\epsilon\,\epsilon; \dot{z}_\varsigma(y) = \dot{\hat{\varsigma}} + \epsilon\,\hat{e} + y\,\dot{\hat{e}}, \tag{12}$$

while, replacing Equation (10) in Equation (12), it follows

$$\dot{z}_\epsilon = \left(k_\epsilon^2 + \varsigma\right)\hat{e} + \varsigma y + k_\epsilon\left(-z_\epsilon + k_\epsilon^2\hat{e} + \hat{e}\,\varsigma + k_\epsilon y\right)$$
$$= -k_\epsilon\,z_\epsilon + \left(k_\epsilon^2 + \varsigma\right)\left\{y + k_\epsilon\,\hat{e} + \dot{\hat{e}}\right\}; \tag{13a}$$

$$\dot{z}_\varsigma = \dot{\hat{\varsigma}} + y\,\dot{\hat{e}} + \hat{e}\left[-z_\epsilon + k_\epsilon^2\hat{e} + k_\epsilon y + \hat{e}\,(-z_\varsigma + \hat{\varsigma} + y\,\hat{e})\right]$$
$$= -\hat{e}\,z_\epsilon - \hat{e}^2 z_\varsigma + \left\{\dot{\hat{\varsigma}} - \left[-y\,\dot{\hat{e}} - k_\epsilon^2\hat{e}^2 - k_\epsilon y\hat{e} - \hat{e}^2\hat{\varsigma} - y\,\hat{e}^3\right]\right\} = -\hat{e}\,z_\epsilon - \hat{e}^2 z_\varsigma + \left\{\dot{\hat{\varsigma}} - \Delta(y, \hat{e}, \hat{\varsigma})\right\}. \tag{13b}$$

The term $\Delta(y, \hat{e}, \hat{\varsigma})$ is a function of known signals. By setting at zero the quantities between braces in both Equations (13a) and (13b), the two scalar adaptation laws for $(\hat{e}, \hat{\varsigma})(t)$ are found as follows

$$\dot{\hat{e}} = -y - k_\epsilon\,\hat{e}, \qquad \dot{\hat{\varsigma}} = -\left[y\,\dot{\hat{e}} + k_\epsilon^2\hat{e}^2 + k_\epsilon y\hat{e} + \hat{e}^2\hat{\varsigma} + y\,\hat{e}^3\right], \tag{14}$$

where, from Equations (6) and (8), $\hat{e}(0) = u\,\omega_0\,\cos(\omega_0\,t + \hat{\varphi}_i(0))$ and $\hat{\varsigma}(0) = \omega_0^2$.

Consequently, Equations (13a) and (13b) become

$$\begin{bmatrix} \dot{z}_\epsilon \\ \dot{z}_\varsigma \end{bmatrix} = \begin{bmatrix} -k_\epsilon & 0 \\ -\hat{e} & -\hat{e}^2 \end{bmatrix} \cdot \begin{bmatrix} z_\epsilon \\ z_\varsigma \end{bmatrix}. \tag{15}$$

The first-row of Equation (15) makes that $z_\epsilon \to 0$ exponentially. However, this is not sufficient to construct an estimation of the state variable ϵ since the overall convergence of the estimation is linked to the convergence of all the error dynamics $[z_\epsilon, z_\varsigma]$. However, once the two adaptation laws have reached their steady-state values, namely $\left[\dot{\hat{e}}, \dot{\hat{\varsigma}}\right] = [0, 0]$, the asymptotic estimates for $[\epsilon_{est}, \varsigma_{est}] = [\epsilon, \omega^2]$ are given by Equation (11) (see the proof in Section 3). Note that ω_0 may be the rough initial estimate of the operating rotor frequency, taken from the rotor actuator datasheet as the steady-state operation value. As mentioned above, the estimates of Equation (11) are two PI, with integral parts $(\alpha_\epsilon)\,\hat{e}$ and $\hat{\varsigma}$, as may be inferred from Equation (14).

2.2.2. ANC-FE Control Design with Analytical Output Noise (y_{an})

Regarding the assumption about y, which in Equations (8)–(14) is regarded as the measured output of the closed-loop system Equations (1a)–(7), it requires a sensor in output able to distinguish and provide only the sinusoidal vibration measurement $(d_e(t, \omega))$. First, it is a logic short-circuit since once the disturbance measurement is available a direct cancellation could be viable. Second, the sensor in turn may be affected by measurement noise, which is source of uncertainty. Then, the ANC-FE control Equations (8)–(14) with measured output noise $(y(t, \omega))$ cannot be directly implemented in closed-loop. For this scope, the value of the output noise y, which is used as a known quantity in the ANC-FE Equations (8)–(14), is reconstructed analytically from Equation (1b). By using position, speed and acceleration measurements of the disc center coordinates $\left(\left[\ddot{x}, \dot{x}, x\right]^T\right)$ and an analytical chasing of the journal center displacement $(\overline{z_j}(\omega_0))$, the analytical output noise (y_{an}) is computed from Equations (1b)–(3) as follows

$$y_{an}\left(t, \overline{z_j}, \hat{\omega}\right) - \left\{ \ddot{x}(t) + \frac{k}{m}\left(x - \overline{z_j}(t, \hat{\omega}(t))\right) + d_w - \vec{u}_x \right\} \tag{16}$$

$$\overline{z_j}(t, \hat{\omega}(t), i)\overline{z_{j0}} + \int_0^t \left[k\overline{e}^3 l^{-1}\left(\tau, \overline{z_j}, \overline{e}, i\right) \left\{ \Sigma\left(\tau, \overline{z_j}, \overline{e}, \hat{\omega}(\tau), i\right) \overline{z_j}(\tau, \hat{\omega}(\tau)) + \frac{1}{\mu_0} x(\tau) \right\} \right] d\tau \tag{17}$$

where: $\overline{z_{j0}}$ is the arbitrary initial condition of the analytical chasing $\overline{z_j}(t, \hat{\omega}(t), i)$ of the journal displacement (z_j); $z_j(t, \omega, i)$ is a function of the frequency through the matrix $\Sigma(t, z_j, e, \omega, i)$ in Equation (3); $\overline{e}$ is the eccentricity computed in the analytical chasing coordinates $(\overline{z_j})$. From Equations (1b) and (16), it follows

$$y_{an}\left(t, \overline{z_j}, \hat{\omega}\right) = \begin{bmatrix} \vdots \\ y_{ani} \\ \vdots \end{bmatrix} \doteq \ddot{d}_e(t, \omega) + \frac{k}{m}\left[-z_j(t, \omega) + \overline{z_j}(t, \hat{\omega}(t))\right] = \ddot{d}_e(t, \omega) + \delta_y\left(t, \overline{z_{j0}}, \omega, \hat{\omega}(t)\right) \equiv y_{an}\left(t, \overline{z_j}, \omega, \hat{\omega}\right) \tag{18}$$

where from (8) $\ddot{d}_{ei}(\omega, t) \propto d_{ei} = y$, so that $y_{ani} \propto \ddot{d}_{ei}(\omega, t)$. The constant proportional factor is the square of unknown frequency $-\omega^2$, which, from (11), is absorbed in the tuning procedure of k_ϵ. Then, y_{an} and the bounded disturbance $d_e(t, \omega)$ are consistent since, eventually, the two sinusoidal vibrations have the same frequency (ω). Indeed, from Equation (18), $y_{an}\left(t, \overline{z_j}, \omega, \hat{\omega}\right)$ is a function of ω. We now state the main assumption in this paper.

Assumption 1. *The analytical chasing error $\delta_y\left(t, \overline{z_{j0}}, \omega, \hat{\omega}(t)\right)$ in Equation (18) is bounded and small enough that $y_{an}\left(t, \overline{z_j}, \hat{\omega}\right)$ is assumed at the same frequency of $\dot{d}_e(t, \omega) = \Omega\, d_e(t, \omega)$.* ▼

From Equations (8) and (16)–(18), according to Remark 2, and under the Assumption 1, the asymptotic adaptation laws Equations (11)–(14) are updated considering the computation for only one component y_{ani}, as follows

$$\epsilon_{est} = \left(k_\epsilon^2 + \varsigma_{est}\right)\hat{e} + k_\epsilon y_{ani};$$

$$\varsigma_{est} = \hat{\varsigma} + y_{ani}\,\hat{e}; \dot{\hat{e}} = -y_{ani} - k_\epsilon\,\hat{e}; \dot{\hat{\varsigma}} = -\left[y_{ani}\,\dot{\hat{e}} + k_\epsilon^2\hat{e}^2 + k_\epsilon y_{ani}\hat{e} + \hat{e}^2\hat{\varsigma} + y_{ani}\hat{e}^3\right],$$

$$\boxed{\omega_{est} = \lim_{t \to \infty} \hat{\omega}(t) = \lim_{t \to \infty} \sqrt{\varsigma_{est}(t)}} \tag{19}$$

with $\quad \epsilon_{est}(0) = \left(k_\epsilon^2 + \varsigma_{est}(0)\right)\hat{e}(0) + k_\epsilon y_{ani}(0); \varsigma_{est}(0) = \hat{\varsigma}(0) + y_{ani}(0)\hat{e}(0);$
$\hat{e}(0) = u\,\omega_0\,\cos(\omega_0\,t + \hat{\varphi}_i(0))\,; \hat{\varsigma}(0) = \omega_0^2.$

Equations (16), (17), and (19), constitute the frequency estimator (FE) equations. Equation (15) holds as function of y_{ani}.

By absorbing Equations (16), (17) and (19) in the ANC-TC Equations (5)–(7), the ANC-FE control, Equations (19)–(20c) with analytical output noise $\left(y_{an}\left(t, \overline{z_j}, \hat{\omega}\right)\right)$, is implementable in closed-loop as follows

$$\begin{cases} \hat{u}_x \triangleq \hat{u}_{x0} + \ddot{\hat{d}}_e(t, \omega_{est}) \\ \hat{u}_{x0} \triangleq \frac{k}{m}x + d_w + \hat{v}_x \\ \hat{v}_x \triangleq -\frac{k}{m}\hat{z}_j + \ddot{x}_r - k_1\dot{\tilde{x}} - k_2\tilde{x} \end{cases} \tag{20a}$$

$$\begin{cases} \ddot{\hat{d}}_e(t,\omega_{est}) \triangleq \omega_{est}^2 \begin{bmatrix} -u & sin(\omega_{est}t + \hat{\varphi}(t)) \\ u & cos(\omega_{est}t + \hat{\varphi}(t)) \end{bmatrix} \\ \hat{\varphi}(t,\omega_{est}) \triangleq \int_0^t \left(\varphi(\tau,\omega_{est}) \begin{bmatrix} k_{\varphi_1} & 0 \\ 0 & k_{\varphi_2} \end{bmatrix} \cdot \left(\ddot{\tilde{x}} + k_1 \dot{\tilde{x}} + k_2 \tilde{x} \right) \right) d\tau \\ \varphi(t,\omega_{est}) \triangleq u \begin{bmatrix} cos(\omega_{est}t + \phi_0) & 0 \\ 0 & sin(\omega_{est}t + \phi_0) \end{bmatrix} \end{cases} \qquad (20b)$$

$$\begin{cases} \hat{z}_j \triangleq -\frac{m}{k}\left(\hat{\eta} + c_\eta \cdot \dot{x} \right) \\ \dot{\hat{\eta}} \triangleq -c_\eta \cdot \hat{v}_x + c_\eta \cdot \hat{\eta} + c_\eta^2 \cdot \dot{x} \end{cases} \quad \Rightarrow \quad \hat{\eta}(0) \doteq 0, \quad \hat{z}_j \triangleq -\frac{m}{k} \cdot c_\eta \cdot \dot{x}(0), \qquad (20c)$$

where, in Equations (20a)–(20c), the asymptotic estimate $\omega_{est} = \hat{\omega}(t)_{t\to\infty}$ (Equation (19)) is considered for the constant ω as the constant frequency parameter in the phase estimator (compare Equations (5)–(7) with Equations (20a)–(20c).

At steady-state, when all the estimates' terms are converged (i.e., $\hat{v}_x \to v_x$, $\hat{z}_j \to z_j$, $\hat{d}_e \to d_e$), the closed-loop system, Equations (1b)–(4) and (19)–(20c), is regulated by $\hat{u}_x \to u_x$ around the reference x_r (Figure 2).

Remark 3. *The idea of frequency estimation, namely Equations (8)–(15), inspired from immersion and invariance (I&I) approach, are borrowed from the algorithm proposed in [32]. Nevertheless, the novelty of applying this technique to a rotordynamic model Equation (1a) [24,30] consists of a further manipulation. Equations (16)–(19) have been introduced with the aim of reconstructing analytically (y_{an}) an unavailable signal, the output noise (y)). The signal chasing Equation (17) is possible by virtue of the closed-form formulation, Equations (1b)–(4), presented in [24], which allows a reliable numerical extraction of the coordinates of the journal disc center to be performed, with a very high determinant value of the matrix Λ, by integration of Equation (1b).*

In particular, the system Equation (8) of dimension 3 $[\varsigma,\epsilon,y_{ani}]$ is immersed in a wider space with 5 dimensions $[\varsigma,\hat{\varsigma},\epsilon,\hat{\epsilon},y_{ani}]$. Then, the observer problem is attracted by the surface at $z_\epsilon = 0$ and $z_\varsigma = 0$, where it remains at steady-state, so that the resultant invariant locus is a manifold (Figure 3). ◊

Remark 4. *It is worth highlighting that Equation (20c) is identical to Equation (7), namely, the reduced-order-observer, which estimates the z_j displacement, acts independently from the other control blocks. The FE controller designed in this paper (Equations (16)–(19)) based on the analytical output y_{an} results as a plug-in estimation block downstream to the ANC-TC Equations (5)–(7) proposed in [24]. This property shall be exploited for control convergence proof.* ◊

Remark 5. *The control ANC-FE Equations (19) and (20a)–(20c) are a general-purpose control with respect to the actual lubricated rotordynamics' operating condition, namely cavitated/uncavitated. Indeed, according to the notations in Equations (1b)–(4), the matrices $l^{-1}\left(t,\bar{z}_j,\bar{e},i\right)$ and $\Sigma\left(t,\bar{z}_j,\bar{e},\hat{\omega}(t),i\right)$ in Equation (17) are functions of the parameter i.* ◊

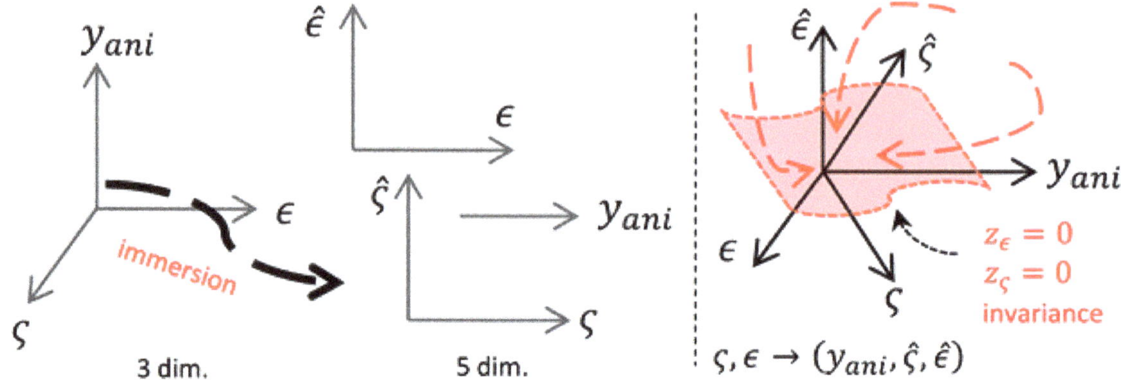

Figure 3. Immersion and Invariance application on analytical output noise (y_{ani}).

3. Closed-Loop ANC-FE Control Convergence Proof

3.1. Proposition Statement

In this section a formal convergence proof of the proposed closed-loop ANC-FE control Equations (19) and (20a)–(20c) is provided through the following proposition statement.

Proposition 1. *Consider the lubricated rotordynamic model Equations (1b)–(4) and $y = d_{ei}(t, \omega)$ Equation (8) as the sinusoidal output noise from the exosystem (2), with unknown constant frequency (ω) and initial phase (φ_0). Consider the reference dynamics $\vec{x}_r = [x_{cr}, y_{cr}]$ for the disc center coordinates, with $\tilde{x} = (x - x_r)$.*

From Equations (9) and (16)–(20c), $\tilde{\varphi} = (\overline{\varphi_0} - \hat{\varphi}(t))$, $\tilde{z}_j = (z_j - \hat{z}_j)$, $z_{\epsilon}(t, y_{ani}) = (-\epsilon_{est} + (k_{\epsilon}^2 + \varsigma_{est})\hat{\epsilon} + k_{\epsilon} y_{ani})$ and $z_{\varsigma}(t, y_{ani}) = (-\hat{\omega}^2(t) + \hat{\varsigma} + y_{ani}\,\hat{\epsilon})$.

Under the Assumption 1, the $\hat{u}_x$ control in Equations (20a)–(20c) makes that for the closed-loop system Equations (1b)–(4) and (19)–(20c):

i. *the equilibrium point $(z_{\epsilon}(y_{ani}), \hat{\epsilon}(y_{ani})z_{\varsigma}(y_{ani})) = (0,0)$ is asymptotically stable;*

ii. *$\epsilon_{est} = (k_{\epsilon}^2 + \varsigma_{est})\hat{\epsilon} + k_{\epsilon} y_{ani}$. is an asymptotic estimate of the analytical output noise derivative $\dot{y}_{ani}$ and $\omega_{est} = \lim_{t \to \infty} \hat{\omega}(t) = \sqrt{\hat{\varsigma} + y_{ani}\,\hat{\epsilon}}$ is an asymptotic estimate of the rotor operating frequency ω; $\square\square$*

iii. *$(\tilde{\varphi}, \tilde{z}_j, \tilde{x}) = (0,0,0)$ is exponentially stable (thesis from [24]). $\square\square\square$*

3.2. Proof

Let $V_1(t, z_{\epsilon}, z_{\varsigma})$ be a suitable Lyapunov function

$$V_1(t, z_{\epsilon}, z_{\varsigma}) = \frac{z_{\epsilon}^2}{2\,k_{\epsilon}} + \frac{z_{\varsigma}^2}{2} \geq 0. \tag{21}$$

Deriving in time Equation (21), from Equation (15)

$$\dot{V}_1(t, z_{\epsilon}, z_{\varsigma}) = -z_{\epsilon}^2 - \hat{\epsilon} z_{\varsigma} z_{\epsilon} - \hat{\epsilon}^2 z_{\varsigma}^2 = -\begin{bmatrix} z_{\epsilon} & \hat{\epsilon} z_{\varsigma} \end{bmatrix} \begin{bmatrix} 1 & \frac{1}{2} \\ \frac{1}{2} & 1 \end{bmatrix} \begin{bmatrix} z_{\epsilon} \\ \hat{\epsilon} z_{\varsigma} \end{bmatrix} \leq 0, \quad \forall (z_{\epsilon}, \hat{\epsilon} z_{\varsigma}). \tag{22}$$

Due to $\dot{V}_1 \leq 0$, $V_1(t, z_{\epsilon}, z_{\varsigma})$ is always nonincreasing in $t \in (0, \infty)$, so that V_1 is bounded and, consequently, $(z_{\epsilon}, z_{\varsigma})$ are bounded. By deriving in time $\dot{V}_1$ Equation (22) and from Equations (15) and (19), it follows

$$\begin{aligned} \ddot{V}_1(t, z_{\epsilon}, z_{\varsigma}) &= -2z_{\epsilon}\dot{z}_{\epsilon} - \dot{\hat{\epsilon}} z_{\varsigma} z_{\epsilon} - \hat{\epsilon}\dot{z}_{\varsigma} z_{\epsilon} - \hat{\epsilon} z_{\varsigma}\dot{z}_{\epsilon} - (\hat{\epsilon}^2) z_{\varsigma}^2 - \hat{\epsilon}^2\left(\dot{z}_{\varsigma}^2\right) \\ &= 2k_{\epsilon} z_{\epsilon}^2 - z_{\varsigma} z_{\epsilon}(-y_{ani} - k_{\epsilon}\,\hat{\epsilon}) - \hat{\epsilon} z_{\epsilon}\left(-\hat{\epsilon} z_{\epsilon} - \hat{\epsilon}^2 z_{\varsigma}\right) + k_{\epsilon}\hat{\epsilon} z_{\varsigma} z_{\epsilon} \\ &\quad - 2\hat{\epsilon} z_{\varsigma}^2(-y_{ani} - k_{\epsilon}\,\hat{\epsilon}) - 2\hat{\epsilon}^2 z_{\varsigma}\left(-\hat{\epsilon} z_{\epsilon} - \hat{\epsilon}^2 z_{\varsigma}\right). \end{aligned} \tag{23}$$

The $\ddot{V}_1$ is bounded since under the Assumption 1 the analytical output noise y_{ani} is bounded and the estimate $\hat{e}(s) = [-y_{ani}(s)/(s + k_\epsilon)]$ is a passive filter of a bounded signal, so it is bounded and also its derivative $\dot{\hat{e}} = (-y_{ani} - k_\epsilon\,\hat{e}\,)$ is bounded in turn. Then, B.2.1 Barbalat's Lemma [40] applies on $\dot{V}_1$, so that $\lim\limits_{t \to \infty} \dot{V}_1(t) = 0$. Consequently, $(z_e, \hat{e}z_\varsigma) \to (0,0)$ globally asymptotically. Now, according to [32] and from Equations (10), (13) and (19), $\epsilon_{est} = (k_\epsilon^2 + \varsigma_{est})\hat{e} + k_\epsilon y_{ani}$ is an asymptotic estimate of analytical output noise derivative $\dot{y}_{ani}$, while $\omega_{est} = \hat{\omega}(t)_{t \to \infty} = \sqrt{\hat{\varsigma} + y_{ani}\,\hat{e}}$ is an asymptotic estimate of the rotor operating frequency ω. This proves the theses (i)–(ii).

The rest of the proof is inherited from [24].

Let $V_2(t, \widetilde{\varphi})$ and $V_3(t, \widetilde{\eta})$ be introduced as two suitable Lyapunov functions

$$V_2(t, \widetilde{\varphi}) = \frac{1}{2}\,\widetilde{\varphi}^T \cdot \widetilde{\varphi} \geq 0, \tag{24}$$

$$V_3(t, \widetilde{\eta}) = \frac{1}{2}\,\widetilde{\eta}^T \cdot \widetilde{\eta} \geq 0. \tag{25}$$

Deriving $V_2(t)$ Equation (24) in time, from Equations (6) and (20b) it follows

$$\dot{V}_2(t) = \widetilde{\varphi}^T \dot{\widetilde{\varphi}} = -\widetilde{\varphi}^T \Phi(t, \omega_{est}) \begin{bmatrix} k_{\varphi 1} & 0 \\ 0 & k_{\varphi 2} \end{bmatrix} \left(\omega^2 \widetilde{d}_e(t, \omega_{est})\right) = -\widetilde{d}_e^T \left[\left(\omega^2 k_\varphi\right) I_{(2\times 2)}\right] \widetilde{d}_e \leq 0, \tag{26}$$

where: $k_{\varphi 1} = k_{\varphi 2} = k_\varphi > 0$; $I_{(2\times 2)}$ is the 2-by-2 identity matrix; the linear parametrization of the estimation error in Equation (6) has been considered, i.e., $\widetilde{d}_e(t, \omega_{est}) = \Phi^T(t, \omega_{est})\widetilde{\varphi}(t, \omega_{est})$ (where the ω_{est} has been considered constant at steady-state so that it is replaced for the actual constant ω). Due to $\dot{V}_2 \leq 0$, $V_2(t)$ does not increase in $t \in (0, \infty)$. Then, V_2 is bounded and so $\widetilde{\varphi}$ is bounded. By time-deriving V_2 again, under the assumption of bounded $\left(\Phi, \dot{\Phi}\right)$, the Barbalat's Lemma in [40] applies on $\dot{V}_2$. Then, $\lim\limits_{t \to \infty} \dot{V}_2(t) = 0$ and $\widetilde{d}_e \to 0$; consequently, $\widetilde{\varphi} \to 0$ asymptotically. To prove the exponential convergence of $\widetilde{\varphi}(t, \omega_{est})$ estimation error the Persistency of Excitation condition [40] is required, provided that $\exists\, T,\, k_T \in \mathbb{R}^+$ such that

$$\int_t^{t+T} \Phi(\tau, \omega_{est})\Phi^T(\tau, \omega_{est})\, d\tau \geq k_T I_{(2\times 2)} > 0\,, \forall t \geq 0\,, \tag{27}$$

which is always trivially fulfilled, given the regressor definition in Equations (6) and (20b). Now, consider, from Equations (7) and (20c), the following definitions of $\widetilde{\eta}, \dot{\widetilde{\eta}}, \widetilde{v}_x, \widetilde{u}_x$

$$\begin{aligned} -\frac{k}{m}(z_j - \hat{z}_j) &= (c_\eta \dot{x} + \eta) - (c_\eta \dot{x} + \hat{\eta})\widetilde{\eta} \\ \dot{\eta} - \dot{\hat{\eta}} &= (-cv_x + c\eta) - (-c\hat{v}_x + c\hat{\eta}) = -c\widetilde{v}_x + c\widetilde{\eta}\widetilde{\eta}\,, \\ v_x - \hat{v}_x &= (u_x - \hat{u}_x) + \omega^2\,\widetilde{d}_e = \widetilde{u}_x + \omega^2\,\widetilde{d}_e\widetilde{v}_x \\ &\quad -\frac{k}{m}\widetilde{z}_j - \omega^2\,\widetilde{d}_e\widetilde{u}_x \end{aligned} \tag{28}$$

with $c_\eta < 0$ (see [24] for detailed derivation of Equation (28)). Deriving $V_3(t)$ Equation (25) in time, from (28) it follows

$$\begin{aligned} \dot{V}_3(t) &= \widetilde{\eta}^T \cdot \dot{\widetilde{\eta}} = c \cdot \widetilde{\eta}^T\widetilde{\eta} + (-c) \cdot \widetilde{\eta}^T \left(\widetilde{u}_x + \omega^2\widetilde{d}_e\right) \leq \frac{c}{2} \cdot \|\widetilde{\eta}\|^2 + \left(-\frac{c}{2}\right)\left(\|\widetilde{u}_x\| + \|\omega^2\widetilde{d}_e\|\right)^2 \\ &= \frac{c}{2} \cdot \|\widetilde{\eta}\|^2 + \left(-\frac{c}{2}\right)\|\omega^2\widetilde{d}_e\|^2 + \left(-\frac{c}{2}\right)\| -\frac{k}{m}\widetilde{z}_j - \omega^2\,\widetilde{d}_e\|^2 + (-c)\| -\frac{k}{m}\widetilde{z}_j - \omega^2\,\widetilde{d}_e\|\|\omega^2\widetilde{d}_e\| \\ &\leq \frac{c}{2} \cdot \|\widetilde{\eta}\|^2 + \left(-\frac{c}{2}\right)\|\omega^2\widetilde{d}_e\|^2\left(-\frac{c}{2}\right)\left(\| -\frac{k}{m}\widetilde{z}_j\| + \|\omega^2\,\widetilde{d}_e\|\right)\left(\| -\frac{k}{m}\widetilde{z}_j\| + \|\omega^2\,\widetilde{d}_e\| + 2\|\omega^2\widetilde{d}_e\|\right) \\ &= -2c\,\|\omega^2\widetilde{d}_e\|^2 - 2c\|\widetilde{\eta}\|\|\omega^2\,\widetilde{d}_e\| \leq c\|\widetilde{\eta}\|^2 - c\,\omega^4\|\widetilde{d}_e\|^2\,. \end{aligned} \tag{29}$$

Equation (29) is obtained considering the triangular inequalities $\widetilde{\eta}^T\left(\widetilde{u}_x + \omega^2\widetilde{d}_e\right) \leq \frac{1}{2}\left(\|\widetilde{\eta}\|^2 + \|\widetilde{u}_x + \omega^2\widetilde{d}_e\|^2\right)$ and $\|\left(-\frac{k}{m}\widetilde{z}_j\right) + (-\omega^2\,\widetilde{d}_e)\|^2 \leq \left(\| -\frac{k}{m}\widetilde{z}_j\| + \| -\omega^2\,\widetilde{d}_e\|\right)^2$; then, the square of $\zeta(t) \triangleq \left(\|\tilde{\eta}\| + \|\omega^2 \tilde{d}_e\| \right)$ has been reconstructed, so that the inequalities $\|\tilde{\eta}\| \|\omega^2 \tilde{d}_e\| \leq -\frac{1}{2}\left(\|\tilde{\eta}\|^2 + \|\omega^2 \tilde{d}_e\|^2 \right)$ have been used.

Following this, $\tilde{d}_e(t)$ is bounded on $[0, \infty)$ so that, according to Lemma A.1 in [41], any globally exponentially convergent observer for the modal disturbance $\tilde{d}_e(t)$ guarantees the exponential convergence for $\tilde{\eta} \to 0$ and, consequently, for $\tilde{z}_j \to 0$ and $\tilde{u}_x = (u_x - \hat{u}_x) \to 0$ (see Equation (28)). This is enforced by the global exponential zero convergence of $\tilde{d}_e(t)$, as has been proved in Equations (26) and (27).

Eventually, by replacing in Equation (1a) the control $\hat{u}_x$ Equation (20a) ($\hat{u}_x$ for u_x), it follows that for $\tilde{u}_x \to 0$

$$\ddot{\tilde{x}} = -k_1 \dot{\tilde{x}} - k_2 \tilde{x} + \frac{k}{m}\tilde{z}_j - \ddot{\tilde{d}}_e = -k_1 \dot{\tilde{x}} - k_2 \tilde{x} - \tilde{u}_x = -k_1 \dot{\tilde{x}} - k_2 \tilde{x} \tag{30}$$

is a Hurwitz system dynamics. Hence, also $\tilde{x} \to 0$ globally exponentially. This proves the thesis (*iii*). ∎∎∎

Remark 6. *It has been proved that, under the Assumption 1, the convergence of asymptotical frequency estimate FE in Equation (19)—with its additional inner loop $(\hat{\omega}(t), y_{an}\left(t, \overline{\tilde{z}}_j, \omega, \hat{\omega}\right))$) in which the FE is elaborated—is standalone with respect to the rest of the control loop: it is a plug-in estimation block, according to Remark 2. In this case, once $\omega_{est} \to \omega$, at steady-state ω_{est} is considered as an internal parameter for the control loop. Then, the frequency estimation phase is a warm-up process for the ANC-TC Equations (5)–(7) [34,35], which through Equation (19) is transformed into Equations (20a)–(20c).* ◊

Remark 7. *From [24,32], as aforementioned at the beginning of Section 2.2.1, the closed-loop system from the input disturbance $d_e(t)$ to the output $x(t)$, including the phase estimator Equation (20b), the journal displacement observer Equation (20c) and the asymptotic frequency estimator Equation (19), results in a regulation loop unifying the parameter and state estimation problems as reduced-order observed-based problems, with adaptive notch filtering characteristics [24].* ◊

4. Results

4.1. Numerical Simulation Setup and Method Description

The proposed control is implemented in the Simulink MATLAB environment with the parameter set as shown in Table 2 and with the control parameters reported in Table 3. The results are presented in four simulation cases for cavitated and uncavitated operating conditions with measured and analytical output noise.

The subscript "*i*" in the notation of the analytical output noise y_{ani} is omitted for editing convenience of the graphs. Hence, y_{an} in all figures is referred to as the only one component y_{ani}, accordingly to Equation (16) and Remark 2.

The signal $\ddot{d}_{ei}(t, \omega)$ and z_j in the figures represent the actual sinusoidal disturbance and the journal center vector coordinates, respectively, whose values are computed from the simulated model differential equations. These signals, $(\ddot{d}_{ei}(t, \omega), z_j)$ would not be available in practice, so their analytical chasing in simulation is reported only for the sake of comparison with respect to the analytical signals.

Table 3. Control Parameters.

Control Gains	Values	Initial Conditions	Values
$c_\eta\,[\mathrm{s}^{-1}]$	-400	$\hat{x}_j(0)$	0
$k_1,\,k_2$	2000	$\hat{y}_j(0)$	0
k_{φ_1}	50.66	$\hat{\varphi}(0)$ (rad)	[4, 4]
k_{φ_2}	50.66	ω_0 (rad/s)	2985
k_ϵ	7.5	$\overline{z_{j0}}$ (m)	$[0,\, -2 \times 10^{-5}]$

Bias and initialization issues can mislead the estimated response measure of performance obtained from a simulation. Several techniques surveyed in [34] can be used to estimate online the length of the warm-up phase in the output data collection of a simulation model, to understand, automatically, when this preliminary phase ends and when the actual numerical simulation starts giving a reliable response. The same issue is tackled in Bayesian parameter estimation [35] where the parameters of the model under study are usually derived from the available data, using optimization and sampling method estimations. It is shown that a warm-up phase occurs (the asymptotical frequency estimation phase) before the closed-loop ANC control can start its exponential hooking of the exosystem output disturbance signal. In practice, such a length cannot be quantified, but for the proposed technique, the warm-up phase length estimation is not required.

The simulations are designed in order to validate numerically the theoretical analytical proof reported in the previous Section 3. In practice, the settling time and trajectory behavior of the phase estimate can be forecast since its trend is exponential, properly, while this does not occur for the frequency estimate, which converges asymptotically. Hence, to highlight the differences in the convergence behaviors of the estimation errors, a further numerical evidence is provided. The Inexact Newton method [36] is applied to the frequency estimation error and phase estimation error successions, while the convergence of the overall disturbance error succession is considered only in the time since it follows from the previous two errors. The iterative Newton method is applied to the error successions in both cavitated and uncavitated conditions, and only in the case of analytical output noise, which is more interesting in practice.

To analyze and evaluate the order of the convergence of the estimation errors, the following point series are defined as the difference between the estimation dynamics terms and the constant target values of each series (ω and φ_0). Denoting the successions of frequency and phase estimation with $\left\{\omega_{est,yan_n}\right\}_n$ and $\left\{\hat{\varphi}_{-,yan_n}\right\}_n$, respectively, sampled at discrete time instants n (1 sample each 5 (ms)), the following

$$e_{\omega_n} = \left\{\omega_{est,yan_n}\right\}_n - \omega \quad e_{\varphi_n} = \left\{\hat{\varphi}_{-,yan_n}\right\}_n - \varphi_0 \tag{31}$$

represent the successions of the frequency and phase estimation errors, respectively. Considering the absolute value of the ratio between the $\{n+1\}$-element of each error succession and the $\{n\}$-element power of $p_{\omega,\varphi}$, as long as the limits of these ratios are constant for $n \to \infty$, namely

$$\lim_{n\to\infty} \frac{\left|\omega_{est,yan_{n+1}} - \omega\right|}{\left|\omega_{est,yan_n} - \omega\right|^{p_\omega}} = constant \quad \lim_{n\to\infty} \frac{\left|\hat{\varphi}_{-,yan_{n+1}} - \varphi_0\right|}{\left|\hat{\varphi}_{-,yan_n} - \varphi_0\right|^{p_\varphi}} = constant \tag{32}$$

then, each succession converges with at least a polynomial behavior.

We can rewrite Equation (32) as follows

$$\left|\omega_{est,yan_{n+1}} - \omega\right| \le k_\omega \cdot \left|\omega_{est,yan_n} - \omega\right|^{p_\omega} \quad \left|\hat{\varphi}_{-,yan_{n+1}} - \varphi_0\right| \le k_\varphi \cdot \left|\hat{\varphi}_{-,yan_n} - \varphi_0\right|^{p_\varphi} \tag{33}$$

parametrized by the coefficient pair $(k_{\omega,\varphi}, p_{\omega,\varphi})$. In Equation (33): $p_{\omega,\varphi}$ represents the order of convergence for each error succession, Equations (31) and (32), while $k_{\omega,\varphi}$ is a proportional factor, denoting the exponential convergence time-constant (when $p_{\omega,\varphi} = 1$).

The following numerical simulations are proposed to: (1a) evaluate the convergence performance of disturbance estimation errors in time; (2) verify that the initial frequency estimation warm-up phase length is acceptable in practice in a real application (namely the asymptotical convergence occurs at $t < \infty$). In all four simulation scenarios, 5% of initial frequency estimation error and about 25% (about 70 deg) of initial phase estimation error are considered.

4.2. Figure Descriptions

Figures 4 and 5 illustrate the frequency estimation performance in time, in the case of the cavitated condition (Figure 4) and in the case of the uncavitated condition (Figure 5). Three curves are plotted: ω denotes the real constant operating frequency; $\omega_{est,yan}$ denotes the asymptotical frequency estimate with the analytical output noise; $\omega_{est,y}$ denotes the asymptotical frequency estimate in the case of measured output noise.

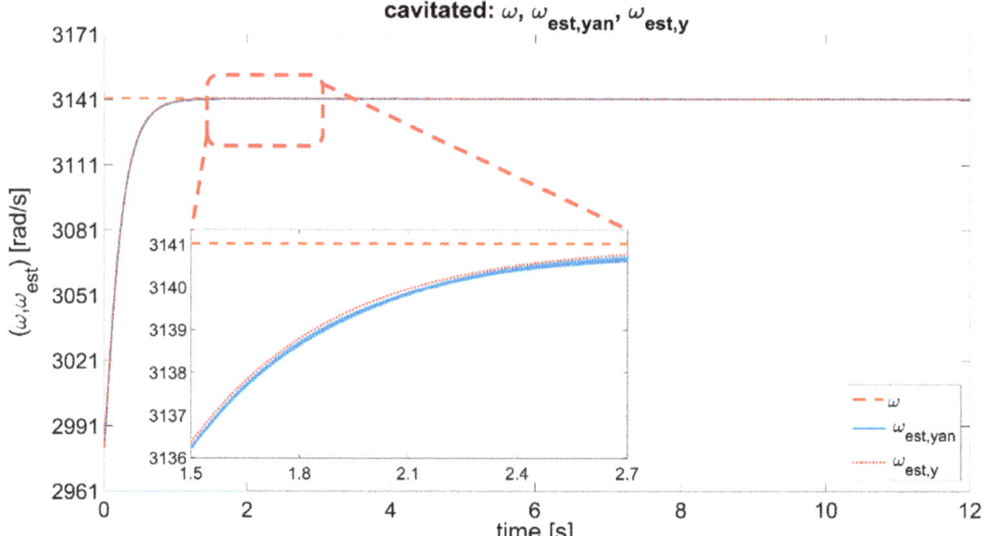

Figure 4. Frequency estimation performance in the case of cavitated condition.

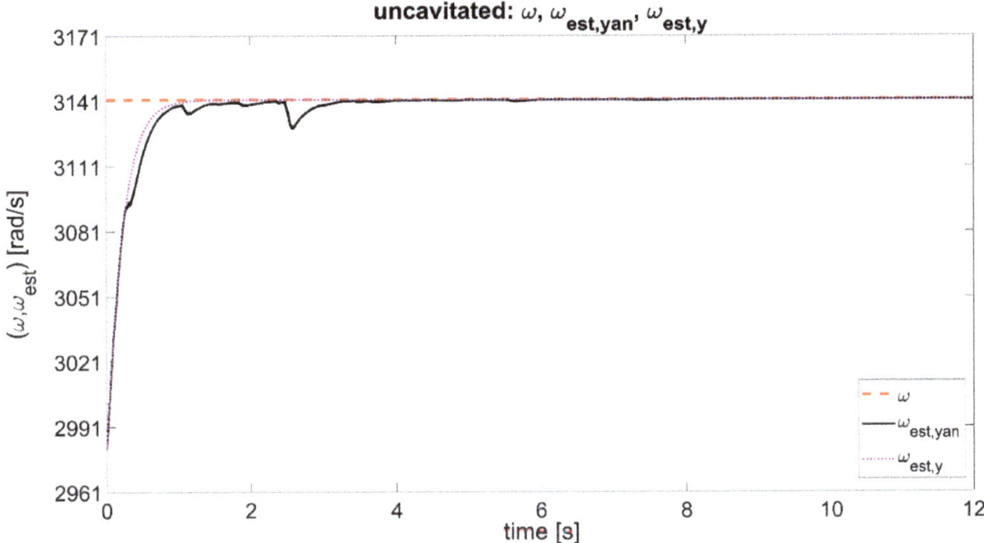

Figure 5. Frequency estimation performance in the case of uncavitated condition.

In Figure 6 the frequency estimation performance in time, considering the analytical output noise y_{an}, is compared in uncavitated and cavitated conditions.

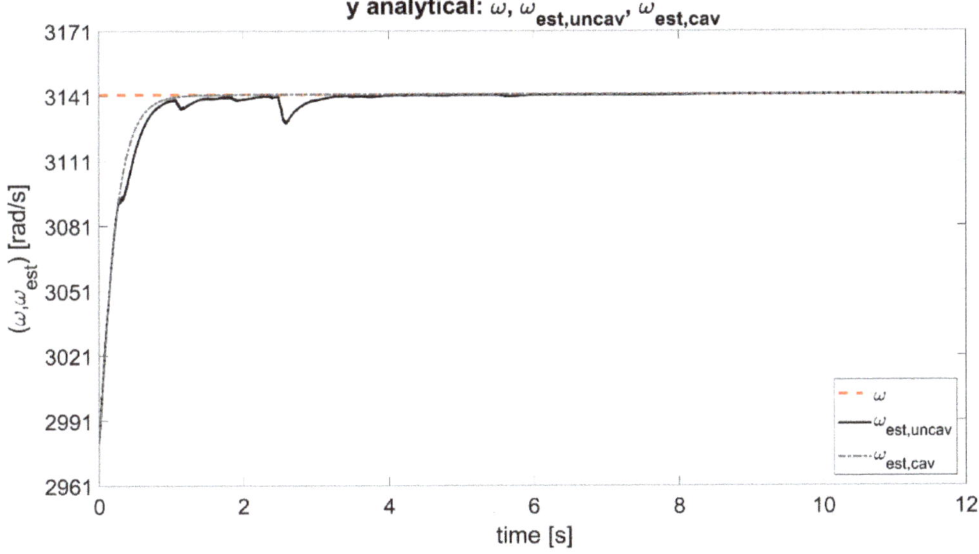

Figure 6. Comparison of frequency estimation performance in uncavitated and cavitated conditions with the analytical output noise y_{an}.

Figure 7 shows the comparison (in offset view) of phase estimation performance in time, in the four simulation cases. The $(\hat{\varphi}_{unc,yan}, \hat{\varphi}_{unc,y})$ represent the phase estimate for the uncavitated case, with analytical and measured output noise, respectively; the $(\hat{\varphi}_{cav,yan}, \hat{\varphi}_{cav,y})$ represent the phase estimate for the cavitated case, with analytical and measured output noise, respectively. In the graph, the warm-up phase edges, and time interval labels, are marked-up in the four cases for illustration purposes.

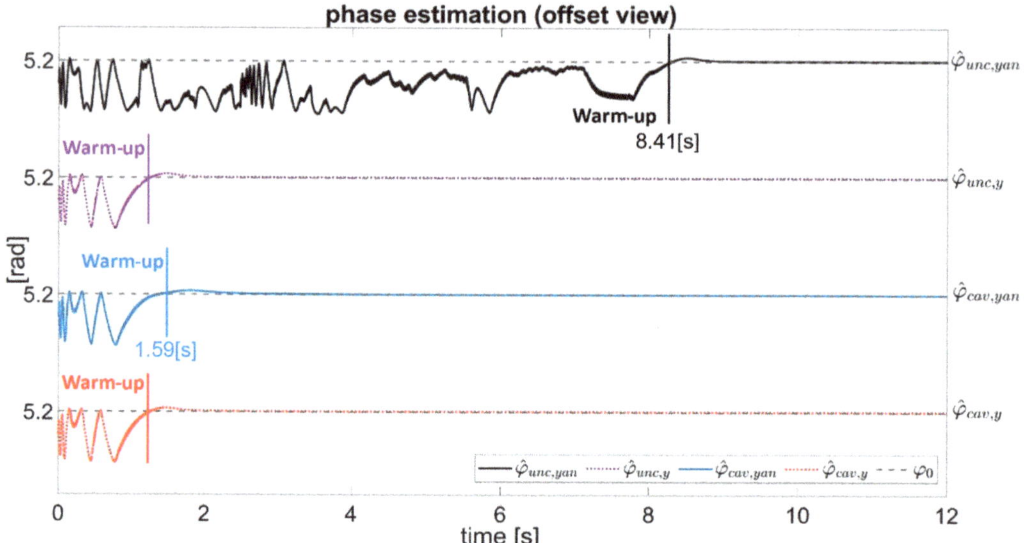

Figure 7. Offset view of phase estimation in the four simulation cases, with warm-up phase edges.

Figure 8 shows the scatter plot of the (k, p)-parametrized successions of frequency estimation error $\left|e_{\omega_{n+1}}\right| = k_\omega \left|e_{\omega_n}\right|^{p_\omega}$ (Figure 8A,B) and phase estimation error $\left|e_{\varphi_{n+1}}\right| = k_\varphi \left|e_{\varphi_n}\right|^{p_\varphi}$ (Figure 8C,D), following Equations (31)–(33). Both cavitated and uncavitated conditions with analytical output noise y_{an} are considered. The timeseries in Figure 8A,B are plotted considering the overall simulation time data. To evaluate whether the phase estimates' convergent behavior is exponential starting from the warm-up phase end (according to Proposition 1), the timeseries in Figure 8C,D are plotted on the two restricted simulation time intervals starting from 8.41 (s) and 1.59 (s) for uncavitated and cavitated cases, respectively, as shown in Figure 7.

The text boxes in Figure 8 report the fitting data statistics obtained applying the iterative Inexact Newton method on the frequency and phase estimation error timeseries, in cavitated and uncavitated conditions with analytical output noise. Such statistics are used to evaluate the order of the convergence $p_{\omega,\varphi}$ in the four cases. The arrows indicate the direction of the succession point distribution from $e_{\omega_{n=0}}$ to $e_{\omega_{n\to\infty}}$, and from $e_{\varphi_{n=0}}$ to $e_{\varphi_{n\to\infty}}$, as defined in Equation (31).

Figure 9 shows the difference between the scalar analytical noise y_{an} and the actual disturbance component $\ddot{d}_{ei}(t, \omega)$ for the uncavitated case (Figure 9A) and for the cavitated case (Figure 9C). For ease of graph interpretation, due to the high frequency of the signals, a restricted time lapse (0–0.0012) (s) was chosen, arbitrarily, on the overall simulation time of 12 (s), in order to highlight the detail of the comparison over a few periods.

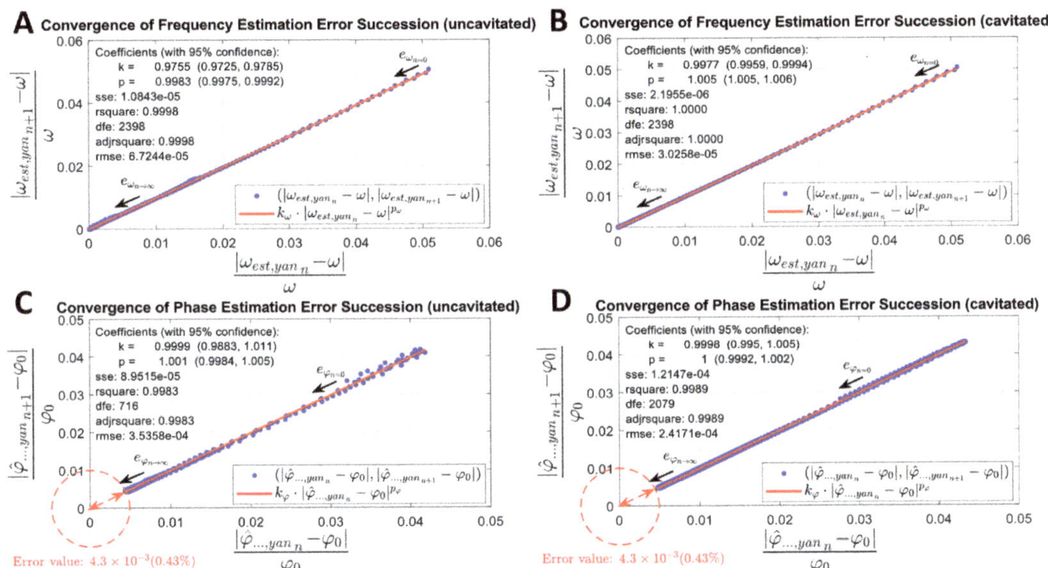

Figure 8. Convergence of frequency estimation error timeseries (**A**,**B**) and phase estimation error timeseries (**C**,**D**) for uncavitated and cavitated condition, with analytical output noise.

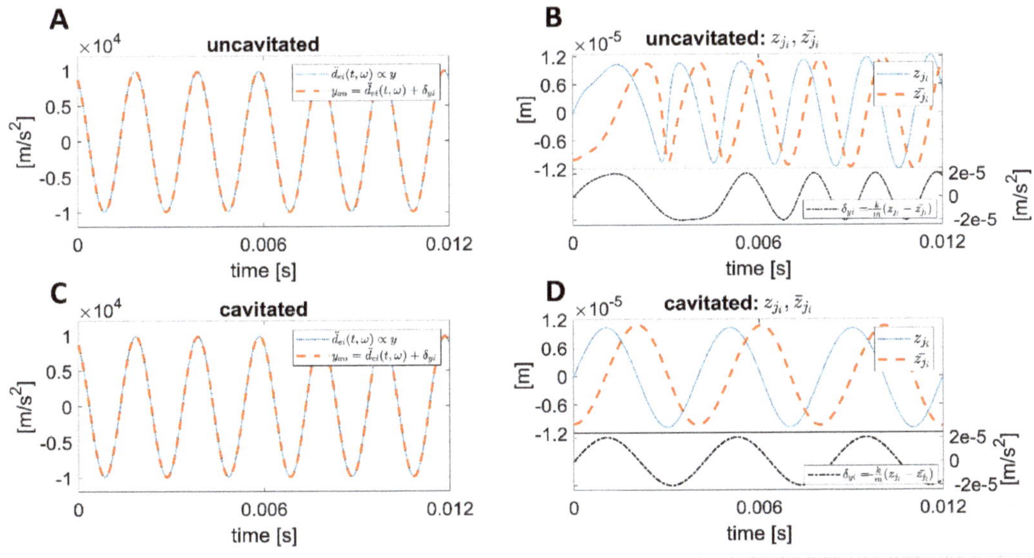

Figure 9. Scalar analytical output noise y_{an} and actual disturbance component $\ddot{d}_{ei}(t,\omega)$ for uncavitated and cavitated conditions (**A**,**C**); "i" component of actual journal displacement z_{j_i} and its analytically reconstruction $\overline{z}_{J_i}$ with the scaled difference δ_{yi} for uncavitated and cavitated conditions (**B**,**D**).

In Figure 9B,D, the behavior of only one component of the journal displacement coordinate vector, z_{j_i}, is shown in time. It is compared with the same component "i" of its analytically reconstructed value $\overline{z}_{j_i}$. Both Figure 9B,D contain a subplot showing the scaled difference $\delta_{yi} = -\frac{k}{m}\left(z_{j_i} - \overline{z}_{j_i}\right)$, where δ_{yi} is the acceleration injected error referred to as only one component of the vector $\delta_y\left(t, \overline{z}_{j0}, \omega, \hat{\omega}(t)\right)$, as defined in Equation (18).

Figure 10 shows the journal displacement observation errors $\widetilde{z}_j$ as defined according to Equations (20c) and (28), in the four simulation cases. The signals $z_j = \begin{bmatrix} x_j, y_j \end{bmatrix}^T$ are not available in practice, but they are shown only for comparison to their corresponding observation $\hat{z}_j = \begin{bmatrix} \hat{x}_j, \hat{y}_j \end{bmatrix}^T$.

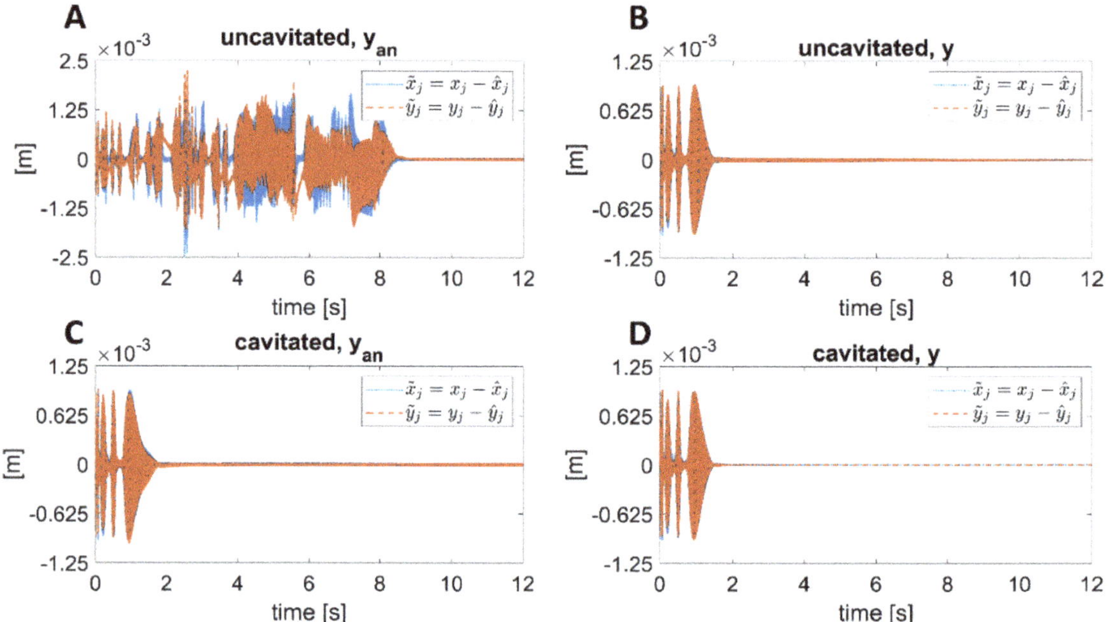

Figure 10. Observation errors of the journal displacement coordinate $\widetilde{z}_j$ with analytical output noise for uncavitated and cavitated conditions (**A**,**C**) and with measured output noise for uncavitated and cavitated conditions (**B**,**D**).

Figure 11 illustrates the hooking of the sinusoidal disturbance $\ddot{d}_{ei}(t, \omega)$ by the estimate $\hat{\ddot{d}}_{ei}(t, \omega_{est})$ in the case of the uncavitated condition with analytical noise. For ease of graph interpretation, due to the high frequency of the signals, the signal hooking is shown on the restricted time interval (7.5–8.5) (s).

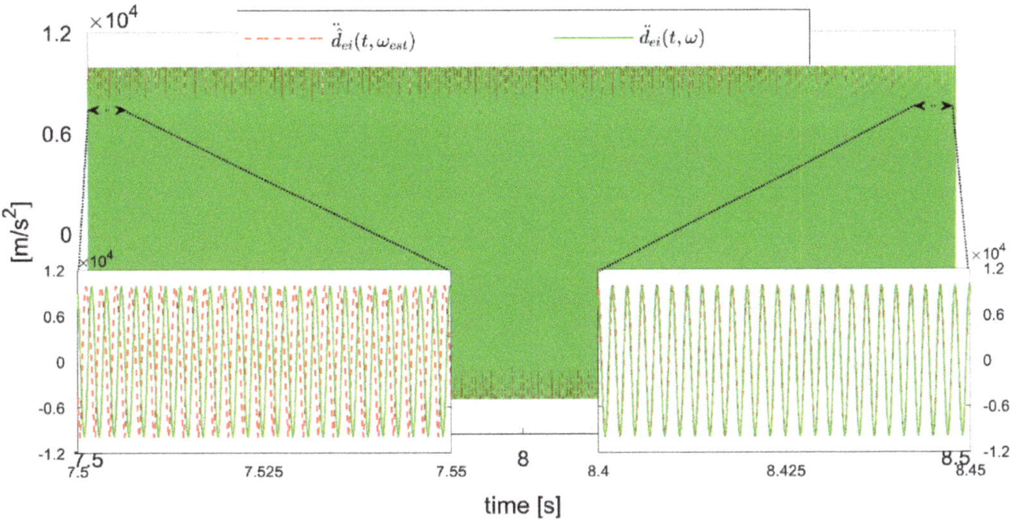

Figure 11. Hooking of the sinusoidal disturbance $\ddot{d}_{ei}(t,\omega)$ by its estimate $\ddot{\hat{d}}_{ei}(t,\omega_{est})$ in the case of uncavitated condition with analytical disturbance (y_{an}).

Figure 12 illustrates the sinusoidal disturbance estimation errors $\ddot{\tilde{d}}_{ei}(t,\omega)$ in the four simulation cases.

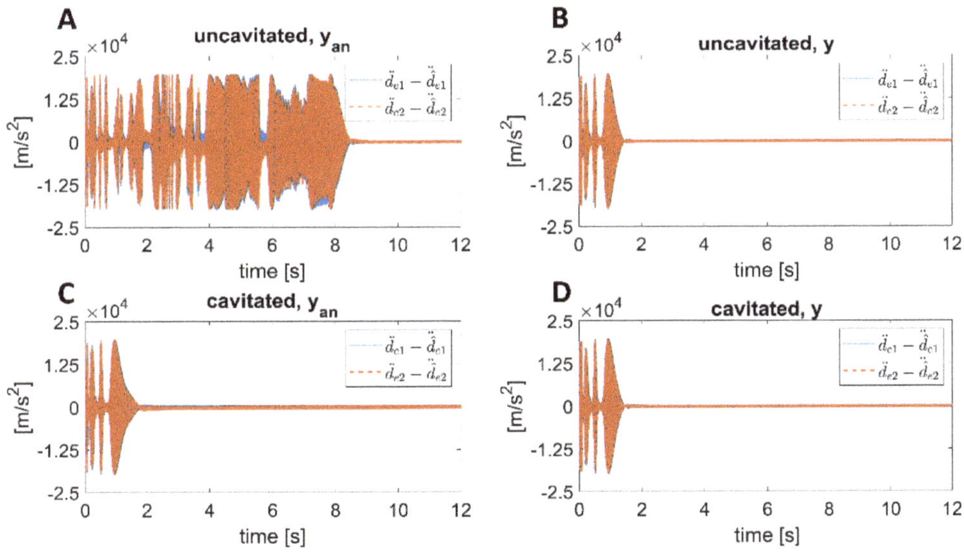

Figure 12. Sinusoidal disturbance estimation errors $\ddot{\tilde{d}}_{ei}(t,\omega)$ are reported for uncavitated and cavitated conditions with analytical output noise (**A**,**C**) and for uncavitated and cavitated conditions with measured output noise (**B**,**D**).

In Figure 13, the control vector $\hat{u}_x = [\hat{u}_{x,1}, \hat{u}_{x,2}]^T$ (Equation (20a)) is plotted in the four simulation cases. A zoomed plot, referred to by the dashed rectangle, highlights the detail of the last transient variation in the control before the steady-state behavior.

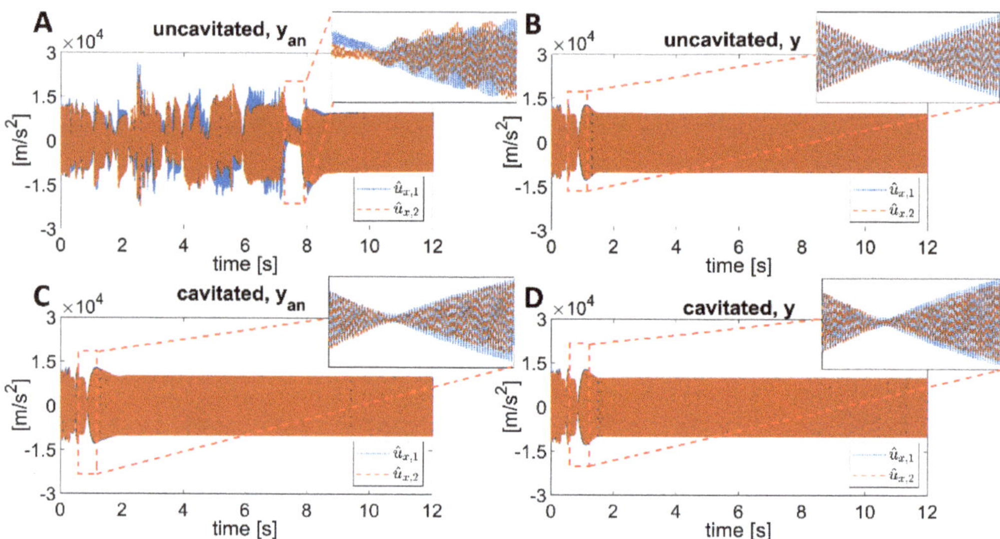

Figure 13. Control vector $\hat{u}_x = [\hat{u}_{x,1}, \hat{u}_{x,2}]^T$ for uncavitated and cavitated conditions with analytical output noise (**A**,**C**) and for uncavitated and cavitated conditions with measured output noise (**B**,**D**).

Figure 14 illustrates the disturbance attenuation on disc center coordinates $[x_c, y_c]$ in the four simulation cases.

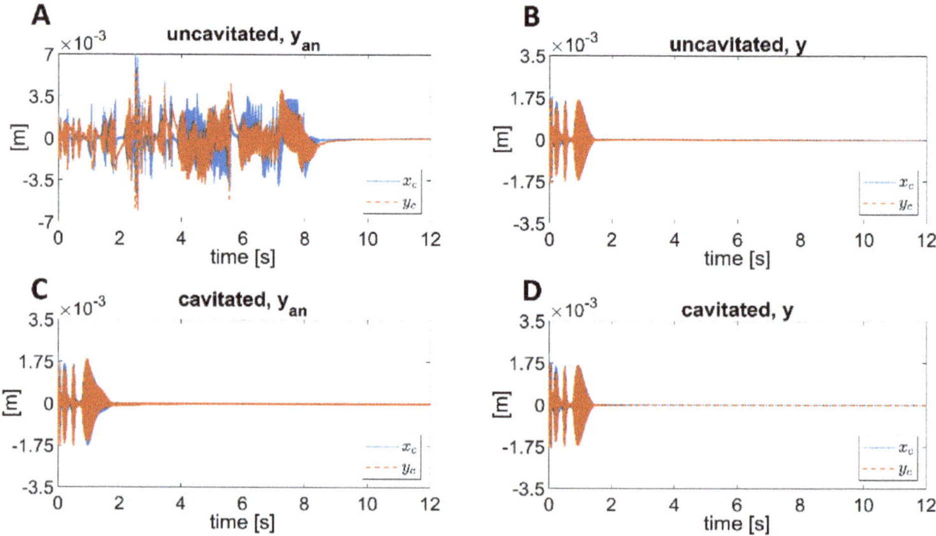

Figure 14. Disturbance attenuation of disc center coordinates (x_c, y_c) for uncavitated and cavitated conditions with analytical output noise (**A**,**C**) and for uncavitated and cavitated conditions with measured output noise (**B**,**D**).

5. Discussions and Conclusions

5.1. Discussions

The closed-loop ANC-FE operates on the output noise date to compute the asymptotical frequency estimate. In the following simulative analysis, both measured and analytical noise reconstruction is considered (in two cavitated and uncavitated operating conditions). Due to the reconstruction error δ_y (Figure 9), the analytical output noise is affected by an intrinsic additive disturbance, which entails a data quality lowering with respect to the measured output noise case. The latter is considered an ideal case, and it is used in this analysis only for the sake of comparison.

The control performance, in terms of settling time, is evaluated in four simulation cases. From Figure 4 it emerges that, from the comparison between the case of analytical output noise $\omega_{est,yan}$ and the case of measured output noise $\omega_{est,y}$, for the cavitated operating condition, the frequency estimation performance is very close. The residual vibration in the analytical case is due to the discrepancy between y and y_{an}, but it has a very low amplitude so that, macroscopically, the two curves cannot be distinguished. The comparison of the frequency estimation for the uncavitated condition is presented in Figure 5, showing that the performance degradation in terms of the settling of time, for analytical output noise $\omega_{est,yan}$ with respect to measured output noise $\omega_{est,y}$, is restrained in 2 s. In Figure 6, both curves refer to the analytical output noise, and it is visible that different convergence performances are registered in cavitated and uncavitated cases. This difference is also recovered by the performance in phase estimation (Figure 7). As expected, the frequency estimation procedure represents a warm-up phase of the overall closed-loop disturbance estimation. In fact, according to Remark 6, the convergence of the asymptotical frequency estimate is standalone with respect to the rest of the control loop. Hence, only once the frequency estimate has reached its steady-state value, the phase estimation starts its exponential convergence ((thesis iii) of Proposition 1) in the last segment of all four curves (Figures 4, 5 and 7). The worst performance is registered in the uncavitated condition with analytical noise (solid black line in Figure 7) since the phase estimation settling time exceeds the other three simulation cases by 6 s.

The numerical analysis of the frequency and phase estimation timeseries gives the fitting results shown in Figure 8, considering 95% confidence in the goodness of fit (GOF) statistics. The order of convergence calculated for the frequency estimation timeseries are $p_{\omega,unc} = 0.9983$ and $p_{\omega,cav} = 1.005$ under uncavitated and cavitated conditions, respectively (Figure 8A,B), with acceptable GOF where the $R^2_{\omega,unc} = 0.9998$ and $R^2_{\omega,cav} = 1.0$. The frequency estimation timeseries are sampled in the overall simulation time (degree of freedom, dfe = 2398). Aiming to show that the phase estimation converges exponentially only after the initial warm-up (according to Proposition 1), in the case of phase estimation (Figure 8C,D), each timeseries sampling is restricted to the interval in which its own warm-up phase ends: 8.41 (s) (dfe = 716) and 1.59 (s) (dfe = 2079) under uncavitated and cavitated conditions, respectively, as shown in Figure 7. The convergence of phase estimation approximates the exponential behavior better than frequency estimation, obtaining $p_{\varphi,unc} = 1.001$ and $p_{\varphi,cav} = 1$ under uncavitated and cavitated conditions, respectively (Figure 8C,D) with acceptable GOF where $R^2_{\varphi,unc} = 0.9983$ and $R^2_{\varphi,cav} = 0.9989$. The final phase estimation error is registered within 0.43%. Overall, with respect to the fitting curve, both frequency and phase estimation error timeseries exhibit less dispersion of the points in the cavitated case than the points sampled in the uncavitated condition.

Figure 9 illustrates that, although an additive disturbance δ_{yi} affects the analytical output y_{an}, due to the disturbance δ_{yi} amplitude, which is bounded and considerably smaller than the output noise itself (nine orders of magnitude less), y_{an} fits the real disturbance $\ddot{d}_{ei}(t,\omega)$ very well in both cavitated and uncavitated conditions (Figure 9A,C). This enforces Assumption 1, and Proposition 1 applies. Figure 10 illustrates the comparison of the reduced-order observer performance (Equation (20c)) in the center journal coordinates' estimation in the four simulation cases. It emerges that the performance degradation in the journal vibration absorption in the two cases with the analytical output (Figure 10A,C) is restrained in a few tenths of millimeters with respect to the ideal case with measured output (Figure 10B,D). The exponential hooking of the disturbance $\ddot{d}_{ei}(t, \omega)$ by the adaptive noise cancellation control $\hat{\ddot{d}}_{ei}(t, \omega_{est})$ (Equation (20b)) is illustrated in Figure 11. For sake of brevity, only the worst case (uncavitated condition with analytical output noise) is reported. Analogous results are obtained in the other three simulation cases. The preliminary phase of the overall disturbance estimations is dictated by the asymptotical convergence of the frequency estimate ω_{est}. From that instant, the same closed-loop behavior of [24] is recovered, with the last segment of the disturbance convergence, which is exponential.

The disturbance adaptation error $\tilde{\ddot{d}}_{ei}(t, \omega)$, reported in Figure 12, in the four cases recovers the observer behavior (Figure 10). Moreover, the behavior and performance of the control input (Figure 13) recover the behaviors of the displacement observation error (Figure 10) and the sinusoidal estimation errors (Figure 11) in the four cases. A zoomed plot is reported to highlight the sinusoidal trend of the two control vector components, oscillating at the same frequency as the rejected disturbance (as shown also in [24]). Figure 13 shows that the control amplitude stabilizes at the same time (respectively, in the four cases) in which all the estimation errors converge. Figure 14 illustrates that, from the comparison of the rotor center coordinate stabilization in the four cases, it emerges that the performance degradation in terms of vibration amplitude, for analytical output noise with respect to the (ideal) measured output noise case, is restrained in a few tenths of millimeters.

Overall, better estimation performance is obtained in the cavitated condition with respect to the uncavitated case. As the main figures of merit for the proposed adaptive closed-loop control, the rotor vibration attenuation time (about 8.5 (s) in the worst case) and the vibration residual amplitude (order of 10^{-1} (mm)) are considered, which are acceptable values for a practical application. Similar results, in terms of estimation error convergence time, are obtained in [17,18] (order of seconds), which presented tests of an experimental setup on thin steel strips [17] (order of millimeters) and simulative in silico validation on a reversible cold strip rolling mill [18].

5.2. Conclusions

The aim of this study was to provide a mathematical model for a novel adaptive noise cancellation (ANC) technique designed to stabilize a flexible rotor shaft supported by two hydrodynamic full (short) journal bearings and affected by a sinusoidal disturbance output noise with unknown frequency. The disturbance frequency corresponds to the rotor operating angular speed (ω), which is driven by an external actuator, so that uncertainties may arise in the frequency actual value due to actuation operating point fluctuations. The adaptation with respect to the frequency estimation (FE) generalizes the ANC to the novel ANC-FE.

As the main novelty in this study, inspired from immersion and invariance (I&I) techniques, an asymptotical frequency estimation (FE) module is designed as a combination of state-observer and asymptotical parameter estimation. The FE operation represents the warm-up phase of the overall adaptive noise cancellation control. The FE module is externalized as an additive plug-in block, which processes the analytical reconstruction of the output data downstream of the ANC closed-loop system (Figure 2).

The mathematical structural proof of the ANC-FE control theoretical formulation, Proposition 1, is provided under Assumption 1, which requires that the analytical output noise reconstruction is a signal at the same frequency of disturbance. It follows that: the frequency estimation convergence is asymptotical; the disturbance phase estimation and the rotor center coordinate stabilization are exponential.

The mathematical results have been validated experimentally in silico by numerical simulations performed in four scenarios: cavitated and uncavitated conditions with analytical and measured output noise. Moreover, a data fitting analysis with the Inexact Newton method (with 95% of confidence) is performed on frequency and phase estimation error point series, in order to also validate numerically Proposition 1, demonstrating that the phase estimate succession approximates the exponential behavior better than frequency estimate succession. The orders of convergence obtained by the frequency estimation timeseries are $p_{\omega,unc} = 0.9983$ and $p_{\omega,cav} = 1.005$; the order of convergence obtained by the phase estimation timeseries is $p_\varphi = 1$ for both uncavitated and cavitated conditions. Considering the analytical output noise as the case of practical interest, the settling time of disturbance rejection and then of the rotor center coordinates stabilization is about 76% less in the cavitated than in the uncavitated condition, 2 (s) and 8.5 (s), respectively.

Simulation shows that the analytical output noise is very close to the measured output noise, enforcing Assumption 1. The warm-up phase length is restrained in less than 10 s, which is acceptable as an initialization process duration in a real application. For future investigations the possible application of the proposed control technique to other dynamical complex systems can be considered.

Author Contributions: Conceptualization, G.A., R.D. and A.R.; methodology, G.A. and R.D.; software, G.A.; validation, G.A. and R.D.; formal analysis, G.A., R.D. and A.R.; investigation, G.A., R.D. and A.R.; data curation, G.A.; writing—original draft preparation, G.A. and R.D.; writing—review and editing, G.A., R.D. and A.R.; visualization, G.A. and R.D.; supervision, A.R. All authors have read and agreed to the published version of the manuscript.

Funding: This research is funded by FARB 2019 University of Salerno (Italy).

Institutional Review Board Statement: Not applicable.

Informed Consent Statement: Not applicable.

Conflicts of Interest: The authors declare no conflict of interest.

References

1. Friswell, M.I.; Penny, J.E.T.; Garvey, S.D.; Lees, A.W. *Dynamics of Rotating Machines*; Cambridge University Press: Cambridge, UK, 2010; ISBN 9780511780509.
2. Lund, J.W. Review of the Concept of Dynamic Coefficients for Fluid Film Journal Bearings. *J. Tribol. ASME US* **1987**, *109*, 37–41. [CrossRef]
3. Hamrock, B.J.; Schmid, S.R. *Fundamental of Fluid Film Lubrication*, 2nd ed.; CRC Press: Boca Raton, FL, USA, 2004; ISBN 0824753712.
4. Salazar, J.G.; Santos, I.F. Active tilting-pad journal bearings supporting flexible rotors: Part I—The hybrid lubrication. *Tribol. Int.* **2017**, *107*, 94–105. [CrossRef]
5. Chen, Y.; Yang, R.; Sugita, N.; Mao, J.; Shinshi, T. Identification of bearing dynamic parameters and unbalanced forces in a flexible rotor system supported by oil-film bearings and active magnetic devices. *Actuators* **2021**, *10*, 216. [CrossRef]
6. Muszynska, A. Stability of whirl and whip in rotor/bearing systems. *J. Sound Vib.* **1988**, *127*, 49–64. [CrossRef]
7. D'Agostino, V.; Guida, D.; Ruggiero, A.; Senatore, A. An analytical study of the fluid film force in finite-length journal bearings. Part I. *Lubr. Sci.* **2001**, *13*, 329–340. [CrossRef]
8. Poritsky, H. Contribution to the theory of oil whip. *Trans. ASME* **1953**, *75*, 1153–1161.
9. Tripathy, D.; Bhattacharyya, K. Analysis of a Hydrodynamic Journal Bearing of Circular Cross Section Lubricated by a Magnetomicropolar Fluid. In Proceedings of the Lecture Notes in Mechanical Engineering; Springer: Singapore, 2022; pp. 1495–1502.
10. Das, S.; Guha, S.K. Non-linear stability analysis of micropolar fluid lubricated journal bearings with turbulent effect. *Ind. Lubr. Tribol.* **2019**, *71*, 31–39. [CrossRef]
11. Bhattacharjee, B.; Chakraborti, P.; Choudhuri, K. Evaluation of the performance characteristics of double-layered porous micropolar fluid lubricated journal bearing. *Tribol. Int.* **2019**, *138*, 415–423. [CrossRef]
12. Harika, E.; Bouyer, J.; Fillon, M.; Hélène, M. Measurements of lubrication characteristics of a tilting pad thrust bearing disturbed by a water-contaminated lubricant. *Proc. Inst. Mech. Eng. Part J. J. Eng. Tribol.* **2013**, *227*, 16–25. [CrossRef]
13. Vania, A.; Pennacchi, P.; Chatterton, S. Dynamic Effects Caused by the Non-Linear Behavior of Oil-Film Journal Bearings in Rotating Machines. In *Proceedings of the Volume 7: Structures and Dynamics, Parts A and B*; ASME: New York, NY, USA, 2012; p. 657.
14. Das, S.; Guha, S.K.; Chattopadhyay, A.K. Linear stability analysis of hydrodynamic journal bearings under micropolar lubrication. *Tribol. Int.* **2005**, *38*, 500–507. [CrossRef]
15. Sukumaran Nair, V.P.; Prabhakaran Nair, K. Finite element analysis of elastohydrodynamic circular journal bearing with micropolar lubricants. *Finite Elem. Anal. Des.* **2004**, *41*, 75–89. [CrossRef]
16. Prabhakaran Nair, K.; Sukumaran Nair, V.P.; Jayadas, N.H. Static and dynamic analysis of elastohydrodynamic elliptical journal bearing with micropolar lubricant. *Tribol. Int.* **2007**, *40*, 297–305. [CrossRef]
17. Marko, L.; Saxinger, M.; Steinboeck, A.; Kugi, A. Cancellation of unknown multi-harmonic disturbances in multivariable flexible mechanical structures. *Automatica* **2022**, *137*, 110123. [CrossRef]

18. Liu, L.; Shao, N.; Deng, R.; Ding, S. Immersion and invariance adaptive decentralized control for the speed and tension system of the reversible cold strip rolling mill. *Int. J. Adapt. Control Signal Process.* **2022**, *36*, 785–801. [CrossRef]
19. Burrows, C.R.; Sahinkaya, M.N. Vibration control of multi-mode rotor-bearing systems. *Proc. R. Soc. Lond. A* **1983**, *386*, 77–94.
20. Ruggiero, A.; D'Amato, R.; Magliano, E.; Kozak, D. Dynamical simulations of a flexible rotor in cylindrical uncavitated and cavitated lubricated journal bearings. *Lubricants* **2018**, *6*, 40. [CrossRef]
21. Zhou, S.; Shi, J. Active balancing and vibration control of rotating machinery: A survey. *Shock Vib. Dig.* **2001**, *33*, 361–371. [CrossRef]
22. Stanway, R.; Burrows, C.R. Active vibration control of a flexible rotor on flexibly-mounted journal bearings. *J. Dyn. Syst. Meas. Control* **1981**, *103*, 383–388. [CrossRef]
23. Zheng, S.; Li, H.; Peng, C.; Wang, Y. Experimental Investigations of Resonance Vibration Control for Noncollocated AMB Flexible Rotor Systems. *IEEE Trans. Ind. Electron.* **2017**, *64*, 2226–2235. [CrossRef]
24. D'Amato, R.; Amato, G.; Wang, C.; Ruggiero, A. A Novel Tracking Control Strategy with Adaptive Noise Cancellation for Flexible Rotor Trajectories in Lubricated Bearings. *IEEE ASME Trans. Mechatron.* **2022**, *27*, 753–765. [CrossRef]
25. Lei, S.; Palazzolo, A. Control of flexible rotor systems with active magnetic bearings. *J. Sound Vib.* **2008**, *314*, 19–38. [CrossRef]
26. Marko, L.; Saxinger, M.; Steinboeck, A.; Kemmetmüller, W.; Kugi, A. Frequency-adaptive cancellation of harmonic disturbances at non-measurable positions of steel strips. *Mechatronics* **2020**, *71*, 102423. [CrossRef]
27. Marino, R.; Tomei, P. Adaptive disturbance rejection for unknown stable linear systems. *Trans. Inst. Meas. Control* **2016**, *38*, 640–647. [CrossRef]
28. Kumar, P.; Tiwari, R. Dynamic analysis and identification of unbalance and misalignment in a rigid rotor with two offset discs levitated by active magnetic bearings: A novel trial misalignment approach. *Propuls. Power Res.* **2020**, *10*, 58–82. [CrossRef]
29. D'Amato, R.; Amato, G.; Ruggiero, A. Adaptive Noise Cancellation-Based Tracking Control for a Flexible Rotor in Lubricated Journal Bearings. In Proceedings of the 2019 23rd International Conference on Mechatronics Technology (ICMT), Salerno, Italy, 23–26 October 2019; pp. 1–5. [CrossRef]
30. Vance, J.M. *Rotordynamics of Turbomachinery*; Wiley: Hoboken, NJ, USA, 1988; ISBN 0471802581.
31. Avramov, K.V.; Borysiuk, O.V. Nonlinear dynamics of one disk asymmetrical rotor supported by two journal bearings. *Nonlinear Dyn.* **2012**, *67*, 1201–1219. [CrossRef]
32. Carnevale, D.; Astolfi, A. A minimal dimension observer for global frequency estimation. *Proc. Am. Control Conf.* **2008**, *2*, 5236–5241. [CrossRef]
33. Carnevale, D. Robust hybrid estimation and rejection of multi-frequency signals. *Int. J. Adapt. Control Signal Process.* **2016**, *30*, 1649–1673. [CrossRef]
34. Hoad, K.; Robinson, S.; Davies, R. Automating warm-up length estimation. *J. Oper. Res. Soc.* **2010**, *61*, 1389–1403. [CrossRef]
35. Ballnus, B.; Schaper, S.; Theis, F.J.; Hasenauer, J. Bayesian parameter estimation for biochemical reaction networks using region-based adaptive parallel tempering. *Bioinformatics* **2018**, *34*, i494–i501. [CrossRef]
36. Ypma, T.J. Local Convergence of Inexact Newton Methods. *SIAM J. Numer. Anal.* **2006**, *21*, 583–590. [CrossRef]
37. Sastry, S.; Bodson, M. *Adaptive Control—Stability, Convergence, and Robustness*; Prentice-Hall: Hoboken, NJ, USA, 1989; p. 201.
38. Marino, R.; Tomei, P. Output Regulation for Unknown Stable Linear Systems. *IEEE Trans. Autom. Control* **2015**, *60*, 2213–2218. [CrossRef]
39. Sun, X.; Member, S.; Wu, M.; Yin, C.; Wang, S.; Tian, X. Multiple-Iteration Search Sensorless Control for Linear Motor in Vehicle Regenerative Suspension. *IEEE Trans. Transp. Electrif.* **2021**, *7*, 1628–1637. [CrossRef]
40. Marino, R.; Tomei, P. *Nonlinear Control Design: Geometric, Adaptive and Robust*; Prentice Hall: London, UK, 1995; Volume 1.
41. Marino, R.; Tomei, P.; Verrelli, C.M. *Induction Motor Control Design*; Springer Science & Business Media: Berlin/Heidelberg, Germany, 2010; ISBN 1849962847.

Article

Complete Balancing of the Six-Bar Mechanism Using Fully Cartesian Coordinates and Multiobjective Differential Evolution Optimization

María T. Orvañanos-Guerrero [1], Mario Acevedo [2], Claudia N. Sánchez [1], Daniel U. Campos-Delgado [3], Amir Aminzadeh Ghavifekr [4], Paolo Visconti [1,5] and Ramiro Velázquez [1,*]

[1] Facultad de Ingeniería, Universidad Panamericana, Aguascalientes 20290, Mexico; torvananos@up.edu.mx (M.T.O.-G.); cnsanchez@up.edu.mx (C.N.S.); paolo.visconti@unisalento.it (P.V.)
[2] Facultad de Ingeniería, Universidad Panamericana, Zapopan 45010, Mexico; macevedo@up.edu.mx
[3] Facultad de Ciencias, Instituto de Investigación en Comunicación Óptica, Universidad Autónoma de San Luis Potosí, San Luis Potosí 78295, Mexico; ducd@fciencias.uaslp.mx
[4] Faculty of Electrical and Computer Engineering, University of Tabriz, Tabriz 5166616471, Iran; aa.ghavifekr@tabrizu.ac.ir
[5] Department of Innovation Engineering, University of Salento, 73100 Lecce, Italy
* Correspondence: rvelazquez@up.edu.mx; Tel.: +52-449-910-6200

Citation: Orvañanos-Guerrero, M.T.; Acevedo, M.; Sánchez, C.N.; Campos-Delgado, D.U.; Ghavifekr, A.A.; Visconti, P.; Velázquez, R. Complete Balancing of the Six-Bar Mechanism Using Fully Cartesian Coordinates and Multiobjective Differential Evolution Optimization. *Mathematics* **2022**, *10*, 1830. https://doi.org/10.3390/math10111830

Academic Editors: Higinio Rubio Alonso, Alejandro Bustos Caballero, Jesus Meneses Alonso and Enrique Soriano-Heras

Received: 23 April 2022
Accepted: 21 May 2022
Published: 26 May 2022

Abstract: The high-speed operation of unbalanced machines may cause vibrations that lead to noise, wear, and fatigue that will eventually limit their efficiency and operating life. To restrain such vibrations, a complete balancing must be performed. This paper presents the complete balancing optimization of a six-bar mechanism with the use of counterweights. A novel method based on fully Cartesian coordinates (FCC) is proposed to represent such a balanced mechanism. A multiobjective optimization problem was solved using the Differential Evolution (DE) algorithm to minimize the shaking force (ShF) and the shaking moment (ShM) and thus balance the system. The Pareto front is used to determine the best solutions according to three optimization criteria: only the ShF, only the ShM, and both the ShF and ShM. The dimensions of the counterweights are further fine-tuned with an analysis of their partial derivatives, volumes, and area–thickness relations. Numerical results show that the ShF and ShM can be reduced by 76.82% and 77.21%, respectively, when importance is given to either of them and by 45.69% and 46.81%, respectively, when equal importance is given to both. A comparison of these results with others previously reported in the literature shows that the use of FCC in conjunction with DE is a suitable methodology for the complete balancing of mechanisms.

Keywords: six-bar mechanism; dynamic balancing; fully Cartesian coordinates; multiobjective optimization; differential evolution; Pareto front

MSC: 91B03; 70B15

1. Introduction

The high-speed motion of the links of a mechanism generates dynamic reactions that are transmitted as vibrations to the fixed frame producing undesirable effects such as noise, wear, fatigue, and energy losses that will eventually limit the mechanism's performance and operating life. In addition to these technical problems, vibrations may induce a constant machine maintenance involving wastes that have an environmental impact and noise pollution that can lead to health problems [1].

The dynamic reactions under discussion are produced by both external and inertial forces. The former are naturally forces generated outside the mechanism such as those induced by an actuator. The latter are due to the large accelerations of the links. Some representative methods for calculating dynamic reactions in mechanisms are those of Uicker [2] and Waldron [3].

The force and the moment that occur at the mechanism's fixed frame are typically known as shaking force (ShF) and shaking moment (ShM), respectively. Thus, a traditional but still open challenge in machine theory is how to effectively remove or minimize the dynamic reactions, i.e., the ShF and ShM, derived from the mechanism's motion. This process is known as complete (or dynamic) balancing.

The dynamic balancing of mechanisms has been widely studied in mechanical engineering. Some recent historical reviews addressing the topic can be found in [1,4–6].

Among the first publications, we can find the work of Fischer introducing in 1902 a method called principal vectors [7]. This method allows for the balancing of the ShF by analyzing each of the links of the mechanism and determining the points, called principal points, in which the static balancing can be reached. The work of Fischer provided the basis for the methods studying the motion of the centers of mass (CoM) of the links in a mechanism. The method of principal vectors was subsequently used by Goryachkin [8], Yudin [9], and Kreutzinger [10].

Another early method proposed for the dynamic balancing of mechanisms was the static substitution of masses. Its aim is to statically replace the mass of the coupler by concentrated masses, thus transforming the problem of mechanism balancing into a simpler problem of balancing the rotating links. The works of Maxwell [11], Smith [12], and Talbourdet [13] are based on this method.

The 1920s were marked by a special interest in the balancing of engines [14,15] and machines related to agriculture [16]. The Lanchester balancer stands out among these works. It is still being used for balancing four-stroke engines.

During the 1940s, some methods based on function approximation were proposed to achieve partial equilibrium. The work of Gheronimus [17] is a representative example. In this work, the balancing conditions were formulated by minimizing the root mean square (RMS) or the maximum values of the ShM. The duplicate mechanism method [18] was also proposed in this decade. It achieves a complete balancing with the addition of axial and mirror symmetric mechanism duplicates.

The balancing methods based on harmonic analysis appeared in the 1960s with the use of the crank-slider mechanism in internal combustion engines [19,20]. Such methods reduce the ShM by balancing certain harmonics of both the ShF and ShM. To carry out this process, the unbalanced forces and moments are divided into Fourier series and then analyzed in parts.

In 1969, Berkof and Lowen [21] proposed a new solution for dynamic balancing using a method called linearly independent vectors. This method consists of formulating an equation considering a vector representation describing the position of the total CoM of the mechanism in conjunction with the equation representing the closed kinematic chain. A system of equations containing linearly independent vectors is then obtained allowing to find the balancing conditions of the mechanism by reducing to zero the time-dependent coefficients. This method was subsequently explored in [22,23].

In the 1970s, the dynamic balancing theory achieved significant advances. Berkof [24] formulated the first method for complete balancing using counterweights and counter-inertias to eliminate the undesired forces and moments of the moving links, respectively. It then became clear that a complete balancing of mechanisms is feasible at the expense of considering complex design modifications and an inevitable increase of the total mass and volume of the mechanism.

Methods proposing a partial balancing were then proposed searching to keep the added mass reasonably small. In [25], Wiederrich and Roth formulated simple general conditions for determining the inertia properties of a four-bar mechanism and achieved its partial balancing. Dresig examined the partial balancing conditions for several 2D mechanims with six and eight-bar linkages [26].

Later approaches involved the use of planetary [27] and auto-balanced systems [28]. More recent ones propose the use of the instantaneous dynamic balancing conditions [29], the trajectory planning of the CoM of the mechanism [30], and the reformulation of the balancing conditions following a Taylor-based approach [29]. Finally, the most recent and novel balancing method encompasses the flexibility of the links of the mechanism [31].

We focus our attention on Berkof's counterweight approach. After the partial balancing methods, optimization techniques were then explored searching a tradeoff between the added mass and the reduction of the ShF and ShM.

In 1998, Segla presented the static balancing optimization of a robotic mechanism [32]. The balancing conditions were first obtained and then, a basic genetic algorithm programmed in Fortran was used for the optimization.

Currently, the most popular optimization techniques for mechanical problems are based on metaheuristic methods: evolutionary [33], differential evolution [34], genetic [35], and Firefly [36] algorithms. These proposals were designed to find heuristics (i.e., partial solutions) that may provide sufficiently good tradeoffs for the dynamic balancing problem.

To our knowledge, all previous works addressing optimization techniques for the dynamic balancing of mechanisms make use of Cartesian coordinates (CC) to obtain the expressions of the dynamic reactions. A major drawback of CC is that they involve trigonometric functions that derive into complex mathematical expressions that are computationally burdensome.

Our research approach differs from the existing literature in that it presents an alternative to CC: fully Cartesian coordinates (FCC), which are also called natural coordinates [37]. By using FCC, the dynamic reactions at the fixed frame can be formulated by means of equations of less complexity (no angular variables). Then, the resulting ShF and ShM expressions are used to optimize the dynamic balancing of mechanisms through the use of counterweights.

Our previous work [38,39] successfully reported the dynamic balancing of a four-bar mechanism and its optimization with Projected Gradient Descent. In this paper, we address the dynamic balancing of a more complex system, a six-bar mechanism, and explore the differential evolution (DE) algorithm as our optimization method.

Parallel mechanisms are increasingly being used in robotic applications [40,41]. The six-bar mechanism is a typical parallel manipulator. This single degree-of-freedom planar linkage is typically used as a variable-speed transmission mechanism where the input crank rotates at constant speed and the output link works as an overrunning clutch mounted on the output shaft [42].

The rest of the paper is organized as follows: In Section 2, the FCC-based mass-matrix definition for the six-bar mechanism with counterweights is introduced together with the expressions of the ShF and ShM. In Section 3, the details of the multiobjective DE optimization are presented. In Section 4, a numerical example is presented to illustrate the proposed approach. In Section 5, results are discussed and compared to others previously reported in the literature. Finally, Section 6 concludes, summarizing the main contributions and giving the future work perspectives of this research.

2. Mechanical Analysis

2.1. FCC-Based Definition of the Mass Matrix of an Element Defined by Three Basic Points

This section details the method used to obtain the mass matrix $\mathbf{M_{3P}}$ of elements defined by three basic points using FCC and the concept of virtual work.

Consider an element defined by three basic points i, j, and k, as shown in Figure 1. The element is located in a global coordinate system (x, y) and in a local coordinate system $(\bar{x}, \bar{y})$ with its origin at point i and the $\bar{x}$ axis directed toward point j.

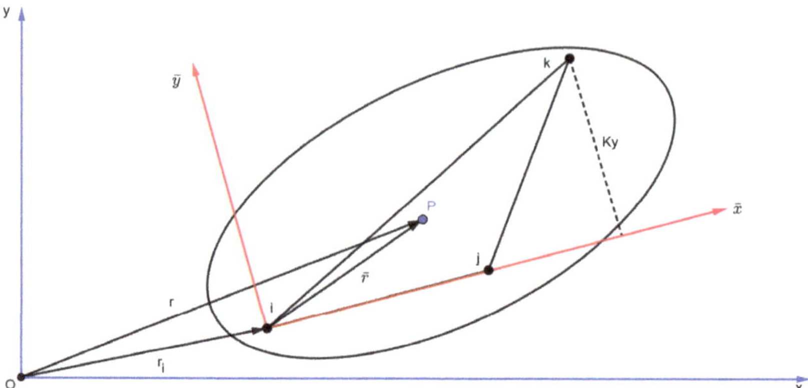

Figure 1. A 2D element with three basic points (i, j, and k).

The location of any point P in this element is defined by a vector $\mathbf{r}$ in the global reference system and a vector $\bar{r}$ in the local coordinate system. In this way, $\mathbf{r}$ can be expressed, according to Equation (1).

$$\mathbf{r} = \mathbf{r_i} + \mathbf{A}\bar{\mathbf{r}} \tag{1}$$

where $\mathbf{A}$ is the rotation matrix. Being the element rigid, the local position of vector $\bar{\mathbf{r}}$ remains constant regardless of the element's motion. Thus, the position of point P can be defined according to Equation (2).

$$\mathbf{r} = \mathbf{r_i} + \mathbf{A}\bar{\mathbf{r}} = \mathbf{r_i} + c_1(\mathbf{r_j} - \mathbf{ri}) + +c_2(\mathbf{r_k} - \mathbf{r_i}) \tag{2}$$

where c_1 and c_2 are the components of vector $\bar{\mathbf{r}}$ in the local coordinate system. The components of vector $\mathbf{r}$ can be expressed in matrix form as shown in Equation (3).

$$\mathbf{r} = \begin{Bmatrix} x \\ y \end{Bmatrix} = \begin{bmatrix} 1 - c_1 - c_2 & 0 & c_1 & 0 & c_2 & 0 \\ 0 & 1 - c_1 - c_2 & 0 & c_1 & 0 & c_2 \end{bmatrix} \begin{Bmatrix} x_i \\ y_i \\ x_j \\ y_j \\ x_k \\ y_k \end{Bmatrix} = \mathbf{Cq} \tag{3}$$

where $\mathbf{q}^\mathbf{T} = \{x_i \quad y_i \quad x_j \quad y_j \quad x_k \quad y_k\}$ is the vector that contains the FCC of the element. Note that matrix $\mathbf{C}$ is constant for a given point P and does not change with the system's motion, thus fulfilling Equations (4) and (5).

$$\dot{\mathbf{r}} = \mathbf{C}\dot{\mathbf{q}} \tag{4}$$

$$\ddot{\mathbf{r}} = \mathbf{C}\ddot{\mathbf{q}} \tag{5}$$

Coefficients c_1 and c_2 in matrix $\mathbf{C}$ can be expressed in terms of the coordinates of points i, j, and k in the local reference frame according to Equation (6).

$$\bar{\mathbf{r}} = c_1(\mathbf{r_j} - \mathbf{r_i}) + c_2(\mathbf{r_k} - \mathbf{r_i}) \tag{6}$$

Since $\bar{\mathbf{r}}_\mathbf{i} = 0$ is located at the local reference origin, Equation (6) can be rewritten as shown in Equation (7).

$$\bar{\mathbf{r}} = [\,\bar{\mathbf{r}}_\mathbf{j} \mid \bar{\mathbf{r}}_\mathbf{k}\,] \begin{Bmatrix} c_1 \\ c_2 \end{Bmatrix} = \bar{\mathbf{X}}\mathbf{c} \tag{7}$$

Vector $\mathbf{c}$ contains the coefficients c_1 and c_2, and matrix $\bar{\mathbf{X}}$ has the components of vectors $\bar{\mathbf{r}}_j$ and $\bar{\mathbf{r}}_k$ in its columns (Equation (8)).

$$\bar{\mathbf{X}} = [\,\bar{\mathbf{r}}_j \mid \bar{\mathbf{r}}_k\,] = \begin{bmatrix} x_j & x_k \\ y_j & y_k \end{bmatrix} = \begin{bmatrix} l_{ij} & K_x \\ 0 & K_y \end{bmatrix} \tag{8}$$

Now, it is possible to define the virtual work W^* generated by the inertial forces (Equation (9)).

$$W^* = -\rho \int_\Omega \dot{\mathbf{r}}^{*T} \ddot{\mathbf{r}} \, d\Omega \tag{9}$$

where ρ is the density of the element's material. Substituting Equations (4) and (5) into Equation (9) yields to the definition of virtual work (Equation (10)):

$$W^* = -\rho \int_\Omega \dot{\mathbf{q}}^{*T} \mathbf{C}^T \mathbf{C} \ddot{\mathbf{q}} \, d\Omega \tag{10}$$

Since vectors $\dot{\mathbf{q}}^{*T}$ and $\ddot{\mathbf{q}}$ are independent of Ω, they can be taken out of the integral as expressed by Equation (11).

$$W^* = -\dot{\mathbf{q}}^{*T} \left(\rho \int_\Omega \mathbf{C}^T \mathbf{C} d\Omega \right) \ddot{\mathbf{q}} \tag{11}$$

On the other hand, taking into account the definition of virtual work proposed in [43] (Equation (12)) and comparing it to Equation (11), the mass matrix $\mathbf{M}_{3P}$ can be expressed by Equation (13).

$$W^* = -\dot{\mathbf{q}}^{*T} \mathbf{M} \ddot{\mathbf{q}} \tag{12}$$

$$\mathbf{M}_{3P} = \rho \int_\Omega \mathbf{C}^T \mathbf{C} d\Omega \tag{13}$$

Further development of the product of $\mathbf{C}^T \mathbf{C}$ in Equation (13) yields to Equation (14).

$$\mathbf{M}_{3P} = \rho \int_\Omega \begin{bmatrix} c_e & 0 & c_f & 0 & c_g & 0 \\ 0 & c_e & 0 & c_f & 0 & c_g \\ c_f & 0 & c_h & 0 & c_i & 0 \\ 0 & c_f & 0 & c_h & 0 & c_i \\ c_g & 0 & c_i & 0 & c_j & 0 \\ 0 & c_g & 0 & c_i & 0 & c_j \end{bmatrix} d\Omega \tag{14}$$

with:

$$c_e = c_1^2 + 2c_1 c_2 - 2c_1 + c_2^2 - 2c_2 + 1 \tag{15}$$

$$c_f = -c_1^2 - c_1 c_2 + c_1 \tag{16}$$

$$c_g = -c_1 c_2 - c_2^2 + c_2 \tag{17}$$

$$c_h = c_1^2 \tag{18}$$

$$c_i = c_1 c_2 \tag{19}$$

$$c_j = c_2^2 \tag{20}$$

Note that Equation (14) involves solving the integrals of Equations (21)–(23).

$$\int_\Omega \rho \, d\Omega = m \tag{21}$$

$$\int_\Omega \rho c\, d\Omega = \int_\Omega \rho \bar{\mathbf{X}}^{-1}\bar{\mathbf{r}}\, d\Omega = \rho \int_\Omega \begin{bmatrix} \frac{1}{l_{ij}} & -\frac{K_x}{K_y l_{ij}} \\ 0 & \frac{1}{K_y} \end{bmatrix}\begin{bmatrix} \bar{x} \\ \bar{y} \end{bmatrix} d\Omega =$$

$$\rho \int_\Omega \begin{bmatrix} \frac{\bar{x}}{l_{ij}} - \frac{\bar{y}K_x}{K_y l_{ij}} \\ \frac{\bar{y}}{K_y} \end{bmatrix} d\Omega = \begin{bmatrix} \frac{m\bar{x}_g}{l_{ij}} - \frac{m\bar{y}_g K_x}{K_y l_{ij}} \\ \frac{m\bar{y}_g}{K_y} \end{bmatrix} \tag{22}$$

$$\int_\Omega \rho c c^{\mathrm{T}}\, d\Omega = \bar{\mathbf{X}}^{-1}\left(\int_\Omega \rho \bar{\mathbf{r}}\bar{\mathbf{r}}^{\mathrm{T}} d\Omega\right)\bar{\mathbf{X}}^{-\mathrm{T}} = \bar{\mathbf{X}}^{-1}\left(\int_\Omega \rho \begin{bmatrix} \bar{x}^2 & \bar{x}\bar{y} \\ \bar{x}\bar{y} & \bar{y}^2 \end{bmatrix} d\Omega\right)\bar{\mathbf{X}}^{-\mathrm{T}} =$$

$$\bar{\mathbf{X}}^{-1}\begin{bmatrix} I_y & I_{xy} \\ I_{xy} & I_x \end{bmatrix}\bar{\mathbf{X}}^{-\mathrm{T}} = \begin{bmatrix} \frac{I_x K_x^2}{K_y^2 l_{ij}^2} - \frac{2 I_{xy} K_x}{K_y l_{ij}^2} + \frac{I_y}{l_{ij}^2} & -\frac{I_x K_x}{K_y^2 l_{ij}} + \frac{I_{xy}}{K_y l_{ij}} \\ -\frac{I_x K_x}{K_y^2 l_{ij}} + \frac{I_{xy}}{K_y l_{ij}} & \frac{I_x}{K_y^2} \end{bmatrix} \tag{23}$$

where m is the total mass of the element, $\bar{\mathbf{r}}$ represents the local coordinates of the center of gravity, and I_x, I_y, and I_{xy} are the moments and products of inertia with respect to local coordinates with origin at the basic point i.

By substituting the integrals of Equations (21)–(23) into Equation (14), we finally obtain the mass matrix $\mathbf{M_{3P}}$ (Equation (24)).

$$\mathbf{M_{3P}} = \begin{bmatrix} e & 0 & f & 0 & g & 0 \\ 0 & e & 0 & f & 0 & g \\ f & 0 & h & 0 & i & 0 \\ 0 & f & 0 & h & 0 & i \\ g & 0 & i & 0 & j & 0 \\ 0 & g & 0 & i & 0 & j \end{bmatrix} \tag{24}$$

with:

$$e = \frac{I_x K_x^2}{K_y^2 l^2} - \frac{2 I_x K_x}{K_y^2 l_{ij}} + \frac{I_x}{K_y^2} - \frac{2 I_{xy} K_x}{K_y l_{ij}^2} + \frac{2 I_{xy}}{K_y l_{ij}} + \frac{I_y}{l_{ij}^2} + \frac{2 K_x m\bar{y}_g}{K_y l_{ij}} + m - \frac{2m\bar{x}_g}{l_{ij}} - \frac{2m\bar{y}_g}{K_y} \tag{25}$$

$$f = -\frac{I_x K_x^2}{K_y^2 l_{ij}^2} + \frac{I_x K_x}{K_y^2 l_{ij}} + \frac{2 I_{xy} K_x}{K_y l_{ij}^2} - \frac{I_{xy}}{K_y l_{ij}} - \frac{I_y}{l_{ij}^2} - \frac{K_x m\bar{y}_g}{K_y l_{ij}} + \frac{m\bar{x}_g}{l_{ij}} \tag{26}$$

$$g = \frac{I_x K_x}{K_y^2 l_{ij}} - \frac{I_x}{K_y^2} - \frac{I_{xy}}{K_y l_{ij}} + \frac{m\bar{y}_g}{K_y} \tag{27}$$

$$h = \frac{I_x K_x^2}{K_y^2 l_{ij}^2} - \frac{2 I_{xy} K_x}{K_y l_{ij}^2} + \frac{I_y}{l_{ij}^2} \tag{28}$$

$$i = -\frac{I_x K_x}{K_y^2 l_{ij}} + \frac{I_{xy}}{K_y l_{ij}} \tag{29}$$

$$j = \frac{I_x}{K_y^2} \tag{30}$$

2.2. Six-Bar Mechanism

Figure 2 shows the six-bar mechanism. This kind of mechanism exhibits one degree of freedom and five mobile links.

Each of the links has a local coordinate system with its origin located at point i and the x axis directed toward point j. Each of the links has a mass m_{bn} and a center of gravity located at local coordinates (x_{bn}, y_{bn}) for $1 \leq n \leq 5$. The distribution of the points is detailed in Table 1.

For this mechanism, it is possible to define a vector **q** representing the positions of its basic points (Equation (31)).

$$\mathbf{q} = [A_x\ A_y\ B_x\ B_y\ C_x\ C_y\ D_x\ D_y\ E_x\ E_y\ F_x\ F_y\ G_x\ G_y]^T \tag{31}$$

By time-deriving **q**, it is possible to obtain a new vector **q̇** representing the velocities of each of the basic points of the mechanism (Equation (32)). Similarly, by time-deriving the velocity vector, the acceleration vector **q̈** can be obtained (Equation (33)).

$$\dot{\mathbf{q}} = [V_{Ax}\ V_{Ay}\ V_{Bx}\ V_{By}\ V_{Cx}\ V_{Cy}\ V_{Dx}\ V_{Dy}\ V_{Ex}\ V_{Ey}\ VF_x\ V_{Fy}\ V_{Gx}\ V_{Gy}]^T \tag{32}$$

$$\ddot{\mathbf{q}} = [A_{Ax}\ A_{Ay}\ A_{Bx}\ A_{By}\ A_{Cx}\ A_{Cy}\ A_{Dx}\ A_{Dy}\ A_{Ex}\ A_{Ey}\ A_{Fx}\ A_{Fy}\ A_{Gx}\ A_{Gy}]^T \tag{33}$$

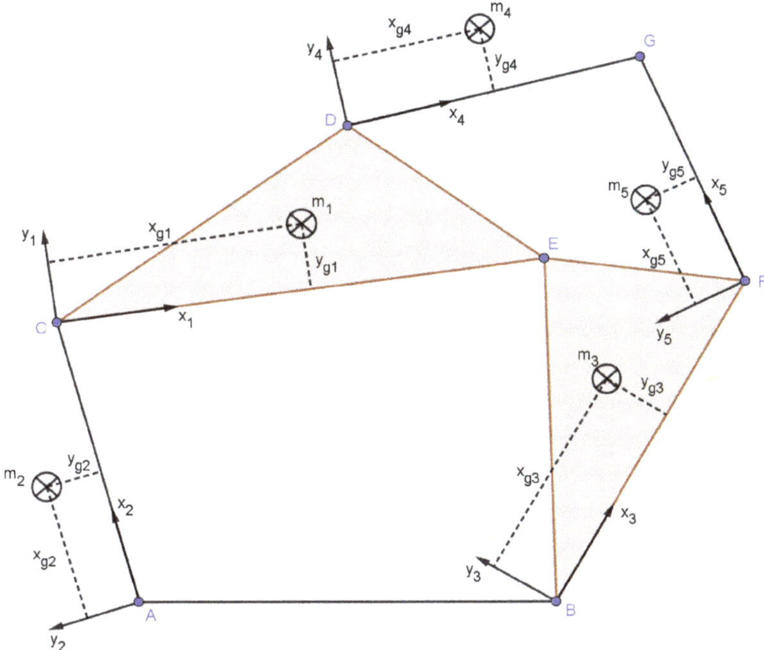

Figure 2. The six-bar mechanism.

Table 1. Point distribution for the six-bar mechanism.

Element	Point i	Point j	Point k
1	C	E	D
2	A	C	
3	B	F	E
4	D	G	
5	F	G	

2.3. Counterweight Addition

To proceed with the dynamic balancing of the six-bar mechanism, a set of cylindrical counterweights is added to its structure (Figure 3).

To simplify the procedure, let us assume that all counterweights are coincident with the basic point *i* of each of the elements of the mechanism. The center of gravity of each counterweight is located at the local coordinates (x_{cn}, y_{cn}) for $1 \leq n \leq 5$. Their thickness is defined by t_{cn} for $1 \leq n \leq 5$.

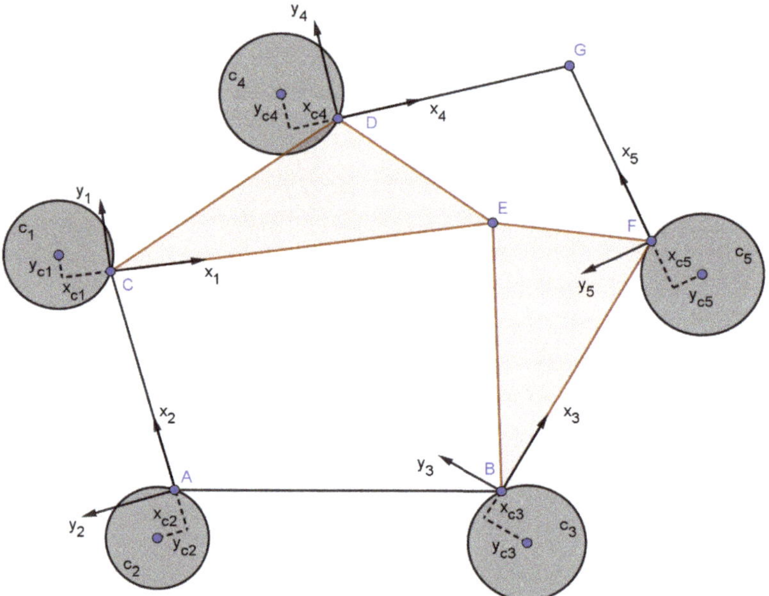

Figure 3. Six-bar mechanism with cylindrical counterweights.

It is then possible to define the mass m_{cn} of each counterweight as a function of its density $\rho_{cn} = \rho$, its thickness t_{cn}, and the location of its CoM (x_{cn}, y_{cn}), as shown in Equation (34).

$$m_{cn} = \rho\, V = \rho \pi t_{cn}(x_{cn}^2 + y_{cn}^2) \tag{34}$$

It is also possible to obtain the mass moments for each of the counterweights with respect to the local coordinate system origin, as shown in Equations (35) and (36).

$$I_{xcn} = \frac{1}{4}m_{cn}(x_{cn}^2 + y_{cn}^2) + my_{cn}^2 = \frac{1}{4}m_{cn}(x_{cn}^2 + 5y_{cn}^2) = \frac{1}{4}\rho \pi t_{cn}(x_{cn}^2 + y_{cn}^2)(x_{cn}^2 + 5y_{cn}^2) \tag{35}$$

$$I_{ycn} = \frac{1}{4}m_{cn}(x_{cn}^2 + y_{cn}^2) + mx_{cn}^2 = \frac{1}{4}m_{cn}(5x_{cn}^2 + y_{cn}^2) = \frac{1}{4}\rho \pi t_{cn}(x_{cn}^2 + y_{cn}^2)(5x_{cn}^2 + y_{cn}^2) \tag{36}$$

The polar moment of inertia of each counterweight I_{zcn} with respect to the local coordinate system origin can be defined by Equation (37).

$$I_{zcn} = I_{xcn} + I_{ycn} = \frac{3}{2}m_{cn}(x_{cn}^2 + y_{cn}^2) = \frac{3}{2}\rho \pi t_{cn}(x_{cn}^2 + y_{cn}^2)^2 \tag{37}$$

Similarly, the product of inertia of each counterweight with respect to the local coordinate system origin can be calculated with Equation (38).

$$I_{xycn} = m_{cn}x_{cn}y_{cn} = \rho \pi t_{cn}(x_{cn}^2 + y_{cn}^2)x_{cn}y_{cn} \tag{38}$$

2.4. Mass Matrix for the Six-Bar Mechanism with Counterweights

To formulate the mass matrix of the balanced mechanism, it is necessary to obtain the mass matrices for each of the linkages (link and counterweight) first. The details of these individual mass matrices can be found in Appendix A.

Given that the mechanism consists of seven basic points, each of them represented by one (x, y) coordinate, the resulting mass matrix will consist of 14 columns and 14 rows.

Equation (39) shows the mass matrix M representing the six-bar mechanism with counterweights.

$$M = \begin{bmatrix}
a_2 & 0 & 0 & 0 & b_2 & c_2 & 0 & 0 & 0 & 0 & 0 & 0 & 0 & 0 \\
0 & a_2 & 0 & 0 & -c_2 & b_2 & 0 & 0 & 0 & 0 & 0 & 0 & 0 & 0 \\
0 & 0 & e_3 & 0 & 0 & 0 & 0 & 0 & g_3 & 0 & f_3 & 0 & 0 & 0 \\
0 & 0 & 0 & e_3 & 0 & 0 & 0 & 0 & 0 & g_3 & 0 & f_3 & 0 & 0 \\
b_2 & c_2 & 0 & 0 & d_2+e_1 & 0 & g_1 & 0 & f_1 & 0 & 0 & 0 & 0 & 0 \\
-c_2 & b_2 & 0 & 0 & 0 & d_2+e_1 & 0 & g_1 & 0 & f_1 & 0 & 0 & 0 & 0 \\
0 & 0 & 0 & 0 & g_1 & 0 & a_4+j_1 & 0 & i_1 & 0 & 0 & 0 & b_4 & c_4 \\
0 & 0 & 0 & 0 & 0 & g_1 & 0 & a_4+j_1 & 0 & i_1 & 0 & 0 & -c_4 & b_4 \\
0 & 0 & g_3 & 0 & f_1 & 0 & i_1 & 0 & h_1+i_3 & 0 & i_3 & 0 & 0 & 0 \\
0 & 0 & 0 & g_3 & 0 & f_1 & 0 & i_1 & 0 & h_1+i_3 & 0 & i_3 & 0 & 0 \\
0 & 0 & f_3 & 0 & 0 & 0 & 0 & 0 & i_3 & 0 & a_5+h_3 & 0 & b_5 & c_5 \\
0 & 0 & 0 & f_3 & 0 & 0 & 0 & 0 & 0 & i_3 & 0 & a_5+h_3 & -c_5 & b_5 \\
0 & 0 & 0 & 0 & 0 & 0 & b_4 & c_4 & 0 & 0 & b_5 & c_5 & d_4+d5 & 0 \\
0 & 0 & 0 & 0 & 0 & 0 & -c_4 & b_4 & 0 & 0 & -c_5 & b_5 & 0 & d_4+d5
\end{bmatrix} \tag{39}$$

where $a_n, b_n, \ldots, j_n$ are the terms of the different mass matrices for linkages n = 1, $\ldots$, 5 (see Appendix A).

2.5. Linear Momentum and Shaking Force

Once vectors $\mathbf{q}$, $\dot{\mathbf{q}}$, and $\ddot{\mathbf{q}}$ have been defined (Equations (31)–(33)), it is possible to calculate the linear momentum L associated to the balanced mechanism (Equation (40)).

$$\begin{bmatrix} L_i \\ L_j \end{bmatrix} = \mathbf{BM\dot{q}} \tag{40}$$

where $\mathbf{B}$ (Equation (41)) is a matrix formed by identity matrices matching the number of basic points in the mechanism.

$$\mathbf{B} = \begin{bmatrix} 1 & 0 & 1 & 0 & 1 & 0 & 1 & 0 & 1 & 0 & 1 & 0 & 1 & 0 \\ 0 & 1 & 0 & 1 & 0 & 1 & 0 & 1 & 0 & 1 & 0 & 1 & 0 & 1 \end{bmatrix}^{\mathbf{T}} \tag{41}$$

By solving Equation (40) and considering that the velocity of the fixed points is always zero ($VA_X = 0$, $VA_Y = 0$, $VB_X = 0$, $VB_Y = 0$), the expressions of the linear momentum ($\mathbf{L_i}$ and $\mathbf{L_j}$) can be obtained.

The ShF_i and ShF_j of the mechanism can be computed by time-deriving the equations $\mathbf{L_i}$ and $\mathbf{L_j}$ (Equation (40)) (expressions are not included in the paper because of their length).

To guarantee the equilibrium, the result of these derivatives must be constant (normally zero) in the analyzed period of time.

2.6. Angular Momentum and Shaking Moment

The use of FCC allows us to express the angular momentum H of the mechanim, as shown in Equation (42):

$$H = \mathbf{q} \times (\mathbf{M\dot{q}}) = \mathbf{rM\dot{q}} \tag{42}$$

where $\mathbf{r}$ is given as a function of the basic points and can be expressed as Equation (43):

$$\mathbf{r} = \begin{bmatrix} -A_Y & A_X & -B_Y & B_X & -C_Y & C_X & -D_Y & D_X & -E_Y & E_X & -F_Y & F_X & -G_Y & G_X \end{bmatrix}^{\mathbf{T}} \tag{43}$$

The ShM can then be calculated by time-deriving H (Equations (44) and (45)).

$$ShM = \frac{dH}{dt} = \mathbf{rM}\left(\frac{d(\dot{\mathbf{q}})}{dt}\right) + \left(\frac{d\mathbf{r}}{dt}\right)\mathbf{M\dot{q}} \tag{44}$$

$$ShM = \frac{dH}{dt} = \mathbf{r}\mathbf{M}\ddot{q} + \dot{\mathbf{r}}\mathbf{M}\dot{q} \tag{45}$$

with:

$$\dot{\mathbf{r}} = [-VA_Y \quad VA_X \quad -VB_Y \quad VB_X \quad -VC_Y \quad VC_X \quad -VD_Y \quad VD_X$$
$$-VE_Y \quad VE_X \quad -VF_Y \quad VF_X \quad -VG_Y \quad VG_X]^{\mathbf{T}} \tag{46}$$

To guarantee the dynamic equilibrium of the mechanism, the ShM must be constant, i.e., the time-derivative of H (Equation (45)) must be zero.

The ShM of the mechanism is finally obtained by solving Equation (45) and considering $VA_X = 0$, $VA_Y = 0$, $VB_X = 0$ y $VB_Y = 0$ (equation is not included in the paper because of its length).

3. Optimization

3.1. Objective Function

Two dimensionless balancing indexes β_i, containing the motion parameters ($\mathbf{q}$, $\dot{\mathbf{q}}$, and $\ddot{\mathbf{q}}$) of the six-bar mechanism, can be used to define the optimization's objective function: β_{ShF} and β_{ShM}.

β_{ShF} (Equation (47)) is defined by the RMS value of the ShF reaction of the optimized mechanism ($rms(^{o}ShF)$) with respect to the RMS value of the original mechanism ($rms(ShF)$), which are both considered over a time period T.

$$\beta_{ShF} = \frac{rms(^{o}ShF)}{rms(ShF)} = \sqrt{\frac{\sum_{k=1}^{N}(^{o}ShF_{ik}^2 + ^{o}ShF_{jk}^2)}{\sum_{k=1}^{N}(ShF_{ik}^2 + ShF_{jk}^2)}} \tag{47}$$

β_{ShM} (Equation (48)) can be calculated in a similar way. Nevertheless, the reaction produced by ShM must also be considered.

$$\beta_{ShM} = \frac{\sum_{k=1}^{N} {}^{o}ShM_k^2}{\sum_{k=1}^{N} ShM_k^2} \tag{48}$$

where ^{o}ShM is the shaking moment of the optimized mechanism and ShM is a constant representing the shaking moment of the unbalanced mechanism.

A multiobjective optimization problem emerges as it is desired to minimize both β_{ShF} and β_{ShM} considering the variables boundaries (i.e., the physical limits for the locations of the CoM (x_{cn} and y_{cn}) and the thickness (t_{cn}) of each counterweight). To solve this problem, a linear combination of the objectives is performed as proposed in Equation (49).

$$f(X) = \gamma * \beta_{ShM} + (1 - \gamma) * \beta_{ShF} \tag{49}$$

where γ is a scalar value that assigns the importance to each optimization objective. Thus, the 15 variables to be optimized are: $x_{c1}, y_{c1}, t_{c1}, x_{c2}, y_{c2}, t_{c2}, x_{c3}, y_{c3}, t_{c3}, x_{c4}, y_{c4}, t_{c4}, x_{c5}, y_{c5}$, and t_{c5}. The boundaries for optimization are defined according to Equation (50).

$$x_{cn}^{min} \leqslant x_{cn} \leqslant x_{cn}^{max}$$
$$y_{cn}^{min} \leqslant y_{cn} \leqslant y_{cn}^{max} \tag{50}$$
$$t_{cn}^{min} \leqslant t_{cn} \leqslant t_{cn}^{max}$$

3.2. Algorithm

Once the objective function has been proposed, an optimization method can be applied. In this paper, we explore Differential Evolution (DE) [44].

Being an evolutionary algorithm, DE uses approaches inspired by the theory of evolution. It optimizes a problem by proposing a population of candidate solutions and creating new candidate solutions with the ones that obtained the best scores. Thus, the new generations are better than the previous ones. Recently, DE has been successfully applied to disruptive fields such as the oil market [45] and genome studies [46].

The DE algorithm proposed for the complete balancing optimization of the six-bar mechanism is presented in Algorithm 1. It was programmed in Python.

Algorithm 1: Differential Evolution (DE).

Input : $N = 255$, $F = 0.8$, $CR = 0.7$, $k_{max} = 100$

1 An initial population is generated $S = \{X_{1k}, X_{2k}, \ldots, X_{Nk}\}$
2 X_{best} = solution with the lowest value in the objective function
3 **for** $k = 0$ *until* k_{max} **do**
4      $Q = \{\}$
5   **for** $i = 1$ *until* N **do**
6   Selection of three random solutions in S $\gamma, \delta, \eta \in \{1, \ldots, N\}, \gamma \neq \delta \neq \eta \neq i$
7          $\hat{X}_{ik} = X_{\gamma k} + F(X_{\delta k} - X_{\eta k})$
8          $\hat{X}_{ik} = Clip(\hat{X}_{ik})$
9   **for** $j = 1$ *until* D **do**
10   R = Random value between 0 and 1 with uniform distribution
11   **if** $R \leq CR$ **then**
12                $Y_{ikj} = \hat{X}_{ikj}$
13   **else**
14                $Y_{ikj} = X_{ikj}$
15   **end**
16   **end**
17   **if** $f(Y_{ik}) < f(X_{ik})$ **then**
18             $Q = Q \cup Y_{ik}$
19   **else**
20             $Q = Q \cup X_{ik}$
21   **end**
22   **If** $f(Y_{ik}) < f(X_{best})$ **then** $X_{best} = Y_{ik}$
23   **end**
24      $S = Q$
25      $k = k + 1$
26 **end**
27 **return** X_{best}

Given that the number of variables to optimize is equal to 15, a population of $N = 15^2 = 225$ solutions is created, identifying the n_{th} solution of generation k with the vector X_{Nk}. Solutions are initialized following a uniform distribution bounded by limits between the allowed ranges for each variable. Three solutions are then selected to perform the random mutation: $X_{\gamma k}$, $X_{\delta k}$, and $X_{\eta k}$ to generate a new solution: $\hat{X}_{ik} = X_{\gamma k} + F(X_{\delta k} - X_{\eta k})$ with F being a random value between 0 and 2.

The crossover is performed with a probability $CR = 0.7$. Then, Y_{ik} and X_{ik} are evaluated in the function to be optimized. To be considered part of the new generation, the one with the best results is chosen. The generations are repeated 100 times, resulting in the solution exhibiting the lowest value.

4. Results

This section presents the numerical results of the dynamic balancing of the six-bar mechanism with FCC and DE optimization. To better visualize the influence of the counterweights on the procedure, a Pareto front analysis is proposed. This tool allows to restrict our attention to the set of best solutions and ease the decision-making process.

Table 2 summarizes the physical parameters of the six-bar mechanism shown in Figure 2. Steel links have been considered with a density of 7800 kg/m^3. The counterweights are considered to be made of brass with a density ρ_{cn} = 8500 kg/m^3.

Table 2. Parameters of the six-bar mechanism used in the example. Those indicated with a '-' are not necessary for the numerical analysis.

Link n	1	2	3	4	5
Mass m_{b_n} [kg]	0.6935	0.1022	0.9636	0.1825	0.1679
Length l_n [m]	0.19	0.14	0.13416408	0.25	0.23
Inertia I_{xb_n} [kg m/s^2]	0.00116161	-	0.00622646	-	-
Inertia I_{yb_n} [kg m/s^2]	0.00556534	-	0.00657336	-	-
Inertia I_{zb_n} [kg m/s^2]	-	0.00066856	-	0.00380360	0.00296204
Inertia I_{xyb_n} [kg m/s^2]	0.00167596	-	0.00522914	-	-
CoM x_{b_n} [m]	0.08	0.07	0.07751702	0.125	0.115
CoM y_{b_n} [m]	0.03333333	0.0	0.06559133	0.0	0.0
K_x [m]	0.05	-	0.09838699	-	-
K_y [m]	0.1	-	0.196677398	-	-

A motor located at point A is responsible for actuating the mechanism. Its rotating speed has been fixed at 500 rpm. Using direct kinematics, it is possible to obtain a sample of the positions, velocities, and accelerations at each of the basic points of the mechanism.

By replacing all known parameters in the equations of the balancing indexes (Equations (47) and (48)), it is possible to define the objective function (Equation (49)). According to the mechanical characteristics of this particular example, the boundaries considered for the optimization are shown in Equation (51):

$$-0.16\,\text{m} \leqslant x_{cn}, y_{cn} \leqslant 0.16\,\text{m}$$
$$0.005\,\text{m} \leqslant t_{cn} \leqslant 0.04\,\text{m}$$

(51)

4.1. Optimization with Five Counterweights

The DE algorithm was executed until 200 valid solutions (i.e., solutions resulting from the objective function optimization with values between 0 and 1) were obtained considering γ as a random value with uniform distribution in the range (0, 1), following the random search of the hyper-parameters presented in [47].

Figure 4a shows the relationship between the β_{ShF} and the β_{ShM} values for all the solutions found. Different colors are used to represent the values used for γ in the objective function f(X) (Equation (49)).

In Figure 4b, the dark points represent the Pareto front while the light ones represent the dominated solutions.

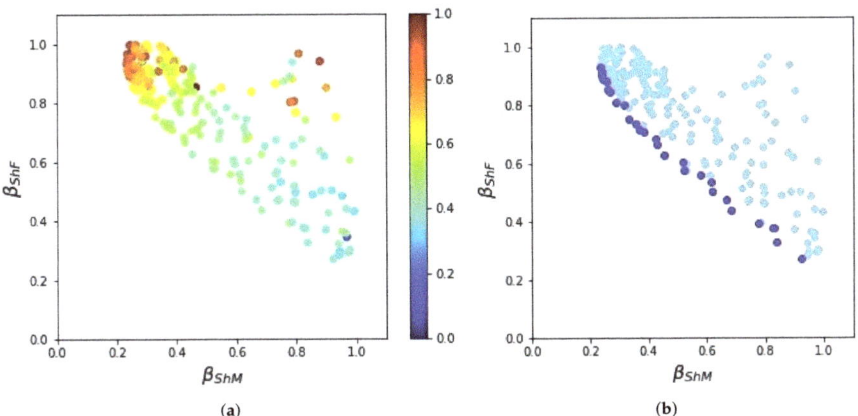

Figure 4. Pareto analysis of the optimization objectives: β_{ShM} and β_{ShF}. (**a**) Optimized β_{ShM} and β_{ShF} according to γ. (**b**) Pareto Front of the optimized objectives.

Among all the solutions found in the Pareto front, it is possible to select the one that is the most appropriate according to the desired type of balancing. To exemplify this statement, let us consider three different solutions from the Pareto front.

In the first one, a greater importance is given to balancing the ShM (β_{ShM} = 0.235917108, β_{ShF} = 0.932850297). In the second one, a greater importance is given to balancing the ShF (β_{ShM} = 0.924195224, β_{ShF} = 0.270900009). Finally, in the third solution, the same importance is given to balancing both the ShF and ShM (β_{ShM} = 0.580111266, β_{ShF} = 0.558041831). These solutions are detailed below.

1. In the first solution, a greater importance is given to balancing the ShM. This is achieved by choosing the minimum value of index β_{ShM} (β_{ShM} = 0. 235917108), which allows us to obtain an improvement of 76.4% without considering any improvement of the ShF (β_{ShF} = 0.932850297). The following variables values correspond to this solution:

$$
\begin{aligned}
x_{c1} &= -0.02611622 & y_{c1} &= -0.033186164 & t_{c1} &= 0.021370695 \\
x_{c2} &= -0.06 & y_{c2} &= -0.012314794 & t_{c2} &= 0.039459546 \\
x_{c3} &= -0.06 & y_{c3} &= 0.016752092 & t_{c3} &= 0.039814955 \\
x_{c4} &= -0.002045161 & y_{c4} &= -0.000151565 & t_{c4} &= 0.005 \\
x_{c5} &= 0.001614171 & y_{c5} &= 0.00639164 & t_{c5} &= 0.005106848
\end{aligned}
$$

2. The second chosen solution in the Pareto front is the one with the minimum value in β_{ShF} (β_{ShF} = 0.270900009), which achieves an improvement of 72.91% in balancing the ShM. This choice assigns no importance to the balancing of the ShM (β_{ShM} = 0.924195224). This solution yields to the following variable values:

$$
\begin{aligned}
x_{c1} &= -0.049437705 & y_{c1} &= -0.04279353 & t_{c1} &= 0.0074319 \\
x_{c2} &= -0.06 & y_{c2} &= -0.001485936 & t_{c2} &= 0.038953174 \\
x_{c3} &= -0.050813691 & y_{c3} &= -0.017724236 & t_{c3} &= 0.039660979 \\
x_{c4} &= 0.001669535 & y_{c4} &= 0.005996916 & t_{c4} &= 0.005 \\
x_{c5} &= 0.000387207 & y_{c5} &= 0.013090811 & t_{c5} &= 0.005
\end{aligned}
$$

3. The third chosen solution is the one in the Pareto front where both indexes are optimized (β_{ShM} = 0.580111266, β_{ShF} = 0.558041831). By using this solution, the ShM is reduced by 41.99% and the ShF is reduced by 44.2%. It corresponds to the following variable values:

$$x_{c1} = -0.06 \qquad y_{c1} = -0.030069707 \qquad t_{c1} = 0.006021448$$
$$x_{c2} = -0.06 \qquad y_{c2} = -0.00383543 \qquad t_{c2} = 0.04$$
$$x_{c3} = -0.06 \qquad y_{c3} = 0.004650961 \qquad t_{c3} = 0.028206906$$
$$x_{c4} = 0.000606455 \qquad y_{c4} = -0.001322538 \qquad t_{c4} = 0.005416754$$
$$x_{c5} = 0.000545683 \qquad y_{c5} = -0.0000185 \qquad t_{c5} = 0.005$$

To measure the method's efficiency, we assessed two parameters: convergence and computing time. Figure 5 shows the process of convergence of the objective function $f(X)$ as a function of generations. Six executions are presented with different γ values. Note that all executions exhibit values higher than 1.0 at the beginning of the execution (i.e., not optimized at all). After the 40th generation, all reach convergence (the closer the value to zero, the better), proving that the DE method actually optimizes the objective function.

Computing time is a major concern in the efficiency of optimization algorithms. To measure it, we executed 50 times the DE optimization procedure on a desktop computer with an Intel Core i7 processor (2.40 GHz), 8 GB RAM, and Windows 10 OS. The average computing time was 15.22 min with a standard deviation of 0.25 min. Being a mechanical design application, the optimization procedure does not require a fast or real-time computing but rather a practical one.

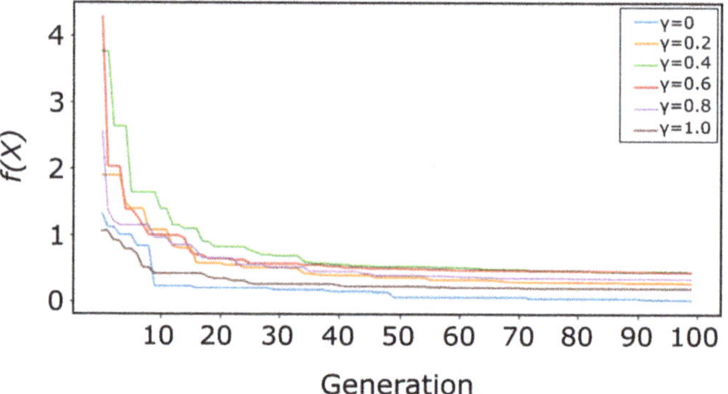

Figure 5. The convergence of the objective function $f(X)$ with the DE method.

4.2. Dimensions of the Counterweights

An analysis is now presented to determine if the proposed boundaries for the optimization are the most suitable or if they should be modified (in case there is the possibility of changing them due to the mechanical constraints of the mechanism and its surrounding space).

Figure 6 shows the box plots of the partial derivatives of the objective function with respect to each of the variables to be optimized $x_n, y_n,$ and t_n for each of the n counterweights $(1 \leq n \leq 5)$ when evaluated at the optimal solutions.

We can state that optimal boundaries have been found when all the values of the partial derivatives of the objective function are close to zero. In the box plots of Figure 6, it can be seen that the partial derivatives of the variables $x_1, y_1, y_2, y_3, x_4, y_4, t_4, x_5, y_5,$ and t_5 are close to zero, implying that the proposed boundaries allow such variables to reach their optimal values.

However, note that variables $t_1, x_2, t_2, x_3,$ and t_3 have partial derivative values that are not close to zero. For example, in t_1 and even more significantly in x_2 and x_3, the values tend to be greater than zero; hence, it can be deduced that better optimization results could be obtained if: (1) the thicknesses of counterweight 1 were smaller than the fixed boundary of 0.005 m (i.e., practically eliminating counterweight 1) and (2) the x-axis positions of counterweights 1 and 3 were less than the limit of -0.16 m (i.e., moving their CoM in their $-x$ local axis).

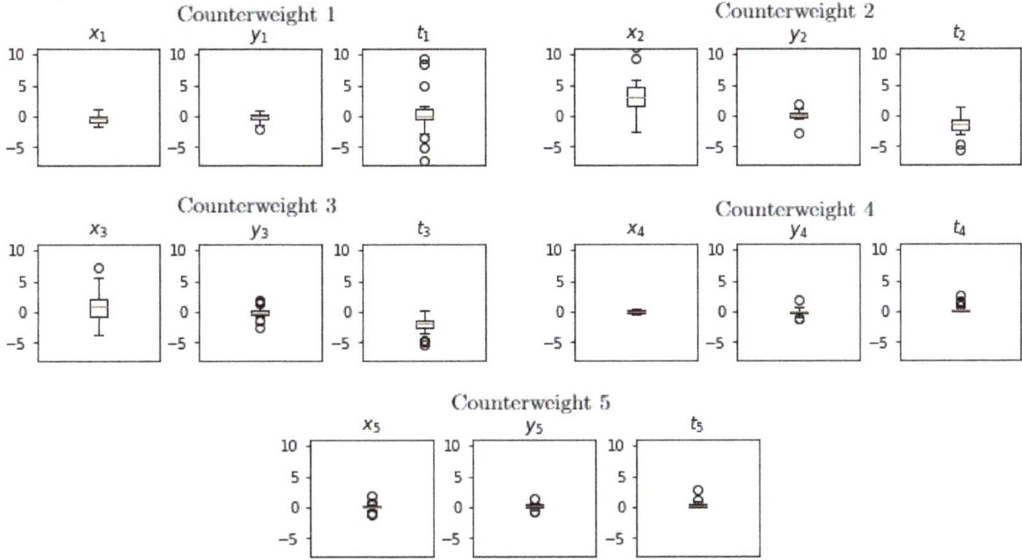

Figure 6. Box plots of partial derivatives of the objective function with respect to each optimization variable when using five counterweights.

On the other hand, the values of the partial derivatives of variables t_2 and t_3 tend to be less than zero, which means that, if possible, it would be convenient to further extend the boundaries of the optimization of these variables to values greater than 0.04 m.

Note that the information obtained from the Pareto front of the partial derivatives of the objective function is extremely useful to make decisions concerning the boundaries of the optimization, allowing us to foresee pertinent changes in the linkages as far as their mechanical conditions allow it.

Figure 7a shows the histograms of the volumes of the counterweights obtained from the different optimization results. Together with the analysis of the counterweights' area and thickness shown in Figure 7b, it is possible to conclude that counterweight 4 can be eliminated: it has a very small volume compared to the other counterweights, and its thickness and area are practically negligible.

In addition, for both counterweights 2 and 3, it can be appreciated that their thickness tends to remain at the upper limit of the optimization, thus confirming the result of the partial derivatives analysis: if the mechanical characteristics of the mechanism allow it, it would be advisable to extend the upper limit of their thicknesses.

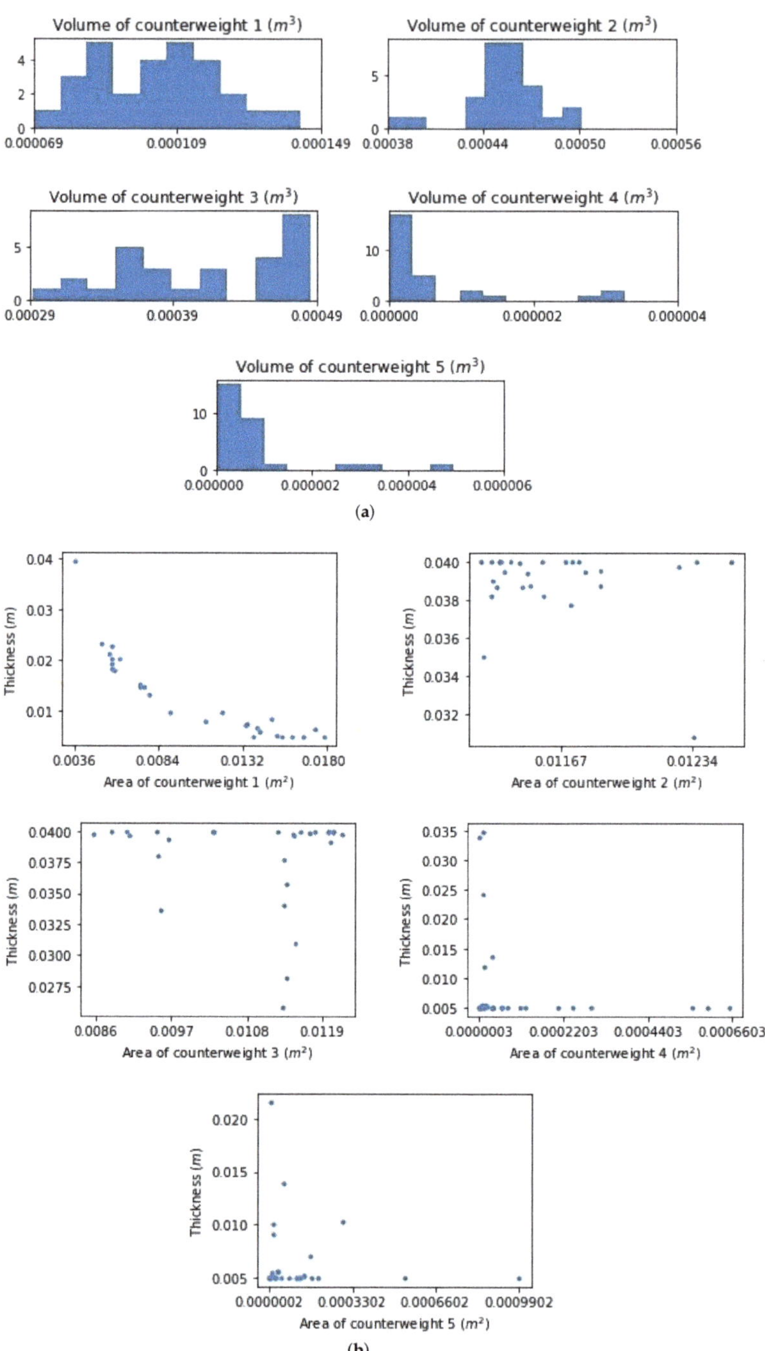

Figure 7. Dimensional analysis for the five counterweights balancing the six-bar mechanism. (**a**) Volumes of the five counterweights. (**b**) Relationship between the area and the thickness for the five counterweights.

4.3. Implementation

The optimization method and the dimensional analysis conclude with the design of a solution that effectively reduces the dynamic reactions of the six-bar mechanism.

Let us consider the third case where both ShF and ShM are optimized. Figure 8 shows the proposed implementation. As suggested by the analysis, counterweight 4 has been eliminated. Note that counterweight 5 is not visible because it exhibits very small dimensions in relation to the mechanism. Yet, it has been considered for the dynamic balancing of the mechanism.

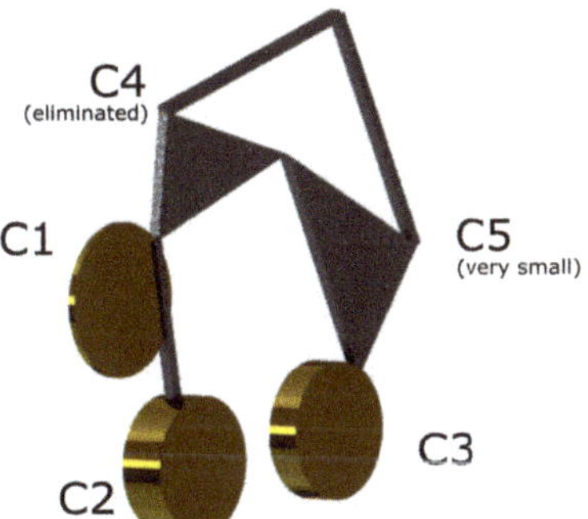

Figure 8. Conceptual implementation of the third solution from the Pareto front for the six-bar mechanism using five counterweights C_i.

4.4. Optimization with Four or Less Counterweights

The previous analysis suggested that not all five counterweights are strictly needed to balance the six-bar mechanism: counterweight 4 is meaningless and counterweight 5 is so small that it can be practically eliminated. When seeking a simple solution, i.e., one that does not involve adding too much volume to the mechanism, it may be useful to reduce the number of counterweights as long as they still provide acceptable results.

We performed the same analysis for four, three, two, and one counterweights. Figure 9 shows a comparison between the Pareto fronts of the different optimization results as a function of the number of counterweights. Note that as expected, similar results are obtained when using five, four, and three counterweights. So, a general conclusion for the implementation would be to use three counterweights.

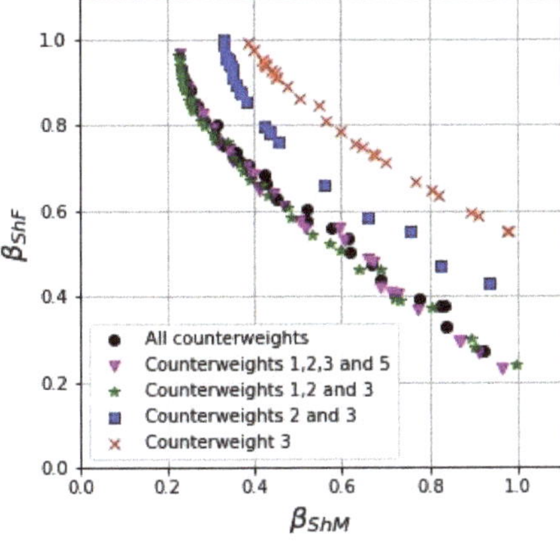

Figure 9. Pareto front comparison of the optimization objectives β_{ShM} and β_{ShF} according to the number of counterweights.

Table 3 summarizes the results obtained. The best results are highlighted in blue. Again, the best choice depends on the desired balancing: only the ShF, only the ShM, or both.

Table 3. Summary of the optimization results for the six-bar mechanism balancing.

Counterweights Cn	ShF Optimization % ShF, % ShM	ShM Optimization % ShF, % ShM	ShF and ShM Optimization % ShF, % ShM
All 5	72.91, 7.59	6.72, 76.41	44.2, 41.99
C1, C2, C3, C4	76.82, 3.59	3.21, 76.97	44.31, 40.73
C1, C2, C3	75.95, 0.5	3.53, 77.21	45.69, 46.81
C2, C3	56.98, 6.52	0.17, 67	41.66, 34.18
C3	44.97, 2.26	1.03, 61	28.69, 30.23

5. Discussion

The implementation of well-designed counterweights allows us to reduce the ShF and ShM by 76.82% and 77.21%, respectively, when importance is given to either of them and by 45.69% and 46.81%, respectively, when equal importance is given to both of them.

Table 4 compares our balancing results with others previously reported in the literature. Note that the comparison is limited to the cases where either the ShF or the ShM are optimized as not all the studies report the joint optimization. Table 4 includes other types of mechanisms such as the crank-slider and the four-bar mechanism, which are indeed simpler structures and thus easier to optimize. Six-bar mechanisms and their balancing are more complex and less frequently found in the literature because, as explained in Section 2, they exhibit a higher number of links with some defined by three basic points.

Table 4. Comparison of balancing results for several types of mechanisms and optimization methods.

Reference, CC or FCC	Mechanism	Optimization Algorithm	% of ShF Optimization	% of ShM Optimization
[48], CC	crank-slider	Differential Evolution	61.42	65.96
[49], CC	crank-slider	Teaching–Learning	48	44
[50], CC	crank-slider	Genetic	46	99
[51], FCC	crank-slider	Differential Evolution	97.76	94.58
[52], CC	four-bar	Genetic	50	68
[36], CC	four-bar	Firefly	86.3	83.39
[38], FCC	four-bar	Gradient Descent	99.70	83.99
[53], CC	six-bar	Genetic	48.5	32.35
This work, FCC	six-bar	Differential Evolution	76.82	77.21

Note that the use of FFC to define the mass matrix of a mechanism and thus obtain the expressions representing the dynamic reactions in its base, in conjunction with an optimization algorithm, is a suitable methodology for the complete balancing of mechanisms. Our previous work in crank-sliders [51] and four-bar [38] mechanisms confirm the efficiency of the proposed approach. The results obtained surpass those of approaches using Cartesian coordinates (CC).

The DE optimization method was successfully applied to solve the balancing problem of a six-bar mechanism. DE is a simple yet robust approach to address multiobjective problems.

Comparison between Pareto fronts has proven to be a useful tool for better visualizing the impact of each counterweight on the dynamic balancing of the mechanism.

The effectiveness of the analysis of the boundaries by means of box plots of the partial derivatives of the variables to be optimized together with the histograms of volumes and relations between area and thickens were also demonstrated. They ease the visualization of possible improvements on the counterweights and allow us to make useful decisions on their implementation.

Balancing six-bar mechanisms is highly appreciated in areas such as mobile robotics, since this type of linkage is used in humanoid robots [54] and bionic legs [55,56] as well as in rehabilitation engineering [57], in which balancing could minimize the reactions caused by the devices' motion, thus improving their performance and usability.

6. Conclusions

This paper has presented a novel approach for the complete (or dynamic) balancing of mechanisms: the use of fully Cartesian coordinates (FCC) in conjunction with an optimization method such as Differential Evolution (DE).

Among the main contributions of this paper is the development of the two-dimensional mass matrix for elements defined by three basic points. To our knowledge, this matrix has not been proposed so far in the literature and it can be applied to a vast number of more complex mechanisms that use type of linkages. By using FCC, this work has demonstrated that even for the most complex mechanisms, it is possible to obtain relatively simple non-trigonometric equations that define the ShF and ShM and to further optimize their dynamic balancing with algorithms such as DE with very good results.

As future work, it is expected to continue exploring FCC together with other meta-heuristic methods to optimize more complex mechanisms in two and three dimensions. Our approach promises to be highly efficient for optimizing the balancing conditions of even more complex mechanisms once their mass matrices have been defined.

Author Contributions: Conceptualization, M.A. and M.T.O.-G.; methodology, M.T.O.-G. and C.N.S.; simulation, M.T.O.-G.; validation, D.U.C.-D., P.V. and R.V.; formal analysis, M.A., R.V. and A.A.G.; investigation, M.T.O.-G., M.A. and R.V.; writing—original draft preparation, M.T.O.-G.; writing—review and editing, R.V. and P.V.; visualization, D.U.C.-D. and A.A.G.; supervision, M.A. and R.V. All authors have read and agreed to the published version of the manuscript.

Funding: This research received no external funding.

Institutional Review Board Statement: Not applicable.

Informed Consent Statement: Not applicable.

Conflicts of Interest: The authors declare no conflict of interest.

Abbreviations

The following abbreviations are used in this manuscript:

CC	Cartesian Coordinates
FCC	Fully Cartesian Coordinates
ShF	Shaking Force
ShM	Shaking Moment
CoM	Center of Mass
DE	Differential Evolution

Appendix A. Mass-Matrix for Individual Linkages

It is necessary to define the mass matrix for each of the linkages in order to assemble the mass matrix of the entire mechanism.

Given that each counterweight is firmly attached to its corresponding link and that their position with respect to the local reference system does not change with the mechanism's motion, it is possible to consider five linkages to be optimized, each of them encompassing a link and a counterweight.

Linkages 1 and 3 have mass matrices consisting of three basic points, while linkages 2, 4, and 5 have mass matrices involving two basic points.

To define the different mass matrices, it is necessary to substitute the terms m, $\bar{x}_g$, $\bar{y}_g$, I_x, I_y, I_{xy}, and I_z in the equations that define the mass matrices of two and three points, considering the contribution of both the link and the counterweight. Thus, for each n linkage, Equations (A1)–(A7) can be written.

$$m_n = m_{bn} + m_{cn} = m_{bn} + \rho \pi t_{cn}(x_{cn}^2 + y_{cn}^2) \tag{A1}$$

$$x_{gn} = \frac{x_{bn}m_{bn} + x_{cn}m_{cn}}{m_{bn} + m_{cn}} \tag{A2}$$

$$y_{gn} = \frac{y_{bn}m_{bn} + y_{cn}m_{cn}}{m_{bn} + m_{cn}} \tag{A3}$$

$$I_{xn} = I_{xbn} + I_{xcn} = I_{xbn} + \frac{1}{4}\rho \pi t_{cn}(x_{cn}^2 + y_{cn}^2)(x_{cn}^2 + 5y_{cn}^2) \tag{A4}$$

$$I_{yn} = I_{ybn} + I_{ycn} = I_{ybn} + \frac{1}{4}\rho \pi t_{cn}(x_{cn}^2 + y_{cn}^2)(5x_{cn}^2 + y_{cn}^2) \tag{A5}$$

$$I_{xyn} = I_{xybn} + I_{xycn} = I_{xybn} + \rho \pi t_{cn}(x_{cn}^2 + y_{cn}^2)x_{cn}y_{cn} \tag{A6}$$

$$I_{zn} = I_{zbn} + I_{zcn} = I_{xbn} + I_{ybn} + I_{xcn} + I_{ycn} \tag{A7}$$

Appendix A.1. Mass Matrix for Linkages 1 and 3

To define the mass matrix of linkage 1, let us consider point C as i, point E as j, and point D as k. Similarly, for linkage 3, point B is considered as i, point F as j, and point E as k.

Figure A1 shows the distances K_{x1}, K_{y1}, K_{x3}, and K_{y3}. Note that for linkage 1, $l_{ij} = l_1$ corresponds to the distance $\bar{CE}$ while for linkage 2, $l_{ij} = l_2$ corresponds to the distance $\bar{BF}$.

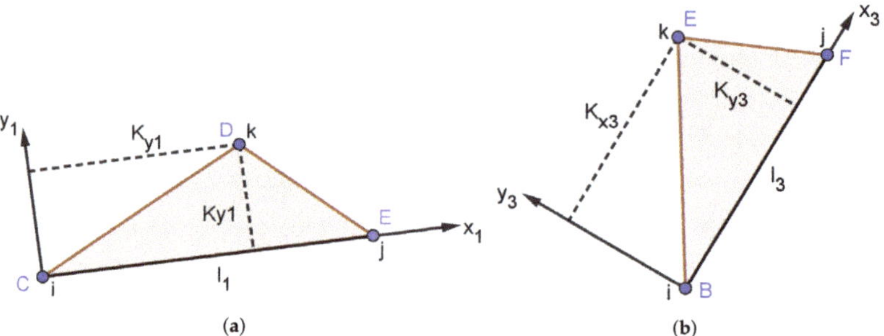

Figure A1. Linkages defined by three basic points. (a) Linkage 1. (b) Linkage 3.

By using the mass matrix $\mathbf{M_{3P}}$ (Equation (24)), the terms of the mass matrix for linkages $n = 1$ and $n = 3$ can be defined according to Equations (A8)–(A13).

$$
\begin{aligned}
e_n = {} & \frac{I_{xbn}K_{xn}^2}{K_{yn}^2 l_n^2} - \frac{2I_{xbn}K_{xn}}{K_{yn}^2 l_n} + \frac{I_{xbn}}{K_{yn}^2} - \frac{2I_{xybn}K_{xn}}{K_{yn}l_n^2} + \frac{2I_{xybn}}{K_{yn}l_n} + \frac{I_{ybn}}{l_n^2} + \frac{\pi K_{xn}^2 \rho_{cn} t_{cn} x_{cn}^4}{4K_{yn}^2 l_n^2} \\
& + \frac{3\pi K_{xn}^2 \rho_{cn} t_{cn} x_{cn}^2}{2K_{yn}^2 l_n^2} y_{cn} + \frac{5\pi K_{xn}^2 \rho_{cn} t_{cn} y_{cn}^4}{4K_{yn}^2 l_n^2} + \frac{2K_{xn} m_{bn} y_{bn}}{K_{yn}l_n} + \frac{2\pi K_{xn}\rho_{cn}}{K_{yn}l_n} t_{cn} x_{cn}^2 y_{cn} \\
& + \frac{2\pi K_{xn}\rho_{cn}}{K_{yn}l_n} t_{cn} y_{cn}^3 - \frac{2\pi K_{xn}\rho_{cn}}{K_{yn}l_n^2} t_{cn} x_{cn}^3 y_{cn} - \frac{2\pi K_{xn}\rho_{cn}}{K_{yn}l_n^2} t_{cn} x_{cn} y_{cn}^3 - \frac{\pi K_{xn}\rho_{cn} t_{cn} x_{cn}^4}{2K_{yn}^2 l_n} \\
& - \frac{3\pi K_{xn}\rho_{cn}}{K_{yn}^2 l_n} t_{cn} x_{cn}^2 y_{cn}^2 - \frac{5\pi K_{xn}\rho_{cn} t_{cn} y_{cn}^4}{2K_{yn}^2 l_n} - \frac{2m_{bn}^2 x_{bn}}{l_n m_{bn} + \pi l_n \rho_{cn} t_{cn} x_{cn}^2 + \pi l_n \rho_{cn} t_{cn} y_{cn}^2} \\
& - \frac{2m_{bn}^2 y_{bn}}{K_{yn} m_{bn} + \pi K_{yn}\rho_{cn} t_{cn} x_{cn}^2 + \pi K_{yn}\rho_{cn} t_{cn} y_{cn}^2} - \frac{2\pi m_{bn}\rho_{cn} t_{cn} x_{bn} x_{cn}^2}{l_n m_{bn} + \pi l_n \rho_{cn} t_{cn} x_{cn}^2 + \pi l_n \rho_{cn} t_{cn} y_{cn}^2} \\
& - \frac{2\pi m_{bn}\rho_{cn} t_{cn} x_{bn} y_{cn}^2}{l_n m_{bn} + \pi l_n \rho_{cn} t_{cn} x_{cn}^2 + \pi l_n \rho_{cn} t_{cn} y_{cn}^2} - \frac{2\pi m_{bn}\rho_{cn} t_{cn} x_{cn}^3}{l_n m_{bn} + \pi l_n \rho_{cn} t_{cn} x_{cn}^2 + \pi l_n \rho_{cn} t_{cn} y_{cn}^2} \\
& - \frac{2\pi m_{bn}\rho_{cn} t_{cn} x_{cn}^2 y_{bn}}{K_{yn} m_{bn} + \pi K_{yn}\rho_{cn} t_{cn} x_{cn}^2 + \pi K_{yn}\rho_{cn} t_{cn} y_{cn}^2} \\
& - \frac{2\pi m_{bn}\rho_{cn} t_{cn} x_{cn}^2 y_{cn}}{K_{yn} m_{bn} + \pi K_{yn}\rho_{cn} t_{cn} x_{cn}^2 + \pi K_{yn}\rho_{cn} t_{cn} y_{cn}^2} \\
& - \frac{2\pi m_{bn}\rho_{cn} t_{cn} x_{cn} y_{cn}^2}{l_n m_{bn} + \pi l_n \rho_{cn} t_{cn} x_{cn}^2 + \pi l_n \rho_{cn} t_{cn} y_{cn}^2} - \frac{2\pi m_{bn}\rho_{cn} t_{cn} y_{bn} y_{cn}^2}{K_{yn} m_{bn} + \pi K_{yn}\rho_{cn} t_{cn} x_{cn}^2 + \pi K_{yn}\rho_{cn} t_{cn} y_{cn}^2} \\
& - \frac{2\pi m_{bn}\rho_{cn} t_{cn} y_{cn}^3}{K_{yn} m_{bn} + \pi K_{yn}\rho_{cn} t_{cn} x_{cn}^2 + \pi K_{yn}\rho_{cn} t_{cn} y_{cn}^2} - \frac{2\pi^2 \rho_{cn}^2 t_{cn}^2 x_{cn}^5}{l_n m_{bn} + \pi l_n \rho_{cn} t_{cn} x_{cn}^2 + \pi l_n \rho_{cn} t_{cn} y_{cn}^2} \\
& - \frac{2\pi^2 \rho_{cn}^2 t_{cn}^2 x_{cn}^4 y_{cn}}{K_{yn} m_{bn} + \pi K_{yn}\rho_{cn} t_{cn} x_{cn}^2 + \pi K_{yn}\rho_{cn} t_{cn} y_{cn}^2} - \frac{4\pi^2 \rho_{cn}^2 t_{cn}^2 x_{cn}^3 y_{cn}^2}{l_n m_{bn} + \pi l_n \rho_{cn} t_{cn} x_{cn}^2 + \pi l_n \rho_{cn} t_{cn} y_{cn}^2} \\
& - \frac{4\pi^2 \rho_{cn}^2 t_{cn}^2 x_{cn}^2 y_{cn}^3}{K_{yn} m_{bn} + \pi K_{yn}\rho_{cn} t_{cn} x_{cn}^2 + \pi K_{yn}\rho_{cn} t_{cn} y_{cn}^2} - \frac{2\pi^2 \rho_{cn}^2 t_{cn}^2 x_{cn} y_{cn}^4}{l_n m_{bn} + \pi l_n \rho_{cn} t_{cn} x_{cn}^2 + \pi l_n \rho_{cn} t_{cn} y_{cn}^2} \\
& - \frac{2\pi^2 \rho_{cn}^2 t_{cn}^2 y_{cn}^5}{K_{yn} m_{bn} + \pi K_{yn}\rho_{cn} t_{cn} x_{cn}^2 + \pi K_{yn}\rho_{cn} t_{cn} y_{cn}^2} + 1 + \frac{5\pi \rho_{cn} t_{cn}}{4l_n^2} x_{cn}^4 + \frac{3\pi \rho_{cn} t_{cn}}{2l_n^2} x_{cn}^2 y_{cn}^2 \\
& + \frac{\pi \rho_{cn} t_{cn} y_{cn}^4}{4l_n^2} + \frac{2\pi \rho_{cn} t_{cn}}{K_{yn}l_n} x_{cn}^3 y_{cn} + \frac{2\pi \rho_{cn} t_{cn}}{K_{yn}l_n} x_{cn} y_{cn}^3 + \frac{\pi \rho_{cn} t_{cn} x_{cn}^4}{4K_{yn}^2} + \frac{3\pi \rho_{cn} t_{cn}}{2K_{yn}^2} x_{cn}^2 y_{cn}^2 \\
& + \frac{5\pi \rho_{cn} t_{cn}}{4K_{yn}^2} y_{cn}^4
\end{aligned}
\tag{A8}
$$

$$
\begin{aligned}
f_n = {} & -\frac{I_{xbn}K_{xn}^2}{K_{yn}^2 l_n^2} + \frac{I_{xbn}K_{xn}}{K_{yn}^2 l_n} + \frac{2I_{xybn}K_{xn}}{K_{yn}l_n^2} - \frac{I_{xybn}}{K_{yn}l_n} - \frac{I_{ybn}}{l_n^2} - \frac{\pi K_{xn}^2 \rho_{cn} t_{cn} x_{cn}^4}{4K_{yn}^2 l_n^2} \\
& - \frac{3\pi K_{xn}^2 \rho_{cn} t_{cn} x_{cn}^2}{2K_{yn}^2 l_n^2} y_{cn}^2 - \frac{5\pi K_{xn}^2 \rho_{cn} t_{cn} y_{cn}^4}{4K_{yn}^2 l_n^2} - \frac{K_{xn} m_{bn} y_{bn}}{K_{yn}l_n} - \frac{\pi K_{xn}\rho_{cn} t_{cn}}{K_{yn}l_n} x_{cn}^2 y_{cn} \\
& - \frac{\pi K_{xn}\rho_{cn} t_{cn}}{K_{yn}l_n} y_{cn}^3 + \frac{2\pi K_{xn}\rho_{cn}}{K_{yn}l_n^2} t_{cn} x_{cn}^3 y_{cn} + \frac{2\pi K_{xn}\rho_{cn}}{K_{yn}l_n^2} t_{cn} x_{cn} y_{cn}^3 + \frac{\pi K_{xn}\rho_{cn} t_{cn} x_{cn}^4}{4K_{yn}^2 l_n} \\
& + \frac{3\pi K_{xn}\rho_{cn} t_{cn} x_{cn}^2}{2K_{yn}^2 l_n} y_{cn}^2 + \frac{5\pi K_{xn}\rho_{cn} t_{cn} y_{cn}^4}{4K_{yn}^2 l_n} + \frac{m_{bn} x_{bn}}{l_n} + \frac{\pi \rho_{cn}}{l_n} t_{cn} x_{cn}^3 + \frac{\pi \rho_{cn}}{l_n} t_{cn} x_{cn} y_{cn}^2 \\
& - \frac{5\pi \rho_{cn} t_{cn}}{4l_n^2} x_{cn}^4 - \frac{3\pi \rho_{cn} t_{cn}}{2l_n^2} x_{cn}^2 y_{cn}^2 - \frac{\pi \rho_{cn} t_{cn} y_{cn}^4}{4l_n^2} - \frac{\pi \rho_{cn} t_{cn} y_{cn}}{K_{yn}l_n} x_{cn}^3 - \frac{\pi \rho_{cn} t_{cn} x_{cn}}{K_{yn}l_n} y_{cn}^3
\end{aligned}
\tag{A9}
$$

$$g_n = \frac{I_{xbn}K_{xn}}{K_{yn}^2 l_n} - \frac{I_{xbn}}{K_{yn}^2} - \frac{I_{xybn}}{K_{yn}l_n} + \frac{\pi K_{xn}\rho_{cn}t_{cn}x_{cn}^4}{4K_{yn}^2 l_n} + \frac{3\pi K_{xn}\rho_{cn}t_{cn}x_{cn}^2}{2K_{yn}^2 l_n}y_{cn}^2$$
$$+ \frac{5\pi K_{xn}\rho_{cn}t_{cn}y_{cn}^4}{4K_{yn}^2 l_n} + \frac{m_{bn}y_{bn}}{K_{yn}} + \frac{\pi\rho_{cn}}{K_{yn}}t_{cn}x_{cn}^2 y_{cn} + \frac{\pi\rho_{cn}}{K_{yn}}t_{cn}y_{cn}^3 - \frac{\pi\rho_{cn}t_{cn}y_{cn}}{K_{yn}l_n}x_{cn}^3$$
$$- \frac{\pi\rho_{cn}t_{cn}x_{cn}}{K_{yn}l_n}y_{cn}^3 - \frac{\pi\rho_{cn}t_{cn}x_{cn}^4}{4K_{yn}^2} - \frac{3\pi\rho_{cn}t_{cn}}{2K_{yn}^2}x_{cn}^2 y_{cn}^2 - \frac{5\pi\rho_{cn}t_{cn}}{4K_{yn}^2}y_{cn}^4 \tag{A10}$$

$$h_n = \frac{I_{xbn}K_{xn}^2}{K_{yn}^2 l_n^2} - \frac{2I_{xybn}K_{xn}}{K_{yn}l_n^2} + \frac{I_{ybn}}{l_n^2} + \frac{\pi K_{xn}^2\rho_{cn}t_{cn}x_{cn}^4}{4K_{yn}^2 l_n^2} + \frac{3\pi K_{xn}^2\rho_{cn}t_{cn}x_{cn}^2}{2K_{yn}^2 l_n^2}y_{cn}^2$$
$$+ \frac{5\pi K_{xn}^2\rho_{cn}t_{cn}y_{cn}^4}{4K_{yn}^2 l_n^2} - \frac{2\pi K_{xn}\rho_{cn}}{K_{yn}l_n^2}t_{cn}x_{cn}^3 y_{cn} - \frac{2\pi K_{xn}\rho_{cn}}{K_{yn}l_n^2}t_{cn}x_{cn}y_{cn}^3 + \frac{5\pi\rho_{cn}t_{cn}}{4l_n^2}x_{cn}^4 \tag{A11}$$
$$+ \frac{3\pi\rho_{cn}t_{cn}}{2l_n^2}x_{cn}^2 y_{cn}^2 + \frac{\pi\rho_{cn}t_{cn}y_{cn}^4}{4l_n^2}$$

$$i_n = - \frac{I_{xbn}K_{xn}}{K_{yn}^2 l_n} + \frac{I_{xybn}}{K_{yn}l_n} - \frac{\pi K_{xn}\rho_{cn}t_{cn}x_{cn}^4}{4K_{yn}^2 l_n} - \frac{3\pi K_{xn}\rho_{cn}t_{cn}x_{cn}^2}{2K_{yn}^2 l_n}y_{cn}^2 - \frac{5\pi K_{xn}\rho_{cn}t_{cn}y_{cn}^4}{4K_{yn}^2 l_n}$$
$$+ \frac{\pi\rho_{cn}t_{cn}y_{cn}}{K_{yn}l_n}x_{cn}^3 + \frac{\pi\rho_{cn}t_{cn}x_{cn}}{K_{yn}l_n}y_{cn}^3 \tag{A12}$$

$$j_n = \frac{I_{xbn}}{K_{yn}^2} + \frac{\pi\rho_{cn}t_{cn}x_{cn}^4}{4K_{yn}^2} + \frac{3\pi\rho_{cn}t_{cn}}{2K_{yn}^2}x_{cn}^2 y_{cn}^2 + \frac{5\pi\rho_{cn}t_{cn}}{4K_{yn}^2}y_{cn}^4 \tag{A13}$$

Appendix A.2. Mass Matrix for Linkages 2, 4, and 5

Linkages 2, 4, and 5 are defined by two basic points. For linkage 2, point A is considered as *i* while point C is considered as *j*; for linkage 4, point D is considered as *i* while point G is considered as *j*; for linkage 5, point F is considered as *i* while point G is considered as *j*. Each of them have their origin at point *i* and the x axis directed toward point *j*. For each linkage, l_n is the distance between points *i* and *j*.

By substituting the corresponding terms in the mass matrix of the elements defined by two basic points M_{2P} (equation can be found in [38]), it is possible to obtain the terms of the mass matrix of each of these elements consisting of a counterweight and a linkage (Equations (A14)–(A17)).

$$a_n = m_{bn} + \pi\rho_{cn}t_{cn}\left(x_{cn}^2 + y_{cn}^2\right) - \frac{1}{l_n\left(m_{bn} + \pi\rho_{cn}t_{cn}(x_{cn}^2 + y_{cn}^2)\right)}$$
$$\left(2.0m_{bn} + 2.0\pi\rho_{cn}t_{cn}\left(x_{cn}^2 + y_{cn}^2\right)\right)\left(m_{bn}x_{bn} + \pi\rho_{cn}t_{cn}x_{cn}\left(x_{cn}^2 + y_{cn}^2\right)\right) \tag{A14}$$
$$+ \frac{1}{l_n^2}\left(I_{bn} + \pi\rho_{cn}t_{cn}x_{cn}^2\left(x_{cn}^2 + y_{cn}^2\right) + \pi\rho_{cn}t_{cn}y_{cn}^2\left(x_{cn}^2 + y_{cn}^2\right) + 0.5\pi\rho_{cn}t_{cn}\left(x_{cn}^2 + y_{cn}^2\right)^2\right)$$

$$b_n = \frac{1}{l_n}\left(m_{bn}x_{bn} + \pi\rho_{cn}t_{cn}x_{cn}\left(x_{cn}^2 + y_{cn}^2\right)\right)$$
$$- \frac{1}{l_n^2}\left(I_{bn} + \pi\rho_{cn}t_{cn}x_{cn}^2\left(x_{cn}^2 + y_{cn}^2\right) + \pi\rho_{cn}t_{cn}y_{cn}^2\left(x_{cn}^2 + y_{cn}^2\right) + 0.5\pi\rho_{cn}t_{cn}\left(x_{cn}^2 + y_{cn}^2\right)^2\right) \tag{A15}$$

$$c_n = \frac{1}{l_n\left(m_{bn} + \pi\rho_{cn}t_{cn}(x_{cn}^2 + y_{cn}^2)\right)}\left(-m_{bn} - \pi\rho_{cn}t_{cn}\left(x_{cn}^2 + y_{cn}^2\right)\right)$$
$$\left(m_{bn}y_{bn} + \pi\rho_{cn}t_{cn}y_{cn}\left(x_{cn}^2 + y_{cn}^2\right)\right) \tag{A16}$$

$$d_n = \frac{1}{l_n^2}\left(I_{bn} + \pi\rho_{cn}t_{cn}x_{cn}^2\left(x_{cn}^2 + y_{cn}^2\right) + \pi\rho_{cn}t_{cn}y_{cn}^2\left(x_{cn}^2 + y_{cn}^2\right) + 0.5\pi\rho_{cn}t_{cn}\left(x_{cn}^2 + y_{cn}^2\right)^2\right) \tag{A17}$$

References

1. Arakelian, V.; Briot, S. *Balancing of Linkages and Robot Manipulators. Mechanisms and Machine Science*; Springer International Publishing: Cham, Switzerland, 2015; Volume 27.
2. Uicker, J.J.; Pennock, G.R.; Shigley, J.E. *Theory of Machines and Mechanisms*, 5th ed.; Oxford University Press: New York, NY, USA, 2016.
3. Waldron, K.J.; Kinzel, G.L. *Kinematics, Dynamics, and Design of Machinery*, 2nd ed.; John Wiley: Hoboken, NJ, USA, 2004.
4. Arakelian, V.; Dahan, M.; Smith, M.A. Historical review of the evolution of the theory on balancing of mechanisms. In *Symposium on History of Machines and Mechanisms Proceedings HMM 2000*; Ceccarelli, M., Ed.; Springer: Dordrecht, The Netherlands, 2000; pp. 291–300.
5. Arakelian, V. Inertia forces and moments balancing in robot manipulators: A review. *Adv. Robot.* **2017**, *31*, 717—726. [CrossRef]
6. Wei, B.; Zhang, D. A review of dynamic balancing for robotic mechanisms. *Robotica* **2021**, *39*, 55–71. [CrossRef]
7. Fisher, O. Über die reduzierten Systeme und die Hauptpunkte der Glieder eines Gelenkmechanismus und ihre Bedeutung für die technische Mechanik. *Z. Für Angew. Math. Und Phys.* **1902**, *47*, 429–466.
8. Goryachkin, V.P. The forces of inertia and their balancing. In *Collection of Scientific Works*; Kolos: Moscow, Russia, 1914; pp. 283–418.
9. Yudin, V. *The Balancing of Machines and Their Stability*; Edition of Academy of Red Army: Moscow, Russia, 1941; 124p.
10. Kreutzinger, R. Über die bewegung des Schwerpunktes beim Kurbelgetriebe. *Getriebetechnik* **1942**, *10*, 397–398.
11. Maxwell, R.L. *Kinematics and Dynamics of Machinery*, 1st ed.; Prentice Hall: Englewood Cliff, NJ, USA, 1960.
12. Smith, M.R.; Maundert, L. Inertia forces in a four-bar linkage. *J. Mech. Eng. Sci.* **1967**, *9*, 218–225. [CrossRef]
13. Talbourdet, G.L.; Shepler, P.R. Mathematical solution of 4-bar linkages-IV. Balancing of linkages. *Mach. Des.* **1941**, *13*, 73–77.
14. Lanchester, F.W. Engine Balancing. *Inst. Automob. Eng.* **1914**, *8*, 195–271. [CrossRef]
15. Root, R.E. *Dynamics of Engine and Shaft*; John Wiley: New York, NY, USA, 1932.
16. Artobolevsky, I.I.; Edelshtein, B.V. *Methods of Inertia Calculation for Mechanisms of Agricultural Machines*; Selkhozizdate: Moscow, Russia, 1935.
17. Gheronimus, Y.L. An approximate method of calculating a counterweight for the balancing of vertical inertia forces. *Mechanisms* **1968**, *3*, 283–288. [CrossRef]
18. Doucet, E. Équilibrage dynamique des moteurs en ligne. *Tech. Automob. Arienne* **1946**, *37*, 30–31.
19. Emod, I.; Jurek, A. Massenausgleich am Kurbelgetriebe von Sechszylinder-viertakt-V-motoren mit 6 Kurbeln und 60 Zylinderwinkeln. *Period. Polytech. Mech. Eng.* **1967**, *3–4*, 205–211.
20. Semenov, M.V. The synthesis of partly balanced plane mechanisms. *Mechanisms* **1968**, *3*, 339–353. [CrossRef]
21. Berkof R.S.; Lowen, G.G. A new method for complete force balancing simple linkages. *J. Eng. Ind.* **1969**, *91B*, 21–26. [CrossRef]
22. Smith, M.R. Dynamic analysis and balancing of linkages with interactive computer graphics. *Comput. Aided Des.* **1975**, *7*, 15–19. [CrossRef]
23. Tepper, F.R.; Lowen, G.G. General theorems concerning full force balancing of planar linkages by internal mass redistribution. *J. Eng. Ind.* **1972**, *94*, 789–796. [CrossRef]
24. Berkof, R.S. Complete force and moment balancing of inline four-bar linkages. *Mech. Mach. Theory* **1973**, *8*, 397–410. [CrossRef]
25. Wiederrich, J.L.; Roth, B. Momentum balancing of four-bar linkages. *J. Manuf. Sci. Eng.* **1976**, *4*, 1289–1295. [CrossRef]
26. Dresig, H.; Jacobi, P. Vollständiger trägheitskraftausgleich von ebenen koppelgetrieben durch anbringen eines zweischlages. *Maschinenbautechnik* **1974**, *23*, 5–8.
27. Feng, G. Complete Shaking Force and Shaking Moment balancing of four types of six-bar linkages. *Mech. Mach. Theory* **1989**, *24*, 275–287. [CrossRef]
28. Kochev, I.S. Active balancing of the frame Shaking Moment in high speed planar machines. *Mech. Mach. Theory* **1992**, *27*, 53–58. [CrossRef]
29. de Jong, J.J.; van Dijk, J.; Herder, J.L. A screw based methodology for instantaneous dynamic balance. *Mech. Mach. Theory* **2019**, *141*, 267–282. [CrossRef]
30. Acevedo, M.; Orvañanos-Guerrero, M.T.; Velázquez, R.; Arakelian, V. An alternative method for Shaking Force balancing of the 3RRR PPM through acceleration control of the center of mass. *Appl. Sci.* **2020**, *10*, 1351. [CrossRef]
31. Meijaard, J.P.; van der Wijk, V. Dynamic balancing of mechanisms with flexible links. *Mech. Mach. Theory* **2022**, *172*, 104784. [CrossRef]
32. Segla, S.; Kalker-Kalman, C.M.; Schwab, A.L. Statical balancing of a robot mechanism with the aid of a genetic algorithm. *Mech. Mach. Theory* **1998**, *33*, 163–174. [CrossRef]
33. Farmani, M.; Jaamialahmadi, A.; Babaie, M. Multiobjective optimization for force and moment balance of a four-bar linkage using evolutionary algorithms. *J. Mech. Sci. Technol.* **2011**, *25*, 2971–2977. [CrossRef]
34. Zamuda, A.; Brest, J.; Boskovic, B.; Zumer V. Differential evolution for multiobjective optimization with self adaptation. In Proceedings of the 2007 IEEE Congress on Evolutionary Computation, Singapore, 25–28 September 2007.

35. Erkaya, S. Investigation of balancing problem for a planar mechanism using genetic algorithm. *J. Mech. Sci. Technol.* **2013**, *27*, 2153–2160. [CrossRef]
36. Bošković, M.; Šalinić, S.; Bulatović, R.; Miodragović, G. Multiobjective optimization for dynamic balancing of four-bar mechanism. In Proceedings of the 6th International Congress of Serbian Society of Mechanics, Mountain Tara, Serbia, 19–21 June 2017.
37. García de Jalón, J. Twenty-five years of natural coordinates. *Multibody Syst. Dyn.* **2007**, *18*, 15–33. [CrossRef]
38. Orvañanos-Guerrero, M.T.; Sánchez, C.N.; Rivera, M.; Acevedo, M.; Velázquez, R. Gradient descent-based optimization method of a four-bar mechanism using Fully Cartesian coordinates. *Appl. Sci.* **2019**, *9*, 4115. [CrossRef]
39. Acevedo, M.; Orvañanos-Guerrero, M.T.; Velázquez, R.; Haro, E. Optimum balancing of the four-bar linkage using Fully Cartesian coordinates. *IEEE Lat. Am. Trans.* **2019**, *17*, 983–990. [CrossRef]
40. Bourbonnais, F.; Bigras, P.; Bonev, I.A. Minimum-time trajectory planning and control of a pick-and-place five-bar parallel robot. *IEEE/ASME Trans. Mechatron.* **2015**, *20*, 740–749. [CrossRef]
41. Wang, D.; Wang, L.; Wu, J.; Ye, H. An experimental study on the dynamics calibration of a 3-DOF parallel tool head *IEEE/ASME Trans. Mechatron.* **2019**, *24*, 2931–2941.
42. Pennock, G.R.; Israr, A. Kinematic analysis and synthesis of an adjustable six-bar linkage *Mech. Mach. Theory* **2009**, *44*, 306–323. [CrossRef]
43. García de Jalón, J.; Bayo, E. *Kinematic and Dynamic Simulation of Multibody Systems: The Real-Time Challenge*; Springer: New York, NY, USA, 1994.
44. Storn, R.; Price, K. Differential evolution-a simple and efficient heuristic for global optimization over continuous spaces. *J. Glob. Optim.* **1997**, *11*, 341–359. [CrossRef]
45. Das, A.K.; Mishra, D.; Das, K.; Mallick, P.K.; Kumar, S.; Zymbler, M.; El-Sayed, H. Prophesying the short-term dynamics of the crude oil future price by adopting the survival of the fittest principle of improved grey optimization and extreme learning machine. *Mathematics* **2022**, *10*, 1121. [CrossRef]
46. Álvarez Gutiérrez, D.; Sánchez Lasheras, F.; Martín Sánchez, V.; Suárez Gómez, S.L.; Moreno, V.; Moratalla-Navarro, F.; Molina de la Torre, A.J. A new algorithm for multivariate genome wide association studies based on differential evolution and extreme learning machines. *Mathematics* **2022**, *10*, 1024. [CrossRef]
47. Bergstra, J.; Bengio, Y. Random search for hyper-parameter optimization. *J. Mach. Learn. Res.* **2012**, *13*, 281–305.
48. Etesami, G.; Felezi, M.E.; Nariman-Zadeh, N. Pareto optimal multi-objective dynamical balancing of a slider-crank mechanism using differential evolution algorithm. *Int. J. Automot. Eng.* **2019**, *9*, 3021–3032.
49. Chaudhary, K.; Chaudhary, H. Optimal design of planar slider-crank mechanism using teaching-learning-based optimization algorithm. *J. Mech. Sci. Technol.* **2015**, *29*, 5189–5198. [CrossRef]
50. Chaudhary, K.; Chaudhary, H. Optimum balancing of slider-crank mechanism using equimomental system of point-masses. *Procedia Technol.* **2014**, *14*, 35–42. [CrossRef]
51. Orvañanos-Guerrero, M.T.; Acevedo, M.; Sánchez, C.N.; Giannoccaro, N.I.; Visconti, P.; Velázquez, R. Efficient balancing optimization of a simplified slider-crank mechanism. In Proceedings of the 2020 IEEE ANDESCON, Quito, Ecuador, 13–16 October 2020.
52. Chaudhary, K.; Chaudhary, H. Shape optimization of dynamically balanced planar four-bar mechanism. *Procedia Comput. Sci.* **2015**, *57*, 519–526. [CrossRef]
53. Belleri B.K.; Kerur, S.B. Balancing of planar six-bar mechanism with genetic algorithm. *J. Mech. Energy Eng.* **2020**, *4*, 303–308. [CrossRef]
54. Hernández, E.; Velázquez, R.; Macías-Quijas, R.; Pissaloux, E.; Giannoccaro, N.I.; Lay-Ekuakille, A. Kinematic computations for small-size humanoid robot KUBO. *ARPN J. Eng. Appl. Sci.* **2017**, *12*, 7311–7320.
55. Xu, K.; Liu, H.; Zhu, X.; Song, Y. Kinematic analysis of a novel planar six-bar bionic leg. *Mech. Mach. Sci.* **2019**, *73*, 13–21.
56. Velázquez, R.; Garzón-Castro, C.L.; Acevedo, M.; Orvañanos-Guerrero, M.T.; Ghavifekr, A.A. Design and characterization of a miniature bio-inspired mobile robot. In Proceedings of the 2021 12th International Symposium on Advanced Topics in Electrical Engineering, Bucharest, Romania, 25–27 March 2021.
57. Shao, Y.; Xiang, Z.; Liu, H.; Li, L. Conceptual design and dimensional synthesis of cam-linkage mechanisms for gait rehabilitation. *Mech. Mach. Theory* **2016**, *104*, 31–42. [CrossRef]

mathematics

Singularities of Serial Robots: Identification and Distance Computation Using Geometric Algebra

Isiah Zaplana [1,*], Hugo Hadfield [2] and Joan Lasenby [2]

[1] Department of Mechanical Engineering, KU Leuven, 3000 Leuven, Belgium
[2] Department of Engineering, University of Cambridge, Cambridge CB2 1PZ, UK; hh409@cam.ac.uk (H.H.); jl221@cam.ac.uk (J.L.)
* Correspondence: isiah.zaplana@kuleuven.be

Abstract: The singularities of serial robotic manipulators are those configurations in which the robot loses the ability to move in at least one direction. Hence, their identification is fundamental to enhance the performance of current control and motion planning strategies. While classical approaches entail the computation of the determinant of either a $6 \times n$ or $n \times n$ matrix for an n-degrees-of-freedom serial robot, this work addresses a novel singularity identification method based on modelling the twists defined by the joint axes of the robot as vectors of the six-dimensional and three-dimensional geometric algebras. In particular, it consists of identifying which configurations cause the exterior product of these twists to vanish. In addition, since rotors represent rotations in geometric algebra, once these singularities have been identified, a distance function is defined in the configuration space $\mathcal{C}$, such that its restriction to the set of singular configurations $\mathcal{S}$ allows us to compute the distance of any configuration to a given singularity. This distance function is used to enhance how the singularities are handled in three different scenarios, namely, motion planning, motion control and bilateral teleoperation.

Keywords: serial robotic manipulators; singularity identification; geometric algebra; rotor group; distance to a singularity

MSC: 15A66; 15A75; 70B15; 70Q05

Citation: Zaplana, I.; Hadfield, H.; Lasenby, J. Singularities of Serial Robots: Identification and Distance Computation Using Geometric Algebra. *Mathematics* **2022**, *10*, 2068. https://doi.org/10.3390/math10122068

Academic Editors: Higinio Rubio Alonso, Alejandro Bustos Caballero, Jesus Meneses Alonso and Enrique Soriano-Heras

Received: 15 April 2022
Accepted: 13 June 2022
Published: 15 June 2022

Publisher's Note: MDPI stays neutral with regard to jurisdictional claims in published maps and institutional affiliations.

1. Introduction

A serial robot manipulator is an open kinematic chain made up of a sequence of rigid bodies, called links, connected by means of actuated kinematic pairs, called joints, that provide relative motion between consecutive links. At the end of the last link, there is a tool or device known as the end-effector. Only two types of joints are considered throughout this work: revolute joints, which only perform rotations, and prismatic joints, which only perform translations. If joint i is revolute (prismatic), the amount it rotates (translates) is encoded by an angle θ_i (a displacement d_i). These scalars are known as the joint variables of the robot.

From a kinematic point of view, the end-effector position and orientation (also known as the pose) can be expressed as a differentiable function $f : \mathcal{C} \to X$, where $\mathcal{C}$ denotes the space of joint variables, called the configuration space of the robot, and X is the space of all positions and orientations of the end-effector with respect to a reference frame, which is usually called the operational space. A serial robot is said to have n degrees of freedom (DoF) if its configuration can be minimally specified by n variables. For a serial robot, the number and nature of the joints determine the number of the DoF. For the task of positioning and orientating its end-effector in the three-dimensional space, the manipulators with more than six DoF are called redundant while the rest are non-redundant.

In order to describe its relative position and orientation, a frame $\{o, x, y, z\}$ is attached to each joint (Figure 1). The relations between consecutive joint frames are conventionally described by homogeneous transformation matrices. In particular, $^{i-1}T_i$ relates frame $\{i\}$ to frame $\{i-1\}$ (the first joint frame is related to a fixed reference frame, known as the world frame). Therefore, the end-effector pose ^0T_n of a robot with n DoF with respect to the world frame can be represented as:

$$^0T_n = {}^0T_1\,{}^1T_2 \cdots {}^{n-1}T_n \tag{1}$$

with

$$^0T_n = \begin{pmatrix} R & p \\ 0 & 1 \end{pmatrix}, \tag{2}$$

where R is a rotation matrix that describes the end-effector orientation with respect to the world frame, while p is a position vector describing the end-effector position with respect to the world frame. This description is equivalent to the one provided by f, known as the kinematic function of the serial robot. Thus, $f(q) = x$, where x denotes the vector describing the end-effector pose and $q = (q_1, \ldots, q_n)$ denotes the vector whose components are the joint variables, also known as the configuration of the robot. Clearly, either $q_i = \theta_i$ if joint i is revolute or $q_i = d_i$ if joint i is prismatic.

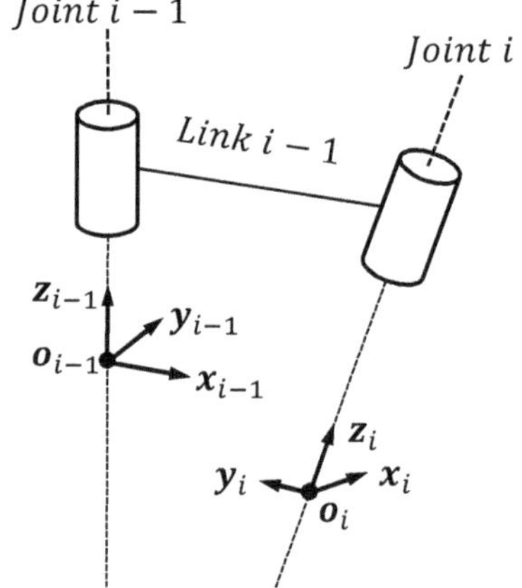

Figure 1. A frame $\{o, x, y, z\}$ is attached to each joint of the serial robot to describe its relative position and orientation.

Deriving the kinematic relation defined by f with respect to time, we obtain another relation:

$$\dot{x} = J(q)\dot{q}, \tag{3}$$

where $\dot{x}$ denotes the end-effector velocity vector, $\dot{q}$; the vector of the joint velocities; and J, the Jacobian matrix of f. If $J = [J_1 \cdots J_n]$, then each column J_i can also be computed as:

$$J_i = \left[\begin{array}{c} z_i \times (o_n - o_i) \\ z_i \end{array} \right] \quad \text{if joint } i \text{ is revolute,}$$

$$J_i = \left[\begin{array}{c} z_i \\ 0 \end{array} \right] \quad \text{if joint } i \text{ is prismatic.} \tag{4}$$

Definition 1. *Given a serial robot with n DoF, a singularity or kinematic singularity is a configuration $q \in \mathcal{C}$ satisfying $\rho(J(q)) < \min\{n, 6\}$, where $\rho(\cdot)$ denotes the rank of the matrix argument. The set of all singular configurations is a subset of $\mathcal{C}$ that is usually denoted by $\mathcal{S}$ and known as the singular set.*

Using the relation (3), it is easy to see that if $q \in \mathcal{C}$ is a singularity of a given serial robot, then the following two statements hold:

- The robot loses at least one degree of freedom or, equivalently, its end-effector cannot be translated along or rotated around at least one Cartesian direction.
- Finite linear and angular velocities of the end-effector may require infinite joint velocities.

In addition, Gottlieb [1] and Hollerbach [2] have independently proven that any serial manipulator with $n > 2$ DoF has singularities. The identification of such singularities is achieved by solving the following non-linear equation:

$$\det(J(q)) = 0 \tag{5}$$

if the robot is non-redundant and

$$\det(J(q)J^T(q)) = 0 \tag{6}$$

if it is redundant.

In general, if the serial robot possesses at least one revolute joint, several coefficients of the Jacobian matrix are non-linear expressions, and thus, neither Equation (5) nor Equation (6) are easy to formulate and solve. However, for manipulators with a spherical wrist, a simplification can be made. For these robots, the axes of their last three joints intersect at a common point, known as the wrist center point, or are parallel (the intersection point and, hence, the wrist center point, is the point at the infinity). Since the origin of the frame attached to the end-effector can be placed at the wrist center point, a zero block appears in $J(q)$ by definition (see Equation (4)). Hence:

$$J(q) = \begin{bmatrix} J_{11}(q) & 0 \\ J_{21}(q) & J_{22}(q) \end{bmatrix}, \tag{7}$$

where $J_{11}(q)$, $J_{21}(q)$ are blocks of order $3 \times (n-3)$, $J_{22}(q)$ is a block of order 3×3 and 0 denotes a block of order 3×3, whose entries are all zero. Now, Equation (5) is simplified to:

$$\det(J(q)) = \det(J_{11}(q))\det(J_{22}(q)), \tag{8}$$

from which the singularities can be obtained as solutions of either $\det(J_{11}(q)) = 0$ or $\det(J_{22}(q)) = 0$. These two equations allow us to decouple the singularities into position and orientation singularities as follows:

- Position singularities $PS = \{q \in \mathcal{C} : \det(J_{11}(q)) = 0\}$;
- Orientation singularities $OS = \{q \in \mathcal{C} : \det(J_{22}(q)) = 0\}$.

Similarly, we can make the same decoupling for redundant robots:

- Position singularities $PS = \{q \in \mathcal{C} : \text{rank}(J_{11}(q)) < 3\}$;
- Orientation singularities $OS = \{q \in \mathcal{C} : \det(J_{22}(q)) = 0\}$.

Remark 1. *The Jacobian matrix $J(q)$ is represented with respect to the world frame (usually located at the base of the robot). However, sometimes it is useful to represent $J(q)$ in a different frame $\mathcal{B}$. To do so, the following identity is used:*

$$J(q)^{\mathcal{B}} = BJ(q), \tag{9}$$

where:

$$B = \begin{bmatrix} R_0^{\mathcal{B}} & 0 \\ 0 & R_0^{\mathcal{B}} \end{bmatrix}, \tag{10}$$

with $R_0^{\mathcal{B}} = (R_{\mathcal{B}}^0)^T$ and where $R_{\mathcal{B}}^0$ denotes the rotation matrix that relates the orientation of $\mathcal{B}$ with respect to the orientation of the world frame.

Singularity identification is one of the fundamental research fields in robot kinematics since, as stated before, it affects the motion of the robot and its performance when executing different tasks. Therefore, such identification is fundamental to enhancing the performance of current control and motion planning strategies by designing approaches to handle them. For instance, current applications of the subject include the handling of singularities for a robust control architecture in human–robot collaboration [3], the planning of singularity-free trajectories [4,5] and the smooth trajectory generation for rotating extensible manipulators or painting robots [6,7]. However, such identification is still problematic. The majority of the current approaches are based on the computation of either $\det(J(q)) = 0$ or $\det(J(q)J^T(q)) = 0$ or on the manipulation of the singular values of $J(q)$ [8–10]; hence, they are not computationally efficient. In addition, there is no efficient way of computing how close an arbitrary configuration is to a given singularity (which is fundamental for defining a threshold from where the strategies to handle them start to work). In this context, geometric algebra turns out to be very useful. In addition, it is currently applied to several problems in robot kinematics and geometry [11–15].

There is not much literature regarding the identification of singularities using geometric algebra, and the majority of the contributions focus on parallel mechanisms. For serial robots, Corrochano & Sobczyk [16] extend the Lie bracket of two vectors defined in any Lie algebra to what they call the superbracket of the lines $\ell_1, \ldots, \ell_6$, $[\ell_1, \ldots, \ell_6]$, where the line ℓ_i denotes the axis of joint i. For serial robots with six DoF, the main idea is to split the superbracket into smaller superbrackets, called bracket monomials, that are equated to zero. The singularities are the solutions of these monomial bracket equations. Following the same idea, Kanaan et al. [17] define the superbracket in a Grassmann–Caley algebra. Since the Lie bracket is well defined in every Grassmann–Caley algebra, the superbracket is also well defined. However, splitting the superbracket into the bracket monomials in this context is not intuitive and, as a consequence, is not always realizable. In addition, there is no standard procedure for computing these brackets' monomials.

For parallel mechanisms, the majority of works [18–22] focus on approaches developed for some particular parallel robots. The main idea consists of computing, for each leg of the mechanism, the exterior product of the twists defined by its joints and equating it to zero. For those legs with less than six actuated joints, combinations of two, three or more legs are considered. The main problem with these approaches is their lack of generalisability. Each approach is designed for the specific parallel robot the authors work with.

Huo et al. [23] present a mobility analysis applying conformal geometric algebra and a singularity analysis using an idea similar to the ones presented in the above-mentioned contributions. A mobility analysis of overconstrained parallel mechanisms is performed using Grassmann–Cayley algebra by Chai et al. [24], while Yang and Li [25] propose a novel identification method for the constraint singularities of parallel robots based on differential manifolds. Finally, Kim et al. [26] apply conformal geometric algebra to the identification of the singularities of a particular type of parallel manipulator, the SPS-parallel manipulator. Several lines and planes are defined using the different joint axes. Then, the relative positions of different combinations of these geometric entities are studied to geometrically find the singularities. However, this method cannot be extended to other classes of parallel or serial robots, nor can it be implemented as an algorithm due to its complex geometrical nature.

In this paper, a novel approach for singularity identification based on the six-dimensional and three-dimensional geometric algebras is introduced. It extends the works developed for parallel robots and reviewed above. In particular, one of the novelties of this method is that it can be applied to both redundant and non-redundant serial robots of any geometry. We first model the twists defined by the joint axes as vectors of the six-dimensional geometric algebra. Then, we manipulate the exterior product of these twists. In addition, this method can be simplified for serial robots with a spherical wrist using, instead of the six-dimensional geometric algebra $\mathcal{G}_6$, the three-dimensional geometric algebra $\mathcal{G}_3$. Once the singularities have been identified and since rotors describe the transformations between arbitrary multivectors in geometric algebra, a distance function D can be defined in the configuration space $\mathcal{C}$ that can be used to determine the distance of any arbitrary configuration $q \in \mathcal{C}$ to a given singularity $q_s \in \mathcal{S}$. This is the first time, to the best of the authors' knowledge, that such a distance function has been defined. It is well-known that there are several indices that can be used to check whether a given configuration is close to a singularity or not. For instance, the Jacobian matrix $J(q)$ allows us to define the manipulability index w_{m} as:

$$w_{\mathrm{m}} = \sqrt{\det(J(q)^T J(q))} = \sigma_n \cdots \sigma_1, \tag{11}$$

where $0 \leq \sigma_1 \leq \cdots \leq \sigma_n$ are the singular values of $J(q)$. Alternatively, we can also define the condition number of $J(q)$, $w_{\mathrm{c}} = \sigma_n / \sigma_1$. Clearly, the former is close to zero when the configuration is close to a singularity, while the value of the latter increases as the robot approaches to a singular configuration. Although there are several approaches based on the use of such indices [27,28], none of them define a distance function, and as stated in [29], they do not provide a realistic measure of how close a singularity is, only whether it is close or not. On the other hand, Yao et al. [30] propose a different index of closeness to singularities for planar parallel robots based on the volume of the workspace. Although interesting, it is still not a distance function, and it cannot be easily applied to serial robots. Similarly, Nawratil [31] defines a distance function for parallel manipulators of the Stewart–Gough type. However, it measures how close a given pose of the end-effector is to a singular pose (i.e., the pose associated with a singular configuration). Hence, such a distance function is not defined in the configuration space $\mathcal{C}$ but in the operational space X. Finally, Bu [32] defines an angle between the velocity vector associated with one of the joints and the manifold generated by the others. Again, such an angle acts as a measure of closeness but not as a distance function, and thus, it does not provide a realistic measure of how close a singularity is.

The rest of the paper is organized as follows: Section 2 presents an overview of geometric algebra that will be useful for understanding the proposed contribution. In Section 3, the novel singularity identification approach and the simplification for serial robots with a spherical wrist are fully developed, while the novel distance function is constructed in Section 4. The application of these results to the Kuka LWR 4+, a redundant serial robot with a spherical wrist, is given in Section 5. Section 6 lists three different applications where both the singularity identification and the novel distance function can be applied in order to illustrate their utility. Finally, the conclusions are given in Section 7.

2. Mathematical Preliminaries: Geometric Algebra

One of the main problems of vector spaces is that linear transformations between them are represented through matrices, which entails a high computational cost when implemented. To overcome these and related problems, geometric algebra provides an excellent framework. It was first introduced by W. K. Clifford [33], whom based his own work on the previous works of H. Grassmann [34] and W. R. Hamilton [35]. Throughout this section, a brief overview of geometric algebra is presented. More detailed treatments of the subject can be found in [36–39].

Definition 2. *Given two vectors $x_1, x_2 \in \mathbb{R}^n$, the outer or exterior product of x_1 and x_2, $x_1 \wedge x_2$, is a new element that can be seen as the oriented area of the parallelogram obtained by sweeping the vector x_1 along x_2. The exterior product is bilinear, associative and anticommutative. In particular, $x \wedge x = 0$ for every $x \in \mathbb{R}^n$.*

The new element defined by the exterior product is called a bivector, and it is defined to have grade two. By extension, the outer product of a bivector with a vector is known as a trivector, is denoted by $x_1 \wedge x_2 \wedge x_3$ and is defined to have grade three. Trivectors can be seen as the oriented volume obtained by sweeping the bivector $x_1 \wedge x_2$ along x_3.

This can be generalized to an arbitrary dimension. Thus,

$$x_1 \wedge x_2 \wedge \cdots \wedge x_k \tag{12}$$

denotes a k-blade, i.e., an element of grade k. Linear combinations of k-blades are known as k-vectors, while linear combinations of k-vectors (for different k) are known as multivectors.

In his work [33], Clifford extends the exterior product by adding a scalar product between vectors, the inner product. He defines the geometric product (also known as the Clifford product) as follows:

$$x_1 x_2 = x_1 \cdot x_2 + x_1 \wedge x_2 \quad (x_1, x_2 \in \mathbb{R}^n). \tag{13}$$

Thus, the geometric product between two vectors has two components: the scalar component given by the inner product and the bivector component given by the exterior product. Clearly, it also inherits the associativity and bilinearity of the exterior product.

When applied to an orthonormal basis $B = \{e_1, \ldots, e_n\}$ of $\mathbb{R}^n$, the geometric product acts as follows:

$$e_i e_j = \begin{cases} 1 & \text{for} \quad i = j \\ e_i \wedge e_j & \text{for} \quad i \neq j \end{cases} \tag{14}$$

Thus, for each $0 \leq k \leq n$, the set of k-vectors is spanned by:

$\boxed{k = 0}$ $\{1\}$ (scalars).
$\boxed{k = 1}$ $\{e_1, \ldots, e_n\}$ (vectors).
$\boxed{k = 2}$ $\{e_i \wedge e_j\}_{1 \leq i < j \leq n}$ (bivectors).
$\boxed{k = 3}$ $\{e_i \wedge e_j \wedge e_k\}_{1 \leq i < j < k \leq n}$ (trivectors).

$$\vdots$$

$\boxed{k = r}$ $\{e_{i_1} \wedge \cdots \wedge e_{i_r}\}_{1 \leq i_1 < \cdots < i_r \leq n}$ (r-vectors).

$$\vdots$$

$\boxed{k = n}$ $\{e_1 \wedge \cdots \wedge e_n\}$ (pseudoscalar).

Then, for each $0 \leq k \leq n$, there are exactly $C(n, k)$ generators for the set of k-vectors, and thus, the set of k-vectors defines a vector space with basis $B_k = \{e_{i_1} \wedge \cdots \wedge e_{i_k}\}_{1 \leq i_1 < \cdots < i_k \leq n}$ and dimension $C(n, k)$.

Definition 3. *Let $\mathbb{R}^n$ denote the real vector space of dimension n. Then, the vector space spanned by the basis*

$$B = \{e_{i_1} \wedge \cdots \wedge e_{i_r}\}_{\substack{1 \leq i_1 < \cdots < i_r \leq n \\ 0 \leq r \leq n}} \tag{15}$$

endowed with the geometric product defined in (13) is an algebra over $\mathbb{R}$ known as the geometric algebra (GA) of $\mathbb{R}^n$. Such an algebra is denoted by $\mathcal{G}_n$ and has the dimension $C(n, 0) + C(n, 1) + \cdots + C(n, n) = 2^n$.

Remark 2. *Since the grading structure of multivectors is a property associated with the exterior product, the elements of $\mathcal{G}_n$ can still be called k-blades, k-vectors and multivectors.*

An important family of linear operators in $\mathcal{G}_n$ are the grade-k projection operators, denoted by $\langle \cdot \rangle_k$ for $0 \leq k \leq n$. Applied to an arbitrary multivector A, $\langle A \rangle_k$ projects onto the grade-k components in A, i.e., it returns the components of A that can be expressed as a linear combination of $\{e_{i_1} \wedge \cdots \wedge e_{i_k}\}_{1 \leq i_1 < \cdots < i_k \leq n}$. Obviously, if A_k denotes a k-vector, then $\langle A_k \rangle_k = A_k$.

Using these operators, general multivectors $A \in \mathcal{G}_n$ can be expressed as:

$$A = \langle A \rangle_0 + \langle A \rangle_1 + \cdots + \langle A \rangle_n. \tag{16}$$

Hence, the set of all k-vectors for a given $1 \leq k \leq n$ is a vector subspace of $\mathcal{G}_n$ denoted by $\langle \mathcal{G}_n \rangle_k$ and spanned by $B_k = \{e_{i_1} \wedge \cdots \wedge e_{i_k}\}_{1 \leq i_1 < \cdots < i_k \leq n}$.

The multivector representation (16) is very useful in defining another important operator in $\mathcal{G}_n$. This linear operator is known as the reversion operator and is denoted by the superscript $\sim$. The reversion is defined over the geometric product of m vectors as:

$$(a_1 \cdots a_m)^{\sim} = a_m \cdots a_1. \tag{17}$$

Applied to k-vectors, we have that:

$$\widetilde{A}_k = (-1)^{\frac{k(k-1)}{2}} A_k \tag{18}$$

due to the anticommutativity of the exterior product. Finally, since reversion is a linear operator, the reverse of an arbitrary multivector is:

$$\widetilde{A} = \left\langle \widetilde{A} \right\rangle_0 + \cdots + \left\langle \widetilde{A} \right\rangle_n = \langle A \rangle_0 + \langle A \rangle_1 - \langle A \rangle_2 + \cdots + (-1)^{\frac{n(n-1)}{2}} \langle A \rangle_n. \tag{19}$$

Notice that the reversion operator corresponds simply to matrix transposition when a matrix representation of the n-dimensional algebra is considered.

Finally, another operator of great interest is the dual operator. Every grade-n element of $\mathcal{G}_n$ is of the form $\alpha(e_1 \wedge \cdots \wedge e_n)$ for a scalar $\alpha \in \mathbb{R}$. For each $\alpha \in \mathbb{R}$, $\alpha(e_1 \wedge \cdots \wedge e_n)$ is known as the volume element E_α of $\mathcal{G}_n$, while the generator $e_1 \wedge \cdots \wedge e_n$ is known as the pseudoscalar of $\mathcal{G}_n$ and is usually denoted by I. Pseudoscalars allow us to define the dual operator, whose action over a k-vector A_k is:

$$A_k^* = I A_k, \tag{20}$$

where A_k^* is an $(n-k)$-vector.

Now, let us go back to the bivectors of $\mathcal{G}_n$ since they will be fundamental in the modelling of the rotations in $\mathbb{R}^n$. An important property of these bivectors is that they always square to a scalar. Therefore, given a bivector B, the unit bivector associated with B is $B' = B/|B^2|$. Unit bivectors of $\mathcal{G}_n$ always square to -1. This allows us to compute the following series:

$$\exp(\alpha B') = \sum_{m=0}^{\infty} \frac{(\alpha B')^m}{m!}, \tag{21}$$

where $\alpha \in \mathbb{R}$. Expanding Equation (21), we have that:

$$\begin{aligned}
\exp(\alpha B') &= 1 + \alpha B' - \frac{\alpha^2}{2} - \frac{\alpha^3 B}{3!} + \cdots \\
&= \left(1 - \frac{\alpha^2}{2} + \cdots\right) + B'\left(\alpha - \frac{\alpha^3}{3!} + \cdots\right) \\
&= \cos(\alpha) + B' \sin(\alpha).
\end{aligned} \tag{22}$$

Equation (22) indicates that $\exp(\alpha B')$ could be related to rotations. Indeed, we have the following result.

Proposition 1. *Let B be a unit bivector and $0 \leq \theta \leq 2\pi$; then, $R = \exp(-(\theta B)/2) = \cos(\theta/2) - B \sin(\theta/2) \in \langle \mathcal{G}_n \rangle_0 + \langle \mathcal{G}_n \rangle_2$ defines a rotation by an angle θ with a rotation plane represented by B. It acts over an element $X \in \mathcal{G}_n$ through the sandwiching product:*

$$X' = R X \widetilde{R}. \tag{23}$$

Such an element R is termed a rotor.

Rotors satisfy the following properties:

(1) $R\widetilde{R} = 1$;

(2) $Rx\widetilde{R} = (-R)x(-\widetilde{R})$ for $x \in \mathbb{R}^n$;

(3) $Rxy\widetilde{R} = Rx\widetilde{R}Ry\widetilde{R}$ for $x, y \in \mathbb{R}^n$;

(4) Given two frames $\{e_1, \ldots, e_n\}$ and $\{f_1, \ldots, f_n\}$, there always exists a rotor R that transforms one into the other, i.e., there exists a rotor R such that $e_i = Rf_i\widetilde{R}$ for $i = 1, \ldots, n$. This rotor is computed as [36,40]:

$$R = \frac{1 + e^1 f_1 + \cdots + e^n f_n}{\|1 + e^1 f_1 + \cdots + e^n f_n\|}, \tag{24}$$

where $\{e^1, \ldots, e^n\}$ denotes the reciprocal frame of $\{e_1, \ldots, e_n\}$.

The first property is the analogous version of the property defining the orthogonal matrices with determinant equal to 1, which are known to represent rotations. The second property proves that both R and $-R$ encode the same rotation, while the third property is known as the geometric covariance of rotors.

In general, rotors define a group $\mathfrak{R}$ with the geometric product as the group product:

$$\mathfrak{R} = \{R \in \langle \mathcal{G}_n \rangle_0 + \langle \mathcal{G}_n \rangle_2 \; : \; R\widetilde{R} = 1\}. \tag{25}$$

Therefore, the product of two different rotors R_1 and R_2 also encodes a rotation. In particular, it is the rotation resulting from the composition of the rotations encoded by R_1 and R_2, respectively. In addition, the second property states that $\mathfrak{R}$ provides a double covering of the rotation group.

Finally, one of the most important geometric algebras is the spatial geometric algebra $\mathcal{G}_3$, whose basis is:

$$\{1, e_1, e_2, e_3, e_{12}, e_{13}, e_{23}, I\}, \tag{26}$$

where $\{e_1, e_2, e_3\}$ is an orthonormal basis of $\mathbb{R}^3$ and $e_{ij} = e_i \wedge e_j$.

3. Identification of Singularities Using Geometric Algebra

Since six degrees of freedom are required to describe the position and orientation of a rigid body in the three-dimensional space, the more natural way of formulating the singularity problem is through the six-dimensional geometric algebra $\mathcal{G}_6$, which extends naturally the three-dimensional algebra $\mathcal{G}_3$ introduced in Section 2. Screw theory [41,42] provides an intuitive and geometrical description of the differential kinematics of serial and parallel manipulators using six-dimensional vectors. Because of this, throughout this chapter, some concepts taken from this theory will be employed. This will provide the initial framework to completely understand the approach introduced in this section.

As stated in the introduction, we are going to work with three-dimensional rigid motions, i.e., three-dimensional orientation-preserving isometries. They form a Lie group, called the special Euclidean group, denoted by $SE(3)$. Its associated Lie algebra is:

$$\mathfrak{se}(3) = \left\{ \hat{\xi} \in \mathcal{M}_4 \; : \; \hat{\xi} = \begin{pmatrix} \Omega & v \\ 0 & 0 \end{pmatrix} \right\}, \tag{27}$$

where Ω is a skew-symmetric matrix of order 3, $v \in \mathbb{R}^3$, and $\mathcal{M}_4$ denotes the vector space of order 4 square matrices with real entries. Since every skew-symmetric matrix Ω can be represented as a vector ω, we can express an element $\hat{\xi} \in \mathfrak{se}(3)$ as a six-dimensional vector $\xi = [\omega \; v]^T$, termed a twist. Therefore, twists are the infinitesimal generators of rigid motions via the exponential map, i.e., $\exp(\hat{\xi}t) = f(t)$ with $f \in SE(3)$. The next theorem is a fundamental result in screw theory.

Theorem 1 (Chasles, 1830 [43]). *Every rigid motion $f \in SE(3)$ can be realized as a rotation around an axis followed (preceded) by a translation along the same axis.*

Definition 4. *A screw motion consists of a rotation around an axis followed (preceded) by a translation along the same axis, the screw axis ℓ. The ratio between the translational and the rotational parts of the motion is known as the pitch and is denoted by h. In particular, if a point is rotated around ℓ by an angle $\theta \neq 0$ and translated along ℓ by an amount d, then $h = d/\theta$. By convention, if $\theta = 0$, $h = \infty$.*

Remark 3. *For infinitesimal motions, if $\theta \neq 0$, then the pitch is defined as $h = \dot{d}/\dot{\theta}$.*

Hence, every rigid motion is a screw motion. Particular cases of screw motions are pure rotations (pure translations) where the translation (rotation) is the identity or, equivalently, $h = 0$ ($h = \infty$). In addition, every screw motion can be characterized by the triple (ℓ, h, q), where q denotes the magnitude of the motion. If $h \neq \infty$, then $\theta = q$ and $d = h\theta$, while if $h = \infty$, then $\theta = 0$ and $d = q$. We call this triple the screw associated with the screw motion, and we denote it using \$.

Proposition 2. *Given a screw \$ $= (\ell, h, q)$ with screw axis ℓ, pitch h and magnitude q, there exists a twist ξ such that the rigid motion it generates is the screw motion associated with \$.*

Proposition 2 states a correspondence between twists and screws that is useful for our purposes. In particular, if p is a point on ℓ and v is its direction unit vector, then $\ell = \{p + v\lambda : \lambda \in \mathbb{R}\}$ and we have that:

$$\xi = \theta \begin{bmatrix} v \\ p \times v + hv \end{bmatrix} \quad \text{for a general screw motion,}$$

$$\xi = \theta \begin{bmatrix} v \\ p \times v \end{bmatrix} \quad \text{for a pure rotation,} \tag{28}$$

$$\xi = d \begin{bmatrix} 0 \\ v \end{bmatrix} \quad \text{for a pure translation.}$$

A twist ξ associated with a magnitude 1 screw \$ is said to be a unit twist. Hence, any twist ξ can be seen as a unit twist multiplied by the magnitude of the associated screw axis:

$$\xi = \theta \xi_U = \theta \begin{bmatrix} v \\ p \times v + hv \end{bmatrix}, \tag{29}$$

where ξ_U is a unit twist. Clearly, ξ is associated with \$ $= (\ell, h, q)$, while ξ_U is associated with \$ $= (\ell, h, 1)$.

Proposition 3. *Let us consider a rigid body performing a screw motion represented by the screw \$ $= (\ell, h, q(t))$, where the magnitude $q(t)$ is a time-dependent variable. Its velocity during the screw motion is given by the associated twist ξ, where, now, the pitch is defined as in Remark 3. In particular:*

$$\xi = \dot{\theta}(t) \begin{bmatrix} v \\ p \times v + hv \end{bmatrix} \quad \text{if } \theta \neq 0,$$

$$\xi = \dot{d}(t) \begin{bmatrix} 0 \\ v \end{bmatrix} \quad \text{if } \theta = 0,$$

$$\tag{30}$$

where, here, $\dot{\theta}(t)$ ($\dot{d}(t)$) is known as the twist amplitude.

Now, let us consider a serial robot with n DoF, where ω, v denote the angular and linear velocity vectors of its end-effector. If Equation (3) is expanded, the following is obtained:

$$\begin{bmatrix} v \\ \omega \end{bmatrix} = J_1(q)\dot{q}_1 + \cdots + J_n(q)\dot{q}_n, \tag{31}$$

where J_i denotes the i-th column of the Jacobian matrix J. Notice that the right side of Equation (31) can be seen as the addition of the twists associated with the joints of the robot, where $\dot{q}_i$ plays the role of the twist amplitude and where the linear and angular parts are interchanged. However, for the sake of formality, let us consider the unit twist ξ_i associated with the i-th joint of the robot (Since, from now on, we are going to work exclusively with unit twists, the subindex U is omitted for simplicity). Then:

$$\xi_i(q)\dot{q}_i = \begin{cases} \begin{bmatrix} z_i \\ z_i \times (o_n - o_i) \end{bmatrix} \dot{q}_i & \text{if joint } i \text{ is revolute} \\[1em] \begin{bmatrix} 0 \\ z_i \end{bmatrix} \dot{q}_i & \text{if joint } i \text{ is prismatic} \end{cases} \tag{32}$$

where, as stated in the introduction, z_i is the direction vector of the joint axis, o_n (o_i) is the origin of the frame attached to the end-effector (i-th joint) and $\dot{q}_i = \dot{\theta}_i$ if joint i is revolute and $\dot{q}_i = \dot{d}_i$ if joint i is prismatic.

Remark 4. *The unit twists $\xi_i(q)$ defined in Equation (32) are represented with respect to the world frame, not with respect to the local frame attached to the previous joint. If the unit twists are defined with respect to a local frame, we need to use the adjoint transformation to represent them with respect to the world frame. In particular, $\xi_i'(q) = Ad_f \xi_i(q)$, where $Ad_f : \mathbb{R}^6 \to \mathbb{R}^6$ is the adjoint transformation associated with the rigid motion f, i.e., the rigid motion transforming the reference frame to the local frame in which the twist is initially represented.*

The following is a key result:

Theorem 2 (Tsai, 1999 [44]). *Given a serial robot with n DoF:*

$$\begin{bmatrix} \omega \\ v \end{bmatrix} = \xi_1(q)\dot{q}_1 + \cdots + \xi_n(q)\dot{q}_n = [\xi_1(q) \ \cdots \ \xi_n(q)]\dot{q}, \tag{33}$$

where, again, ω, v denote the angular and linear velocity vectors of the robot's end-effector and $\dot{q} = (\dot{q}_1, \ldots, \dot{q}_n)$.

The main advantage of the screw-based Jacobian matrix defined in Equation (33) is that it allows a geometrical identification of the singularities. Moreover, if an approach based on geometric algebra is used, an intuitive geometrical and computer-friendly algebraic identification of the singularities is possible. For that purpose, let us consider the geometric algebra $\mathcal{G}_6$, where, for every $i = 1, \ldots, n$, the unit twist $\xi_i(q)$ can be modelled as a vector. Indeed, we make the identification $\xi_i(q) = [\xi_{i_1} \ \cdots \ \xi_{i_6}]^T$ with the vector $x = \xi_{i_1} e_1 + \cdots + \xi_{i_6} e_6 \in \mathcal{G}_6$, where $e_1, \ldots, e_6$ are the basis vectors of $\mathcal{G}_6$.

The following gives the main result of this section.

Theorem 3. *Let $\xi_i(q)$ denote the unit twist defined by the i-th joint expressed as a vector of $\mathcal{G}_6$. Then:*

$$\xi_1(q) \wedge \cdots \wedge \xi_6(q) = \det([\xi_1(q) \ \cdots \ \xi_6(q)])e_1 \wedge \cdots \wedge e_6. \tag{34}$$

Theorem 3 can be seen as a particular case of a more general result:

Theorem 4. *Let $a_1, \ldots, a_n$ be a set of n vectors of $\mathcal{G}_n$. Then:*

$$a_1 \wedge \cdots \wedge a_n = \det([a_1 \ \cdots \ a_n])e_1 \wedge \cdots \wedge e_n \tag{35}$$

Theorem 4 can be easily deduced from Equation (4.143) in [36] (p. 108):

$$F(I) = \det(F)I, \tag{36}$$

where F is a linear transformation in $\mathcal{G}_n$, $F : \mathcal{G}_n \rightarrow \mathcal{G}_n$, and, as stated in Section 2, I denotes the pseudoscalar of $\mathcal{G}_n$.

In particular, Theorem 4 is true for any set of six vectors $a_1, \ldots, a_6$ of $\mathcal{G}_6$, which proves Theorem 3. Now, the following corollary of Theorem 3 allows us to characterize the singularities of any serial robot of 6 DoF.

Corollary 1. *Given a serial robot with 6 DoF and associated unit twists* $\xi_1(q), \ldots, \xi_6(q)$, *then* $q \in \mathcal{S}$ *if, and only if,* $\xi_1(q) \wedge \cdots \wedge \xi_6(q) = 0$.

Proof. Taking the dual of Equation (34), the following identity is obtained:

$$(\xi_1(q) \wedge \cdots \wedge \xi_6(q))^* = \det([\xi_1(q) \ \cdots \ \xi_6(q)]) \tag{37}$$

and, therefore, the singularities of the serial robot are those configurations $q \in \mathcal{C}$ verifying that:

$$(\xi_1(q) \wedge \cdots \wedge \xi_6(q))^* = 0. \tag{38}$$

Now, since for a given non-zero multivector $M \in \mathcal{G}_n$, $M^* = 0$ if, and only if, $M = 0$, Equation (38) can be simplified to:

$$\xi_1(q) \wedge \cdots \wedge \xi_6(q) = 0. \tag{39}$$

Thus, $q \in \mathcal{S}$ if, and only if, $\xi_1(q) \wedge \cdots \wedge \xi_6(q) = 0$. $\square$

Remark 5. *Notice that what has also been proven in Corollary 1 is that the outer product of n vectors of* $\mathcal{G}_n$ *is zero if, and only if, the n vectors are linearly dependent.*

In addition, Corollary 1 allows us to re-define the singular set as:

$$\mathcal{S} = \{q \in \mathcal{C} : \xi_1(q) \wedge \cdots \wedge \xi_6(q) = 0\}. \tag{40}$$

Remark 6. *What Theorem 3 states is that, for instance, if two unit twists* ξ_1 *and* ξ_2 *satisfy* $\xi_1 \wedge \xi_2 = 0$, *then they represent the same twist, and hence, they generate the same screw motion. This means that if such a screw motion is a pure translation, then the translational axes are either parallel or coincident, while if the screw motion is a pure rotation, the rotational axes are coincident (Since the twists contain the term* $(z_i \times (o_6 - o_i))$ *for* $i = 1, 2$, *they cannot be parallel). Regarding the kinematic singularities of serial robots, this implies that two prismatic joints whose axes are either parallel or coincident give rise to a singularity and, equivalently, that two revolute joints whose axes are coincident give rise to a singularity. This is, in fact, in agreement with what is known about kinematic singularities since two parallel revolute joint axes do not give rise to a singularity. Obviously, the same geometrical interpretation can be made for three, four or more unit twists satisfying that their outer product is zero.*

With respect to redundant serial robots, it is clear that, for $n > 6$, $\xi_1(q) \wedge \cdots \wedge \xi_n(q) = 0$ for any $q \in \mathcal{C}$. Hence, Corollary 1 on its own does not allow us to characterize the singularities of redundant robots. However, this problem can be easily overcome by studying all the possible combinations of six unit twists in $\{\xi_1(q), \ldots, \xi_n(q)\}$. We denote the set of all combinations of six elements that can be drawn from $\{1, \ldots, n\}$ by S. Clearly, S has $C(n, 6) = \binom{n}{6}$ elements of the form $\{i_1, \ldots, i_6\}$, where $1 \leq i_1 < \quad < i_6 \leq n$ and $1 \leq i \leq C(n, 6)$.

Theorem 5. *Given a serial robot with n DoF and associated unit twists* $\xi_1(q), \ldots, \xi_n(q)$, *then* $q \in \mathcal{S}$ *if, and only if, for each* $1 \leq i \leq C(n,6)$:

$$\xi_{i_1}(q) \wedge \cdots \wedge \xi_{i_6}(q) = 0, \tag{41}$$

where $\{i_1, \ldots, i_6\}$ *is the i-th element of S.*

Proof. It follows that, for $q \in \mathcal{C}$:

$$\xi_{i_1}(q) \wedge \cdots \wedge \xi_{i_6}(q) = 0 \text{ for every } 1 \leq i \leq C(n,6)$$

$$\overset{(1)}{\Longleftrightarrow} \det([\xi_{i_1}(q) \ \cdots \ \xi_{i_6}(q)]) = 0 \text{ for every } 1 \leq i \leq C(n,6) \tag{42}$$

$$\overset{(2)}{\Longleftrightarrow} \rho([\xi_1(q) \ \cdots \ \xi_n(q)]) < 6$$

where (1) uses Equations (37) and (2) uses the fact that all the minors of order 6 of the matrix $[\xi_1(q) \ \cdots \ \xi_n(q)]$ have null determinants. Clearly, $\rho([\xi_1(q) \ \cdots \ \xi_n(q)]) < 6$ if, and only if, $\rho(J(q)) < 6$, which, in turn, is equivalent to $q \in \mathcal{S}$ (by Definition 1). $\square$

The computation of either Equation (39) for non-redundant robots or Equation (41) for redundant ones is computationally more efficient than the computation of either $\det(J(q)) = 0$ or $\det(J(q)J^T(q)) = 0$. The main reason for this lies in the computational complexity of the operations needed to obtain the expressions (39) or (41) with respect to the complexity of the operations needed for obtaining $\det(J(q)) = 0$ or $\det(J(q)J^T(q)) = 0$. It is clear that the outer product of n vectors of $\mathcal{G}_n$ behaves like the addition and product of real numbers, and hence, it has complexity $O(n) + O(n^2)$, while the determinant has complexity $O(n^3)$ or $O(n^4)$ depending on the algorithm used. In addition, for redundant robots, there are two main operations: the product between $J(q)$ and $J^T(q)$ and the determinant of the product matrix. This implies that, for this case, the complexity increases to $O(n^3) + O(n^4)$.

Special Case: Serial Robots with a Spherical Wrist

Similarly to what happens with the Jacobian matrix J, a simplification can be achieved for robots that have a spherical wrist. As stated in Section 1, the singularities of these robots can be decoupled into position and orientation singularities. Position singularities involve the first $n - 3$ joints and are computed by studying the rank of the following matrix:

$$J_p = \begin{bmatrix} s_1 & \cdots & s_{n-3} \end{bmatrix}, \tag{43}$$

where $s_i = z_i \times (o_n - o_i)$ if joint i is revolute and $s_i = z_i$ if joint i is prismatic. On the other hand, orientation singularities involve the last three joints and are computed through the determinant of the following matrix:

$$J_o = \begin{bmatrix} z_{n-2} & z_{n-1} & z_n \end{bmatrix}. \tag{44}$$

Now, let us consider the three-dimensional geometric algebra $\mathcal{G}_3$. As proven in Theorem 4 for $n = 3$, $a_1 \wedge a_2 \wedge a_3 = \det([a_1 \ a_2 \ a_3])e_1 \wedge e_2 \wedge e_3$ for any three vectors $a_1, a_2, a_3 \in \mathbb{R}^3$. Hence, analogously to what has been performed before, the following characterization for the position and orientation singularities can be deduced.

Theorem 6. *Given a serial robot with n DoF and a spherical wrist, if either* $z_i \times (o_n - o_i)$ *or* z_i *are denoted by* s_i *for* $i = 1, \ldots, n - 3$, *then:*

- $q \in \mathcal{C}$ *is a position singularity if, and only if,* $s_{i_1}(q) \wedge s_{i_2}(q) \wedge s_{i_3}(q) = 0$ *for each* $1 \leq i \leq C(n-3,3)$, *where* $\{i_1, i_2, i_3\}$ *is the i-th combination of three elements drawn from* $\{1, \ldots, n-3\}$.

- $q \in \mathcal{C}$ is an orientation singularity if, and only if,

$$z_{n-2}(q) \wedge z_{n-1}(q) \wedge z_n(q) = 0. \tag{45}$$

Proof. The proof is completely analogous to the proof of Corollary 1 and Theorem 5. $\square$

Remark 7. *Since the last three joint axes either intersect at a single point or are parallel, there is only one orientation singularity, namely, when these three joint axes are coplanar. This can also be easily deduced from Equation (45). A schematic representation of such singularity, also called wrist singularity, is depicted in Figure 2.*

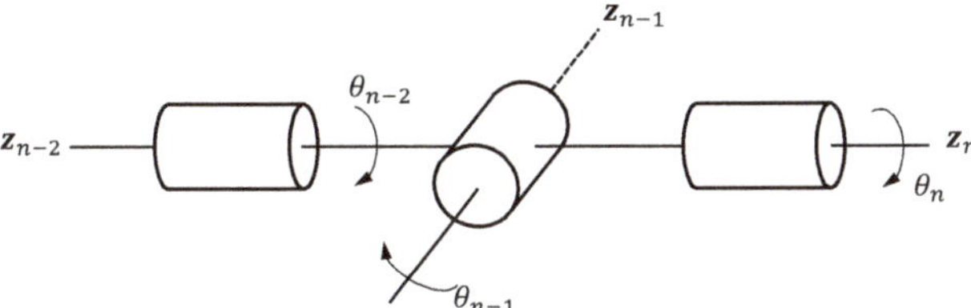

Figure 2. Schematic representation of the wrist singularity.

4. Distance to Singularities

Let $q_1, q_2 \in \mathcal{C}$ be two arbitrary configurations of a serial robot with n DoF and let $\xi_1, \ldots, \xi_n$ be the unit twists associated with its joints. Then, there exist $R_1(q_1, q_2), \ldots, R_n(q_1, q_2)$, where, for each $1 \leq i \leq n$, $R_i(q_1, q_2)$ is a configuration-dependent rotor in the six-dimensional geometric algebra $\mathcal{G}_6$ such that (Figure 3):

$$\xi_i(q_2) = R_i(q_1, q_2)\xi_i(q_1)\widetilde{R}_i(q_1, q_2). \tag{46}$$

The reason why these rotors exist is simple: unit twists are modelled as vectors in $\mathcal{G}_6$ and there always exists a rotor relating any pair of vectors in any geometric algebra $\mathcal{G}_n$. In particular, there is always a rotor relating the same unit twist ξ in two different configurations q_1, q_2.

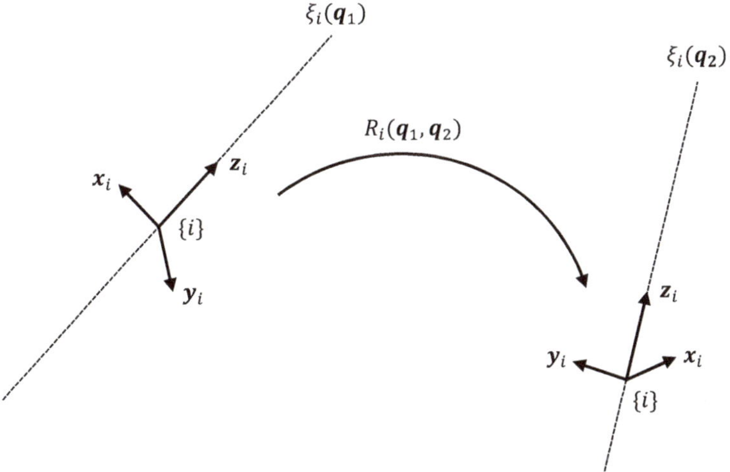

Figure 3. Rotor R_i relating the twist ξ_i in two different configurations q_1 and q_2.

Now, let $q_s \in \mathcal{S}$ denote a singularity of a serial robot. As explained in the previous section, if the serial robot has a spherical wrist, then q_s only involves a maximum of two or three joints and, therefore, two or three unit twists. If, conversely, the robot has no spherical wrist, then it can involve a maximum of six joints. Let us suppose, without loss of generality, that a given singularity q_s involves the joints $i_1, \ldots, i_r$ with associated unit twists $\xi_{i_1}(q_s), \ldots, \xi_{i_r}(q_s)$ for $2 \leq r \leq 6$. Then, for any configuration $q \in \mathcal{C}$, there exist $R_{i_1}(q, q_s), \ldots, R_{i_r}(q, q_s)$ such that:

$$\xi_{i_j}(q_s) = R_{i_j}(q, q_s)\xi_{i_j}(q)\widetilde{R}_{i_j}(q, q_s) \text{ for each } 1 \leq j \leq r. \tag{47}$$

The notation chosen for these rotors expresses a configuration dependence that is not a functional dependency, i.e., there is no analytical expression for these rotors with q as a variable.

Now, it is clear that $R_{i_j}(q, q_s) = 1$ if, and only if, $q = q_s$ for every $j = 1, \ldots, r$. However, since for each j, $R_{i_j}(q, q_s)$ does not define a function on q, a distance function cannot be defined. However, the measure of how close a given configuration q is to a singularity can be set as:

$$q \approx q_s \iff R_{i_j}(q) \approx 1 \text{ for every } j = 1, \ldots, r. \tag{48}$$

Example 1. *Let $q_s \in \mathcal{S}$ be a singularity of a serial robot that only involves the second and third joints. Then, for any configuration $q \in \mathcal{C}$, there exist $R_2(q, q_s)$ and $R_3(q, q_s)$ such that:*

$$\begin{aligned} \xi_2(q_s) &= R_2(q, q_s)\xi_2(q)\widetilde{R}_2(q, q_s), \\ \xi_3(q_s) &= R_3(q, q_s)\xi_3(q)\widetilde{R}_3(q, q_s). \end{aligned} \tag{49}$$

Therefore, q is close to q_s if, and only if:

$$\left.\begin{aligned} R_2(q, q_s) &\approx 1 \\ R_3(q, q_s) &\approx 1 \end{aligned}\right\} \tag{50}$$

and is singular if, and only if:

$$\left.\begin{aligned} R_2(q, q_s) &= 1 \\ R_3(q, q_s) &= 1 \end{aligned}\right\} \tag{51}$$

where, in general, $R_2(q, q_s) \neq R_3(q, q_s)$.

These rotors can be constructed in many different ways. The easiest way consists of considering, for each $1 \leq i \leq n$, the frame $\{i\}$ attached to joint i and constructed from ξ_i. This three-dimensional frame varies with the configuration q. Hence, for two different configurations q_1 and q_2, there are two frames $\{i\}$ attached to joint i. As shown in Section 2, we can recover the three-dimensional rotor that transforms one of the frames into the other. Since each frame $\{i\}$ depends continuously on the configuration q, the rotor $R_i(q)$ is a continuous function defined as follows:

$$\begin{aligned} R_i : \mathcal{C} &\to \mathfrak{R} \\ q &\mapsto R_i(q) \end{aligned} \tag{52}$$

Thus, these configuration-dependent rotors exhibit a functional dependency on the configuration, which allows us to define a distance function. Such distance is based on the norm of a multivector $X \in \mathcal{G}_n$, defined by the relation:

$$\|X\|^2 = \left\langle X\widetilde{X} \right\rangle_0. \tag{53}$$

To prove that $\|\cdot\|$ is a norm, the following two lemmas are necessary.

Lemma 1. *For any given multivector* $X \in \mathcal{G}_n$, $\left\langle X\widetilde{X} \right\rangle_0 \in \mathbb{R}^+ = [0, +\infty)$.

Proof. According to Equation (19), it follows that:

$$\widetilde{X} = \left\langle \widetilde{X} \right\rangle_0 + \left\langle \widetilde{X} \right\rangle_1 + \cdots + \left\langle \widetilde{X} \right\rangle_n. \tag{54}$$

Then:

$$X\widetilde{X} = \sum_{i=0}^{n} \sum_{j=0}^{n} \langle X \rangle_i \left\langle \widetilde{X} \right\rangle_j. \tag{55}$$

Now, note that, for each $i = 1, \ldots, n$, $\langle X \rangle_i$ is a i-vector, i.e., it only contains terms of grade i. The geometric product of two k-vectors (with different k) is stated as follows [36] (p. 103):

$$A_r B_s = \langle A_r B_s \rangle_{|r-s|} + \langle A_r B_s \rangle_{|r-s|+2} + \cdots + \langle A_r B_s \rangle_{r+s}, \tag{56}$$

Therefore, it is clear that $\left\langle \langle X \rangle_i \left\langle \widetilde{X} \right\rangle_j \right\rangle_0 = 0$ for $i \neq j$. Thus:

$$\left\langle X\widetilde{X} \right\rangle_0 = \sum_{i=0}^{n} \left\langle \langle X \rangle_i \left\langle \widetilde{X} \right\rangle_i \right\rangle_0. \tag{57}$$

Now, each $\langle X \rangle_i$ can be expanded as follows:

$$\langle X \rangle_i = \sum_{j=1}^{C(n,i)} \alpha_j(i) e_{j_1} \cdots e_{j_i}, \tag{58}$$

where, for every $1 \leq j \leq C(n,i) = \binom{n}{i}$, $\alpha_j(i) \in \mathbb{R}$ and $e_{j_1} \cdots e_{j_i}$ are the base elements of $\langle \mathcal{G}_n \rangle_i$. Therefore:

$$\left\langle \widetilde{X} \right\rangle_i = \sum_{j=1}^{C(n,i)} \alpha_j(i) e_{j_i} \cdots e_{j_1} \tag{59}$$

and, thus:

$$
\begin{aligned}
\langle X \rangle_i \left\langle \widetilde{X} \right\rangle_i &= \sum_{j=1}^{C(n,i)} \alpha_j(i) e_{j_1} \cdots e_{j_i} \sum_{j=1}^{C(n,i)} \alpha_j(i) e_{j_i} \cdots e_{j_1} \\
&= \sum_{j=1}^{C(n,i)} \sum_{k=1}^{C(n,i)} \alpha_j(i) \alpha_k(i) e_{j_1} \cdots e_{j_i} e_{k_i} \cdots e_{k_1},
\end{aligned}
\tag{60}
$$

where, clearly, $\left\langle e_{j_1} \cdots e_{j_i} e_{k_i} \cdots e_{k_1} \right\rangle_0 = \delta_{jk}$ with δ_{jk} the Kronecker delta. Then:

$$\left\langle \langle X \rangle_i \left\langle \widetilde{X} \right\rangle_i \right\rangle_0 = \sum_{j=1}^{C(n,i)} \alpha_j^2, \tag{61}$$

which, for every $1 \leq i \leq n$, is a positive scalar. This implies that the sum of Equation (57) is also a positive scalar. $\square$

Lemma 2. *Given three strictly positive real numbers* $a_1, a_2, a_3 \in \mathbb{R}^+ \setminus \{0\}$, *the following properties hold:*

- $\sqrt{a_1 + a_2 - a_3} \leq \sqrt{a_1} + \sqrt{a_2}$;
- $\sqrt{a_1 + a_2 + a_3} \leq \sqrt{a_1} + \sqrt{a_2}$ *if, and only if,* $a_3 \leq 2\sqrt{a_1 a_2}$.

Proof. Both properties can be obtained by a straightforward computation. $\square$

Proposition 4. *The function* $\|\cdot\| : \mathcal{G}_n \to \mathbb{R}^+$ *defined by the identity* $\|X\|^2 = \left\langle X\widetilde{X} \right\rangle_0$ *is a norm in* $\mathcal{G}_n$, *i.e.*,

(i) $\|X\| \geq 0$ *for all* $X \in \mathcal{G}_n$. *In particular,* $\|X\| = 0$ *if, and only if,* $X = 0$.
(ii) $\|\lambda X\| = |\lambda| \|X\|$ *for all* $X \in \mathcal{G}_n$ *and* $\lambda \in \mathbb{R}$.
(iii) $\|X + Y\| \leq \|X\| + \|Y\|$ *for all* $X, Y \in \mathcal{G}_n$ *(usually known as the triangle inequality).*

Proof.

(i)   Given a multivector X, identity $\|X\|^2 = \left\langle X\widetilde{X} \right\rangle_0$ is equivalent to:

$$\|X\| = \pm\sqrt{\left\langle X\widetilde{X} \right\rangle_0}. \tag{62}$$

Thus, it is clear by Lemma 1 that the positive branch of Equation (62) is well defined and that $\|X\| \geq 0$. In particular, if $\|X\| = 0$, then:

$$\sqrt{\left\langle X\widetilde{X} \right\rangle_0} = 0 \implies \left\langle X\widetilde{X} \right\rangle_0 = 0 \implies \sum_{i=0}^{n} \left\langle \langle X \rangle_i \left\langle \widetilde{X} \right\rangle_i \right\rangle_0 = 0, \tag{63}$$

where all the terms of the last equation are positive by Lemma 1 and, thus, all of them are equal to zero. Now, note that each addend is the geometric product of an i-vector with its reverse. Therefore, if such product is zero, the corresponding i-vector must be zero. Since all the terms are zero, all the i-vectors that form X are zero, and thus, X is zero.

(ii) If $\lambda \in \mathbb{R}$ and $X \in \mathcal{G}_n$, then:

$$\begin{aligned}
\|\lambda X\| &= \sqrt{\left\langle (\lambda X)(\lambda \widetilde{X}) \right\rangle_0} = \sqrt{\left\langle \lambda^2 X\widetilde{X} \right\rangle_0} \\
&\overset{(1)}{=} \sqrt{\lambda^2 \left\langle X\widetilde{X} \right\rangle_0} = |\lambda| \sqrt{\left\langle X\widetilde{X} \right\rangle_0} = |\lambda| \|X\|,
\end{aligned} \tag{64}$$

where (1) uses the linearity of the grade-0 projection operator (as stated in Section 2).

(iii) Given two different multivectors X and Y, they can be expanded as linear combinations of the basis elements of $\mathcal{G}_n$ as follows:

$$\begin{aligned}
X &= \sum_{i=0}^{2^n} \alpha_i e_{j_1} \cdots e_{j_i}, \\
Y &= \sum_{i=0}^{2^n} \beta_i e_{j_1} \cdots e_{j_i}.
\end{aligned} \tag{65}$$

Now, it follows that:

$$X + Y = \sum_{i=0}^{2^n} (\alpha_i + \beta_i) e_{j_1} \cdots e_{j_i} \tag{66}$$

and hence:

$$\begin{aligned}
\|X + Y\| &= \sqrt{\langle (X+Y)(X+Y)^{\sim} \rangle_0} \overset{(1)}{=} \sqrt{\sum_{i=0}^{2^n} (\alpha_i + \beta_i)^2} \\
&= \sqrt{\underbrace{\sum_{i=0}^{2^n} \alpha_i^2}_{A} + \underbrace{\sum_{i=0}^{2^n} \beta_i^2}_{B} + 2\underbrace{\sum_{i=0}^{2^n} \alpha_i \beta_i}_{C}},
\end{aligned} \tag{67}$$

where (1) uses Lemma 1, while A, B and C are just notations given to simplify the different manipulations. Since $A, B > 0$ (If either A, B are equal to zero, then either $X = 0$ or $Y = 0$, which will make the condition $\|X + Y\| \leq \|X\| + \|Y\|$ trivial):

$$\sqrt{A + B + C} \overset{(1)}{\leq} \sqrt{A} + \sqrt{B} = \sqrt{\sum_{i=0}^{2^n} \alpha_i^2} + \sqrt{\sum_{i=0}^{2^n} \beta_i^2} = \|X\| + \|Y\|, \tag{68}$$

where (1) uses the first or second property of Lemma 2 depending on whether $C < 0$ or $C > 0$. It only remains to check if $C > 0$, $2C \leq 2\sqrt{AB}$. Indeed, the previous inequality is equivalent to that $C^2 \leq AB$. Now:

$$AB = \sum_{i=0}^{2^n} \alpha_i^2 \sum_{i=0}^{2^n} \beta_i^2 = \sum_{i=0}^{2^n} \sum_{j=0}^{2^n} \alpha_i^2 \beta_j^2 = \sum_{i=0}^{2^n} \alpha_i^2 \beta_i^2 + \sum_{\substack{i=0 \\ j \neq i}}^{2^n} \sum_{j=0}^{2^n} \alpha_i^2 \beta_j^2,$$

$$C^2 = \left(\sum_{i=0}^{2^n} \alpha_i \beta_i \right)^2 = \sum_{i=0}^{2^n} \alpha_i^2 \beta_i^2 + \sum_{\substack{i=0 \\ j \neq i}}^{2^n} \sum_{j=0}^{2^n} \alpha_i \beta_i \alpha_j \beta_j \tag{69}$$

and, thus, $C^2 \leq AB$ turns to:

$$\sum_{i=0}^{2^n} \alpha_i^2 \beta_i^2 + \sum_{\substack{i=0 \\ j \neq i}}^{2^n} \sum_{j=0}^{2^n} \alpha_i \beta_i \alpha_j \beta_j \leq \sum_{i=0}^{2^n} \alpha_i^2 \beta_i^2 + \sum_{\substack{i=0 \\ j \neq i}}^{2^n} \sum_{j=0}^{2^n} \alpha_i^2 \beta_j^2, \tag{70}$$

which is equivalent to:

$$\sum_{\substack{i=0 \\ j \neq i}}^{2^n} \sum_{j=0}^{2^n} \alpha_i \beta_i \alpha_j \beta_j \leq \sum_{\substack{i=0 \\ j \neq i}}^{2^n} \sum_{j=0}^{2^n} \alpha_i^2 \beta_j^2 \tag{71}$$

which, in turn, is equivalent to:

$$\begin{aligned}
0 &\leq \sum_{\substack{i=0 \\ j \neq i}}^{2^n} \sum_{j=0}^{2^n} \alpha_i^2 \beta_j^2 - \sum_{\substack{i=0 \\ j \neq i}}^{2^n} \sum_{j=0}^{2^n} \alpha_i \beta_i \alpha_j \beta_j \\
&= \frac{1}{2} \sum_{\substack{i=0 \\ j \neq i}}^{2^n} \sum_{j=0}^{2^n} \alpha_i^2 \beta_j^2 + \frac{1}{2} \sum_{\substack{i=0 \\ j \neq i}}^{2^n} \sum_{j=0}^{2^n} \alpha_i^2 \beta_j^2 - \sum_{\substack{i=0 \\ j \neq i}}^{2^n} \sum_{j=0}^{2^n} \alpha_i \beta_i \alpha_j \beta_j \\
&= \frac{1}{2} \sum_{\substack{i=0 \\ j \neq i}}^{2^n} \sum_{j=0}^{2^n} (\alpha_i \beta_j - \alpha_j \beta_i)^2.
\end{aligned} \tag{72}$$

Since this last inequality is always true, the triangle inequality is also true.
$\square$

Now, a distance function D can be defined for rotors.

Theorem 7. *The function $D : \mathfrak{R} \times \mathfrak{R} \to \mathbb{R}^+$ defined by the identity $D(R_1, R_2) = \|R_1 - R_2\|$ is a distance in $\mathfrak{R}$, i.e.,*

(i)   *$D(R_1, R_2) \geq 0$ for all $R_1, R_2 \in \mathfrak{R}$. In particular, $D(R_1, R_2) = 0$ if, and only if, $R_1 = R_2$;*
(ii)   *$D(R_1, R_2) = D(R_2, R_1)$ for all $R_1, R_2 \in \mathfrak{R}$;*
(iii)   *$D(R_1, R_3) \leq D(R_1, R_2) + D(R_2, R_3)$ for all $R_1, R_2, R_3 \in \mathfrak{R}$.*

Proof. The proof is straightforward and uses the fact that $\|\cdot\|$ is a norm. Given two different rotors R_1 and R_2:

(i) $D(R_1, R_2) = \|R_1 - R_2\| \geq 0$. In particular:

$$D(R_1, R_2) = 0 \iff \|R_1 - R_2\| = 0 \overset{(1)}{\iff} R_1 - R_2 = 0 \iff R_1 = R_2, \tag{73}$$

where (1) uses the first property of a norm.

(ii)   We have that:

$$\begin{aligned}
D(R_1, R_2) &= \|R_1 - R_2\| = \sqrt{\langle (R_1 - R_2)(R_1 - R_2)^{\sim} \rangle_0} \\
&= \sqrt{\langle (R_2 - R_1)(R_2 - R_1)^{\sim} \rangle_0} = \|R_2 - R_1\| = D(R_2, R_1).
\end{aligned} \tag{74}$$

(iii)   Given a third rotor R_3, we have that:

$$\begin{aligned}
D(R_1, R_3) &= \|R_1 - R_3\| = \|R_1 - R_2 + R_2 - R_3\| \\
&\overset{(1)}{\leq} \|R_1 - R_2\| + \|R_2 - R_3\| = D(R_1, R_2) + D(R_2, R_3),
\end{aligned} \tag{75}$$

where (1) uses the third property of a norm.

$\square$

As stated before, the end-effector pose of a serial robot and the pose of each one of its joints are described by the configuration-dependent rotors $R(q)$ and $R_i(q)$, respectively. Thus, one can be tempted to extend the distance function D to $\mathcal{C}$ as follows:

$$\begin{aligned}
&D : \mathcal{C} \times \mathcal{C} \to \mathbb{R}^+ \\
&D(q_1, q_2) = \|R(q_1) - R(q_2)\|
\end{aligned} \tag{76}$$

This function verifies all the requirements of a distance function with the exception of:

$$D(q_1, q_2) = 0 \iff q_1 = q_2. \tag{77}$$

The reason is simple: a given pose of the end-effector can be associated with up to 16 different configurations if the serial robot is non-redundant and an infinite number if it is redundant. In particular, this means that $R(q_1) = R(q_2)$ with $q_1 \neq q_2$. However, this problem can be overcome as follows:

- For each joint i, denote by $\mathcal{C}_i$ the configuration space of the subchain formed by the first i joints. It is clear that, if the robot has n degrees of freedom, $\mathcal{C}_i \subset \mathcal{C}$ for every $1 \leq i \leq n$. Then, the following set of functions can be defined:

$$\begin{aligned}
&D_i : \mathcal{C}_i \times \mathcal{C}_i \to \mathbb{R}^+ \\
&D_i(q_1, q_2) = \|R_i(q_1) - R_i(q_2)\|
\end{aligned} \tag{78}$$

  where, as stated before, R_i is the rotor that describes the pose of joint i. Again, these functions are not distance functions for the same reason as D (Equation (76)) is not a distance function.

- The function:

$$\begin{aligned}
&D : \mathcal{C} \times \mathcal{C} \to [0, +\infty) \\
&D(q_1, q_2) = D_1(q_{1_1}, q_{2_1}) + \cdots + D_n(q_{1_n}, q_{2_n})
\end{aligned} \tag{79}$$

  where q_{1_i} (q_{2_i}) denotes the first i coordinates of the configuration vector q_1 (q_2), defines a distance function in $\mathcal{C}$.

Proof. Since, for each $1 \leq i \leq n$, D_i satisfies the requirements (ii) and (iii) of a distance function, it is clear that D also satisfies them. In addition, $D_i(q_{1_i}, q_{2_i}) \geq 0$ for each $1 \leq i \leq n$ and $q_{1_i}, q_{2_i} \in C_i$. Therefore, $D(q_1, q_2) \geq 0$ for arbitrary $q_1, q_2 \in C$. Finally, if $D(q_1, q_2) = 0$, then, since any term of Equation (79) is a positive scalar, it can be deduced that $D_i(q_{1_i}, q_{2_i}) = 0$ for every $1 \leq i \leq n$. Thus, q_1 and q_2 not only have the same end-effector pose, but the same pose for of each of its joints, which clearly implies that $q_1 = q_2$. $\square$

This distance function can be restricted to $\mathcal{S}$ just by considering the joints involved in a given singularity q_s.

Definition 5. *Let $q_s \in \mathcal{S}$ be a singularity of a serial robot that involves joints $i_1, \ldots, i_r$. Then, the function $D : C \times \mathcal{S} \to \mathbb{R}^+$ is defined by the expression:*

$$D(q, q_s) = D_{i_1}(q_{i_1}, q_{s_{i_1}}) + \cdots + D_{i_r}(q_{i_r}, q_{s_{i_r}}), \tag{80}$$

where, for each $i_1 \leq k \leq i_r$, D_k is the function defined in (78) and is a distance function in C.

5. Application to the Serial Robot Kuka Lwr 4+

To show the advantages of the proposed method, an illustrative example is developed in this section, making use of the Kuka LWR 4+, an anthropomorphic robotic arm with seven degrees of freedom and a spherical wrist. It is schematically depicted in Figure 4. Since it has a spherical wrist, its singularities can be decoupled into position and orientation singularities. Hence, Theorem 6 can be applied in order to find out such singularities. The computations with the vectors of $\mathcal{G}_3$ have been carried out using the Clifford Multivector Toolbox of MATLAB [45].

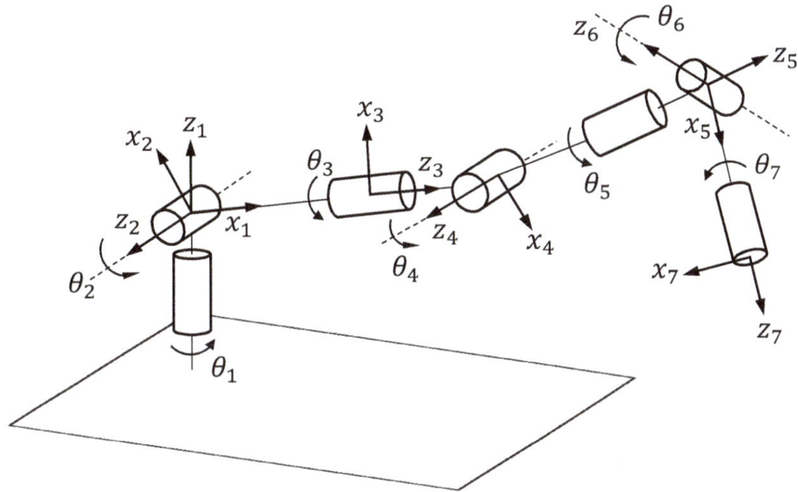

Figure 4. Schematic representation of the Kuka LWR 4+.

With respect to the position singularities, the following system of $C(4, 3) = 4$ equations should be solved:

$$\left.\begin{array}{l}
(z_1 \times (o_7 - o_1)) \wedge (z_2 \times (o_7 - o_2)) \wedge (z_3 \times (o_7 - o_3)) = 0 \\
(z_1 \times (o_7 - o_1)) \wedge (z_2 \times (o_7 - o_2)) \wedge (z_4 \times (o_7 - o_4)) = 0 \\
(z_1 \times (o_7 - o_1)) \wedge (z_3 \times (o_7 - o_3)) \wedge (z_4 \times (o_7 - o_4)) = 0 \\
(z_2 \times (o_7 - o_2)) \wedge (z_3 \times (o_7 - o_3)) \wedge (z_4 \times (o_7 - o_4)) = 0
\end{array}\right\} \tag{81}$$

where, as computed in [46], we have that:

$$z_1 \times (o_7 - o_1)$$

$$= \begin{bmatrix} -400c_2s_1 - 390s_4(c_1s_3 + c_3s_1s_2) - 390c_2c_4s_1 \\ 400c_1c_2 - 390s_4(s_1s_3 - c_1c_3s_2) + 390c_1c_2c_4 \\ 0 \end{bmatrix},$$

$$z_2 \times (o_7 - o_2)$$

$$= \begin{bmatrix} -c_1(400s_2 + 390c_4s_2 - 390c_2c_3s_4) \\ -s_1(400s_2 + 390c_4s_2 - 390c_2c_3s_4) \\ A_1 \end{bmatrix},$$

$$\tag{82}$$

$$z_3 \times (o_7 - o_3)$$

$$= \begin{bmatrix} c_2s_1(390c_4s_2 - 390c_2c_3s_4) - s_2(390s_4(c_1s_3 + c_3s_1s_2) + 390c_2c_4s_1) \\ -s_2(390s_4(s_1s_3 - c_1c_3s_2) - 390c_1c_2c_4) - c_1c_2(390c_4s_2 - 390c_2c_3s_4) \\ c_1c_2(390s_4(c_1s_3 + c_3s_1s_2) + 390c_2c_4s_1) + c_2s_1(390s_4(s_1s_3 - c_1c_3s_2) - 390c_1c_2c_4) \end{bmatrix},$$

$$z_4 \times (o_7 - o_4)$$

$$= \begin{bmatrix} (390c_4s_2 - 390c_2c_3s_4)(c_1c_3 - s_1s_2s_3) - c_2s_3(390s_4(c_1s_3 + c_3s_1s_2) + 390c_2c_4s_1) \\ (390c_4s_2 - 390c_2c_3s_4)(c_3s_1 + c_1s_2s_3) - c_2s_3(390s_4(s_1s_3 - c_1c_3s_2) - 390c_1c_2c_4) \\ A_2 \end{bmatrix},$$

where

$$\begin{aligned} A_1 &= c_1(400c_1c_2 - 390s_4(s_1s_3 - c_1c_3s_2) + 390c_1c_2c_4) + s_1(400c_2s_1 \\ &\quad + 390s_4(c_1s_3 + c_3s_1s_2) + 390c_2c_4s_1), \\ A_2 &= (c_1c_3 - s_1s_2s_3)(390s_4(s_1s_3 - c_1c_3s_2) - 390c_1c_2c_4) - (390s_4(c_1s_3 + c_3s_1s_2) \\ &\quad + 390c_2c_4s_1)(c_3s_1 + c_1s_2s_3), \end{aligned}$$

and where $c_i = \cos(\theta_i)$ and $s_i = \sin(\theta_i)$. However, in order to simplify these expressions, the system of Equation (81) is expressed with respect to the frame attached to the fourth joint of the Kuka LWR 4+. To do so, a relation analogous of relation (9) is applied. Here, instead of pre-multiplying by the corresponding rotation matrix, the system of Equation (81) is multiplied by the three-dimensional rotor R that performs the rotation between the frame attached to the end-effector and the frame attached to the fourth joint. For instance, the first equation of the system (81) becomes:

$$R(z_1 \times (o_7 - o_1)) \wedge (z_2 \times (o_7 - o_2)) \wedge (z_3 \times (o_7 - o_3))\widetilde{R} = 0, \tag{83}$$

which, using the geometric covariance property for rotors introduced in Section 2, becomes:

$$R(z_1 \times (o_7 - o_1))\widetilde{R} \wedge R(z_2 \times (o_7 - o_2))\widetilde{R} \wedge R(z_3 \times (o_7 - o_3))\widetilde{R} = 0. \tag{84}$$

Therefore, the system of Equation (81) becomes:

$$\left. \begin{aligned} a_1 \wedge a_2 \wedge a_3 &= 0 \\ a_1 \wedge a_2 \wedge a_4 &= 0 \\ a_1 \wedge a_3 \wedge a_4 &= 0 \\ a_2 \wedge a_3 \wedge a_4 &= 0 \end{aligned} \right\} \tag{85}$$

where

$$a_1 = (-10c_2s_3(40c_4 + 39))e_1 + (400c_2s_3s_4)e_2 + (400c_2c_3 + 390s_2s_4 + 390c_2c_3c_4)e_3,$$
$$a_2 = (10c_3(40c_4 + 39))e_1 + (-400c_3s_4)e_2 + (10s_3(40c_4 + 39))e_3,$$
$$a_3 = (390s_4)e_3,$$
$$a_4 = (-390)e_1.$$
(86)

Now, the system of Equation (85) becomes:

$$\left.\begin{array}{r} 0 = 0 \\ 40c_2s_4 + 39s_2c_3s_4^2 + 39c_2c_4s_4 = 0 \\ c_2s_3s_4^2 = 0 \\ c_3s_4^2 = 0 \end{array}\right\}$$
(87)

which clearly has two different solutions:

- $s_4 = 0$ or, equivalently, $q_4 = 0$.
- $c_2 = c_3 = 0$ or, equivalently, $q_2 = \pm\frac{\pi}{2}$ and $q_3 = \pm\frac{\pi}{2}$.

  These two solutions correspond to the position singularities of the Kuka LWR 4+. With respect to the orientation singularities, there is only one equation to solve:

$$z_5 \wedge z_6 \wedge z_7 = 0.$$
(88)

Again, the expression of each z_i for $i = 5, 6, 7$ can be simplified by expressing those vectors with respect to the frame attached to the fourth joint. Thus, Equation (88) becomes:

$$e_2 \wedge (-s_5e_1 - c_5e_3) \wedge (c_5s_6e_1 + c_6e_2 - s_5s_6e_3)$$
$$= (-s_5e_2 \wedge e_1 - c_5e_2 \wedge e_3) \wedge (c_5s_6e_1 + c_6e_2 - s_5s_6e_3)$$
$$\overset{(1)}{=} -s_5^2s_6e_1 \wedge e_2 \wedge e_3 - c_5^2s_6e_1 \wedge e_2 \wedge e_3 = -s_6e_1 \wedge e_2 \wedge e_3 = 0,$$
(89)

where (1) uses the anticommutativity of the outer product. Clearly, the last expression of Equation (89) is zero if, and only if, $s_6 = 0$ or, equivalently, if, and only if, $q_6 = 0$. Thus, the Kuka LWR 4+ only has one orientation singularity (the wrist singularity, as explained in Remark 7).

Finally, the distance function defined in Definition 5 can be applied to any of the already obtained singular configurations. Let us consider, for instance, the position singularity $q_4 = 0$. Then, the distance between an arbitrary configuration $q \in C$ and this singularity is given by the expression:

$$D(q, q_s) = \|R_4(q) - R_4(q_s)\|,$$
(90)

where q_s denotes the singular configuration $q_4 = 0$ and R_4 is the rotor defining the pose of the fourth joint of the Kuka LWR 4+.

In particular, R_4 can be found as explained in Section 2. Indeed, if $\{e_1, e_2, e_3\}$ denotes the orthogonal basis defined by the world frame and $\{f_1, f_2, f_3\}$ (respectively, $\{f_1', f_2', f_3'\}$), the orthogonal basis defined by the frame attached to the fourth joint under the effect of configuration q (respectively, singular configuration q_s), then:

$$R_4(q) = \frac{1 + e^1f_1 + e^2f_2 + e^3f_3}{\|1 + e^1f_1 + e^2f_2 + e^3f_3\|},$$
(91)

$$R_4(q_s) = \frac{1 + e^1f_1' + e^2f_2' + e^3f_3'}{\|1 + e^1f_1' + e^2f_2' + e^3f_3'\|},$$

where $\{e^1, e^2, e^3\}$ is the reciprocal frame of $\{e_1, e_2, e_3\}$. Since $\{e_1, e_2, e_3\}$ is also an orthonormal set of vectors, such a reciprocal frame is:

$$
\begin{aligned}
e^1 &= e_1, \\
e^2 &= e_2, \\
e^3 &= e_3.
\end{aligned}
\tag{92}
$$

Thus, Equation (91) turns to:

$$
R_4(q) = \frac{1 + e_1 f_1 + e_2 f_2 + e_3 f_3}{\|1 + e_1 f_1 + e_2 f_2 + e_3 f_3\|},
$$

$$
\tag{93}
$$

$$
R_4(q_s) = \frac{1 + e_1 f_1' + e_2 f_2' + e_3 f_3'}{\|1 + e_1 f_1' + e_2 f_2' + e_3 f_3'\|}.
$$

Evaluating Equation (93) for the KUKA LWR 4+, we obtain:

$$
R_4(q) = \frac{a_1 + a_2 e_1 \wedge e_2 + a_3 e_1 \wedge e_3 + a_4 e_2 \wedge e_3}{\sqrt{a_1^2 + a_2^2 + a_3^2 + a_4^2}},
$$

$$
\tag{94}
$$

$$
R_4(q_s) = \frac{b_1 + b_2 e_1 \wedge e_2 + b_3 e_1 \wedge e_3 + b_4 e_2 \wedge e_3}{\sqrt{b_1^2 + b_2^2 + b_3^2 + b_4^2}},
$$

where $\{e_1 \wedge e_2, e_1 \wedge e_3, e_2 \wedge e_3\}$ are the basic bivectors of $\mathcal{G}_3$ and

$$
\begin{aligned}
a_1 &= c_2 s_3 + c_4 s_1 s_3 - c_4 c_3 c_1 s_2 + s_3 s_4 c_1 + s_4 s_1 s_2 c_3 + s_4 c_1 c_2 + c_2 c_4 s_1, \\
a_2 &= c_2 s_1 s_4 - c_4 c_1 s_3 - c_4 c_3 s_1 s_2 - c_1 c_2 c_4 + s_4 s_1 s_3 - s_4 s_2 c_1 c_3, \\
a_3 &= s_2 s_4 + c_2 c_3 c_4 + c_3 s_1 + c_1 s_2 s_3, \\
a_4 &= c_4 s_2 - c_2 c_3 s_4 - c_1 c_3 + s_1 s_2 s_3, \\
b_1 &= c_2 s_3 + s_1 s_3 - c_3 c_1 s_2 + c_2 s_1, \\
b_2 &= -c_1 s_3 - c_3 s_1 s_2 - c_1 c_2, \\
b_3 &= c_2 c_3 + c_3 s_1 + c_1 s_2 s_3, \\
b_4 &= s_2 - c_1 c_3 + s_1 s_2 s_3.
\end{aligned}
\tag{95}
$$

Therefore, by Proposition 4 and the decomposition used in the Proof of Lemma 1, the distance of an arbitrary configuration q to the position singularity $q_4 = 0$ is given by:

$$
D(q, q_s) = \sqrt{(a_1' - b_1')^2 + (a_2' - b_2')^2 + (a_3' - b_3')^2 + (a_4' - b_4')^2},
\tag{96}
$$

where

$$
a_i' = \frac{a_i}{\sqrt{a_1^2 + a_2^2 + a_3^2 + a_4^2}} \quad \text{and} \quad b_i' = \frac{b_i}{\sqrt{b_1^2 + b_2^2 + b_3^2 + b_4^2}}.
\tag{97}
$$

6. Handling of Singularities

Once the set of singular configurations $\mathcal{S}$ has been identify, several methods can be applied to handle the singularities. The detailed treatment of this topic is beyond the scope of this work. However, in order to show the possibilities of the distance function proposed in Section 4, we comment on three different situations, namely, motion planning, motion control and bilateral teleoperation. In each one of these situations, the distance function defined in Definition 5 plays an important role for handling the singularities.

6.1. Singularity Handling in Motion Planning

Motion planning consists of programming collision-free motions for a given robotic manipulator from a start position to a goal position among a collection of static obstacles. The subset of robot configurations that do not cause collision with such obstacles is termed the free-of-obstacles configuration space, and it is denoted by $\mathcal{C}_{\text{free}}$. The main methods used for motion planning can be grouped into three categories:

- Potential field methods, where a differentiable real-valued function $U : \mathcal{C} \to \mathbb{R}$, called the potential function, is defined. Such a function has an attractive component that pulls the trajectory towards the goal configuration and a repulsive component that pushes the trajectory away from the start configuration and from the obstacles.
- Sampling-based multi-query methods, where a roadmap is constructed over $\mathcal{C}_{\text{free}}$. The nodes represent free-of-obstacles configurations, while the edges represent feasible local paths between those configurations. Once the roadmap is constructed, a search algorithm finds out the best solution trajectory by selecting and joining the local paths through an optimization process.
- Sampling-based single-query methods, where tree-structure data are constructed by searching new configurations (nodes) in $\mathcal{C}_{\text{free}}$ and connecting them through local paths (edges). Its main difference with respect to the multi-query methods is that, while the multi-query methods work in two steps (construction of the roadmap and searching of a solution trajectory), in the single-query methods, both steps are taken together. Each new configuration added to the set of nodes is connected by a local path and evaluated in order to check its feasibility.

For any method of these three categories, the distance function D defined in Definition 5 can be applied to construct solution trajectories that also avoid the singularities. Indeed:

- For a potential field method, it is sufficient to add a repulsive component that pushes the trajectory, not only away from obstacles, but also away from singularities. To do so, the most efficient method is to define, for each singularity q_s, a quadratic repulsive component as follows:

$$
U_{r,q_s}(q) = \begin{cases} \dfrac{\kappa}{2}\left(\dfrac{1}{D(q,q_s)} - \dfrac{1}{D_0}\right)^2 & \text{if } D(q,q_s) \leq D_0 \\ 0 & \text{if } D(q,q_s) > D_0 \end{cases} \tag{98}
$$

where D_0 is set as a threshold for the distance D and $\kappa \in \mathbb{R}$.
- For a sampling-based method with multiple queries, it is sufficient to remove from the roadmap those nodes associated with singular configurations. During the construction of the roadmap, each configuration $q \in \mathcal{C}$ is evaluated to determine whether q is free-of-obstacles or not. Similarly, the idea is to evaluate each $q \in \mathcal{C}$ in order to determine whether q is close to a singularity or not. To speed up the process, both evaluations can be carried out together:
  (1) Select a value $D_0 > 0$ that will work as a threshold.
  (2) Given a discretization of the configuration space $\mathcal{C}$, each q of this discretization is evaluated to check whether:
      * It is free-of-obstacles;
      * It is far from any singularity. This can be achieved simply by evaluating whether $D(q,q_s) > D_0$ or $D(q,q_s) \leq D_0$;
  (3) If q is free-of-obstacles and far from any singularity, then it can be added to the set of nodes of the roadmap.
- For a sampling-based method with a single query, the approach is completely analogous to the one used for methods with multiple-queries due to the similarities between both categories.

6.2. Singularity Handling in Motion Control

Motion control consists of making the end-effector of a robot follow a time-varying trajectory specified within the manipulator workspace. A typical Inverse Dynamics Control scheme (depicted as a block diagram in Figure 5) can be described as:

- An input, i.e., the desired or target configuration q_d together with its velocity $\dot{q}_d$;
- A controller based on the dynamical model of the robot:

$$\tau = M(q)\ddot{q} + C(q, \dot{q})\dot{q} + g(q) \tag{99}$$

where $M(q)$ denotes the inertia matrix of the robot, $C(q, \dot{q})$ denotes the matrix of Coriolis and centrifugal forces and $g(q)$ denotes the gravity vector.
- An output, i.e., the vector of torques τ, that is sent to the robot to perform the desired motion;
- The robot executes the motion and updates the vectors q and $\dot{q}$;
- The robot sends such updated vectors to the controller (also known as the feedback of the system).

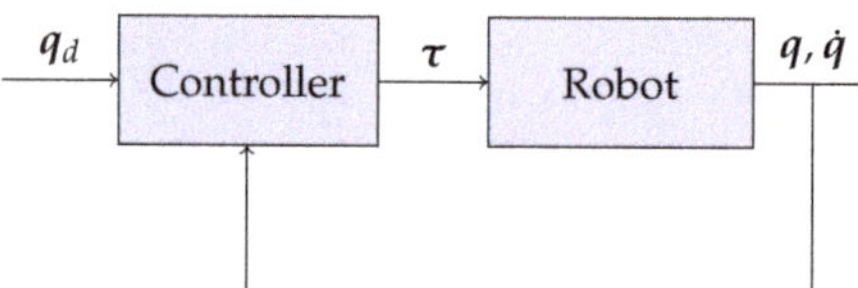

Figure 5. Standard motion control scheme.

To handle the singularities, a restriction can be defined inside the controller:

(1) The target configuration q_d is entered into the controller.
(2) q_d is checked in order to determine whether it is close to a singularity or not:
   - Select a threshold value $D_0 > 0$.
   - Evaluate the condition $D(q_d, q_s) > D_0$ for each singularity q_s.
   - If the evaluation returns yes, then τ can be computed from q_d using the dynamical model (Equation (99)) and sent to the robot. Otherwise, q_d is substituted by $q_d + D_0 q_d$ and evaluated again.

A block diagram of this scheme is depicted in Figure 6.

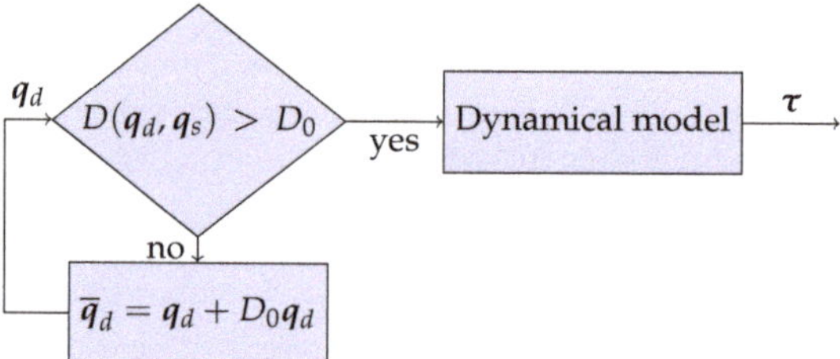

Figure 6. Proposed control scheme in presence of singularities.

6.3. Singularity Handling in Bilateral Teleoperation

Teleoperated robotic systems are characterized by a robot that executes the movements/actions commanded by a human operator. Any high-level or planning decision is made by a human user, while the robot is responsible for their mechanical implementation [47]. Teleoperation systems are often, at least conceptually, split into two parts: a local manipulator and a remote manipulator. The first one refers to the device moved by the human operator, while the second refers to the robot or robot system that performs the action.

According to the information flow direction, the teleoperation may be unilateral or bilateral. In unilateral teleoperation, the local manipulator sends position or force data to the remote manipulator and only receives, as feedback, visual information from the remote scene. However, in bilateral teleoperation, the position or force data are also sent from the remote manipulator in addition to the visual information.

In a bilateral teleoperation system, some strategies for handling of kinematic singularities can make use of the distance function D defined in Definition 5. For instance, the following scheme could be applied:

(1)   Select a value $D_0 > 0$ that will work as a threshold.
(2)   The local manipulator sends a pose (force) p (f) to the remote manipulator.
(3)   The controller of the remote manipulator obtains the associated configuration q.
(4)   The distance $D(q, q_s)$ is computed for each $q_s \in \mathcal{S}$.
(5)   If, for some q_s, $D(q, q_s) < D_0$, then the remote manipulator controller computes a reaction force f_s in the same direction of the motion but with inverse sense.
(6)   The remote manipulator sends this force f_s to the local manipulator.
(7)   The human operator will not be able to move the local manipulator in such a direction implied by f_s and, thus, the singularity q_s will never be reached.

7. Conclusions

This paper proposes a novel singularity identification method for arbitrary serial robots based on the six-dimensional geometric algebra $\mathcal{G}_6$. For non-redundant serial robots, we take the six unit twists $\xi_1, \ldots, \xi_6$ associated with the joints, and we model them as vectors of $\mathcal{G}_6$. Hence, the problem reduces to find the configurations causing the exterior product $\xi_1(q) \wedge \cdots \wedge \xi_6(q)$ to vanish since, as proven in Corollary 1, $\xi_1(q) \wedge \cdots \wedge \xi_6(q) = 0$ if, and only if, $q \in \mathcal{S}$. Analogously, for a redundant robot with n DoF, we consider the $C(n, 6)$ different combinations of six unit twists taken from $\{\xi_1, \ldots, \xi_n\}$, and we find the configurations causing all the exterior products of the form $\xi_{j_1}(q) \wedge \cdots \wedge \xi_{j_6}(q)$ for $1 \leq j \leq C(n, 6)$ to vanish.

For serial robots with a spherical wrist, a simplification is possible. For these manipulators, the singularities are of two types: position singularities and orientation singularities. The former are identified as the configurations causing the exterior products $s_{i_1}(q) \wedge s_{i_2}(q) \wedge s_{i_3}(q)$ to vanish for $1 \leq i \leq C(n - 3, 3)$, where s_{i_j} is the linear velocity component of the unit twist ξ_{i_j} and is modelled as a vector of $\mathcal{G}_3$, while the latter are identified as the configuration causing the exterior product $z_{n-2}(q) \wedge z_{n-1}(q) \wedge z_n(q)$ to vanish, where z_i is the i-th joint axis and, again, is modelled as a vector of $\mathcal{G}_3$. Thus, the simplification consists of evaluating the exterior product of three vectors in $\mathcal{G}_3$ instead of six vectors in $\mathcal{G}_6$.

Once the singularities are identified, a distance function is defined such as its restriction to the singular set $\mathcal{S}$, defined in Definition 5, is also a distance function that allows us to check how far an arbitrary configuration q is to a singularity. This distance function exploits the fact that between any two vectors $x, y \in \mathcal{G}_n$, there always exists a rotor R such that $y = Rx\tilde{R}$.

The advantages of the strategy introduced in this work are clear. First, it is a computer-friendly approach that avoids the computation of the determinant of an order $6 \times n$ (for non-redundant robots) or $n \times n$ (for redundant robots) matrix and the Jacobian matrix J. In addition, the novel distance function defined in Definition 5 can be used to improve the performance of current control schemes or motion planning algorithms, which, as seen in the introduction, is still a hot research topic in robotics.

Author Contributions: Conceptualization, I.Z. and J.L.; methodology, I.Z.; investigation, I.Z. and H.H.; writing—original draft preparation, I.Z. and H.H.; writing—review and editing, I.Z., H.H. and J.L.; supervision, J.L. All authors have read and agreed to the published version of the manuscript.

Funding: This research received no external funding.

Institutional Review Board Statement: Not applicable.

Informed Consent Statement: Not applicable.

Data Availability Statement: Not applicable.

Conflicts of Interest: The authors declare no conflict of interest.

References

1. Gottlieb, D. Robots and Topology. In Proceedings of the IEEE International Conference on Robotics and Automation (ICRA), San Francisco, CA, USA, 7–10 April 1986; pp. 1689–1691.
2. Hollerbach, J. Optimum kinematic design for a seven degree of freedom manipulator. In *Robotics Research: The Second International Symposium*; Hanafusa, H., Inoue, H., Eds.; MIT Press: Cambridge, MA, USA, 1985; pp. 215–222.
3. Carmichael, M.; Khonasty, R.; Aldini, S.; Liu, D. Human Preferences in Using Damping to Manage Singularities During Physical Human-Robot Collaboration. In Proceedings of the IEEE International Conference on Robotics and Automation (ICRA), Paris, France, 31 May–31 August 2020; pp. 10184–10190.
4. Lopez-Franco, C.; Diaz, D.; Hernandez-Barragan, J.; Arana-Daniel, N.; Lopez-Franco, M. A Metaheuristic Optimization Approach for Trajectory Tracking of Robot Manipulators. *Mathematics* **2022**, *10*, 1051. [CrossRef]
5. Thananjeyan, B.; Tanwani, A.; Ji, J.; Fer, D.; Patel, V.; Krishnan, S.; Goldberg, K. Optimizing Robot-Assisted Surgery Suture Plans to Avoid Joint Limits and Singularities. In Proceedings of the International Symposium on Medical Robotics (ISMR), Atlanta, GA, USA, 3–5 April 2019; pp. 1–7.
6. Dupac, M. Smooth trajectory generation for rotating extensible manipulators. *Math. Methods Appl. Sci.* **2018**, *41*, 2281–2286. [CrossRef]
7. Wang, X.; Zhang, D.; Zhao, C.; Zhang, H.; Yan, H. Singularity analysis and treatment for a 7R 6-DOF painting robot with non-spherical wrist. *Mech. Mach. Theory* **2018**, *126*, 92–107. [CrossRef]
8. Ratajczak, J.; Tchoń, K. Normal forms and singularities of non-holonomic robotic systems: A study of free-floating space robots. *Syst. Control. Lett.* **2020**, *138*, 104661. [CrossRef]
9. Almarkhi, A.; Maciejewski, A. Singularity Analysis for Redundant Manipulators of Arbitrary Kinematic Structure. In Proceedings of the 16th International Conference on Informatics in Control, Automation and Robotics (ICINCO), Prague, Czech Republic, 29–31 July 2019; pp. 42–49.
10. Sharifi, H.; Black, W. Identification Algorithm to Determine the Trajectory of Robots with Singularities. *arXiv* **2019**, arXiv:1911.06632.
11. Zhu, G.; Wei, S.; Zhang, Y.; Liao, Q. A Novel Geometric Modeling and Calculation Method for Forward Displacement Analysis of 6-3 Stewart Platforms. *Mathematics* **2021**, *9*, 442. [CrossRef]
12. Hadfield, H.; Wei, L.; Lasenby, J. The forward and inverse kinematics of a Delta Robot. In *Advances in Computer Graphics*; Magnenat-Thalmann, N., Stephanidis, C., Wu, E., Thalmann, D., Sheng, B., Kim, J., Papagiannakis, G., Gavrilova, M., Eds.; Springer International Publishing: Berlin/Heidelberg, Germany, 2020; pp. 447–458.
13. Thiruvengadam, S.; Tan, J.; Miller, K. A Generalised Quaternion and Clifford Algebra Based Mathematical Methodology to Effect Multi-stage Reassembling Transformations in Parallel Robots. *Adv. Appl. Clifford Algebr.* **2021**, *31*, 39. [CrossRef]
14. Breuils, S.; Tachibana, K.; Hitzer, E. New Applications of Clifford's Geometric Algebra. *Adv. Appl. Clifford Algebr.* **2022**, *32*, 1–17. [CrossRef]
15. Hitzer, E.; Lavor, C.; Hildenbrand, D. Current survey of Clifford geometric algebra applications. *Math. Methods Appl. Sci.* 2022, *in press* . [CrossRef]
16. Corrochano, E.; Sobczyk, G. Applications of Lie algebras and the algebra of incidence. In *Geometric Algebra with Applications in Science and Engineering*; Corrochano, E., Sobczyk, G., Eds.; Birkhäuser Boston: Boston, MA, USA, 2001; pp. 252–277.
17. Kanaan, D.; Wenger, P.; Caro, S.; Chablat, D. Singularity analysis of lower mobility parallel manipulators using Grassmann–Cayley algebra. *IEEE Trans. Robot.* **2009**, *25*, 995–1004. [CrossRef]
18. Tanev, T. Singularity analysis of a 4-DOF parallel manipulator using geometric algebra. In *Advances in Robot Kinematics: Mechanisms and Motion*; Lennarčič, J., Roth, B., Eds.; Springer: Dordrecht, The Netherlands, 2006; pp. 275–284.
19. Chai, X.; Xiang, J. Mobility Analysis of Limited-Degrees-of-Freedom Parallel Mechanisms in the Framework of Geometric Algebra. *ASME J. Mech. Robot.* **2016**, *8*, 41005.
20. Yao, H.; Chen, Q.; Chai, X.; Li, Q. Singularity analysis of 3-RPR parallel manipulators using geometric algebra. *Adv. Appl. Clifford Algebr.* **2017**, *27*, 2097–2113. [CrossRef]
21. Chai, X.; Li, Q. Analytical mobility analysis of Bennett linkage using geometric algebra. *Adv. Appl. Clifford Algebr.* **2017**, *27*, 2083–2095. [CrossRef]
22. Ma, J.; Chen, Q.; Yao, H.; Chai, X.; Li, Q. Singularity analysis of the 3/6 Stewart parallel manipulator using geometric algebra. *Math. Methods Appl. Sci.* **2018**, *41*, 2494–2506. [CrossRef]
23. Huo, X.; Sun, T.; Song, Y. A geometric algebra approach to determine motion/constraint, mobility and singularity of parallel mechanism. *Mech. Mach. Theory* **2017**, *116*, 273–293. [CrossRef]

24. Chai, X.; Li, Q.; Ye, W. Mobility analysis of overconstrained parallel mechanism using Grassmann-Cayley algebra. *Appl. Math. Model.* **2017**, *51*, 643–654. [CrossRef]

25. Yang, S.; Li, Y. Classification and analysis of constraint singularities for parallel mechanisms using differential manifolds. *Appl. Math. Model.* **2020**, *77*, 469–477. [CrossRef]

26. Kim, J.; Jeong, J.; Park, J. Inverse kinematics and geometric singularity analysis of a 3-SPS/S redundant motion mechanism using conformal geometric algebra. *Mech. Mach. Theory* **2015**, *90*, 23–36. [CrossRef]

27. Huo, L.; Baron, L. The joint-limits and singularity avoidance in robotic welding. *Ind. Robot.* **2008**, *35*, 456–464. [CrossRef]

28. Yahya, S.; Moghavvemi, M.; Mohamed, H. Singularity avoidance of a six degree of freedom three dimensional redundant planar manipulator. *Comput. Math. Appl.* **2012**, *64*, 856–868. [CrossRef]

29. Siciliano, B.; Sciavicco, L.; Villani, L.; Oriolo, G. *Robotics: Modelling, Planning and Control*; Springer Publishing Company: Berlin/Heidelberg, Germany, 2008.

30. Yao, H.; Li, Q.; Chen, Q.; Chai, X. Measuring the closeness to singularities of a planar parallel manipulator using geometric algebra. *Appl. Math. Model.* **2018**, *57*, 192–205. [CrossRef]

31. Nawratil, G. Singularity distance for parallel manipulators of Stewart Gough type. In *Advances in Mechanism and Machine Science*; Uhl, T., Ed.; Springer International Publishing: Berlin/Heidelberg, Germany, 2019; pp. 259–268.

32. Bu, W. Closeness to singularities of robotic manipulators measured by characteristic angles. *Robotica* **2016**, *34*, 2105–2115. [CrossRef]

33. Clifford, W.; Smith, H.; Tucker, R. *Mathematical Papers by William Kingdon Clifford-Edited*; Macmillan: London, UK, 1882.

34. Grassmann, H. *Ausdehnungslehre-Extension Theory (English Re-Edition)*; American Mathematical Society: Providence, RI, USA, 2000.

35. Hamilton, W.R. *Elements of Quaternions*; Longmans, Green: London, UK, 1866.

36. Doran, C.; Lasenby, A. *Geometric Algebra for Physicists*; Cambridge University Press: Cambridge, UK, 2003.

37. Dorst, L.; Fontijne, D.; Mann, S. *Geometric Algebra for Computer Science: An Object-Oriented Approach to Geometry*; Morgan Kaufmann Publishers Inc.: Burlington, MA, USA, 2007.

38. Perwass, C. *Geometric Algebra with Applications in Engineering*; Springer Publishing Company, Incorporated: Berlin/Heidelberg, Germany, 2009.

39. Hildenbrand, D. *Foundations of Geometric Algebra Computing*; Springer Publishing Company, Incorporated: Berlin/Heidelberg, Germany, 2012.

40. Lavor, C.; Xambó-Descamps, S.; Zaplana, I. *A Geometric Algebra Invitation to Space-Time Physics, Robotics and Molecular Geometry*; SRMA/Springerbriefs; Springer: Berlin/Heidelberg, Germany, 2018.

41. Murray, R.; Li, Z.; Shankar-Sastry, S. *A Mathematical Introduction to Robotic Manipulation*; CRC Press: Boca Raton, FL, USA, 1994.

42. Davidson, J.; Hunt, K. *Robots and Screw Theory: Applications of Kinematics and Statics to Robotics*; Oxford University Press: Oxford, UK, 2004.

43. Chasles, M. Note sur les propriétés générales du système de deux corps semblables entr'eux. *Bull. Sci. Math. Astron. Phys. Chemiques* **1830**, *14*, 321–326.

44. Tsai, L. *Robot Analysis: The Mechanics of Serial and Parallel Manipulators*; John Wiley and Sons: Hoboken, NJ, USA, 1999.

45. Sangwine, S.; Hitzer, E. Clifford Multivector Toolbox (for MATLAB). *Adv. Appl. Clifford Algebr.* **2017**, *27*, 539–558. [CrossRef]

46. Zaplana, I.; Claret, J.; Basanez, L. Kinematic analysis of redundant robotic manipulators: applications to Kuka LWR 4+ and ABB Yumi. *Rev. Iberoam. Autom. Inform. Ind.* **2018**, *15*, 192–202. [CrossRef]

47. Basañez, L.; Suárez, R. Teleoperation. In *Springer Handbook of Automation*; Nof, S., Ed.; Springer: Berlin/Heidelberg, Germany, 2009; pp. 449–468.

 *mathematics*

Article

Mathematical Analysis of a Low Cost Mechanical Ventilator Respiratory Dynamics Enhanced by a Sensor Transducer (ST) Based in Nanostructures of Anodic Aluminium Oxide (AAO)

Jesús Alan Calderón Chavarri [1,2,*], Carlos Gianpaul Rincón Ruiz [2], Ana María Gómez Amador [3], Bray Jesús Martin Agreda Cardenas [2], Sebastián Calero Anaya [2], John Hugo Lozano Jauregui [2,4], Alexandr Toribio Hinostroza [4] and Juan José Jiménez de Cisneros y Fonfría [2]

[1] Angewandte Nanophysik, Institut Für Physik, Technische Universität Ilmenau, 98693 Ilmenau, Germany
[2] Engineering Department, Pontificia Universidad Católica del Perú, Lima 15088, Peru; rinconr.carlos@pucp.pe (C.G.R.R.); b.agreda@pucp.edu.pe (B.J.M.A.C.); sebastian.calero@pucp.pe (S.C.A.); john.lozanoj@pucp.edu.pe (J.H.L.J.); juanjose.cisneros@pucp.pe (J.J.J.d.C.y.F.)
[3] Departament of Mechanical Engineering, Universidad Carlos III de Madrid, 28911 Madrid, Spain; amgomez@ing.uc3m.es
[4] Mechatronic Department, Northen (Artic) Federal University Named after M.V. Lomonosov, Arkhangelsk 163002, Russia; alexandrtoribio@mail.ru
* Correspondence: alan.calderon@pucp.edu.pe

Citation: Chavarri, J.A.C.; Ruiz, C.G.R.; Gómez Amador, A.M.; Cardenas, B.J.M.A.; Anaya, S.C.; Lozano Jauregui, J.H.; Hinostroza, A.T.; Jiménez de Cisneros y Fonfría, J.J. Mathematical Analysis of a Low Cost Mechanical Ventilator Respiratory Dynamics Enhanced by a Sensor Transducer (ST) Based in Nanostructures of Anodic Aluminium Oxide (AAO). *Mathematics* 2022, 10, 2403. https://doi.org/10.3390/math10142403

Academic Editor: James M. Buick

Received: 8 April 2022
Accepted: 19 June 2022
Published: 8 July 2022

Publisher's Note: MDPI stays neutral with regard to jurisdictional claims in published maps and institutional affiliations.

Abstract: Mechanical ventilation systems require a device for measuring the air flow provided to a patient in order to monitor and ensure the correct quantity of air proportionated to the patient, this device is the air flow sensor. At the beginning of the COVID-19 pandemic, flow sensors were not available in Peru because of the international supply shortage. In this context, a novel air flow sensor based on an orifice plate and an intelligent transducer was developed to form an integrated device. The proposed design was focused on simple manufacturing requirements for mass production in a developing country. CAD and CAE techniques were used in the design stage, and a mathematical model of the device was proposed and calibrated experimentally for the measured data transduction. The device was tested in its real working conditions and was therefore implemented in a breathing circuit connected to a low-cost mechanical ventilation system. Results indicate that the designed air flow sensor/transducer is a low-cost complete medical device for mechanical ventilators that is able to provide all the ventilation parameters by an equivalent electrical signal to directly display the following factors: air flow, pressure and volume over time. The evaluation of the designed sensor transducer was performed according to sundry transducer parameters such as geometrical parameters, material parameters and adaptive coefficients in the main transduction algorithm; in effect, the variety of the described results were achieved by the faster response time and robustness proportionated by transducers of nanostructures based on Anodic Aluminum Oxide (AAO), which enhanced the designed sensor/transducer (ST) during operation in intricate geographic places, such as the Andes mountains of Peru.

Keywords: air flow medical sensor; emergency air flow sensor; low-cost air flow sensor; nanostructures; COVID-19

MSC: 65K10

1. Introduction

Monitoring has become indispensable for mechanical ventilators in order to provide feedback to medical staff regarding treatment. Moreover, ventilation parameters can be used for internal control tasks such as adaptive feedback which improves the equipment performance and response to disturbances. Thus, appropriated values of physical variables and artificial breathing could be achieved by monitoring the ventilation curves in real time [1]. In this sense, a flow sensor is a device used for air flow measurement, and its working concept is commonly based on an induced pressure drop across an orifice which can be measured and related to the air flow value in a breathing circuit over time [2].

Design considerations for air flow measurement in industrial applications are well established in the steady-state condition [3]; nevertheless, the air flow provided by the developed low-cost mechanical ventilator has a particular periodic waveform in a transient condition. Typical medical air flow sensors measure the pressure drop using static pressure taps connected to a transducer. However, this represents a risk of leaking polluted air from the patient, which is a harmful risk given the current pandemic context. To counter this, modern medical air flow sensors are connected to electrical transducers to generate electrical values analogous to values of static pressure [4]. Having a sensor connected to an electrical transducer is quite important for the measured variables, as their electrical equivalent values can be stored in the main control system of the ventilator. On the other hand, any kind of virus can be reduced, something that cannot be achieved with typical airflow connectors. Furthermore, air flow sensors cannot provide air pressure and volume and other parameters, which require monitoring, are measured using an additional sensor and calculated at the processing stage.

In this study, a novel air flow sensor/transducer was proposed as an integrated design that is able to measure the air flow in a breathing circuit in the time domain and estimate the air pressure and volume parameters with the mathematical model developed and calibrated experimentally for a specific low-cost mechanical ventilator based on cams [1]. This proposed medical device directly provides the equivalent electrical signals of air pressure, volume and flow after a transduction stage required due to the nonlinearity of the air flow and pressure drop [5].

During the COVID-19 pandemic, many low-cost mechanical ventilation systems based on the compression of an airbag were developed [6], but they were required to measure the generated air flow rate to the patient [7]. However, air flow sensors were not easy to resource in Peruvian markets due to the international supply shortage during this period of the pandemic. Moreover, the design and fabrication for this medical application requires certain considerations, such as those proposed in this research.

2. Materials and Methods

Broadly, the methodology consists of design, fabrication, experimental study and experimental validation stages. The design was an iterative process, thus in this paper only the final version which was improved according to the experimental results is shown. Figure 1 shows an overview of the research flow. In the design stage, sensor architecture and functions were established. The transduction algorithm correlates the theoretical pressure difference and ventilation curves shown in the display. A Computational Fluid Dynamics (CFD) analysis supports an exhaustive study of air airflow measurement in different flow conditions as well as with different orifice plate dimensions. A sensor prototype and equipment were specially designed for the experimental study. The experimental data obtained were used for the mathematical characterization of the sensor in dynamic [8], steady and real working conditions. Design improvement and model calibration were carried out based on the CFD simulation and experimental characterization [9].

Finally, the design was improved and fabricated for the experimental validation where the device measurement, connected to a mechanical ventilator, was compared to the measurement of a Fluke air flow meter.

2.1. Device Description

Figure 2 shows the sensor/transducer concept. It consists of an orifice plate sensor, a pressure transducer and a controller. The crossing air flow through the orifice plate produces a pressure difference related to the air flow value. Then, the pressure transducer converts the pressure signals into their electric equivalent signals which are sent to the controller for data treatment in order to determine and display the pressure, volume and flow over time using an electronic screen [10].

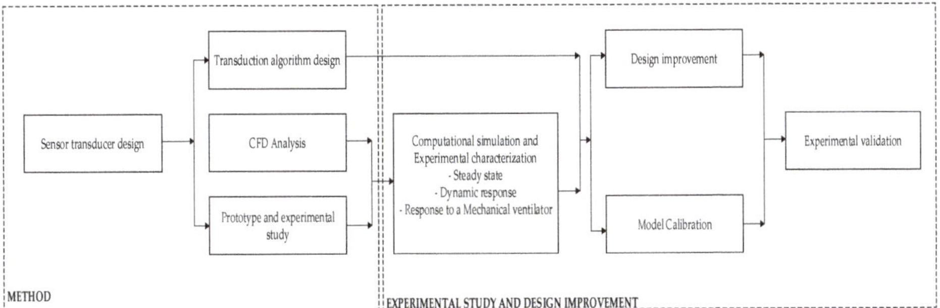

Figure 1. Research overview.

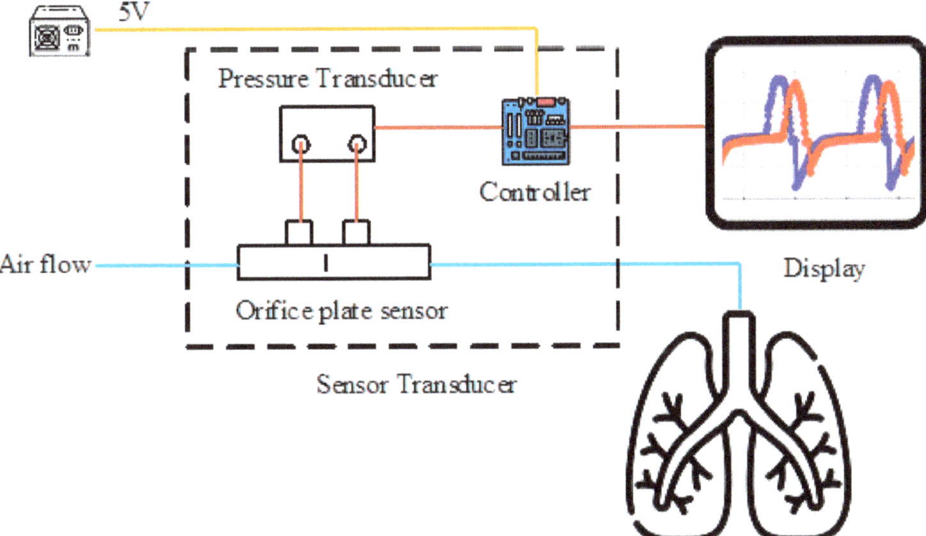

Figure 2. Scheme of the designed air flow sensor transducer.

2.2. Device Design

Figure 3 shows the integrated air flow sensor/transducer developed for this particular transient flow measurement. It consists of a compact air flow sensor based on an orifice plate and a transduction stage of the measured pressure difference over time. The transduction stage and pressure difference measurement were integrated in this device in order to prevent leaks or contaminated air flow. The device measures the air flow through it and calculates (using a microcontroller Arduino UNO) the breathing variables transduced as electrical values that are depicted on the electronic screen and sent to the main control system of the mechanical ventilator.

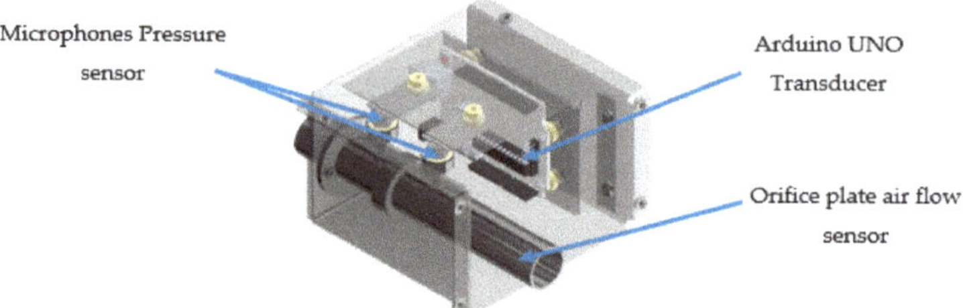

Figure 3. Design of an integrated air flow sensor/transducer for a breathing circuit.

Figure 4 shows the compact air flow sensor developed for the integrated design, which is composed of an upstream and downstream pipe welded to an orifice plate. Dimensions are established based on the analytic and experimental results, which guarantee the pressure difference measurement over time for this particular medical application with a mechanical ventilation system OxygenIP.PE, whose nominal working conditions correspond to a transient flow [11].

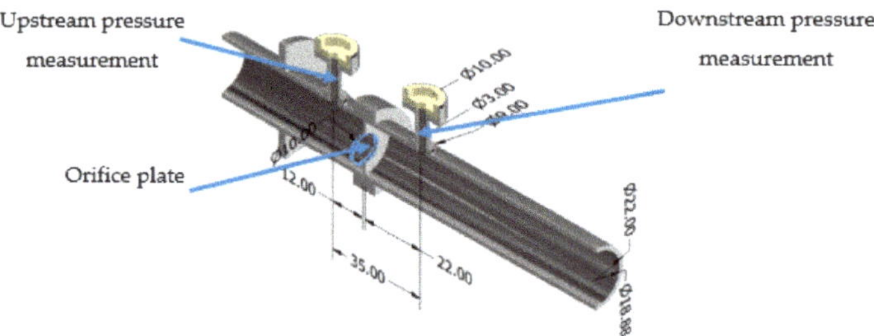

Figure 4. Compact air flow sensor developed for the air flow sensor/transducer.

2.3. Mathematical Modelling

The design of the flow sensor was based on the mathematical model analysis as a consequence of the theoretical equations, which describe the flow behavior through an orifice plate. Moreover, the analysis was enhanced due to the correlation with polynomial models that provided adaptive coefficients for the optimal measurements, for which the process is described in more detail in the following paragraphs.

In Figure 5, a scheme to explain the theoretical model of the flow inside the airflow circuit is depicted. In which f is the air flow from a source, the section area of the hose is a, and the flow reduces its section area b while crossing the orifice plate. Thus, a pressure difference can be measured between points P1 and P2, moreover v_1 and v_2 represent the fluid flow speed of the air flow in the direction depicted by the red arrow [11,12].

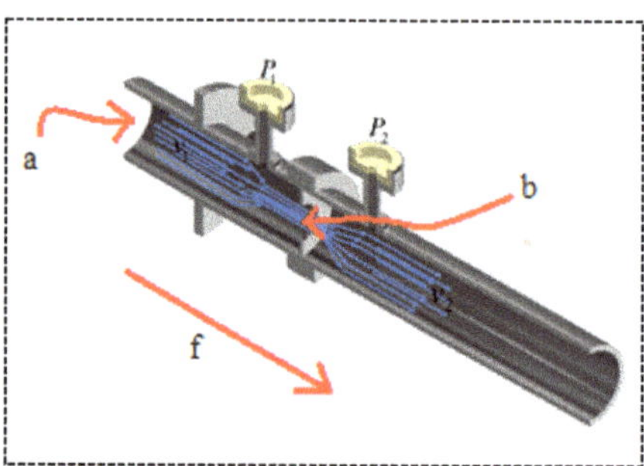

Figure 5. Scheme of the orifice plate theoretical model.

In order to find the pressure, airflow and volume equations, the Stokes equations were analyzed using Equation (1), in which, "f" is the function as the consequence of airflow and "μ" and "D" are the average velocity and dispersion coefficient, respectively [13,14].

$$\frac{\partial f}{\partial t} + \mu \frac{\partial f}{\partial x} = D \frac{\partial f^2}{\partial x^2} \tag{1}$$

This expression can be reduced to the Bernoulli model, and with the analysis of the energy balance, Equation (2) can be obtained [13], in which ρ is the fluid flow density, p_1 and p_2 are the initial and final referential pressure through the fluid flow path, and v_1 and v_2 are the fluid flow speed.

$$\frac{1}{2}\rho v_1{}^2 + p_1 = \frac{1}{2}\rho v_2{}^2 + p_2 \tag{2}$$

By conducting a static pressure analysis, it is possible to obtain Equation (3), where ϱ is the volumetric flow rate [13], A_1 and A_2 are the cross sections area, and the initial and the final static pressure are p_1 and p_2, which are proportionate to the difference of pressure ΔP.

$$\Phi = A_2 \sqrt{\frac{2\Delta P}{\varrho\left(1 - \left(\frac{A_2}{A_1}\right)^2\right)}} \tag{3}$$

From here, it is further possible to reduce to the equation by considering the "discharge coefficient C_d" so that Equation (4) is obtained [13].

$$\Phi = C_d A_2 \sqrt{\frac{2\Delta P}{\varrho\left(1 - \left(\frac{A_2}{A_1}\right)^2\right)}} \tag{4}$$

Finally, a theoretical model can be derived from the expected breathing curve as shown by Equation (5) [13], in which Ri has the geometrical and material information of the fluid flow sensor. Equation (5) proposes the relation among the volumetric flow rate Φ with the pressure difference of ΔP, and τ is the response time.

$$\Phi = \left(\frac{\Delta P}{R_i}\right)\left(1 - e^{\frac{t}{\tau}}\right) \tag{5}$$

2.4. Computational Fluid Dynamics (CFD) Analysis

Sensor Case Setup

The geometry was modelled after the actual state bench design proposal (with an extended outlet to overcome reverse flow issues). Following Bridgeman's steps [15], the sensor case was modelled as axisymmetric and the pressure sensing ports' geometries were disregarded to save computational resources and time. Because of this, better and more structured meshes were obtained. Five geometries were generated in total (one for each orifice diameter). The meshes were then generated with an element size of 0.00025 m, all with good or excellent mesh metrics results, according to Ansys [16]. Figure 6 shows the model for the 10 mm orifice plate of the sensor case, in which A and B keep the distance for the static pressure orifices.

The main equations that govern this analysis are provided by the Navier–Stokes equations, mainly the x-y momentum and continuity equations. The energy equation was not included in the analysis since heat transfer was not one of the concerns of this study. The SST k-ε turbulence model was selected for this study because it combines the following 2 separate models: the k-ε and k-ε turbulence models. The first model is better at capturing near-the-wall turbulence effects, and the second model is superior at capturing turbulence effects away from the walls.

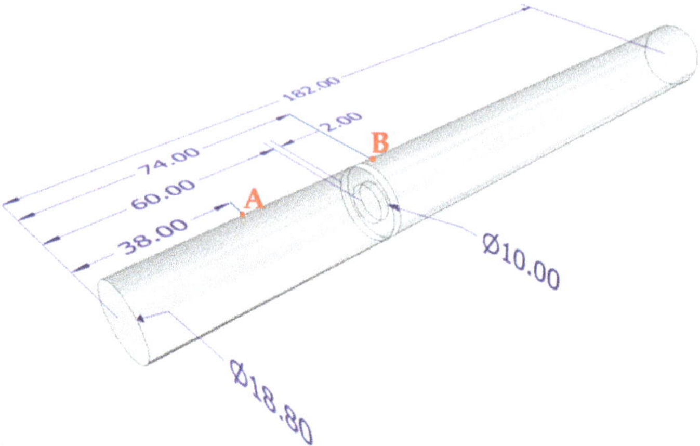

Figure 6. Sensor case geometry for 10 mm orifice plate.

The following initial and border conditions were considered:

- Air assumed as incompressible fluid [17];
- Density = 1.18 kg/m^3;
- Atmospheric pressure at the outlet (101,325 Pa);
- Dynamic Viscosity = 1.7984 $\times$ 10^{-5} kg/(m-s);
- No slip condition at the walls.

A mesh sensitivity analysis was performed for 3 different meshes. The first one contained 6962 elements; the second, 27,843; and the third, 43,045. The populations corresponded to meshes with element sizes of 0.0005 m, 0.00025 m, and 0.0002 m, respectively. These 3 different meshes were run as steady state cases for the 4 mm orifice and 100 L/min airflow. The results were then graphed.

Figure 7 shows that the results began to converge towards a single pressure difference value. If the mesh refining continued, a single pressure difference value would be reached. Since the geometry was simple, a highly populated mesh was not necessary. This is shown in Figure 6, where almost double the number of elements only resulted in an increase of 0.31% of the pressure difference, as well as a considerable increase of almost 6 times the required number of iterations before the residuals reached convergence. Therefore, to save computational resources and time, but at the same time still obtain acceptable results, a mesh with an element size of 0.00025 m was selected for the simulations. The mesh used for the 10 mm orifice model is shown in Figure 8.

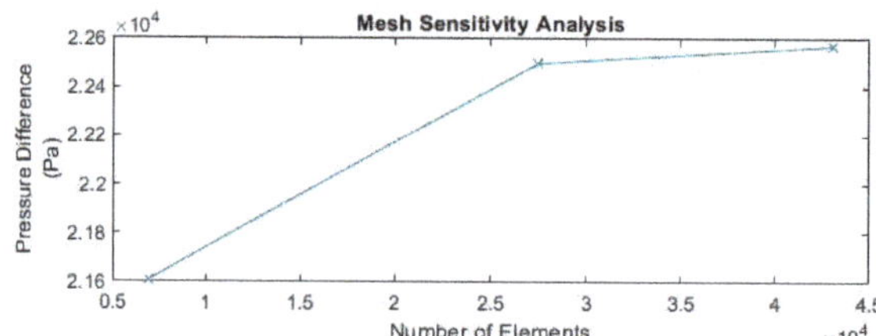

Figure 7. Mesh Sensitivity Analysis.

Figure 8. Mesh for 10 mm orifice model.

2.5. Transduction Stage Design

After studying the theoretical model with the Navier–Stokes equations and analyzing the sensor case via simulation methodologies, experiments were conducted to develop the experimental model of the designed sensor/transducer. Equations (6) and (7) represent a non-linear system, where $u(t)$ is the input signal (for example pressure from cam over pivot), $y(t)$ is the response (such as for example airflow crossing airbag to a connected hose), $x(t)$ is an internal variable which allows the connection between input signals and response, f and h represent the non-linear characteristics of the system such as functions, while θ represents physical parameters of the system and t the time. The function f contains the mechanic parameters, fluid parameters and electrical parameters for the correlation between the theoretic and experimental model of the designed sensor [1,13,18].

$$\frac{dx(t)}{dt} = f(x(t), u(t), \theta) \tag{6}$$

$$y(t) = h(x(t), u(t), \theta) \tag{7}$$

It is possible to propose the mathematical model of the system (sensor/transducer) with Equations (6) and (7) described above. Moreover, it is necessary to find the optimal signal (input variable) that provides a predicted response. Hence, Equation (8) depends on the trajectory solution as set point R_s and $r(k_i)$ is the set point of sample time k_i

$$R_s{}^T = (111 \ldots 11111111) r(k_i) \tag{8}$$

Equation (9) has a dependence of matrices Rs and ΔU. The optimal predicted solution for a desired value in the system is obtained based on the cost function "J" among optimal predicted response with this "trajectory solution Y" such as the optimal predicted solution ΔU, which is described using Equation (9). The procedure to calculate ΔU is given by Equations (10)–(14), furthermore, R is a diagonal matrix to tune the desired optimal response matrix.

$$J = (R_s - Y)^T (R_s - Y) + \Delta U^T R \Delta U \tag{9}$$

On the other hand, the "Optimal predicted response Y" has a dependence of "matrices F, φ and the state vector $X(ki)$" as shown by Equation (10)

$$Y = FX(k_i) + \varphi \Delta U \tag{10}$$

Moreover, using the costing function "J" the optimal estimation was analyzed, as it is described by Equation (11) after replacing Equation (10) in Equation (9).

$$J = (R_s - FX(k_i))^T (R_s - FX(k_i)) - 2\Delta U^T \varphi^T (R_s - FX(k_i)) + \Delta U^T \left(\varphi^T \varphi + R \right) \Delta U \tag{11}$$

Looking for the minimal error, Equation (11) was derived as the dependence of "ΔU" in Equation (12).

$$\frac{\partial J}{\partial \Delta U} = -2\varphi^T (R_s - FX(k_i)) + 2 \left(\varphi^T \varphi + R \right) \Delta U \tag{12}$$

Therefore, the minimal value of "J" was achieved using Equation (13).

$$\frac{\partial J}{\partial \Delta U} = 0 \tag{13}$$

This means that the optimal "ΔU" is given by Equation (14), which helps to obtain the optimal and predicted "Y".

$$\Delta U = \left(\varphi^T \varphi + R \right)^{-1} \varphi^T (R_s - FX(k_i)) \tag{14}$$

On the other hand, the matrix φ can help to obtain information of the material sensor regarding the thin film, which is subject to the air pressure. Moreover, the geometry (nanostructures) of the thin film can be stored over the matrix. Therefore, Equation (15) provides the correlation between the optimal predictive solution of the transduction with the matrix adaptive coefficients, which keep information of the material and geometrical characteristic of the designed sensor.

Furthermore, the periodical responses from the sensor/transducer are provided by Equation (15) though a transfer function analysis, thereby making it possible to identify the first order behavior in the steady state. Therefore, C_s represents the answer variable for the first order, N_s is the input excitation signal, t is the response time and K is the proportional gain.

$$\frac{C_s}{N_s} = \frac{K}{ts + 1} \tag{15}$$

Equation (15) is reduced in the time domain and Equation (16) is obtained, which keeps the model of Equation (5) in a steady state. In this context, the excitation signal is ΔP that looks for the response variable fluid flow Φ, and R is the constant derived from the first order solution for Equation (14) in Equation (10). Hence, it was possible to correlate stability parameters with the geometrical and material properties of the designed sensor.

$$\Phi = \left(\frac{\Delta P}{R}\right) \tag{16}$$

For this reason, the transducer adapts the signal received from the sensor according to the value measured of the physical variable to another equivalent (for example, the pressure difference transduced to its equivalent as an electrical signal, which is from "cmH2O" to "mV"). In this context, Figure 9 represents the static curve given in the section "A, B, C" for the coordinate system "**X1, Y1**". However, the static curve can help to identify the linear response regions, by working in linear regions or with non-linear mathematical models.

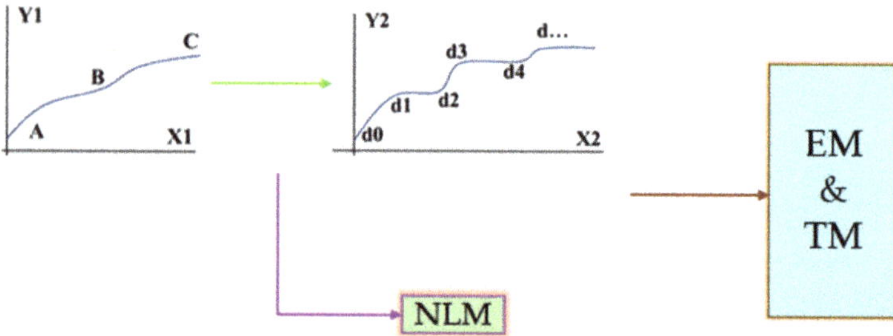

Figure 9. Scheme to represent the sensor/transducer data processing.

It should be noted that depending on the response time of the system (the system is considered as a research plant), it can lose information in the linearization process, due to its non-linear behavior, which is the reason why many authors suggest working in linear zones despite such risks (losing information in linearization process). In addition, the process of discretizing the signal information may also be lost, such as the dependence on the response time of the system and the sampling time (important parameter to discretize a signal).

Based on the conditions described in the previous paragraph, a transducer is necessary to transform the pressure difference into its electrical equivalent (in mV) in order to obtain the value for the air flow and also the displaced air volume. Theoretically, using Navier–Stokes models, the following relationship of the three physical variables is achieved: "pressure, flow and volume", where the numerical solutions of the differential equations are also prone to loss of information as a dependence of the methodologies of numerical approximations in the solutions. However, the disadvantage of a completely theoretical correlation is that the modeling of disturbances is frequently not achieved, which is solved by either approximation modeling or by correlation with experimental data.

Therefore, the transients of the plant's dynamic system (while it is first-order) for the sections represented by d_0, d_1; d_2, d_3; d_4, d ... and with similar slopes only in the linear trends "AB" or "BC" are located in the coordinates **"X2, Y2"**. In such sections, the response times of the plant (the designed sensor/transducer) can be obtained.

Moreover, for each section, d_1, d_2; d_3 d_4, and "d ... , henceforth" the characteristics of overshoots, settling time, damping and parameters that indicate the stability of the designed system can be maintained.

The static and dynamic curves are algebraically interpreted as models of differential equations (polynomial models). Consequently, the correlation is obtained between the theoretical modelling and modelling based on the experimental data analysis results in the final model of the designed "sensor/transducer" system, that implies a response "Experimental Model (EM) and Theoretical Model (TM)" from linear (linearized) or completely non-linear (NLM) models.

Indeed, the system formed by the sensor/transducer represented by "ST" receives a mechanical signal "Ms1" (fluid mechanical signal for this case, which is pressure difference) and transforms it into the electrical equivalent "Es" or a response signal "Ms2" (fluid mechanics variable as volume or airflow equivalent signal). This conversion (transduction) requires characterization and calibration steps, which are depicted by the scheme summarized in Figure 10. From the static measurement data analysis, is possible to recognize the linear and nonlinear operations in the ST system. This information is quite important according to the design of the algorithm, which provides the final information of the ST measurement, such as the breathing variables of pressure, volume, and airflow.

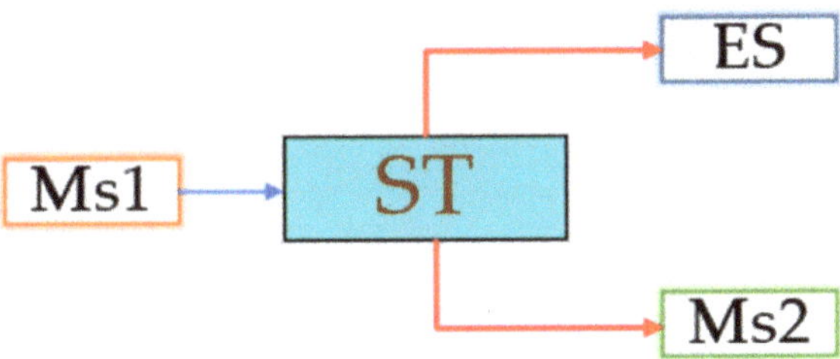

Figure 10. Scheme of a sensor/transducer system.

In the mathematical model of the sensor/transducer as a system, the transducer is composed by electronic devices that convert the fluid-mechanic variable of "pressure difference" to its electrical equivalent, as is the case for the variables "airflow and volume". Furthermore, the characterization of each physical variable to be measured is necessary, with electro/mechanical devices that are able to measure the desired physical variables being monitored, which is showed in Figure 11.

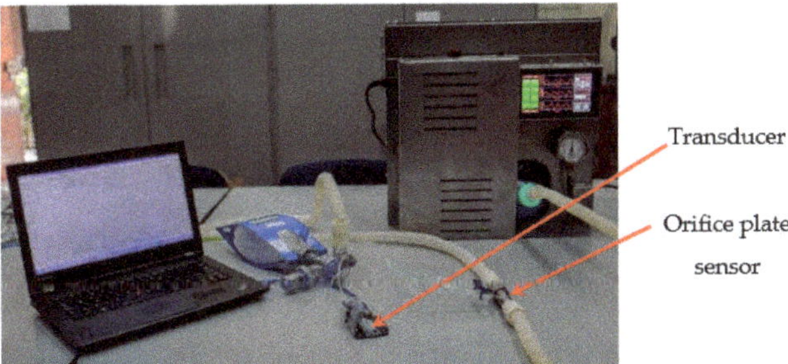

Figure 11. Setup of the designed sensor/transducer system.

3. Experimental Section

3.1. Orifice Plate Diameter Study

The working concept of the orifice plate is to generate a pressure difference which is related to the flow passing through it [1]. In this sense, different diameters could be considered; however, an optimum range of diameters presents a balance between the produced pressure difference sufficiently high to be used on the transduction stage, and sufficiently low to not to produce significant air flow or pressure drop in the breathing circuit.

Figure 12 (by views of subfigures a–c) shows the variable orifice-plate flow sensor it is manufactured by two pipes joined together with a ferrule clamp to have a quick release mechanism. Downstream of the ferrule clamp, an orifice-plate is welded to the flange. In this study, 5 parts were fabricated with different diameters: 4, 6, 8, 10, 12 mm. Distances of the pressure taps to the orifice plate were established according to the design considerations of an orifice plate air flow sensor [3]. However, the distance of the downstream pressure tap was not able to be less than half radius because of welding limitations for the geometry.

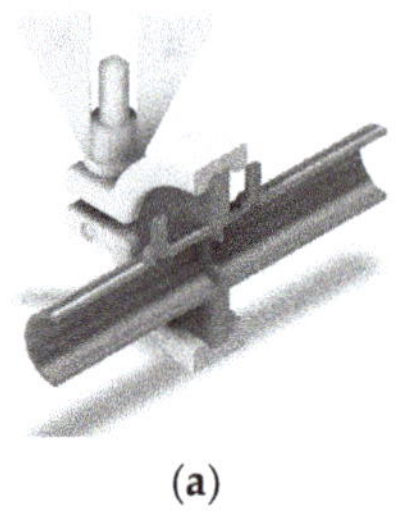

(a)

(b)

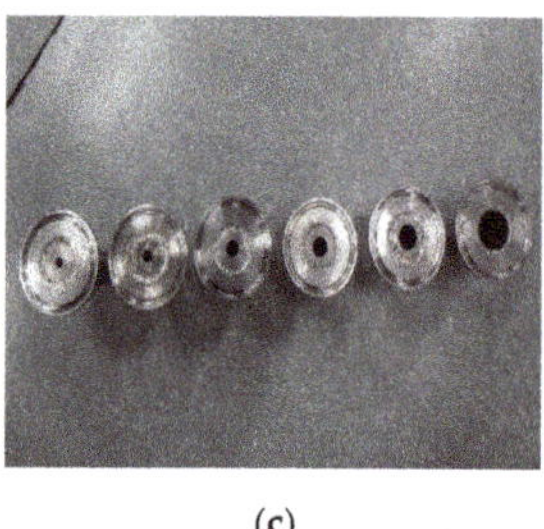

(c)

Figure 12. Variable Orifice-plate flow sensor.

3.1.1. Steady State and Dynamic Response

The variable orifice plate air flow sensor was tested for the air flow measurement in a steady state condition. Figure 13a shows the experimental setup designed for this test, where the variable orifice plate was installed in the air flow test bench, an air compressor providing the air flow through the variable orifice plate sensor was included, a pressure difference transducer based on microphones connected to an Arduino UNO microcontroller board was integrated and an algorithm was designed and programmed for this transduction stage. Figure 13b shows a schematic representation of the experimental setup and the interaction of its components.

3.1.2. Measurement in a Mechanical Ventilator

The variable orifice plate of the air flow sensor was installed in a breathing circuit connected to a mechanical ventilation system based on cams "OxygenIP.PE" [1]. This experiment corresponded to the real working conditions of the air flow sensor/transducer which was designed. A medical gas flow meter Fluke VT650 was connected to the circuit to obtain the real air flow over time, which is showed by the Figure 14a.

The artificial lung Fluke Accu Lung (a precision test lung) was used in order to simulate the ventilation parameters, which is showed by the Figure 14b as part of the experiments setup. This device is able to simulate 3 lung stages, from a healthy lung up to a damaged lung. A healthy lung corresponds to a compliance of 50 mL/cm H2O and a resistance of 5 cm H2O-s/L, while a damaged lung corresponds to a compliance of 10 mL/cm H2O and a resistance of 50 cm H2O-s/L. In this sense, for each orifice diameter inside the sensor, the experiment was run four times, twice for each lung parameters and twice for the minimum and maximum cams, XS and XL, respectively.

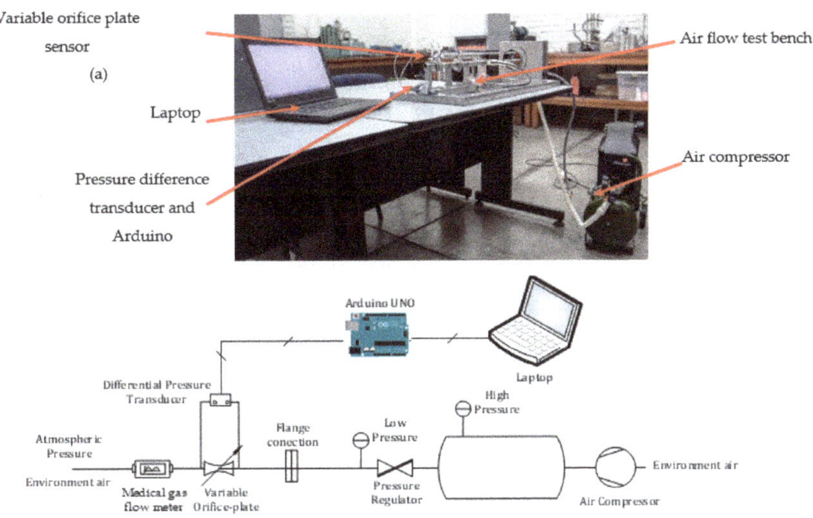

Figure 13. Schematic representation of experimental setup for steady state air flow.

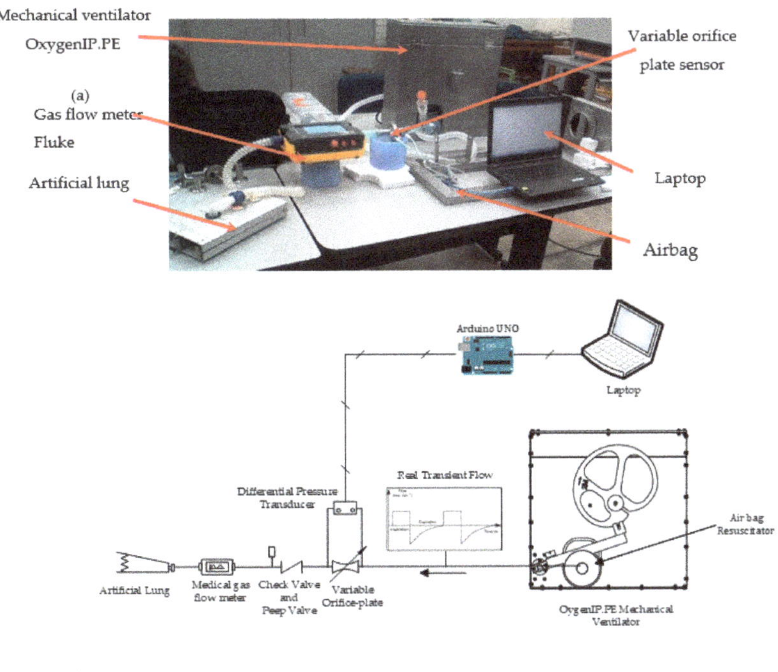

Figure 14. Scheme of experimental setup for the sensor transducer connected to a Mechanical Ventilator.

3.2. Experimental Validation of the Sensor/Transducer Device

In the Figure 15, the designed sensor connected to the mechanical ventilator (Appendix A contains more details) is depicted; Figure 15a shows the designed sensor is shown to be connected to the air flow calibrator; and Figure 15b shows the air flow calibrator that needs to be connected to the personal computer in order to validate the expected measurements, and all the setup is showed by Figure 15c.

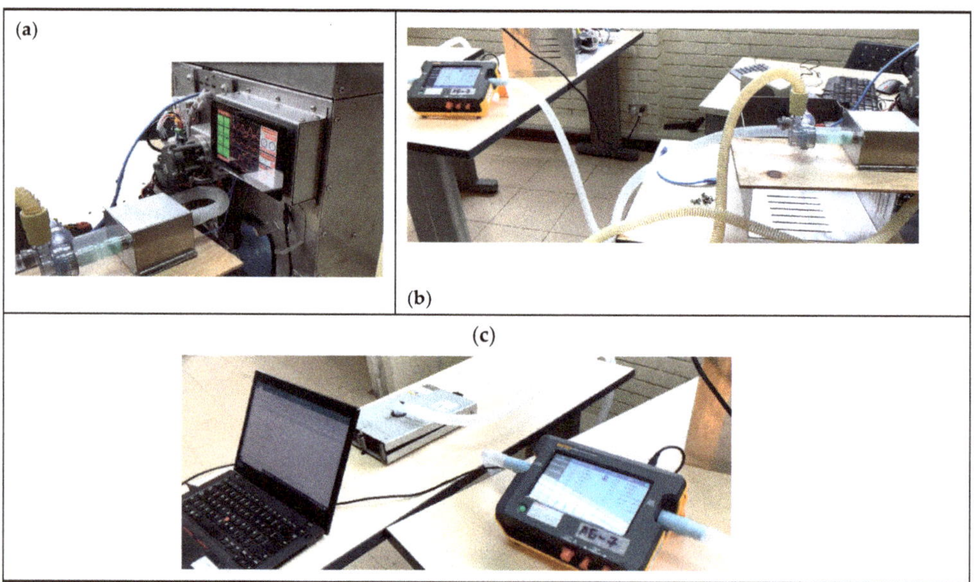

Figure 15. Setup of the sensor/transducer device connected to OxygenIP.PE Mechanical ventilator.

4. Results

4.1. CFD Simulation Results

4.1.1. Steady State Results

The following results were obtained for the steady state simulations for a 100 L/min airflow in a 10 mm diameter orifice. Figure 16a shows how the velocity increased as the area reduced (orifice throat), which is explained by the flow's intention of conserving the flow rate (mass conservation and incompressibility assumption). On the other hand, Figure 16b shows how the static pressure reduced, which occurred in response to the velocity's increase. This is explained by the flow's intention of conserving its momentum. It must also be noted that the behavior of these contours matched the results presented by Karthik [19].

4.1.2. Profile of Mechanical Ventilator Curves

A certain (expected) proportionality between the airflow and difference pressure curves is shown in Figure 17a. It is widely known that a higher pressure drop corresponds to a more abundant flow. This is explained because in turbulent flows the pressure loss is proportional to the flow velocity squared. This diameter alternative presents a maximum pressure drop of 274.28 Pa for a corresponding airflow of 67.74 L/min. Next, the steady and transient state analysis results were compared and plotted. Figure 17b shows an excellent similarity between the transient and steady state curves for the plotted points. Although a margin is observed between both curves, a good estimate of the transient behavior of the sensor can be modeled using steady state approximations. This behavior was expected, since Funk et al. [20] concluded in their work that steady state approximations can be used in transient situations with excellent or valid results for most engineering applications.

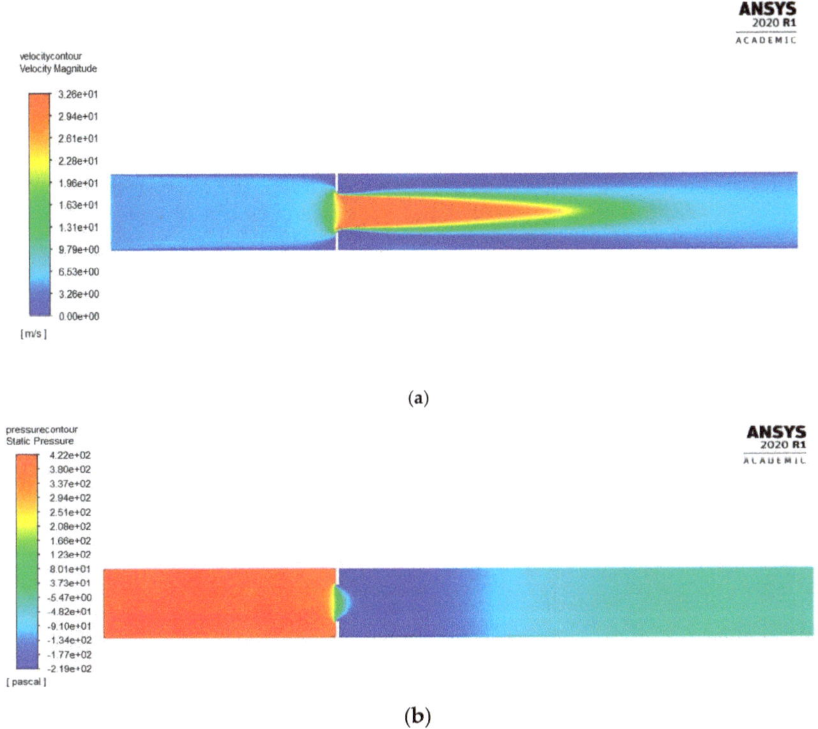

Figure 16. (**a**) Velocity Contours for a 100 L/min airflow, 10 mm diameter orifice; (**b**) Static pressure contours for a 100 L/min airflow, 10 mm diameter orifice.

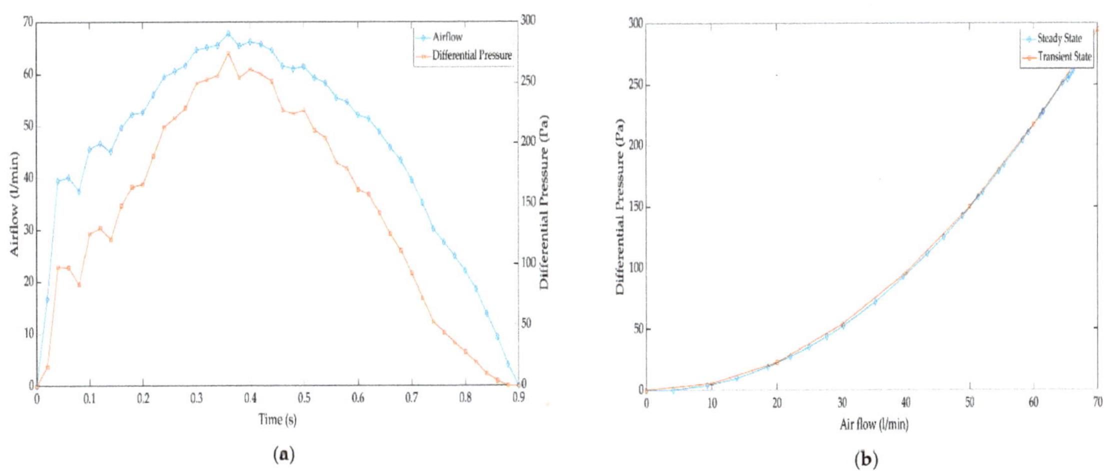

Figure 17. (**a**) Airflow profile and DP profile over time; (**b**) Steady vs. transient state results for a 10 mm diameter orifice.

4.2. Experimental Results

4.2.1. Steady State

In the following paragraphs, the results of experimental tests to verify the results of the proposed ST in steady state are provided. The analyses were made by comparisons with a Fluke instrumentation equipment, because the steady state information helped to obtain an understanding of the response of the designed ST under disturbances, such as unexpected overshoots or electromagnetic noise. In that context, the tests were run for the 4 mm diameter sensor/transducer and a steady-state inlet pressure of 5 mBar, 10 mBar and 20 mBar provided by an air compressor. A graphical comparison between the airflow plots is shown in Figure 18. The "Fluke" curve provides information of the airflow measured by the "Fluke" analyzer, and the "ST" curve shows the airflow as a response of the designed sensor/transducer ST.

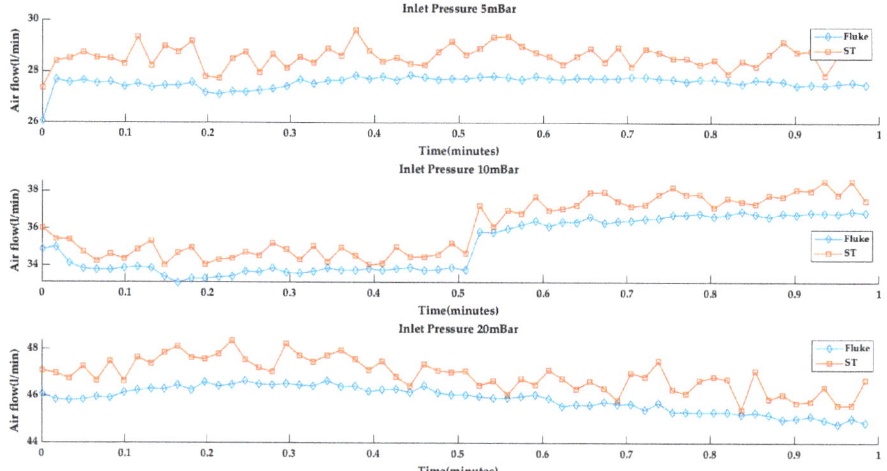

Figure 18. Flow and Flow ST for diameter 4 mm (Inlet pressure: 5, 10, 20 mBar).

The pressure difference curves (in mV) are plotted together in Figure 19. The "ST" curve provides information related to the pressure measured by the sensor/transducer in mV.

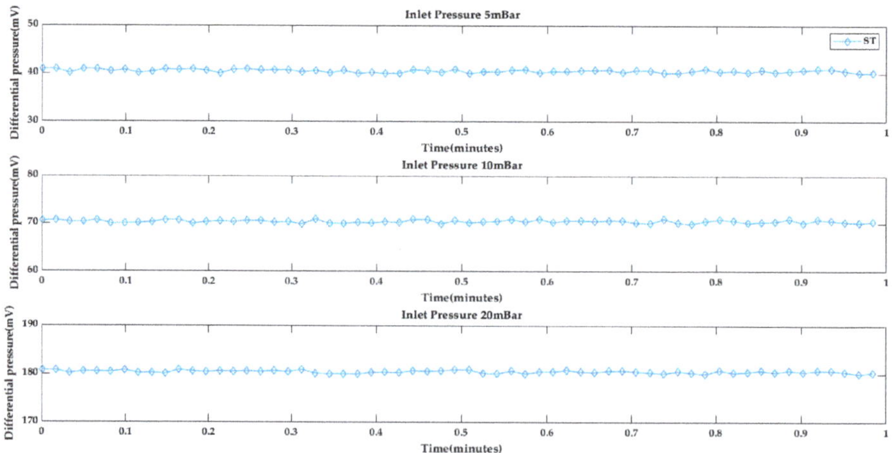

Figure 19. Pressure measured by ST in mV, for diameter 4 mm (Inlet pressure: 5, 10, 20 mBar).

The pressure difference curves are plotted together next in Figure 20. Each curve provides information of the pressure measured by the sensor/transducer in Pascal.

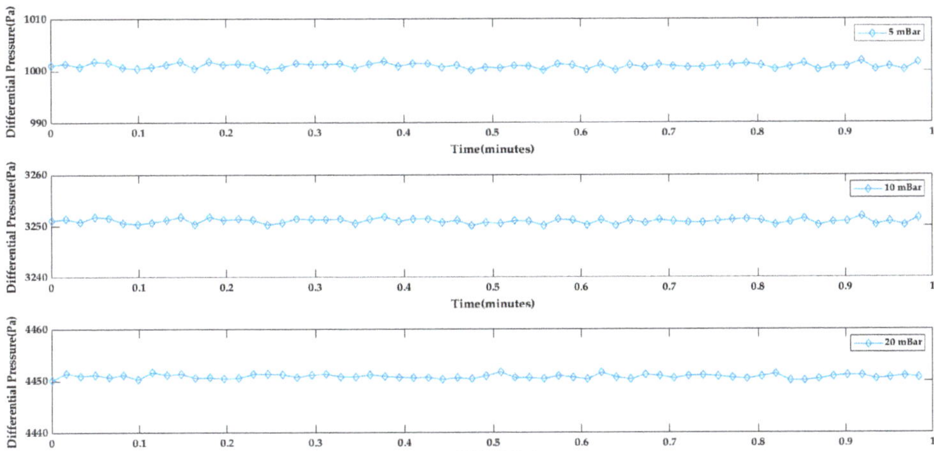

Figure 20. Pressure measured by ST (Pa), for diameter 4 mm (Inlet pressure: 5, 10, 20 mBar).

The tests were run for the 8 mm diameter sensor/transducer and a steady-state inlet pressure of 10 mBar, 20 mBar and 30 mBar was provided by an air compressor, for which the plotted airflow curves are shown in Figure 21. The "Fluke" curve provides information on the airflow measured by the "Fluke" analyzer, and the "ST" curve shows the airflow as the response of the designed sensor/transducer ST.

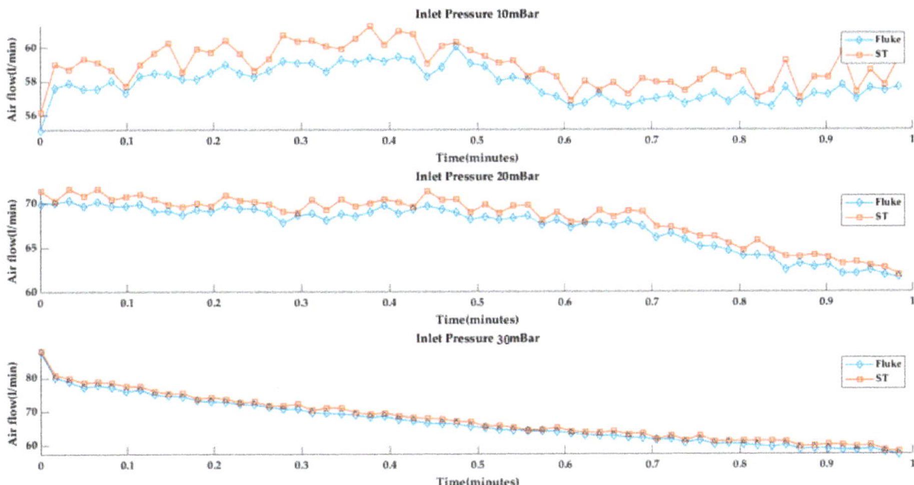

Figure 21. Flow and Flow ST for diameter 8 mm (Inlet pressure: 5, 10, 30 mBar).

The transduced pressure difference curves in mV are shown in Figure 22. Each curve provides information on the pressure measured by the sensor/transducer in mV.

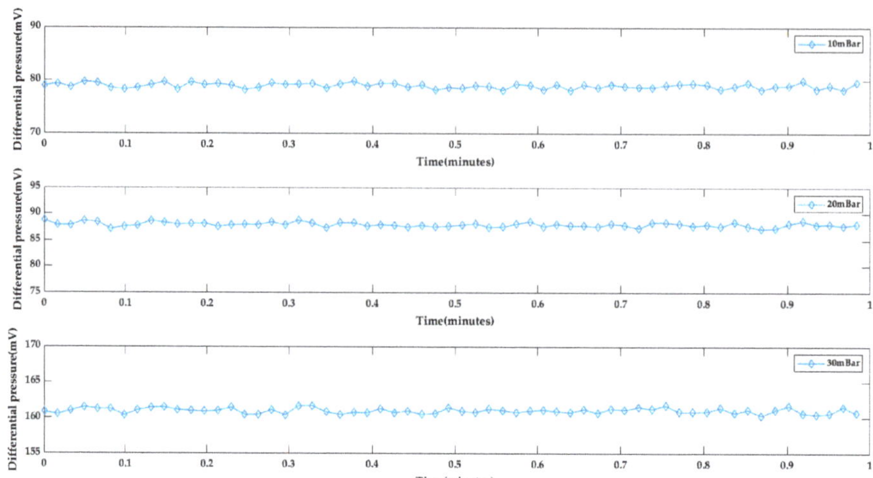

Figure 22. Pressure measured by ST (mV), for diameter 8 mm (Inlet pressure: 5, 10, 20 mBar).

On the other hand, the pressure difference curves are shown in Figure 23. Each curve provides information on the pressure measured by the sensor/transducer in Pascal.

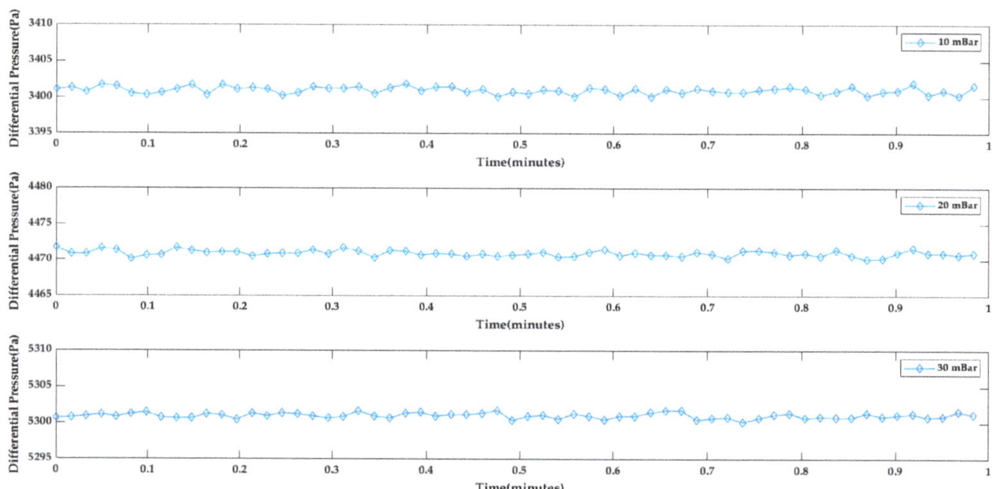

Figure 23. Pressure measured by ST (Pa), for diameter 8 mm (Inlet pressure: 5, 10, 20 mBar).

4.2.2. Open and Close Loop

Figure 24 shows the behavior of the airflow obtained by the designed sensor transducer (ST curve) in comparison with the airflow reference pattern (Fluke curve measured by the Fluke equipment). It is possible to understand that the ST system can measure changes in airflow when the tester prototype is closed or open to evaluate the steady state behavior of the designed ST. Nevertheless, it was possible to develop better responses of the system such as the dependence of the diameter of the ST, and the best responses (less error percent between ST and Fluke measurements) were achieved with orifice diameters of 8, 10 and 12 mm.

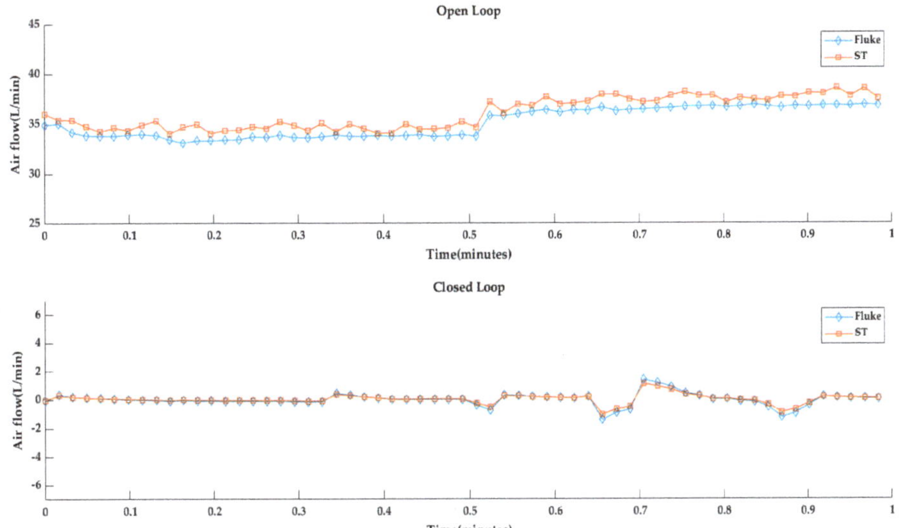

Figure 24. Airflow comparison in open loop and closed loop for the ST evaluation.

Figure 25 shows the pressure difference measured by the designed ST for a closed loop (upper subfigure) and open loop (lower subfigure). The ST system can detect changes or tendencies to maintain a steady state as a consequence of an open or closed loop of the tester prototype; therefore, it was possible to measure this pressure difference with the designed ST.

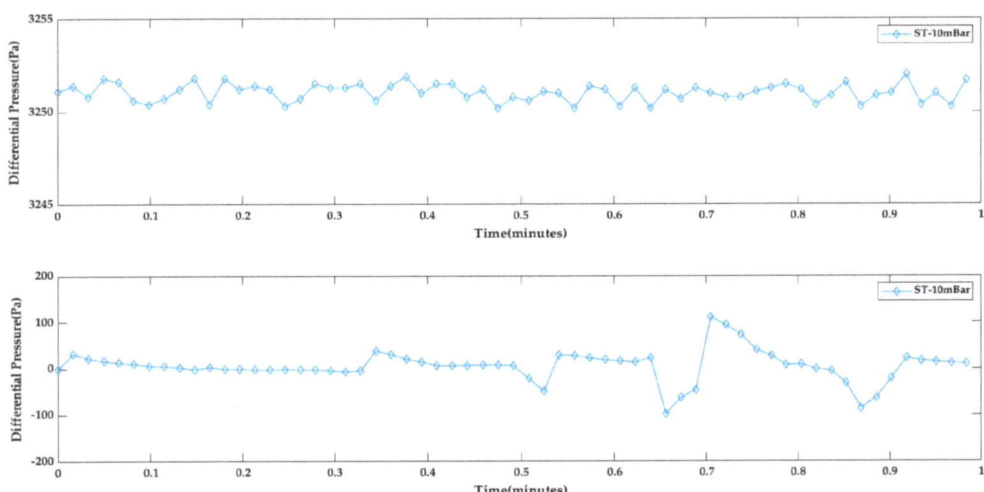

Figure 25. Pressure difference comparison in open loop and closed loop for the ST evaluation.

4.2.3. Dynamic Response

Figure 26 presents the dynamical response of the system ST to a pressure excitation signal given by 10, 20, 30 mBar obtained through the airflow regulated by the air compressor of the tester prototype, with an orifice diameter of 4 mm. It was found that ST cannot provide a faster response in the presence of transient changes (red curves for airflow, pressure difference and volume).

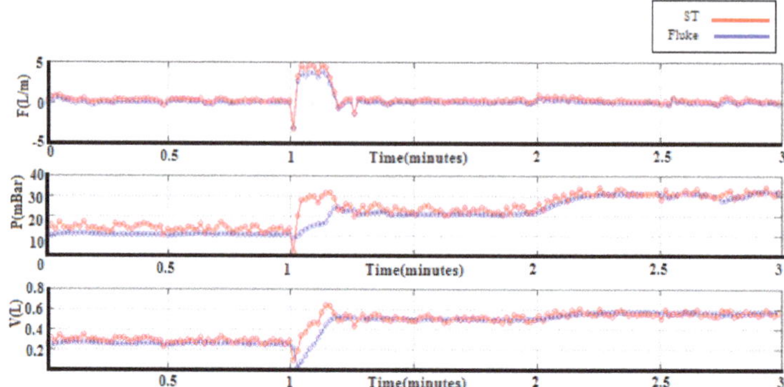

Figure 26. Flow, pressure difference and volume by dynamic analysis.

This can be explained by the geometrical characteristics of the ST and its nonlinear behavior in diameters from 4 mm to 6 mm; however, the ST system can provide a better response in the presence of disturbances or transient changes when the orifice diameter is between 8, 10 and 12 mm. The reference corresponds to the measurements of the Fluke equipment (blue curve for airflow, pressure difference and volume).

For the 8 mm diameter sensor/transducer and the mechanical ventilator excitation signals, the ventilation curves are shown in Figure 27. In every subfigure, the medical gas flow meter Fluke VT 650 proportionated the blue color curves in low frequency speed and red color for high frequency speed, which provide information on the volume, pressure and airflow (subfigures I–III), and the ST curves provide the volume, pressure, airflow the green color curves in low frequency speed and yellow color for high frequency speed (subfigures I–III). During the 2 min of measurement, the ST demonstrated a faster and robust sensor behavior (for 8 mm of diameter) as a consequence of the dynamical evaluation made by a low-cost mechanical ventilator based on cams OxygenIP.PE.

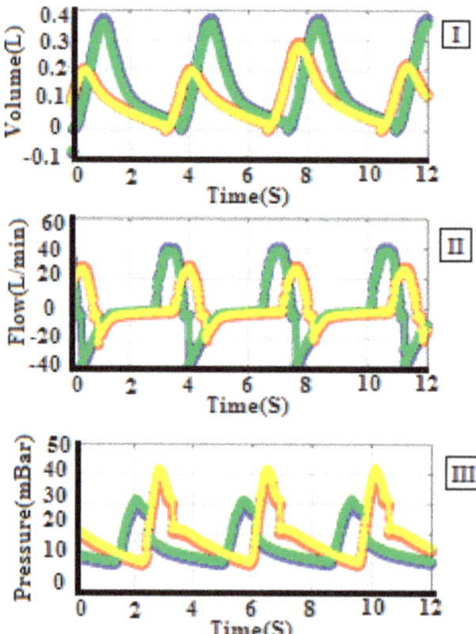

Figure 27. The dynamical behavior of the sensor/transducer connected to a mechanical ventilator low cost based on cams OxygenIP.PE.

Figure 28 shows the airflow curves for the 8 mm diameter sensor/transducer and the mechanical ventilator excitation signals. In every subfigure, the Fluke curves (blue color) provide information on the airflow measured by the medical gas flow meter Fluke VT 650 and the ST curves (red color) represent the airflow measured by the designed sensor/transducer.

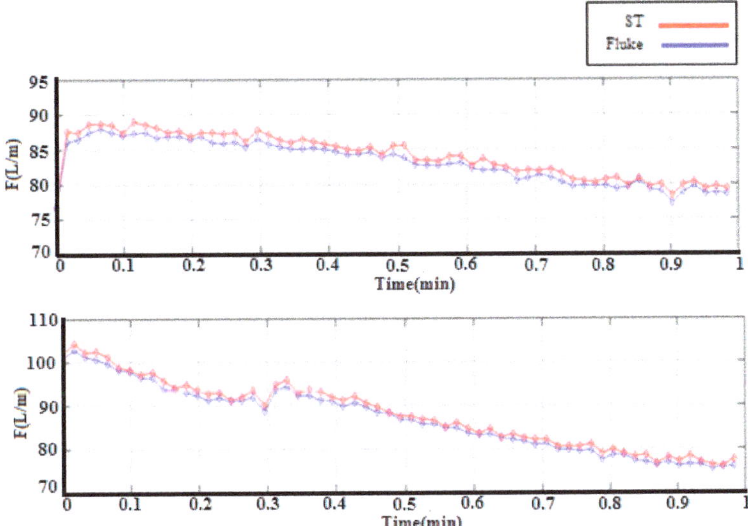

Figure 28. The dynamical behavior of the sensor/transducer connected to an air compressor in airflow range of work between 75 L/min and 105 L/min.

During the 1 min of measurement, the ST shows a faster and robust sensor behavior as a consequence of the dynamical evaluation made using an air compressor in the range of work from 75 L/min to 105 L/min.

5. Discussion

5.1. Data Interpretation Analysis

In order to interpret the experimental data with the designed algorithm, it was necessary to establish a correlation with the theoretical model to enhance the adaptive parameters of the polynomial model analyzed in Section 2.5.

Therefore, the experimental analysis of the designed polynomial model was correlated with Equation (18) in the theoretical model, as this equation gives the pressure difference "ΔP" as dependence on the airflow "Φ", geometrical parameters "A_2, A_1" and flow density "ρ".

$$\Delta P = \left(\frac{\Phi}{A2}\right)^2 \left(\frac{\rho}{2}\right)\left(1 - \left(\frac{A_2}{A_1}\right)^2\right) \tag{17}$$

The theoretical model only produces a quadratic relationship between the airflow with the pressure difference, while "A_2" tends to decrease, the pressure difference and airflow tend to increase.

Since the airflow for the breathing analysis had an "increasing and decreasing behavior", to obtain an appropriate measurement with the designed ST, the static and dynamic behavior of the sensor was studied using an adaptive analysis of the polynomial model.

Therefore, the static behavior was provided as a dependence of "airflow, pressure difference, and diameter". Figure 29 depicts a flow and two points crossing its axis, for which P1, V1, Y1 and P2, V2, Y2 are the pressure, flow speed, and position at points 1 and 2, respectively.

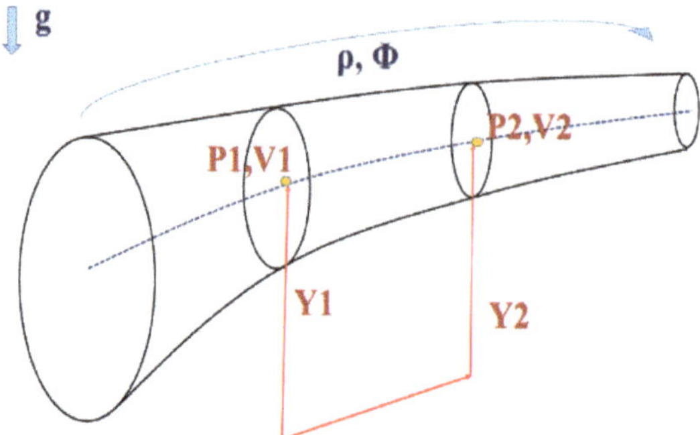

Figure 29. Schematic representation of airflow transmission through points 1 and 2.

The static curve gives the relation among two variables in order to identify a linear and nonlinear range of work; therefore, it is possible to understand its dynamic behavior.

The dynamical analysis of the airflow attempts to obtain the linear answer of the system so that the ST can make estimations of the physical variables volume, and pressure as a consequence of the pressure difference, which was achieved by the correlation between the experimental and theoretical analyses described in previous chapters. Therefore, the curves "A, B, and C" represent the airflow curves that were expected to be achieved in the characterization of the interpretation data with the ST, which is depicted by the Figure 30.

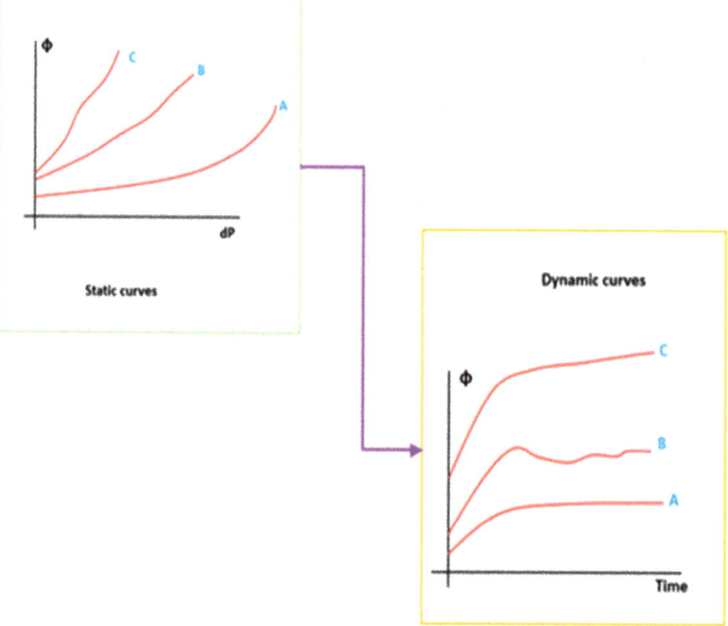

Figure 30. Schematic representation of the airflow dynamic and static behavior.

Two types of studies were performed, one for the static responses of the ST system, and another for the dynamic ones. The first was conducted using a tester prototype with a compressor as an air source, which is depicted in Figure 31 and was used for the static and dynamic tests. The pressure difference measured by the sensor was calibrated in order to provide the same response as the air flow meter Fluke. Finally, the other group of static and dynamic tests was performed with the mechanical ventilator as the air source. Hence, the pressure difference registered provided information to understand the dynamics of the breathing variables (pressure, volume and airflow) in order to understand the behavior of the sensor/transducer as part of the ventilation circuit.

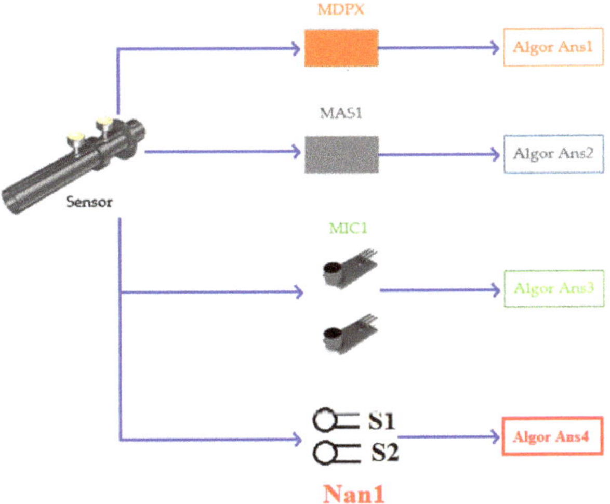

Figure 31. Block diagram testing transducers.

The designed ST depends on the diameter and pressure difference; for this reason, it was necessary to choose which transducer to use to obtain the electrical equivalent of the pressure difference. Finally, the evaluated alternatives were as follows: transducer MAS1, the integrated circuit MDPDX10P, the transducer constructed with two small microphones sourced from the ARDUINO company, and a transducer based on nanostructures. Different alternatives are shown in Figure 31.

The measured pressure difference by the designed ST was central to obtaining the airflow after a transduction process, which was obtained by MDPX and MAS1, making the electrical equivalent conversion from air pressure difference signal to airflow signal in Volts. The electrical transduction was performed with two microphones measuring decibels and correlating with pressure by converting the response to volts. The electrical equivalence values help to analyze the statistics of the physical variables "air pressure, volume and airflow" as a consequence of the designed algorithms by polynomial (as they are depicted by "Algor Ans1", "Algor Ans2", and "Algor Ans3" in Figure 31) correlations between the transduction signals. However, in order to enhance the robustness and response time of the designed ST, we modified the samples that receive the pressure difference by samples based on nanostructures of AAO, as is depicted in S1, S2 in Figure 31, which also send the transduction information through the algorithm depicted by "Algor Ans4".

The best estimation of the breathing physical variables "airflow, pressure, and volume" was chosen because of the pressure difference transduction techniques that were studied in the previous chapters. These microphone-based transducers obtained estimated variables with a maximum error of 1% in a steady state, while the MAS1-based transducer estimated variables with a 0.95% error in a steady state. In turn, the MDPX-based transducer presented a steady-state error of around 2%. However, the cost of the microphone-based transducer was around 70% cheaper than the most expensive transducer, the MAS1. For this reason, the microphone-based transducer was selected as the main transducer for the designed ST.

When performing the main analysis of the transducer algorithm, we sought to identify the adaptive coefficients from the experimental data obtained by the transducer and the polynomial identification estimate using pressure difference "DP", the pressure "P", the airflow "F", and the volume "V". If the adaptive estimation was predicted according to the reference variable with an error of less than 1% in the steady state, then the ST provided information on the breathing variables, which is depicted in Figure 32.

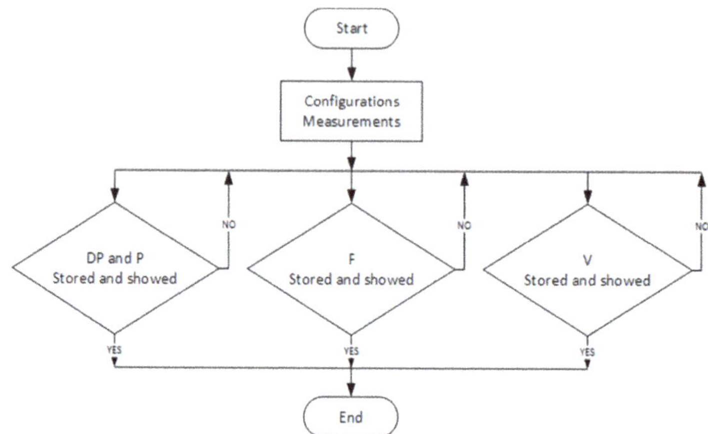

Figure 32. Flow diagram of the ST measurement and data storing.

Therefore, it was necessary to define the input variables in order to correlate the estimated responses under the dependencies of the references variable. Figure 33 shows the two input variables of the system from the data interpretation analysis, which are given by the pivot displacement in sexagesimal degrees (red curve color curve in subfigure III of Figure 33). The pressure difference in its electrical equivalent (mV) was considered as a second input or excitation variable (blue color curve in subfigure I of Figure 33). Moreover, the pressure difference measured by the nanosensor based on nanostructures, as shown in subfigure II, revealed that the nanostructure sensor maintains a rapid response time and robustness under disturbances of the pivot (green color curve in subfigure II of Figure 33).

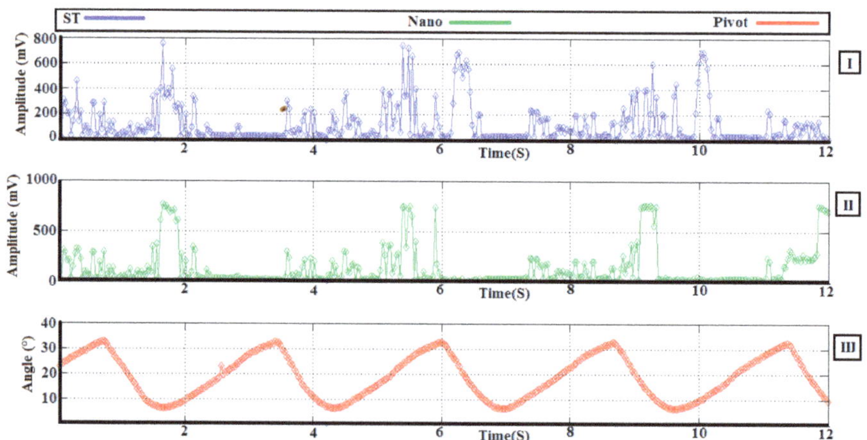

Figure 33. Motor angle displacement and microphones answer.

As a consequence of the input excitation signals, it was possible to achieve and identify the adaptive coefficients which were evaluated in high and low operation because the frequency range of the requirements is between 15 Hz and 20 Hz.

Therefore, Figure 34 shows the result of the ST as a consequence of the electrical transduction from the pressure difference to the airflow values in millivolts for cams of extra small, medium, and extra-large sizes. The obtained data are presented in Figure 34 in subfigures I–III, in which blue indicates high frequency and red curves low frequency.

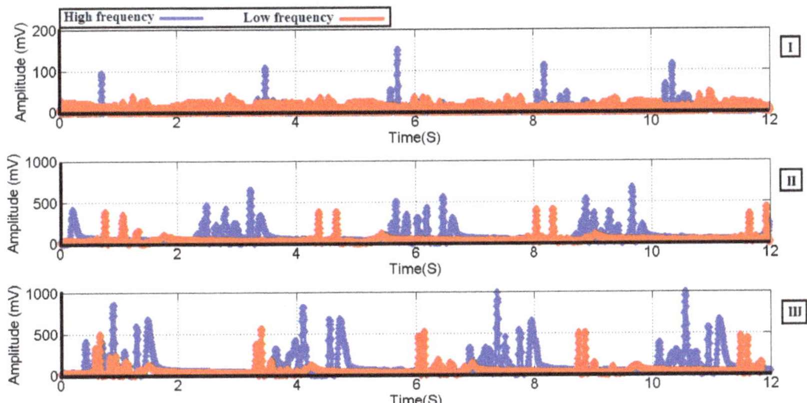

Figure 34. Microphones in high and low operation.

On the other hand, it is apparent that for the ST response as a consequence of the electrical transduction from the pressure difference to the airflow values in millivolts for cams of extra small, medium, and extra-large sizes, the information achieved is shown in Figure 34 for the subfigures I–III, in which the blue color indicates high frequency and red represents low frequency. With this information, it is possible to identify a linear range of work.

The ventilator works as an intermediatory for the signal received from the rotor position sensor (RS1(AS5047) and RS2) used to measure the rotor speed of the motor and control it with a driver (DR1) that has its own control system with its own actuator and sensor with an electrical current to the motor (M).

This is the reason why a controller (as central control unit) also requires the signal from a rotor position sensor to measure the angular displacement of the pivot to provide an estimation of the air volume, and finally, the airflow sensor (FS1), which has its own transduction algorithm to provide information to medical doctors through the touchscreen panel, as depicted in Figure 35.

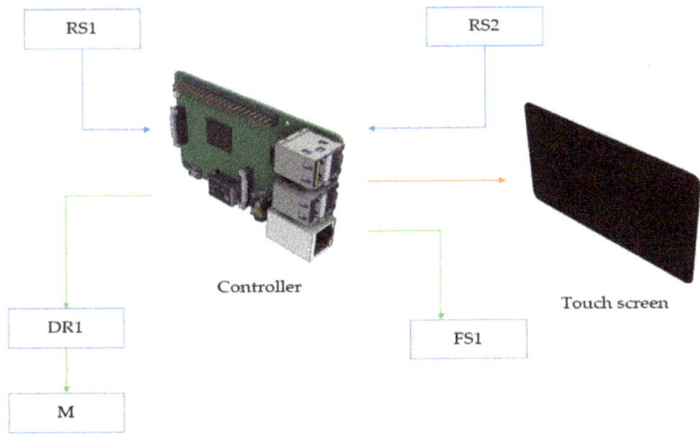

Figure 35. Block diagram of the control system for online operation.

The ST achieved good performance for the diameter of 8 mm due to its optimal response based on appropriated magnitudes and tracking for the mechanical ventilator curves. In the "data interpretation analysis" the ST with an orifice diameter of 10 mm was used. Despite the nonlinear response, the airflow was obtained because of the good performance of the ST algorithm in transducing the pressure difference signal to the airflow variable for a larger range of work (low and high speed of the ventilator).

The range of work of the test was limited to a high frequency of 20 Hz and a low frequency of 15 Hz. The transduction was achieved with the microphone sensors used to obtain the pressure difference as is shown in Figure 36I, where the blue curve provides information about the system working at high speeds and the red curve for low-speed operation. Figure 36I shows that at high speeds it was possible to obtain an increased amplitude (50 percent). Both signals were achieved in millivolts due to the transducer (small microphones) having greater sensitivity at higher speeds.

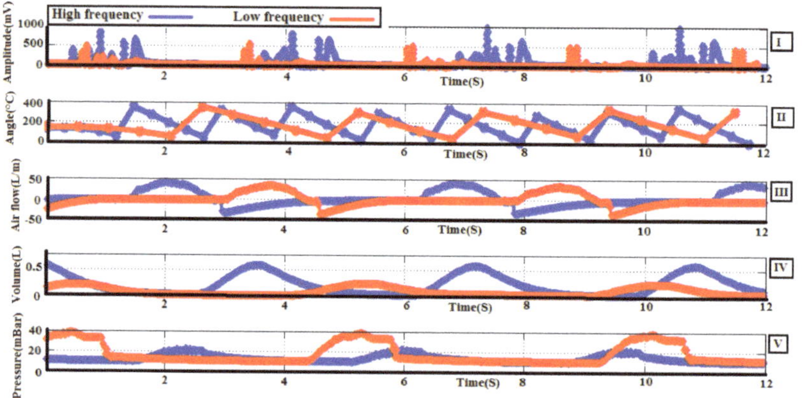

Figure 36. Microphones answer in high and low operation and small and big cam.

Furthermore, it was necessary to define the input variables to correlate the estimated responses under the dependencies of the reference variable (analyzed in previous chapters). Figure 36II shows the input variables of the system from the data interpretation analysis, which are given by the motor displacement in sexagesimal degrees, with the blue curve indicating high speed and red curve representing low speed. It is necessary to remember that motor displacement causes pivot displacement and it is enough to use only one sensor to achieve the angle displacement during the polynomial correlation variable analysis.

Therefore, the main variable obtained from the ST is the airflow rate, given by the transduction operation supported by the predictive estimations for the other breathing variables "volume and pressure" obtained by the polynomial analysis (described in chapters lines above) from the airflow variable.

Figure 36III–V presents the airflow rate, volume, and pressure, respectively, in which the blue color curves are the airflow, volume, and pressure in high-frequency operation, and the red color curves are the airflow, volume, and pressure in low-frequency operation, which were obtained as a consequence of the predictive/adaptive algorithm; however, it is necessary to keep in mind the importance of the previous data analysis during the characterization and calibration of the ST to obtain the presented results.

It was necessary to develop a multivariable control algorithm to control the respiratory variables (air flow, pressure and volume), hence the pressure difference transduced in volume, air flow and pressure as respiratory variables required a multivariable control algorithm, which was able to control the rotor speed of the mechanical ventilator motor, the air pressure difference and the air flow, as the implicit function, pressure and volume.

5.2. Rotor Speed Control of the Mechanical Ventilator Motor

The rotor speed control of the mechanical ventilator motor was achieved by an adaptative predictive model, for which a system identification of the motor physical parameters was necessary. However, unexpected disturbances caused by vibrations during the cam movement can reduce the performance of the control. Therefore, it was necessary to design a PID (Proportional, Integral and Derivative) control as part of the identification system to achieve the physical parameters of the mechanical ventilator motor.

Hence, the control inside the identification of the motor parameters was obtained by the classic PID controller equation in the Laplace (S) domain, in which KPp is the proportional constant, KIp is the integral constant, and KDp is the derivative constant.

$$PID(S) = KP_P + \frac{KI_p}{S} + KD_p S \tag{18}$$

The plant (mechanical ventilator motor) was analyzed in two sections, namely the electrical section and the mechanical section which are described in the following paragraphs.

Rotor Control Position of the Mechanical Ventilator Motor as a Function of Electrical Current

The rotor control position of the mechanical ventilator motor is depicted by the block diagrams in Figure 37. Inp(S) is the electrical current signal as the input excitation signal, Yans(s) is the position response that is necessary to obtain the speed of the motor, Cont(S) is the controller during the identification system, Plant(S) is the transfer function for the plant that is the electrical motor composed by its electrical subsystem and its mechanical subsystem, and Sens(S) is the transfer function for the position sensor.

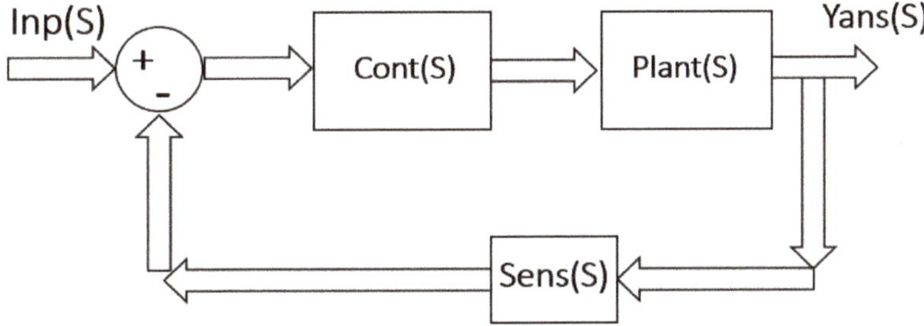

Figure 37. Block diagram for the rotor position control of the mechanical ventilator.

Using the algebra analysis from the block diagram depicted above, the following is obtained:

$$Inp(S) - Sens(S)Y(S)Cont(S)MS(S) = Y(S) \tag{19}$$

From which following equation is achieved:

$$\frac{Yans(S)}{Inp(S)} = \frac{Cont(S)Plant(S)}{1 + Cont(S)Plant(S)Sens(S)} \tag{20}$$

In Equation (20), the transfer function for the rotor position sensor is generalized, but in spite of that, the response time of the position sensor is very short in comparison to the speed of the motor. Hence, its transfer function was reduced as $Sens(S) = Ks$. Therefore, the characteristic equation is given by Equation (21), in which m is the mass of the rotor motor, k_i is the electrical current coefficient of the motor and k_y is the displacement coefficient of the motor.

$$\left(KP_P + \frac{KI_p}{S} + KD_p S\right)\left(\frac{K_i}{m\,S^2 + K_y}\right)Ks + 1 = 0 \tag{21}$$

The reduction of the equation above is given by Equation (22).

$$\left(KD_p\, S^2 + KP_p S + KI_p\right) K_I K_S + S(mS^2 + K_y) = 0 \tag{22}$$

Equation (22) is organized in Equation (23) as a polynomial in descending order.

$$S^3 + \frac{\left(KD_p K_i K_s\right)}{m} S^2 + \frac{\left(KP_p K_i K_s + K_y\right)}{m} S + \frac{KI_p K_i K_s}{m} = 0 \tag{23}$$

The controller's parameters need to be obtained by different methodologies such as a stability analysis and a comparison with the theoretical model of dynamic systems given by Equation (24) [21], in which ω_0 is the natural frequency for the system, ϵ is the damping effect, and α is the coefficient as the auxiliary connector between Equations (23) and (24).

$$S^3 + \omega_0(2\epsilon + \alpha)S^2 + \omega_0^2(1 + 2\epsilon\alpha)S + \alpha\,\omega_0^3 = 0 \tag{24}$$

Therefore, the controller parameters KP_p and KI_p, can be obtained by the comparison of the coefficients from Equations (23) and (24), from which Equations (25)–(27) are proposed as functions of K_i, K_s, KD_p, m, ω_0 and ϵ.

$$\frac{\left(KD_p K_i K_s\right)}{m} = \omega_0(2\epsilon + \alpha) \tag{25}$$

$$\frac{\left(KP_p K_i K_s + K_y\right)}{m} = \omega_0^2(1 + 2\epsilon\alpha) \tag{26}$$

$$\frac{KI_p K_i K_s}{m} = \alpha\,\omega_0^3 \tag{27}$$

Moreover, α is obtained from Equation (27) and represented by Equation (28)

$$\alpha = \frac{KI_p K_i K_s}{m\omega_0^3} \tag{28}$$

The controller parameter KI_p is obtained by analyzing Equations (25)–(28), and presented in Equation (29)

$$KI_p = KD_p\omega_0^2 - \frac{2\epsilon m\omega_0^3}{K_i K_s} \tag{29}$$

In a similar context, the controller parameter KP_p is obtained with Equations (25)–(29), and presented in Equation (30).

$$KP_p = \frac{2\epsilon K_i K_s KD_p\omega_0 + m\omega_0^2\left(1 - 4\epsilon^2\right) - K_y}{K_i K_s} \tag{30}$$

Therefore, Equations (29) and (30) are the integral and proportional controller parameters that determine the controller parameter KD_p, which can be analyzed as a reference parameter (analyzed by stability of the dynamic system) to find the PID control for the motor of the mechanical ventilator. However, the identification parameters of the motor require a controller with slow reactions to provide sufficient time to identify the parameters of the system, and the controller selected for the identification of the physical parameters of the mechanical ventilator motor was a PI controller, hence the derivative parameter KD_p becomes null, and Equation (30) can be obtained from Equation (29), thereby providing the integral parameter of the PI controller.

$$KI_p = -\frac{2\epsilon m\omega_0^3}{K_i K_s} \tag{31}$$

Finally, Equation (32) is the proportional parameter of the PI controller obtained from Equation (30) when the derivative parameter is null.

$$KP_p = \frac{m\omega_0{}^2\left(1 - 4\epsilon^2\right) - K_y}{K_i K_s} \tag{32}$$

5.3. Model Predictive Control Analysis

While the above results indicate good control, they lack a fast response which could be obtained using a predictive model, such as an Optimal Predictive Control model described in more detail in the following (Appendix B).

The optimal excitation signal in order to find the optimal response is given by [1,13]

$$\Delta S = \left(\phi^T\phi + A\right)^{-1}\phi^T\left(A_S - LX(k_i)\right) \tag{33}$$

Therefore, using the last equation, the optimal control response for the pressure difference can be achieved, as well as the rotor speed and implicitly the volume, airflow and pressure. Moreover, the optimal response control can be enhanced by the specific weight matrix "W" that must be included in Equation (A5). Equation (33) has similarities with Equation (A9) but the difference is given by the adaptive matrix coefficient "ϕ", because for this analysis the matrix depends only on the requirement control strategies while maintaining its dependence on "φ" (the geometrical and material coefficients of the designed ST).

As a consequence, the main control algorithm was designed using the adaptive cascade model, which is depicted in Figure 38 in which "U1" is the input variable (pressure difference measured by the designed ST) and the internal response variable is the air flow "U2", which is the input variable used to obtain the pressure "Y1" and volume "Y2", moreover the internal control variable of the rotor speed has a correlation with the desired airflow required by the mechanical ventilator.

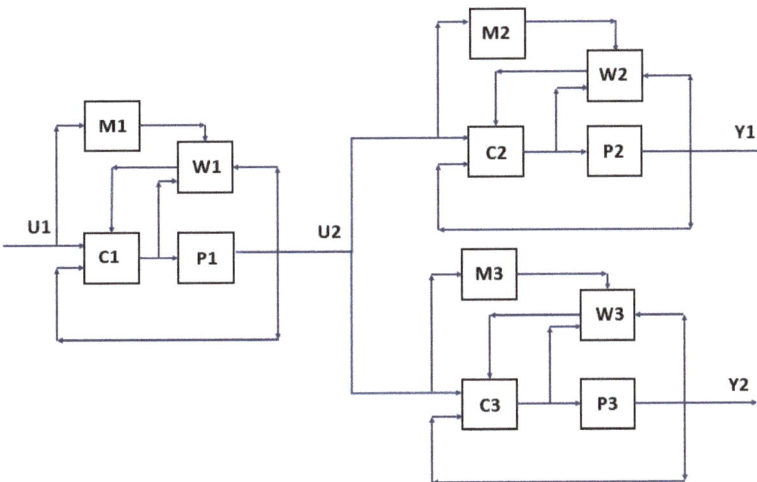

Figure 38. Control adaptive cascade model for the ventilator control system.

To evaluate the performance of the designed ST in comparison with its enhancement based on AAO nanostructures, experiments of the mechanical ventilator controlled by the adaptive cascade algorithm in Huancayo (3250 m over sea level) were conducted. It can be seen in Figure 39I that the curves as well as the controlled air flow (implicit) measured by the designed ST (blue colour curve) and the air flow measured by the designed ST enhanced by nanostructures (red colour curve) were obtained with a steady state error of 1.4% of the ST and 0.8% for the ST enhanced by nanostructures.

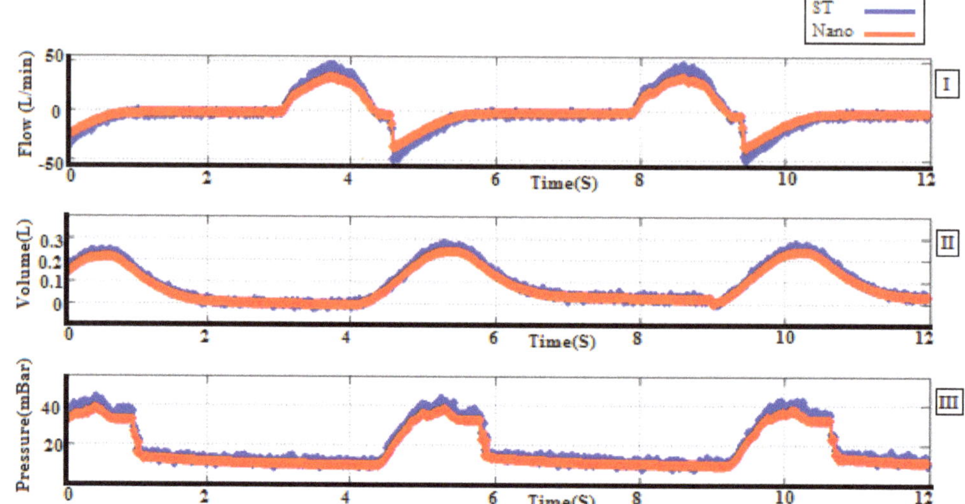

Figure 39. Control of respiratory variables made in Huancayo, Peru.

In Figure 39II the curves of the controlled volume (implicit) and measured by the designed ST (blue colour curve) as well as the volume measured by the designed ST enhanced by nanostructures (red colour curve) are presented, with a steady state error of 1.5% of the ST and 0.7% of the ST enhanced by nanostructures.

Furthermore, in Figure 39III the curves of the controlled pressure (implicit) and measured by the designed ST (blue colour curve) as well as the pressure measured by the designed ST enhanced by nanostructures (red colour curve) are shown, with a steady state error of 1.8% for the ST and 0.8% for the ST enhanced by nanostructures.

Therefore, the designed ST of nanostructures (TiO_2 nanotubes) based on Anodic Aluminium Oxide (AAO) provided better results in robustness, response time and resolution, which supported measurements from the approximated range of operation between 0.985 atm (Lima) and 1 atm (Huancayo).

6. Conclusions

A sensor/transducer (ST) was designed to measure the pressure difference through an orifice plate and to obtain the airflow rate, volume and pressure over time to register the performance of mechanical ventilators. The geometrical characteristics of the proposed sensor based on an orifice plate were researched in order to identify the appropriate parameters to measure the physical variables. The diameter has an important influence on the performance of the sensor; in this sense, a diameter of 8 mm provides the optimal response, in terms of magnitudes and tracking, for the mechanical ventilator curves without significantly increasing the pressure inlet for the breathing circuit which is provided by the mechanical ventilator. A diameter in the range of 4 mm to 6 mm can achieve the pressure difference of airflow, but with a nonlinear response and for the range between 10 mm and 12 mm, the ST can measure the physical variables in linear responses for breathing values (it can be used to measure airflow shared by several patients connected to one mechanical ventilator as a future application). Therefore, it was suggested to use the ST with a diameter of 8 mm due to its linear response and robustness under disturbances.

The mathematical model of the system was designed and evaluated. Consequently, an algorithm was also designed to obtain the airflow, volume and pressure from the pressure difference, which were obtained by the transduction process of the transducer joined to the designed sensor. Moreover, the ST system was analyzed in dynamical and transient behavior for ranges of work that depended on geometrical parameters and physical values of airflow, pressure, and volume of mechanical ventilators for artificial human breathing, This is a novel proposed sensor design such as other new proposals [22–24] because of the multiple variable correlation also variables considered like disturbances (temperature and vibration) and considering the effect of nanostructures in this objective.

For a given orifice diameter, a correlation was observed between the pressure difference and the airflow. This correlation was not linear, but instead showed a parabolic tendency. Secondly, similar results and tendencies were observed between the steady-state and transient-state simulations. Therefore, it can be concluded that a steady-state approach (and its corresponding assumptions) can be considered to validly describe the transient behavior of this orifice plate sensor.

For a given flow, the smaller the orifice diameter, the larger the pressure difference induced in the orifice plate. This could happen because, for a given flow, a smaller orifice diameter requires the flow to pass through the opening. This increases the flow velocity, but at the cost of reducing the flow pressure at the outlet, resulting in greater pressure differences in the orifice plate. Therefore, a bigger inlet pressure would be necessary to help the flow overcome the pressure loss in the pipes. It can also be concluded that a plate with a smaller orifice diameter would be the best for this sensor. Taking this into account, lower pressure drops would occur in the plate, which would not considerably affect the flow, while at the same time being large enough to be detected by the ST and still being in the measuring range.

The designed ST enhanced by nanostructures of anodic aluminium oxide can provide a faster and robust response for geographic conditions where the atmospheric pressure level is different, as was the case in this research (tests in Lima and Huancayo). The nanostructure samples used for the designed sensors have the potential to fix adaptive coefficients to improve the operating work and enhance the response system (breathing variables) in case of disturbances.

Author Contributions: Conceptualization, J.A.C.C., C.G.R.R. and J.J.J.d.C.y.F.; methodology, J.A.C.C. and C.G.R.R.; software, J.A.C.C., J.H.L.J., C.G.R.R., B.J.M.A.C., S.C.A. and J.J.J.d.C.y.F.; validation, J.A.C.C. and A.T.H.; formal analysis, J.A.C.C., C.G.R.R. and S.C.A.; investigation, J.A.C.C., C.G.R.R. and A.M.G.A.; resources, J.A.C.C., C.G.R.R. and S.C.A.; data curation, J.A.C.C. and J.H.L.J.; writing—original draft preparation, J.A.C.C. and C.G.R.R.; writing—review and editing, J.A.C.C., C.G.R.R., A.M.G.A. and J.J.J.d.C.y.F.; visualization, J.A.C.C. and C.G.R.R.; Transduction mathematical and algorithm design and nanostructure design J.A.C.C. Supervision, J.A.C.C., J.H.L.J., C.G.R.R. and J.J.J.d.C.y.F.; project administration, J.J.J.d.C.y.F. and A.M.G.A.; funding acquisition, J.J.J.d.C.y.F. All authors have read and agreed to the published version of the manuscript.

Funding: This work was supported by CONCYTEC-FONDECYT based on the contest "PROYECTOS ESPECIALES: Modalidad-Escalamiento de proyectos COVID-19" [74169], to Pontificia Universidad Católica del Perú, to Protofy and MODASA.

Institutional Review Board Statement: Not applicable.

Informed Consent Statement: Not applicable.

Data Availability Statement: Not applicable.

Acknowledgments: The authors would like to express their special gratitude to the Protofy team, who shared their knowledge and experience of the OxyGEN project development. Furthermore, the authors would like to acknowledge the OxygenIP.PE team, including Andrea Portal, Juan José Leyton and Romel Rosales who devised and assisted in the design of the components inside the mechanical ventilator, Broni Huamaní and Giovani Berrospi who assisted in the transduction stage, experimental tests and sensors assembly, Joaquin Gonzales who developed the electronic display for displaying the ventilation parameters for medical staff and Cristian Rimac who assisted in the schematic representations. Furthermore, we express our gratitude to the Alexandr Hinostroza, because of his medical analysis suggestions during the design and analysis of the experimental data.

Gratitude is also expressed to the researchers of the Metrology Laboratory PUCP, due to their support in the experimental analysis, as well as acknowledgement to the Mechanical Department of PUCP due to the support given to use measurement equipment for the experiments.

Conflicts of Interest: The authors declare no conflict of interest. The funders had no role in the design of the study; in the collection, analyses, or interpretation of data; in the writing of the manuscript, or in the decision to publish the results.

Appendix A

This appendix describes the special equipment designed for the validation and mathematical characterization of the air flow sensor in different working conditions of air flow and the low-cost mechanical ventilator based on cams where the air flow sensor was installed.

Appendix A.1. Steady State Air Flow Test Bench

In order to determine the optimal orifice diameter for the medical flow sensor in stationary conditions, a test bench was specially designed for this experiment (Figure A1). The holders of the sensor can be seen in the diagonal view (Figure A1a) and the side view (Figure A1b) of the test bench, which are quite important due to keep the correct position of the sensor during the experiments. It consists of a pipe where an air flow is induced by an air compressor, a fixed orifice plate to the pipe for flow measurement and a variable orifice-plate flow sensor. The fixed orifice plate was calibrated with a medical gas flow meter Fluke VT650 in a validation laboratory implemented in the COVID19-pandemic context at the Pontificia Universidad Católica del Perú. This fixed orifice-plate can help to determine the real flow through the variable orifice-plate flow sensor. In this sense, the orifice diameter can be changed and the relation between real flow and pressure difference can be obtained.

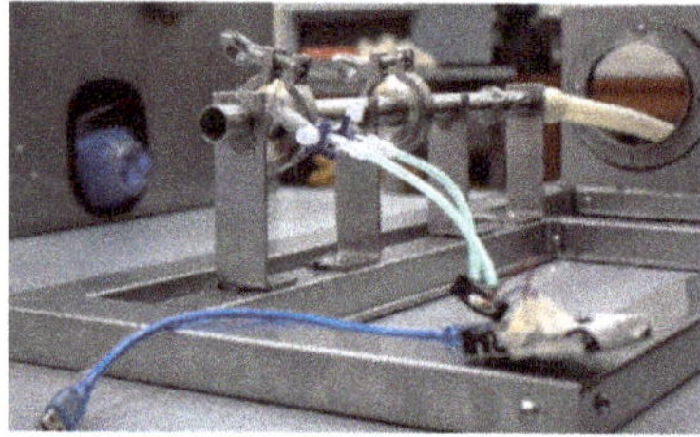

(a)

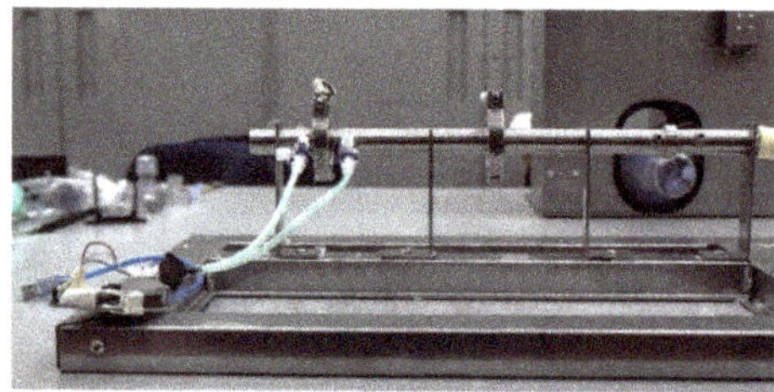

(b)

Figure A1. Air flow test bench designed for model validation.

Appendix A.2. Mechanical Ventilation System Oxygen IP.PE

OxygenIP.PE is a low-cost functional mechanical ventilation prototype based on cams [1] OxyGEN-IP, which was developed by Protofy [6]. A redesign was carried out according to the special requirements and available technology in Peru; however, the working concept (compression mechanism) established by Protofy was maintained due to the satisfactory results [25]. Figure A2 shows the OxygenIP.PE mechanical ventilator. Its working concept consists of an air bag compression mechanism based on cams to produce a pressure increment inside the cam and then air flow to the patient through a breathing circuit. Moreover, it can work with five different cams depending on the air volume requirement for a particular patient.

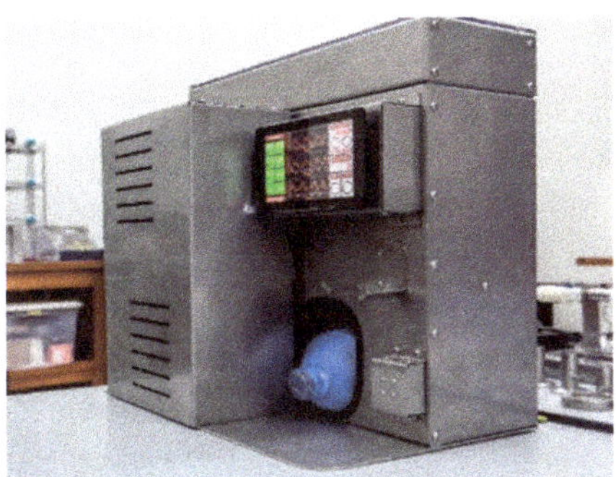

Figure A2. Mechanical ventilation system OxygenIP.PE, where the novel air low sensor/transducer is implemented.

Appendix B

In this research, it was necessary to work with strategies of Model Predictive Control (MPC) to join both subsystems (mechanical and electrical) for the main algorithm, for which an internal identification system was used, while adaptive control coefficients/weights searched for the right control. Therefore, the nonlinear function "g" and internal variables "$z(t)$" due to excitation "$v(t)$" as a function of time "t" were important.

$$\frac{dx(t)}{dt} = g(z(t)v(t), \theta) \tag{A1}$$

On the other hand, "$m(t)$" correlated with "$z(t)$" and "$v(t)$" through a nonlinear function "n".

$$m(t) = n(z(t), v(t), \theta) \tag{A2}$$

Additionally, "A_S" represents trajectory and the input excitation "O" is included depending of the sample time ki.

$$A_s^T = (111\ldots11111)O(k_i) \tag{A3}$$

So, the costing function "J" was analyzed to achieve the optimal desired response [21].

$$J = (A_S - Y)^T(A_S - Y) + \Delta S^T A \Delta S \tag{A4}$$

In which the expected response is

$$Y = LX(k_i) + \phi \Delta S \tag{A5}$$

Otherwise,

$$J = (A_S - LX(k_i))^T\left(A_S - LX(k_i)\right) - 2\Delta S^T \phi^T (A_S - LX(k_i)) + \Delta U^T (\phi^T \phi + A)\Delta S \tag{A6}$$

where "L" and "ϕ" are matrices that contain all the physical parameters of the system (as joining matrices above for the identified result) [21].

$$\frac{\partial J}{\partial \Delta S} = -2\phi^T (A_S - LX(k_i)) + 2(\phi^T \phi + R)\Delta S \tag{A7}$$

So

$$\frac{\partial J}{\partial \Delta S} = 0 \tag{A8}$$

From which the optimal excitation signal in order to find the optimal response is given by [1,21]

$$\Delta S = \left(\phi^T \phi + A\right)^{-1} \phi^T (A_S - LX(k_i)) \tag{A9}$$

References

1. Calderón Ch, J.A.; Rincón, C.; Agreda, M.; Jiménez de Cisneros, J.J. Design and Analysis of a Mechanical Ventilation System Based on Cams. *Heliyon* **2021**, *7*, e08195. [CrossRef] [PubMed]
2. Freescale Semiconductor Literature Distribution Center, Inc. *Ventilator/Respirator Hardware and Software Design Specification*; Freescale Semiconductor Literature Distribution Center, Inc.: Denver, CO, USA, 2011; pp. 11–20.
3. Brzeski, P.; Lazarek, M.; Perlikowski, P. Experimental study of the novel tuned mass damper with inerter which enables changes of inertance. *J. Sound Vib.* **2017**, *404*, 47–57. [CrossRef]
4. Fowles, G.; Boyes, W. *Measurement of Flow*, 4th ed.; Elsevier: Amsterdam, The Netherlands, 2010; ISBN 978075068308.
5. Biselli, P.J.C.; Nóbrega, R.S.; Soriano, F.G. Nonlinear flow sensor calibration with an accurate syringe. *Sensors* **2018**, *18*, 2163. [CrossRef] [PubMed]
6. OxyGEN Project. Available online: https://www.oxygen.protofy.xyz/download (accessed on 23 November 2020).
7. Chang, J.; Acosta, A.; Aspiazu, J.B.; Reategui, J.; Rojas, C.; Cook, J.; Nole, R.; Giampietri, L.; Pérez-Buitrago, S.; Casado, F.L.; et al. Masi: A mechanical ventilator based on a manual resuscitator with telemedicine capabilities for patients with ARDS during the COVID-19 crisis. *HardwareX* **2021**, *9*, e00187. [CrossRef] [PubMed]
8. Jamróz, P. Interaction between the Standard and the Measurement Instrument during the Flow Velocity Sensor Calibration Process. *Processes* **2021**, *9*, 1792. [CrossRef]
9. Włodarczak, S.; Ochowiak, M.; Doligalski, M.; Kwapisz, B.; Krupińska, A.; Mrugalski, M.; Matuszak, M. Flow Rate Control by Means of Flow Meter and PLC Controller. *Sensors* **2021**, *21*, 6153. [CrossRef] [PubMed]
10. Koirala, N.; McLennan, G. Mathematical Models for Blood Flow Quantification in Dialysis Access Using Angiography: A Comparative Study. *Diagnostics* **2021**, *11*, 1771. [CrossRef] [PubMed]
11. Bisgaard, J.; Tajsoleiman, T.; Muldbak, M.; Rydal, T.; Rasmussen, T.; Huusom, J.K.; Gernaey, K.V. Automated Compartment Model Development Based on Data from Flow-Following Sensor Devices. *Processes* **2021**, *9*, 1651. [CrossRef]
12. Hammer, A.; Roland, W.; Zacher, M.; Praher, B.; Hannesschläger, G.; Löw-Baselli, B.; Steinbichler, G. In Situ Detection of Interfacial Flow Instabilities in Polymer Co-Extrusion Using Optical Coherence Tomography and Ultrasonic Techniques. *Polymers* **2021**, *13*, 2880. [CrossRef] [PubMed]
13. Landau, L.D.; Leiftshitz, E.M. *Fluid Mechanics, Institute of Physical Problems*; U.S.S.R. Academy of Sciences: Moscow, Russia, 1987.
14. Aldoghaither, A.; Liu, D.-Y.; Laleg-Kirati, T.M. *Modulating Functions Based Algorithm for the Estimation of the Coefficients and Differentiation Order for a Space-Fractional Advection-Dispersion Equation*; Society for Industrial and Applied Mathematics: Philadelphia, PA, USA, 2015.
15. Bridgeman, D.; Tsow, F.; Xian, X.; Forzani, E. A new differential pressure flow meter for measurement of human breath flow: Simulation and experimental investigation. *AIChE J.* **2016**, *62*, 956–964. [CrossRef] [PubMed]
16. Launder, B.E. MAN—ANSYS Fluent User's Guide Releasde 15.0. *Knowl. Creat. Diffus. Util.* **2013**, *15317*, 724–746.
17. Paz, C.; Suárez, E.; Concheiro, M.; Porteiro, J. CFD transient simulation of a breathing cycle in an oral-nasal extrathoracic model. *J. Appl. Fluid Mech.* **2017**, *10*, 777–784. [CrossRef]
18. CLÍNIC BARCELONA. Available online: https://www.clinicbarcelona.org/noticias/el-dispositivo-de-ventilacion-de-emergencia-desarrollado-por-clinic-germans-trias-i-pujol-y-ub-con-protofy-xyz-recibe-la-aprobacion-de-la-aemps-para-hacer-un-estudio-clinico (accessed on 15 May 2020).
19. Karthik, G.S.Y.; Kumar, K.J.; Seshadri, V. Prediction of Performance Characteristics of Orifice Plate Assembly for Non-Standard Conditions Using CFD. *Int. J. Eng. Tech. Res.* **2015**, *3*, 2321–2869.
20. Funk, J.E.; Wood, D.J.; Chao, S.P. The transient response of orifices and very short lines. *J. Fluids Eng. Trans. ASME* **1972**, *94*, 483–489. [CrossRef]
21. Calderón, J.A.; Barriga, E.B.; Mas, R.; Chirinos, L.; Barrantes, E.; Alencastre, J.; Tafur, J.C.; Melgarejo, O.; Lozano, J.H.; Heinrich, B.; et al. Magnetic Bearing Proposal Design for a General Unbalanced Rotor System enhanced because of using sensors/actuators based in nanostructures. *E3S Web Conf.* **2019**, *95*, 01002. [CrossRef]
22. Халилов, И.А.; Керимов, С.Х.; Багирова, С.А.; Гаджиева, ф., III. Синтез кулачкового механизма с учетом условий передачи сил и контактной прочности. (Khalilov, I.A.; Kerimov, S.K.; Bagirova, S.A.; Gadzhieva, F.S. Synthesis of a Cam Mechanism Taking into Account the Conditions for the Transfer of Forces and Contact Strength). 2017. Available online: http://web.iyte.edu.tr/~{}gokhankiper/ISMMS/Khalilov.pdf (accessed on 6 April 2022).
23. Liang, L.; Qin, K.; El-Baz, A.S.; Roussel, T.J.; Sethu, P.; Giridharan, G.A.; Wang, Y. A Flow Sensor-Based Suction-Index Control Strategy for Rotary Left Ventricular Assist Devices. *Sensors* **2021**, *21*, 6890. [CrossRef] [PubMed]
24. Algarni, M. Optimization of Nano-Additive Characteristics to Improve the Efficiency of a Shell and Tube Thermal Energy Storage System Using a Hybrid Procedure: DOE, ANN, MCDM, MOO, and CFD Modeling. *Mathematics* **2021**, *9*, 3235. [CrossRef]
25. Melaibari, A. Free Vibration of FG-CNTRCs Nano-Plates/Shells with Temperature-Dependent Properties. *Mathematics* **2022**, *10*, 583. [CrossRef]

mathematics

Article

Mechanical Model and FEM Simulations for Efforts on Biceps and Triceps Muscles under Vertical Load: Mathematical Formulation of Results

Emilio Lechosa Urquijo [1], Fernando Blaya Haro [2], Juan David Cano-Moreno [2,*], Roberto D'Amato [2] and Juan Antonio Juanes Méndez [1]

[1] Facultad de Medicina, Departamento Anatomía Humana, Universidad de Salamanca, 37007 Salamanca, Spain; elechosa@gmail.com (E.L.U.); jajm@usal.es (J.A.J.M.)
[2] Escuela Técnica Superior de Ingeniería y Diseño Industrial (ETSIDI), Departamento de Ingeniería Mecánica, Química y Diseño Industrial, Universidad Politécnica de Madrid (UPM), Ronda de Valencia 3, 28012 Madrid, Spain; fernando.blaya@upm.es (F.B.H.); r.damato@upm.es (R.D.)
* Correspondence: juandavid.cano@upm.es

Abstract: Although isometric contractions in human muscles have been analyzed several times, there are no FEA models that allow us to use the same modeled joint (the elbow under our case) in different conditions. Most elbow joints use 3D elements for meshing. Representing the muscles in the joint is quite useful when the study is focused on the muscle itself, knowing stress distribution on muscle, and checking damage in muscle in a detailed manner (tendon–muscle insertion, for example). However, this technique is not useful for studying muscle behavior at different positions of the joint. This study, based on the mechanical model of the elbow joint, proposes a methodology for modelling muscles that will be studied in different positions by meshing them with 1D elements. Furthermore, the methodology allows us to calculate biceps and triceps efforts under load for different angles of elbow joint aperture. The simulation results have been mathematically modelled to obtain general formulations for these efforts, depending on the load and the aperture angle.

Keywords: FEM analysis; biomechanics; effort mathematical models; NMR reconstruction; reverse engineering; elbow joint

MSC: 37M05

Citation: Lechosa Urquijo, E.;
Blaya Haro, F.; Cano-Moreno, J.D.;
D'Amato, R.; Juanes Méndez, J.A.
Mechanical Model and FEM
Simulations for Efforts on Biceps and
Triceps Muscles under Vertical Load:
Mathematical Formulation of Results.
Mathematics **2022**, *10*, 2441. https://
doi.org/10.3390/math10142441

Academic Editors: Higinio Rubio
Alonso, Alejandro Bustos Caballero,
Jesus Meneses Alonso and
Enrique Soriano-Heras

Received: 21 June 2022
Accepted: 11 July 2022
Published: 13 July 2022

Publisher's Note: MDPI stays neutral
with regard to jurisdictional claims in
published maps and institutional affiliations.

1. Introduction

Over the years, the finite element method (FEM or FEA model) has become the preferred tool [1,2] by researchers for the study of analyzes of practically any type. The main reason for this preference is the accuracy [3,4] in the solution and the flexibility to simulate almost any physical event that may be explained by mathematical equations.

One of the fields in which the finite element method has been used on numerous occasions has been biomechanics [5–11], with applications to specific studies such as joint analysis, prosthesis design and muscle behavior analysis.

Li et al. [5] used the FEM method to run a simulation of fluid structure interactions coupling the lattice Boltzmann method and the FEM, and this new study can be used on blood flow and the heart, or urine interaction with bladder. Della Rosa et al. used the FEM method to validate the new fixator with application in fractures [6]. Zhang et al. developed a FEM analysis of a lumbar spine implanted with a lordotic cage [7] at L3-L4, and the results were validated by in vitro testing. The stress distribution on the cage and the range of motion (ROM) of L3-L4 were used to assess the stability of the implant. Denozière and Ku, in their study, proposed a three-dimensional model of a two-level ligamentous lumbar segment [8], simulated by static analyses with the FEM software to predict that mobility after arthrodesis at the upper level was reduced in all rotational degrees of freedom by an average of approximately 44%, relative to healthy normal discs. Samani et al., proposed a breast biomechanical model using a FEM formulation [9], specially focused on the modeling of breast tissue deformation which takes place in breast imaging procedures.

The study proposed by Martinez [12] is focused on the characterization of the muscle and its response under different conditions. Muscle length changes and the internal stress distribution for different electrical stimulations were studied. Weiss et al. [13] described strategies for addressing technical aspects of the 3D computational modelling of ligaments with FEM analysis. Islan et al. [14], studied the behavior of the glenohumeral joint under different postures that represent the routine of a violinist from an ergonomic point of view (RULA method) and using a 3D FEM model to study this movement for a high number of cycles. Sachenkov et al. [15] studied the movement of the femur in the hip, using 1D models for the muscles. Martins et al. [16] presented a study with FEM techniques applied to the pelvic floor, using the Hill model for muscles as a 1D element, extrapolating it to 2D and 3D models, working with isometric and isotonic contractions. Tang et al. [17] studied muscle fatigue with a 3D muscle model in different activation situations. Syomin et al. [18] performed a numerical simulation of contraction of the left ventricle approximated by the axisymmetric body.

Perreault and Heckman evaluated the ability of the Hill model for muscles to describe muscle force responses for naturally and electrically stimulated muscles [19]. Alonso et al. [20] proposed a static physiological optimization model instead of the dynamic optimization model. For this purpose, they used a modeling of muscles by means of 1D elements along with the line of action of the same. Holzbaur et al. proposed a different biomechanical model representing the upper extremity, including shoulder, forearm and arm [21]. The muscles are replaced by a 1D element with a force generation parameter based on experimental data. Park et al. [22] proposed a torque estimation method at the joint of an index finger in the human hand while pinching, the study is supported by electromyography and Hill muscle model. Soechting and Flander used a simplified muscle model [23] to predict torques during arm movement, comparing their prediction with the EMG results measured in 20 subjects. The muscle model is also based on the Hill muscle model. Zajac developed a mathematical model to analyze the muscle tendon, 'musculotendon' actuator [24] based only on the ration between tendon length and muscle fiber length, both at rest.

In the study and FEM analysis of human joints, bones are considered rigid elements, their geometry does not vary when the joint moves, the same happens with cartilages, and they can be approximated to rigid, nondeformable parts [25]. On the other hand, when meshing a muscle in 3D elements into a joint, this mesh is only valid for a certain position, and if there is need to study a different position in the joint, the muscles should be remeshed once again. This problem disappears by using a 1D element, rod-type, to mesh the muscles in the joint. Ligaments can be modeled as springs, 1D elements with two nodes. According to this, a model meshed with 3D elements for rigid bodies and 1D element for muscles elements, can adopt different positions just by moving the elements of the rigid bodies to the new positions.

In the case of 1D elements with just two nodes, if those nodes move in the space, the element will move as well. As ligaments are defined from two different nodes of two different bones, if the bones move to another position, the ligament will adopt the new position. This allows us to reuse the same meshed model in several configurations. For this reason, the final objective of this study is discovering the equation that would allow one to create a new finite element type that could be used in biomechanics analysis where muscles forces are involved in *"isometric contractions"*. This study focuses on the analysis of human muscle behavior in order to define a new equation that could predict the muscle force that would appear in a certain muscle under known conditions. The human joint under study will be the elbow. The proposed FEM analysis is focused only on the biceps and triceps muscles and the muscle elbow joint considered in this study are (Figure 1):

- Long biceps
- Short biceps

- Brachii
- Long supinator
- Long triceps
- Medium triceps
- Lateral triceps

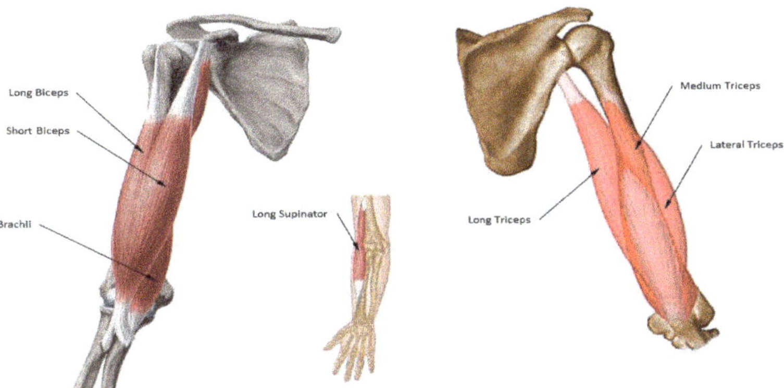

Figure 1. Flexors (**left**) and extensors (**right**) elbow muscles identification.

Since muscles can only work under traction (they do not transmit compression loads), different models must be proposed for a separate study of the triceps and biceps muscles. Several analyses, on different aperture angles of the elbow, will be solved looking for the effort needed on referred muscles, biceps or triceps, to balance 150 N applied on the wrist in an "*isometric contraction*". The scope of the study includes the following aspects:

- Analyzing the elbow under load at different angles;
- Obtaining efforts in different muscles involved in the joint for different positions;
- Defining mathematical models to predict efforts in muscles depending on:
  - Joint angle or muscle length;
  - Applied load on the wrist.

2. Materials and Methods

The proposed methodology will allow one to calculate elbow muscle effort. Furthermore, a post-processing of the results is proposed in order to obtain the formulation of efforts in the studied muscles when the load applied changes. These efforts are referred to isometric load cases. An isometric load involves muscle contraction against resistance in which the length of the muscle remains constant. The research can be divided into two main parts:

1. Finite element models of the mechanism of elbow joint. The 3D anatomical model developed is based on two diagnostic imaging techniques. CT for the bones and MRI for the muscles. From this three-dimensional model, different FEM models were generated, for biceps and triceps muscles studies.
2. Mathematical models predicting efforts in muscles studied. From FEM models, isometric muscle efforts will be obtained for a vertical load applied on wrist, 150 N including forearm mass, at different angles of aperture in the joint, from 44 to 164 degrees. These efforts will be extrapolated for any load applied on the wrist.

The first part is based on previous studies [25]. This model is shown in Figure 2, and it will be described in the more detail in next section.

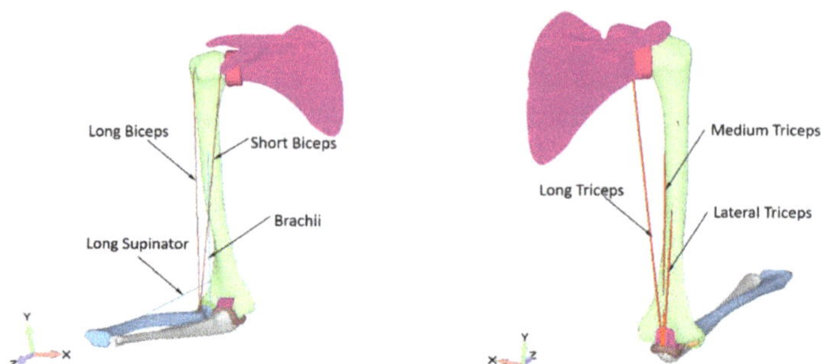

Figure 2. FEA model, muscles identification, flexors (**left**) and extensors (**right**).

2.1. Finite Element Model

For the creation of the FEM model, a 3D CAD model of the elbow joint is necessary, from which the mesh of the FEM model will be created.

Said CAD model is obtained starting from the images obtained by means of CT and MRI, of the elements of the joint (bones and muscles). Thanks to the use of software for generating 3D CAD models from point clouds, Geomagic Desing X, we can generate the CAD model of solid bodies.

The FEA model was implemented in Siemens FEMAP 10.2 as preprocessor and post-processor, and NX Nastran 7.1, as solver. FEA software was described in detail in previous research [25]. Figure 3 shows the two developed FEA models for the biceps and triceps FEM analysis.

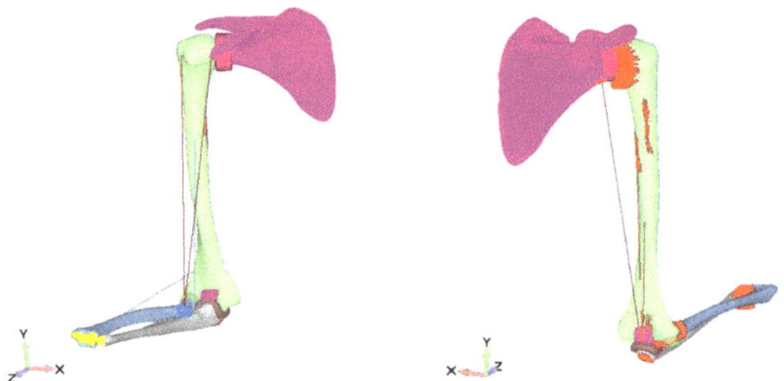

Figure 3. FEA model, biceps study (**left**) and triceps study (**right**), 74° aperture angle.

2.1.1. FEA Model Creation Meshing Process, Element Types

Different types of elements have been used for different components:

- **Bones.** They are meshed using 3D elements (tetrahedral type, 2nd order, 10 nodes). Teo et al. [26] use 3D elements for meshing neck and head parts of the joints in their study. In the same way, Donahue et al. [27] use 3D elements with axisymmetric condition, to model knee parts involved in the study.
- **Cartilages.** They are meshed using 3D elements (tetrahedral type, 2nd order, 10 nodes). Donahue et al. [27] and also Abidin et al. [28] model cartilages using 3D elements.
- **Bone cartilage junction.** Since the cartilage and bone form a solidary union between both parts, during the meshing process, a coincidence of the nodes of the bone and the cartilage in the contact area has been carried out. By ensuring that the bone and cartilage nodes coincide in that area, it is possible to ensure that the transmission of efforts between both parts is correct.

- **Contact between cartilages**. It has been modeled using rigid elements that join the nodes of both cartilage meshed parts involved in the joint, as explained in previous studies [25], using an RBE2 element. In the RBE2 elements used, the degrees of rotation freedom have been released so that only forces are transmitted in the X, Y or Z axes.
- **Muscles**. These are meshed using 1D elements (the type is discussed in this study). Previous studies have considered muscles in this manner, as 1D element acting along the imaginary axis of the muscle. Alonso et al. [20] model muscles in the leg to study human march in the same manner, 1D elements aligned with the axis of the muscles; Sachenkov et al. [29] use the same approximation for muscles, Parekh [30] approximates muscles to a 1D FEM element aligned with the muscle axis. This is a simplification, but it is used by several authors because fusiform muscles only have two insertion points and low pennation angles. Thus, fibers are aligned so with this imaginary axis which is formed by joining both insertion points.
- **Tendons**. These are meshed in combination with muscles, as a unique element musculotendon (MTU). The same 1D element is considered for tendons.
- **Tendon insertion in bone**. It has been modeled by adding a rigid element (RBE2), joining the end node of the 1D MTU element with several nodes in the bone, and thus distributing the reaction force in the area of the insertion of the tendons, as explained in previous research [25]. Of the restrictions that an element with these characteristics can apply, only the displacements in the X, Y or Z axes have been applied, freeing the turns around them. In this way, and since the MTU only transmits axial forces, we eliminate any possible bending moments that may affect the insertion of the tendon in the bone.

The RBE2 elements constitute constraint equations between the nodes to which the element is connected, they are characterized by having a node, called independent, and whose degrees of freedom are independent. The rest of the nodes connected to the RBE2 element are called dependent, and in them it must be fulfilled that the displacements of the independent node must be the same as in the dependent nodes.

That is why they are commonly known as rigid elements, since the distance between the component nodes remains unchanged.

In Figure 4, an example of an RBE2 element can be found, with an independent node (i) and four dependent nodes (j1 ... j4).

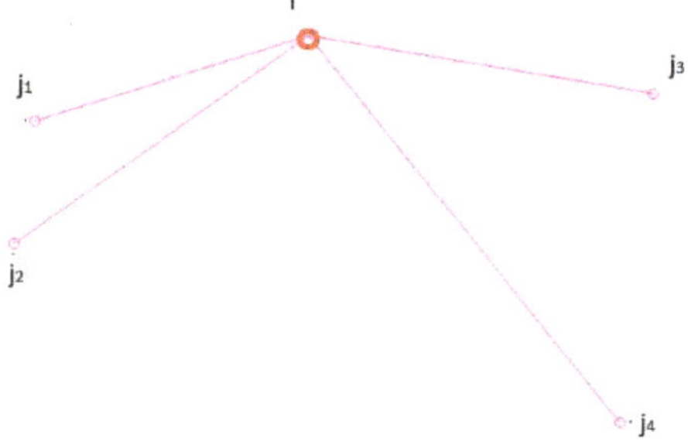

Figure 4. RBE2 element example: 1 Independent node and 4 dependent nodes.

The force balancing equation on an RBE2 is such that the reactions that appear on dependent nodes of the element must balance the force applied in the independent node:

$$\sum_{dependent} \{Reactions\ forces\} = F_{independent} \tag{1}$$

In this way, using an element of this type to simulate tendon insertion in the bone, the stress that appears in the 1D element that simulates MTU in the insertion zone is distributed, eliminating stress concentrations.

Figure 5 shows the application of this insert model.

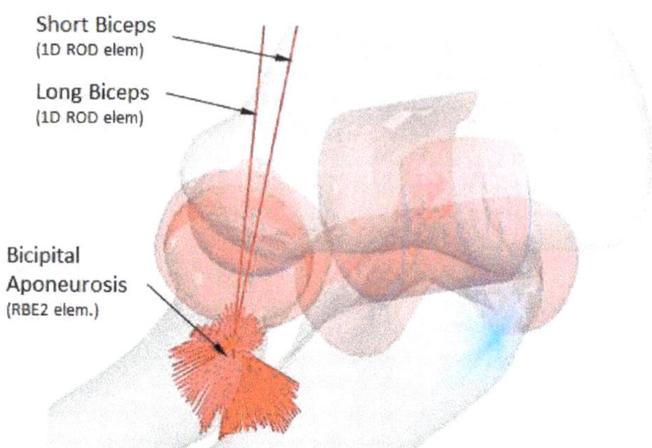

Figure 5. Biceps tendon insertion using RBE2 element.

2.1.2. FEA Model Creation Muscles Modeling

Given that the proposed study is focused on knowing the forces developed by each of the muscles of the elbow joint, in different load situations, the modeling of the muscle must be such that it allows one to obtain these values.

As mentioned above, the objective is to be able to use a 1D element that allows us to reuse the mesh created to solve the FEM simulation without the need to re-mesh the entire model for each position of the joint.

The behavior of the element used in our study to simulate the muscle in the joint must be such that:

- It works only in traction.
- The force of the element used must be generated along the axis that joins the two nodes that define it; it must not have transverse forces.
- It must not present deformations or displacements that alter the isometric character of the contraction studied. Therefore, the deformations have to be low.
- The 1D-type elements existing in FEM analysis tools are described below, also indicating the reasons that lead us to discard them as valid for our purpose.
- Constant spring stiffness. In this case, the results are not valid, since a spring delivers the force along a curve according to the equation $F = K{\cdot}x$, where K is its stiffness (N/mm) and x is its elongation (mm). The use of this element would mean that a muscle would deliver a single force value for a given length, which is not true. A muscle at a given length is capable of delivering a force between 0 and the maximum force that muscle fibers are capable of developing.
- Variable spring stiffness. For the same reason as before, although in this case the maximum force could adjust according to the length of the muscle, it does not serve the purpose of the proposed study. The muscle can deliver a force between 0 and the maximum force for a given length. In the case of using a spring element, there is also the drawback of not always being able to balance the joint and the present study would no longer be isometric.
- Beam element. With this element, it is possible to obtain the necessary force in the joint to balance it. Beam elements are characterized in that they present axial, shear, and bending forces, and the muscles only transmit axial tensile forces.
- Rod element. The use of this element meets the study requirements; it only transmits axial forces, and we can decide on a rigidity in it such that the joint does not present large displacements and therefore it is no longer considered an isometric contraction.

In search of the element that best reflects the behavior of the muscle within the joint, the ROD element, given that it only presents axial forces, is the one that best suits the needs of the proposed mathematical study. However, the said element can work under tension and/or compression, which would alter the results. That is why two different FEM models are generated for the study of the biceps or triceps muscles. One in which the triceps muscles are re-moved, and another in which the biceps muscles are removed. In this way, and considering the direction of the applied load, it can be ensured that the analyzed muscles work only under traction.

Several tests have been carried out to define the characteristics of the ROD element that will be used, maintaining the objective of minimizing its stiffness and keeping the element deformations at low values so as not to lose the isometric load condition in the analyzes.

For these reasons, the elements used in the FEA models are defined as:

- Rod element (1D element);
- Constant cross section: 100 mm^2;
- Material parameters:
  - Young modulus: 2000 MPa;
  - Poisson coefficient: 0.43.

This type of element transmits only axial forces. Its constitutive equation is shown below.

$$\begin{bmatrix} F \\ M \end{bmatrix} = \begin{bmatrix} K_x & 0 \\ 0 & K_T \end{bmatrix} \cdot \begin{bmatrix} \varepsilon_x \\ \alpha_x \end{bmatrix} \tag{2}$$

where

- F is the axial force in the element;
- M is the moment along the axis of the element;
- K_x is the axial stiffness of the element;
- K_T is the rotational stiffness along the axis of the element;
- ε_x is the length variation of the element;
- α_x is the angle rotated along the element axis;

To ensure that no torsional stresses are produced in the element, it is sufficient to assign a value of $K_T = 0$.

2.1.3. Material Properties

The meshing process of the 3D bodies that comprise the model, bones and cartilage, has been carried out automatically using the meshing tool that includes Femap. The geometric characteristics of the elements used are common to both bones and cartilage; they are:

- Second-order tetrahedral elements, 10 nodes;
- Nominal element size 1.5 mm.

The materials considered for each body have the following characteristics, taken from the CES Edu Pack [31].

The use of an isotropic and linear material for cartilage has previously been used in those cases in which the duration of the applied load is less than the time constant of the viscoelastic properties of cartilage, around 1500 s. In this case, given that it is considered an isometric contraction, in which there is no displacement in the joint and the application time is considered well below the 1500 s indicated, it is possible to consider it as isotropic, linear, and elastic with the properties indicated in Table 1. Donahue et al. [27] considered isotropic, linear and elastic tissue for cartilages in the knee joint in their study of tibio-femoral contact in the human knee joint, based on the time required to apply the load compared to the time constant for viscoelastic properties of cartilages, for the same reason. Abidin et al. [28] consider cartilage as an isotropic, linear and elastic material in their study on a human knee joint model and verification.

Table 1. Tissue properties for bones and cartilages.

	Young's Modulus (Mpa)	Poisson	Density (kg/m^3)
Bone	17,200	0.41	1790
Cartilage	8	0.49	1150

The use of an isotropic material for bone modelling is known and has previously been used, Teo et al. [26], meshed head and spine using 3D elements with bone tissue as isotropic, linear, and elastic.

2.1.4. Mesh Resume

The summary of the FEM mesh is as follows:

- Number of elements: 548,139;
- Number of nodes: 821,755;
- Number of second-order tetrahedral elements: 546,035;
- Number of 1D ROD elements: 4 (for biceps analysis) 3 (for triceps analysis);
- Number of rigid elements RBE2: 2100.

A study of the quality of the mesh, based on the Jacobian value of the elements, shows us how good its quality is for our objective. Figure 6 shows the elements that are below the limit value of 0.6 (good quality for the considered analysis).

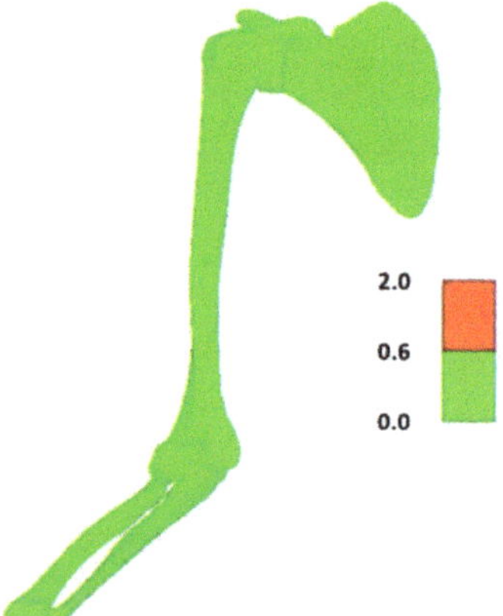

Figure 6. Mesh quality. Jacobian criteria for evaluating mesh quality. Value represented is (1-measured value on element/ideal value on element). Maximum threshold defined as 0.6.

2.1.5. Boundary Conditions

For the boundary conditions, the scapula is considered fixed in the area (pink color on the below image) where the muscles of the scapula would be inserted. These muscles (see Figure 7) are:

1. Levator scapula muscle;
2. Rhomboid minor muscle;
3. Rhomboid major muscle.

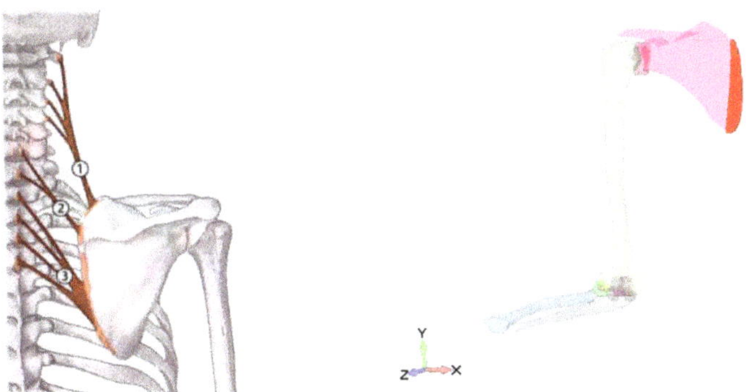

Figure 7. Boundary condition, muscles fixing scapula (**left**), constrain in FEA model, (**right**), scapula fixed.

2.1.6. Load Application

In order to be able to validate results of the analysis, the load considered must be easy to replicate in a controlled environment; that is why we have defined the load as the mass of a dumbbell of 15 kg (approximately, 150 N) for the biceps analysis. In the case of triceps loading, the simulation is more similar to the exercise done using a guided load machine.

The application point of this load is the wrist (the hand is not modeled) and although this can affect slightly to the results, due to the difference in distance of the load applied in the joint, this can be solved varying the load as needed when applied in the hand.

All models have the same load applied, at the same node, and in the same vertical direction (see Figure 8).

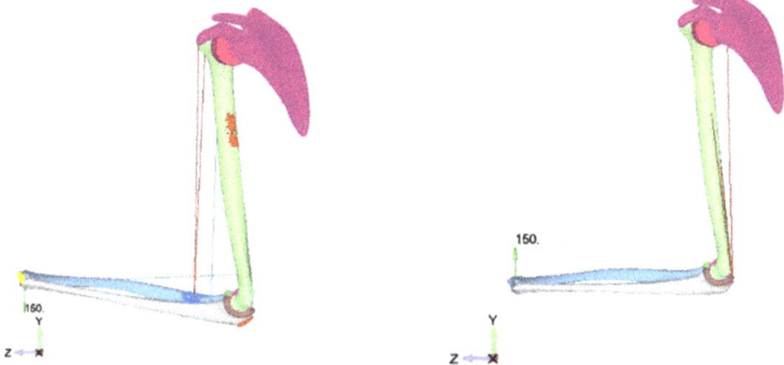

Figure 8. Applied load (150 N) in wrist, vertical direction, biceps analysis (**left**) and triceps analysis (**right**).

The load is applied in vertical direction, positive axis (ascending) when analyzing the triceps and vertical direction, negative axis (descending) when analyzing the biceps. All load cases are considered as static load cases, isometric contraction of the muscle, and, to avoid compression loads on muscles, two different models have been developed; one of them includes only biceps muscles, with descending load, and the other one includes only triceps muscles, with ascending load.

In the model, gravity force has also been considered. Thus, the mass of the arm is considered in the analysis.

To do so, the density of elements has been modified so that the values of the upper arm and forearm mass are [32]:

- Forearm mass: 2.5 kg;
- Upper arm mass: 4.3 kg.

All load cases are considered as static load cases; due to the isometric contraction of the muscle, there must be no displacement in the joint.

2.1.7. Design of Experiment

The FEM model is considered linear and static. No non-linearity has been considered, neither in the materials nor in the restrictions of the model. That is why the general equation of the calculation is so important and can be written as follows:

$$[R] = [K] \cdot [U] \tag{3}$$

where

$[R]$ = global loading matrix;
$[K]$ = global stiffness matrix;
$[U]$ = global displacement matrix.

Equation (3) can be solved directly using the Gauss elimination method, there is no iterative process, and the solution is unique.

The resolution of the problem is carried out through several linear static analyses, each one of them representing a different load condition in the joint, and the result of them is the necessary effort in each muscle of the joint to balance the applied force on the wrist.

As mentioned above, there are two different models, thus guaranteeing that the muscles work only in traction, to study the biceps and the triceps muscles.

In each of the models, the configuration of the joint will be modified by varying its angle between 44 and 164°, at intervals of 10°.

The applied load will always be 150 N.

As a reference for the reading of the results, the fully extended arm corresponds to 164 degrees and the fully flexed arm corresponds to 44 degrees (see Figure 9).

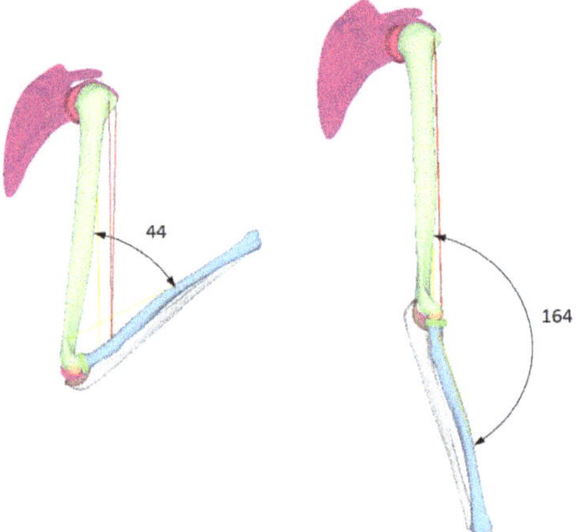

Figure 9. FEA model, minimum (**left**) and maximum (**right**) studied angles.

2.1.8. FEA model Solution

The FEA model has been solved on a workstation with the following characteristics.

- Windows 10, 64-bit, operating system;
- Intel(R) Core (TM) i9-10900X CPU @ 3.70 GHz Processor (20 cores);
- 128 GB RAM memory;
- 1 TB SSD disk drive.

The solution time for any configuration is approximately 20 min.

As contacts in cartilages have been replaced by rigid elements, RBE2 type with free rotation on both nodes, the solution time has been reduced from 67 h to 20 min.

As described in a previous study [23] the results, comparing the model with contacts and the model with rigid elements instead, are similar in both cases, and therefore the use of rigid elements improves the efficiency of the analysis.

2.2. Mathematical Approach

In this work, several mathematical models are derived from the results of the FEM simulation. First, there are grade 6 polynomial curves, which represents the efforts of the five muscles involved for the simulated cases, described in previous section, for a vertical force of 150 N applied on wrist. Muscle effort is defined as the total force that appeared in the 1D element used for each muscle in FEM models. These models make it possible to obtain this effort for each muscle by following an equation for each muscle, with the following general formulation:

$$Muscle_{effort} = \sum_{i=0}^{i=6} a_i \cdot \alpha^i \qquad (4)$$

Once the forces needed in muscles have been calculated to balance the joint, with different angles and with a load of 150 N applied at the wrist, a mathematical approach will be proposed for obtaining these efforts for other loads, due to the mechanism type considered in the elbow joint, the applied forces and torque analysis done in the joint. The value of the force exponent is 1 in all cases, so there is a linear relation between the force applied and the muscle reaction force [33]. According to this, the greater the force applied to the wrist is, the greater the muscle force must be, and both values are proportional. Figure 10 and the following equations justify this simplification in the analysis, where:

- W_s means the weight of the forearm;
- $W_s{}'$ means the equivalent force for W_s applied at the wrist. The moment created by W_s or $W_s{}'$ is the same referred to O point;
- W_q means the load applied in the wrist;
- α means the angle between biceps axis and the normal to the radius longitudinal axis;
- β means the angle between the vertical and the longitudinal axis of the radius;
- a, b and c are geometrical dimensions.

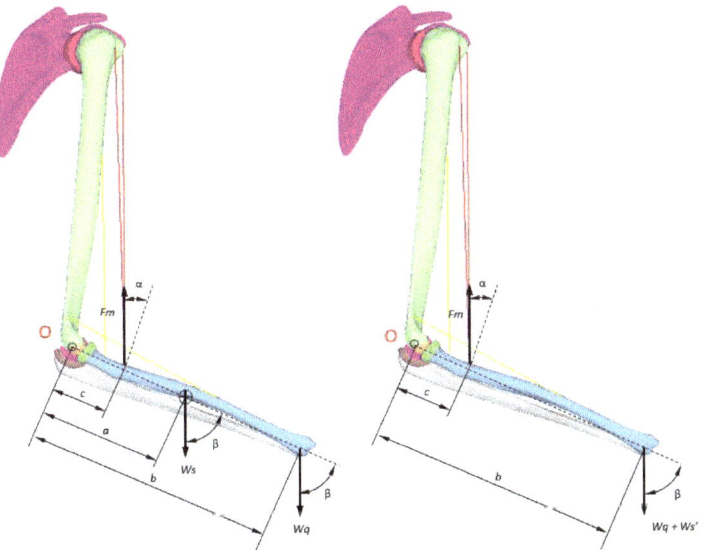

Figure 10. Forces scheme in elbow joint.

The left image considers the proper mass of the arm located in its center of gravity. Balancing equation on that forces scheme, referred to O point scheme, is:

$$Fm = \frac{(W_s \cdot a + W_q \cdot b) \cdot \sin \beta}{c \cdot \cos \alpha} \tag{5}$$

Instead of using W_s, we can use W_s', applied on the wrist (right image). This W_s' should comply:

$$W_s \cdot a \cdot \sin \beta = W_s' \cdot b \cdot \sin \beta \tag{6}$$

So, the balancing equation would be the following:

$$F_m = (W_q + W_s') \frac{b \cdot \sin \beta}{c \cdot \cos \alpha} \tag{7}$$

As seen in Equation (7), the relation between F_m and $W_q + W_s'$ is linear, and constant for a certain angle only depending on b and c and α and β angles.

According to this, the forces in the muscles when the load is different from the 150 N applied in this study (including the W_s' in this 150 N) can be obtained by the relation between the external load applied and the reaction in the muscle.

$$\frac{W_q + W_s'}{F_m} = \frac{F_{ext}}{R_{musc}} = \frac{c \cdot \cos \alpha}{b \cdot \sin \beta} = constant(\alpha, \beta) = \frac{150 \text{ N}}{F_m(150 \text{ N})} \tag{8}$$

In summarizing, forces in muscles depend on:

- Muscle length (or elbow angle);
- Force applied to wrist.

By applying these mathematical approaches, several polynomial equations have been obtained, one per muscle, which establishes the relation between force in muscle as:

$$F_m = F_m(W, \alpha) \tag{9}$$

where:

- F_m is the force in the muscle (N);
- α is the angle of the elbow;
- $W = W_q + W_s'$, is the load applied on the wrist (N).

In all mathematical models of muscle efforts, the effort could be obtained for any angle, α, and for any vertical load, f_v. If the forearm mass, m_{fa}, is considered and g is the gravity acceleration (9.81 m/s^2) W_s' can be expressed as follows:

$$W_s' = W_s \cdot \frac{a}{b} = m_{fa} \cdot \left(\frac{a}{b} \cdot g\right) \tag{10}$$

Thus, for obtaining the effort:

$$Effort(f_v) = F_m\left(f_v + m_{fa} \cdot \left(\frac{a}{b} \cdot g\right), \alpha\right) \tag{11}$$

General equations will be obtained describing the effort in each muscle depending on the angle of the joint and the load applied. These equations and their surface representations will be the results of this research. Both of them, the equations and surfaces, have been obtained using the curve fitting tool from MATLAB. It has been noted that these effort models for the five involve muscles are only used to study isometric muscular contractions, when muscle length does not vary.

3. Results

The model analyzed includes all muscles in the elbow joint, but the post-processing results are only applied to the biceps and triceps muscles.

3.1. FEM Simulations, 3D Results

Figure 11 shows the results of running the design of the experiment described here-inbefore for the 13 cases of elbow angle for the biceps model. The figure represents the Von Mises stress in the involved bones due to the action of the 150 N as a vertical force (descendent) on the wrist. Each of the images represents the stress map resulting from the study of an isometric contraction in the elbow joint at the angle indicated in the image for a load of 150 N applied to the wrist, vertically and downward.

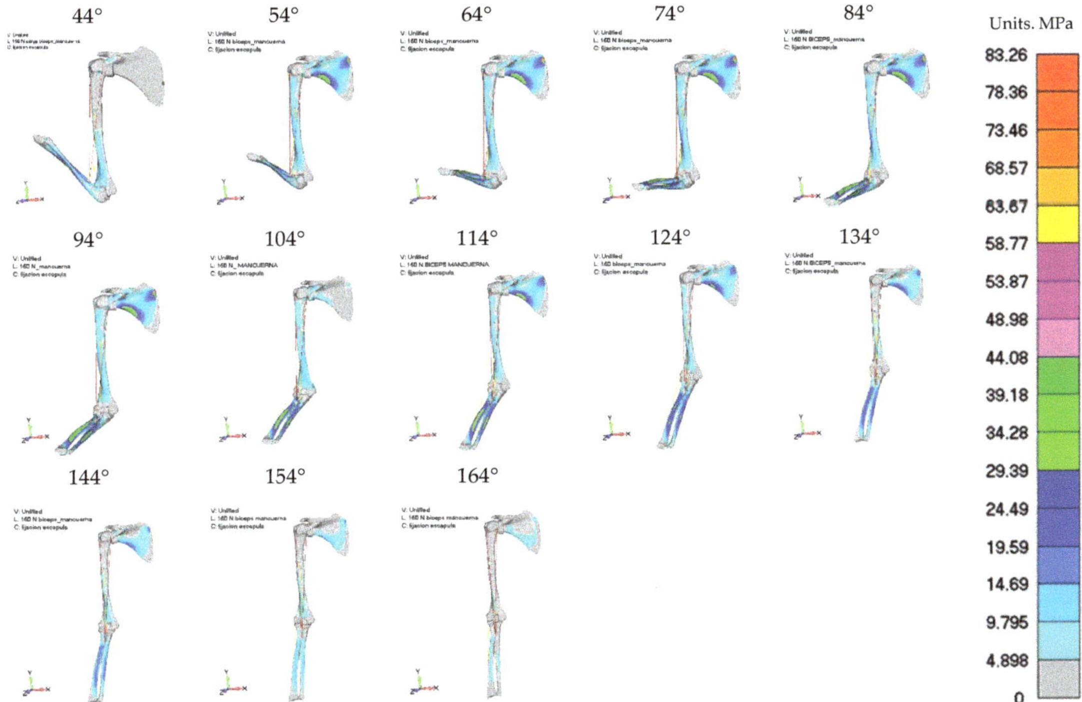

Figure 11. Different elbow angle configuration: bone Von Mises stress distribution under biceps loading (150 N on wrist) cases. Figures are included in more quality in Appendix A.

Similarly, Figure 12 shows the Von Mises stresses on the bones due to the FEM simulation of 150 N of vertical load (ascendant) on the wrist. Each of the images represents the stress map resulting from the study of an isometric contraction in the elbow joint at the angle indicated in the image for a load of 150 N applied to the wrist, vertically and upward.

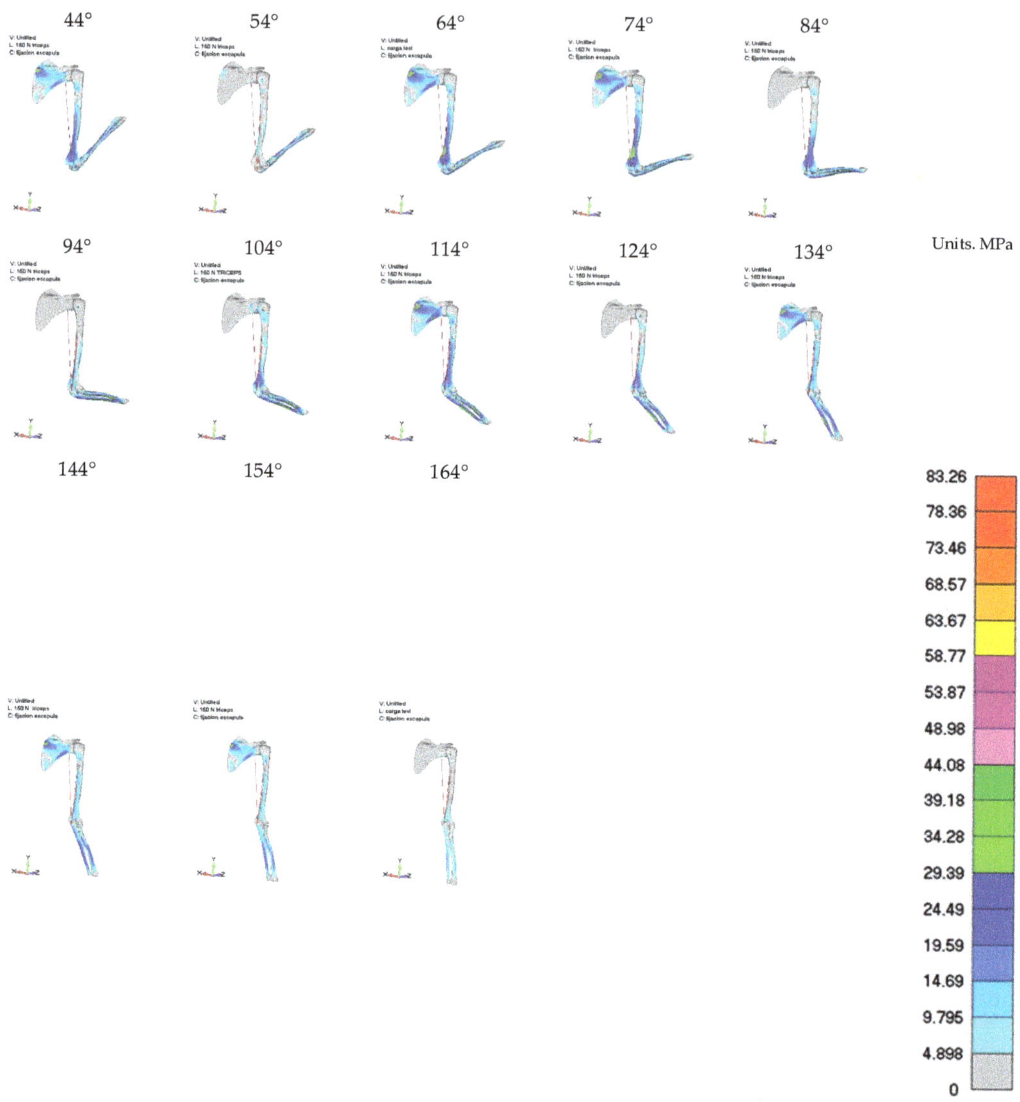

Figure 12. Different elbow angle configuration: bone Von Mises stress distribution under triceps loading (150 N on wrist) cases. Figures are included in more quality in Appendix A.

3.2. Numerical Axial Efforts on Main Different Muscles

This subsection collects axial efforts on biceps and triceps involved muscles for the 26 FEM simulations (13 for biceps and 13 for triceps). These efforts correspond to the reaction obtained in the muscle itself, and length corresponds to the length of the muscle (the whole muscle length including tendons) measured in the FEA model, in that certain elbow angle. Table 2 has these results if the form of force and the corresponding length for this applying force for all muscles studied:

- Biceps muscles (long and short biceps);
- Triceps muscles (lateral, medial and long triceps).

Table 2. Axial efforts (N) in main simulated muscles, 150 N applied on wrist. Isometric contractions at each angle value considered.

Elbow Angle (°)	Long Biceps Force (N)	Long Biceps Length (mm)	Short Biceps Force (N)	Short Biceps Length (mm)	Lateral Triceps Force (N)	Lateral Triceps Length (mm)	Medial Triceps Force (N)	Medial Triceps Length (mm)	Long Triceps Force (N)	Long Triceps Length (mm)
44	244	373	96	310	148	217	70	211	269	284
54	330	375	136	312	128	216	144	209	312	282
64	341	383	152	320	212	215	115	208	340	281
74	341	391	161	328	208	215	119	208	347	281
84	340	399	168	336	235	214	150	207	340	280
94	333	407	170	344	263	214	186	207	347	279
104	302	415	181	351	233	213	164	207	331	279
114	301	422	159	359	233	213	176	206	307	278
124	274	429	143	365	213	211	170	204	271	276
134	261	435	95	371	177	207	150	201	226	273
144	185	439	87	375	131	203	120	197	175	269
154	121	442	46	379	79	199	82	192	119	264
164	52	445	1	381	24	194	40	188	58	260

3.3. Mathematical Models for Muscle Efforts

3.3.1. Efforts for 150 N of Vertical Load

The obtained R-square values are 99.3%, 98.2%, 99.9%, 93.6%, and 97.2%, respectively (see Table 3). Showing results in graph form, one graph for biceps muscles and the other for triceps muscles, a tendency can be observed in both of them. Figure 13 shows the efforts of biceps muscles at each angle (dots) and the mathematical curves (dash point line) that approximate them. The effort in muscle (y-axis on the graph) is related to the elbow angle (x-axis on the graph). The red color is associated with long biceps and the blue color is associated with short biceps. These curves represent the total effort of each muscle involved. Thus, long biceps, for a constant vertical load on the wrist, including forearm mass, reach the maximum effort (340 N) between 60–90°, while for short biceps the maximum effort (170 N) is achieved for the range of 90–110°.

Table 3. The equations developed as polynomial curves of grade 6th, based on the general formulation of Equation (4), has the following constant values for each muscle involved.

Muscle	a_0	a_1	a_2	a_3	a_4	a_5	a_6	SSE	R^2	RMSE
Long biceps	−2323 (−3730, −916.1)	136.6 (55.39, 217.8)	−2.7 (−4.477, −0.9232)	0.02579 (0.007278, 0.04431)	−0.0001193 (−0.0002116, −2.692 × 10^{-5})	2.116 × 10^{-7} (3.433 × 10^{-8}, 3.889 × 10^{-7})	0	694.5	0.993	10.76
Short biceps	−409 (−1729, 911.5)	24.12 (−52.09, 100.3)	−0.4361 (−2.104, 1.232)	0.004284 (−0.01309, 0.02166)	−2.182 × 10^{-5} (−0.0001085, 6.486 × 10^{-5})	4.28 × 10^{-8} (−1.236 × 10^{-7}, 2.092 × 10^{-7})	0	367.3	0.982	7.82
Long Triceps	−1550 (−3477, 377.3)	110.4 (−23.88, 244.7)	−2.702 (−6.433, 1.028)	0.03508 (−0.01795, 0.0881)	−0.0002503 (−0.0006584, 0.0001577)	9.18 × 10^{-7} (−7 × 10^{-7}, 2.536 × 10^{-6})	−1.354 × 10^{-9} (−3.945 × 10^{-9}, 1.236 × 10^{-9})	102.3	0.999	4.13
Medium Triceps	−6448 (−1.38 × 10^4, 908.4)	442 (−70.71, 954.7)	−11.9 (−26.14, 2.344)	0.1638 (−0.0386, 0.3662)	−0.001217 (−0.002775, 0.0003408)	4.643 × 10^{-6} (−1.534 × 10^{-6}, 1.082 × 10^{-5})	−7.144 × 10^{-9} (−1.703 × 10^{-8}, 2.743 × 10^{-9})	1491	0.936	15.76
Lateral Triceps	4661 (−3110, 1.243 × 10^4)	−311.3 (−852.8, 230.3)	8.397 (−6.645, 23.44)	−0.1143 (−0.3281, 0.09946)	0.0008424 (−0.0008029, 0.002488)	−3.214 × 10^{-6} (−9.738 × 10^{-6}, 3.309 × 10^{-6})	4.975 × 10^{-9} (−5.468 × 10^{-9}, 1.542 × 10^{-8})	1664	0.972	16.65

Analogously, Figure 14 shows the efforts of the triceps muscles (dots) and the mathematical curves (dash point line) which approximate them. The effort in the muscle (y-axis in the graph) is related to the elbow angle (x-axis in the graph). In the figure, the red color is associated with medium triceps, the green color is associated with long triceps, and the blue color is associated with the lateral triceps. These curves can be used to know efforts on the three muscles involved in the triceps FEM model. Additionally, as an example, if a muscle had limited effort for any reason, we could know which positions of the elbow joint are not recommendable to use. Therefore, for an isometric contraction with 150 N as a vertical load on the wrist (including the forearm mass), for a maximum effort of 200 N in the lateral triceps, it is necessary to avoid placing the elbow joint between 65° and 130 degrees approximately.

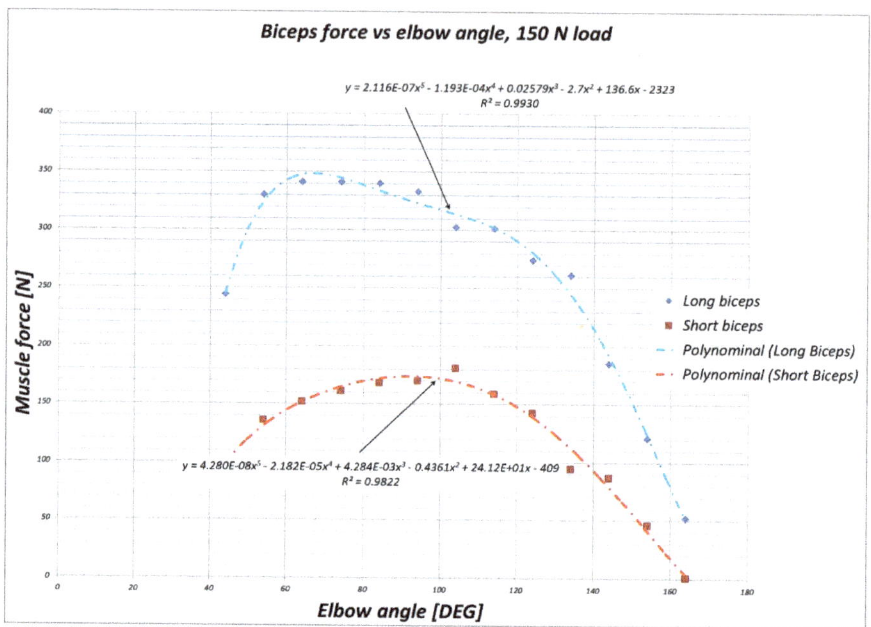

Figure 13. Calculated (dots) and tendency polynomial equation (dash point line) comparison, for biceps forces (*y*-axis). Isometric contractions on the elbow at different positions, from 44 to 164 degrees (*x*-axis) and 150 N are loaded on the wrist.

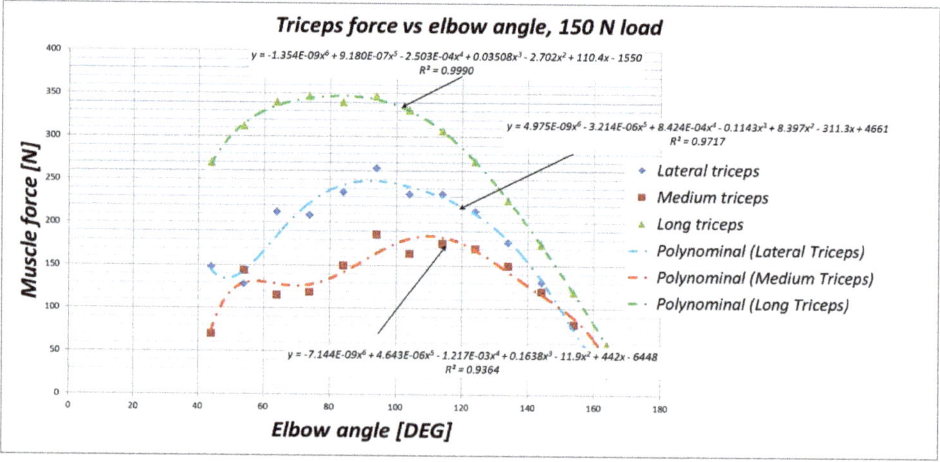

Figure 14. Comparison of the (continuous line) calculated and (dash point line) comparison, for triceps forces (*y*-axis). Isometric contractions on the elbow at different positions, from 44 to 164 degrees (*x*-axis) and 150 N are loaded on the wrist.

3.3.2. General Formulation

Applying the mathematical approach described above, equations of efforts for the five muscles involved will be presented, for any wrist load and elbow angle.

The approximation is done according to a fourth-grade polynomial with the general equation:

$$f(\alpha, w) = p00 + p10\cdot\alpha + p01\cdot w + p11\cdot w\cdot\alpha + p20\cdot\alpha\char94 2 + p21\cdot w\cdot\alpha\char94 2 + p30\cdot\alpha\char94 3 \\ + p31\cdot w\cdot\alpha\char94 3 + p40\cdot\alpha\char94 4 \tag{12}$$

where α is the angle of the elbow and w is the load applied to the wrist.

All coefficients have been calculated for a 95% confidence bound.

The surface of Figure 15 shows the tendency of the effort appearing in the short biceps muscle as a result of applying a vertical load on the wrist. As an example, for a constant vertical load of 750 N (including forearm mass), we could limit the short biceps effort to 200 N if we put the elbow joint angle between 150° to 164°.

<table>
<tr>
<td>

$p00 = 77.91\ (-184.3, 340.2)$
$p10 = -3.618\ (-14.77, 7.534)$
$p01 = -0.8343\ (-1.183, -0.4855)$
$p20 = 0.06135\ (-0.1079, 0.2306)$
$p11 = 0.04654\ (0.03494, 0.05813)$
$p30 = -4.245 \times\ 10{-}04\ (-0.001526, 0.0006774)$
$p21 = -2.803 \times 10{-}04\ (-0.0003988, -0.0001619)$
$p40 = 1.022 \times\ 10{-}06\ (-1.595 \times 10{-}06, 3.64 \times 10{-}06)$
$p31 = 1.695 \times\ 10{-}07\ (-2.081 \times 10{-}07, 5.471 \times 10{-}07)$

</td>
<td>

Goodness of fit:
SSE: 9.822e+04
R-square: 0.9916
Adjusted R-square: 0.9913
RMSE: 22.98

</td>
<td>

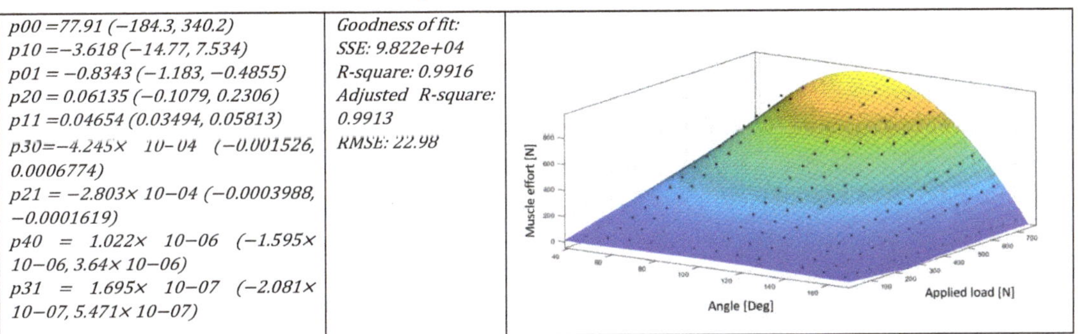

</td>
</tr>
</table>

Figure 15. General formulation of short biceps efforts in function of wrist load (W) and elbow angle (α) and surface reconstruction of the proposed polynomial.

In the range of the example of the previous figure, 150–164°, we could find an effort on long biceps bigger than in short biceps, reaching around 600 N. Figures 16–19 collect the results of the surface reconstruction for the rest of studied muscles.

<table>
<tr>
<td>

$p00 = -2019\ (-2461, -1578)$
$p10 = 94.81\ (76.03, 113.6)$
$p01 = 0.04386\ (-0.5435, 0.6312)$
$p20 = -1.544\ (-1.829, -1.259)$
$p11 = 0.05606\ (0.03653, 0.07558)$
$p30 = 0.01047\ (0.00861, 0.01232)$
$p21 = -0.0003463\ (-0.0005458, -0.0001468)$
$p40 = -2.516 \times\ 10{-}05\ (-2.957 \times 10{-}05, -2.075 \times 10{-}05)$
$p31 = 1.082 \times 10{-}07\ (-5.277 \times 10{-}07, 7.442 \times 10{-}07)$

</td>
<td>

Goodness of fit:
SSE: 2.786e+05
R-square: 0.9938
Adjusted R-square: 0.9935
RMSE: 38.7

</td>
<td>

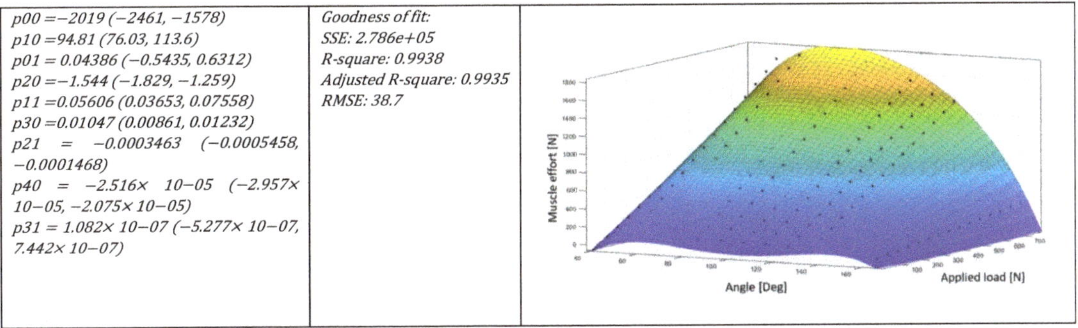

</td>
</tr>
</table>

Figure 16. General formulation of long biceps efforts based on wrist load (W) and elbow angle (α) and surface reconstruction of the proposed polynomial.

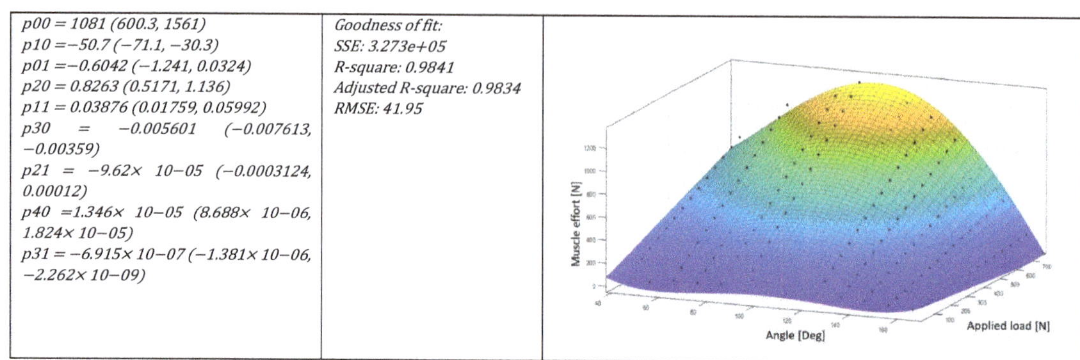

p00 = 1081 (600.3, 1561) *p10 = −50.7 (−71.1, −30.3)* *p01 = −0.6042 (−1.241, 0.0324)* *p20 = 0.8263 (0.5171, 1.136)* *p11 = 0.03876 (0.01759, 0.05992)* *p30 = −0.005601 (−0.007613, −0.00359)* *p21 = −9.62× 10−05 (−0.0003124, 0.00012)* *p40 =1.346× 10−05 (8.688× 10−06, 1.824× 10−05)* *p31 = −6.915× 10−07 (−1.381× 10−06, −2.262× 10−09)*	*Goodness of fit:* *SSE: 3.273e+05* *R-square: 0.9841* *Adjusted R-square: 0.9834* *RMSE: 41.95*

Figure 17. General formulation of lateral triceps efforts based on wrist load on the wrist (W) and elbow angle (α) and surface reconstruction of the proposed polynomial.

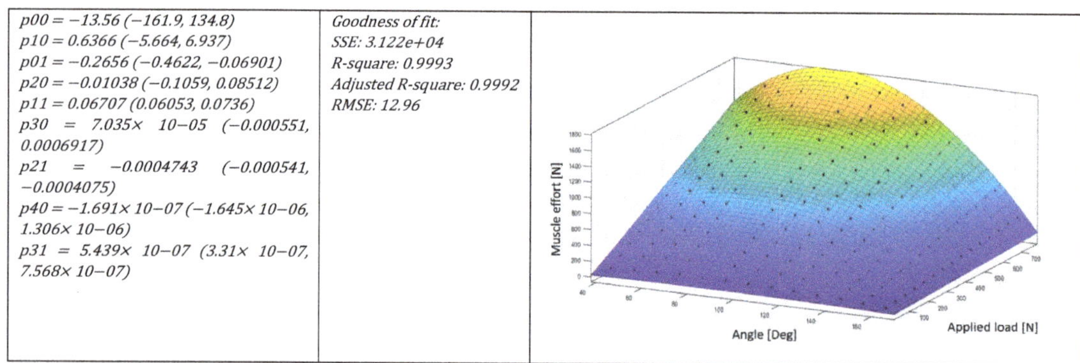

p00 = −13.56 (−161.9, 134.8) *p10 = 0.6366 (−5.664, 6.937)* *p01 = −0.2656 (−0.4622, −0.06901)* *p20 = −0.01038 (−0.1059, 0.08512)* *p11 = 0.06707 (0.06053, 0.0736)* *p30 = 7.035× 10−05 (−0.000551, 0.0006917)* *p21 = −0.0004743 (−0.000541, −0.0004075)* *p40 = −1.691× 10−07 (−1.645× 10−06, 1.306× 10−06)* *p31 = 5.439× 10−07 (3.31× 10−07, 7.568× 10−07)*	*Goodness of fit:* *SSE: 3.122e+04* *R-square: 0.9993* *Adjusted R-square: 0.9992* *RMSE: 12.96*

Figure 18. General formulation of long triceps efforts based on wrist load (W) and elbow angle (α) and surface reconstruction of the proposed polynomial.

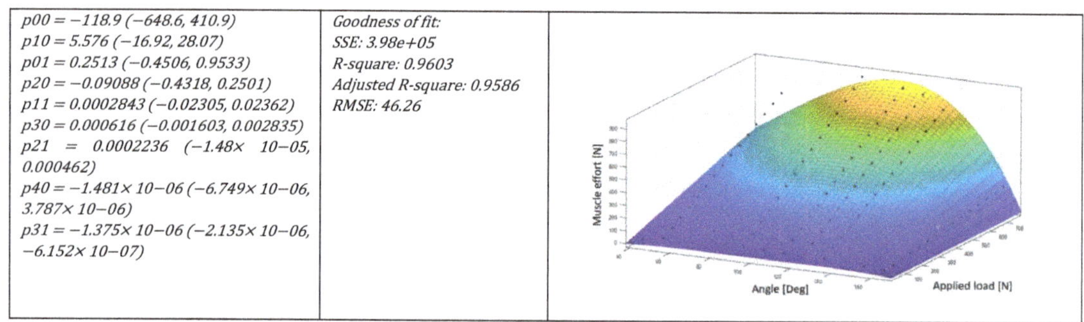

p00 = −118.9 (−648.6, 410.9) *p10 = 5.576 (−16.92, 28.07)* *p01 = 0.2513 (−0.4506, 0.9533)* *p20 = −0.09088 (−0.4318, 0.2501)* *p11 = 0.0002843 (−0.02305, 0.02362)* *p30 = 0.000616 (−0.001603, 0.002835)* *p21 = 0.0002236 (−1.48× 10−05, 0.000462)* *p40 = −1.481× 10−06 (−6.749× 10−06, 3.787× 10−06)* *p31 = −1.375× 10−06 (−2.135× 10−06, −6.152× 10−07)*	*Goodness of fit:* *SSE: 3.98e+05* *R-square: 0.9603* *Adjusted R-square: 0.9586* *RMSE: 46.26*

Figure 19. General formulation of the medium triceps efforts based on wrist load (W) and elbow angle (α) and surface reconstruction of the proposed polynomial.

4. Discussion

In this study, an analytical calculation method is proposed to obtain the forces in the flexor muscles (long and short biceps) and extensors (long, lateral, and middle triceps) in any configuration of the opening angle of the elbow joint and to any load applied to the wrist. The final objective of this study is to allow the creation of a new type of finite element that can be included in a musculoskeletal model that can be analyzed in different positions and configurations without the need to repeat the model meshing process.

There are several studies along the same lines. Therefore, Martinez [12] is focused on the analysis of a certain muscle in a certain load situation, analyzing what happens in the muscle itself. Islan et al. [14], are focused on the muscles involved in the shoulder joint, under static condition predicting the fatigue and cumulative damage to the muscles. Alonso et al. [20] study the forces in the muscles involved in the human march under certain load conditions, weight and step frequency. Park et al. [22] develop a predictive method for muscle forces based on electromyography techniques. Teo et al. [26] are focused on the effect of a collision on the neck muscles. As can be seen, none of them addresses the problem from our perspective. The use of the finite element method is widely recognized as a calculation tool for structural, static, and/or dynamic problems. However, characterization of the muscle, so that it is easy to include it in an FEM model has not been addressed until now.

The main drawbacks presented by the studies carried out so far, in order to include the muscle in a valid finite element model to analyze the behavior of a muscle within a joint in various positions of this, are as follows:

- Studies have focused on one joint position, obtaining results related to muscle effort that are difficult to extrapolate to different positions. Ramírez [12] obtains the results for the anterior tibial muscle under study conditions, but moving the results to other load situations should be difficult to solve
- The equations obtained to characterize the muscle are difficult to implement in a FEM model. Sachenkov et al. [29] use 1D elements to simulate muscles, but there is no single equation for any muscle to implement it in another analysis. The same happens with Alonso et al. [20], where the result is based on a static and physiological optimization that integrates the forces of any Hill MTU unit.
- Several of these studies use 3D elements to represent the muscle, which allows a better understanding of the internal behavior of the muscle itself (distribution of stresses and strains, deformations, etc.) but does not allow the created mesh to be reused in other positions of the joint. Therefore, Martínez [12] studies the anterior tibia muscle in a certain position, and if the muscle length changes, a new mesh should be developed. Another example is shown in research by Islan et al. [14], where they analyze the shoulder muscles of a violin player holding the instrument, and so, if the shoulder position changes, the mesh will do as well, remeshing the muscles into their new position.

With this work, and in a first approximation, we have obtained the equations that govern the behavior of the flexor and extensor muscles of the elbow, only in isometric contractions and referring to this behavior to its response in the form of force provided, throughout the flexion path of the joint for any load applied to the wrist.

For example, to calculate the estimated force that must appear in each of the joint muscles to support a load applied to the wrist of 235 N with the elbow forming an angle of 119 degrees, we need to substitute the values of α and w in Equation (12), depending on which muscle we want to analyze.

Changing the indicated values in Equation (12) would obtain the following estimated forces in the muscles.

Vertical downward load, strength in Biceps muscles.

- Long biceps: 477 N;
- Short biceps: 245 N.

Upward vertical load, strength in the triceps muscles.

- Long triceps: 438 N;
- Middle triceps: 231;
- Side triceps: 304.

In addition to what has been said, the equations obtained to characterize the muscles in their response are easy to implement mathematically, which will undoubtedly help in the future in the possibility of creating a new type of finite element that responds to these behavioral equations obtained.

With the creation of this new type of finite element to represent muscles, a very interesting possibility is opened to analyze the behavior of the joint in different positions, with variable loads, different load cycles, etc. notably reducing the study work since the meshing process of the model is eliminated by being able to reuse the mesh for the desired positions.

Regarding the analysis of the results obtained, it is worth mentioning the following important aspects of the study.

Applying the proposed methodology and using the finite element method, we can obtain the forces that must appear in the muscles involved in the elbow joint to keep it in balance, in any position between 44 and 164 degrees of opening, with a load applied to the wrist of 150 N (Table 2, Figures 13 and 14).

Through this same analysis it is possible to obtain the tensions that appear in the insertions of the tendons of the joint, which allows us to foresee damage before certain load situations (Figures 11 and 12).

This same analysis allows us to obtain a map of tensions in the rigid tissues of the joint, bones, and cartilage. This allows a detailed study of damage to the elements mentioned (Figures 11 and 12).

There are no previous studies that carry out studies similar to the proposed study, so a direct comparison with what has already been done is not possible. However, the trend observed in muscle effort during joint movement is similar to what other studies have published, [34], where the results of the torque in elbow vs. elbow angle shape are consistent with our results and [35] where the relation between the torque of the elbow and angle of the elbow is similar to our expressions, although these studies measure the moment in the joint (or what is the same, the sum of the moments generated by each muscle) and not the independent force in each muscle the results presented in these investigations are similar to those presented in this study in terms of the evolution of the moment in the joint, or of the force in the muscle in our case, along the angle of the joint.

The correlation also presents high R values, which confirms the validity of the analysis (Figures 13 and 14).

Through the application of classical mechanics and the balance of forces and moments in the joint, we can extrapolate the results obtained for a specific load, 150 N in our case, to different loads applied to the joint (Figures 15–19).

Finally, it should be mentioned that the equations that predict the force in each of the muscles in the joint are obtained as a function of the opening angle of the elbow and the applied load.

5. Conclusions

The use of a simplified model of the muscle, as a 1D element with axial forces only (ROD type element), into a FEA model allows one to calculate the forces that must be generated in the muscles involved in the joint.

It is also possible to predict the force required in each muscle for different load situations.

On the basis of the same analysis, the stresses that appear in other elements of the joint, bones, tendon insertions, or cartilage, can be obtained. With the obtained characteristic equations, different lines of research are opened focused on a better characterization of the muscles. It seems necessary to evolve the equations obtained in this study to obtain those that give the force of the muscle as a function of the speed of contraction. Or, based on studies conducted to assess muscle fatigue, they could be modified to predict it in a moving joint.

The evolution of this study should follow the following stages:

- Validation of the theoretical model obtained through testing with individuals with whom it is possible to measure the forces in cases of biceps and/or triceps load that, by means of comparison with this study, allows for knowing its validity.

- Modification of the equations obtained in this study in order to obtain their evolution for concentric or eccentric contractions.
- Creation of a new type of 1D element for application in finite element models that allows the behavior of a muscle to be characterized according to the equations obtained for any type of contraction, isometric, concentric, or eccentric.

Once these phases are completed, the work could be extrapolated to other types of muscle (flat, or peniform) in order to characterize more muscles that allow us to study new joints according to the proposed methodology.

One limitation of the presented study is the approximation of the muscle to the 1D element. This implies that following this methodology cannot obtain information regarding what happens inside the muscle.

Another one is that the analysis is performed for a certain size of elbow joint, and so the FEA model created is valid for this joint size; in any other case (children, older people, etc.), the FEA model should be resized, and the same methodology should be applied.

Author Contributions: Conceptualization, F.B.H. and J.A.J.M.; data curation, E.L.U., J.D.C.-M. and R.D.; formal analysis, E.L.U., F.B.H. and R.D.; investigation, E.L.U., F.B.H. and J.A.J.M.; methodology, F.B.H. and R.D.; project administration, F.B.H.; supervision, F.B.H., R.D. and J.A.J.M.; writing—original draft, E.L.U. and J.D.C.-M.; writing—review and editing, R.D. All authors have read and agreed to the published version of the manuscript.

Funding: This research received no external funding.

Institutional Review Board Statement: Not applicable.

Informed Consent Statement: Not applicable.

Data Availability Statement: Not applicable.

Conflicts of Interest: The authors declare no conflict of interest.

Appendix A

In this appendix, the sub-Figures included in Figures 11 and 12 are going to be collected with higher quality.

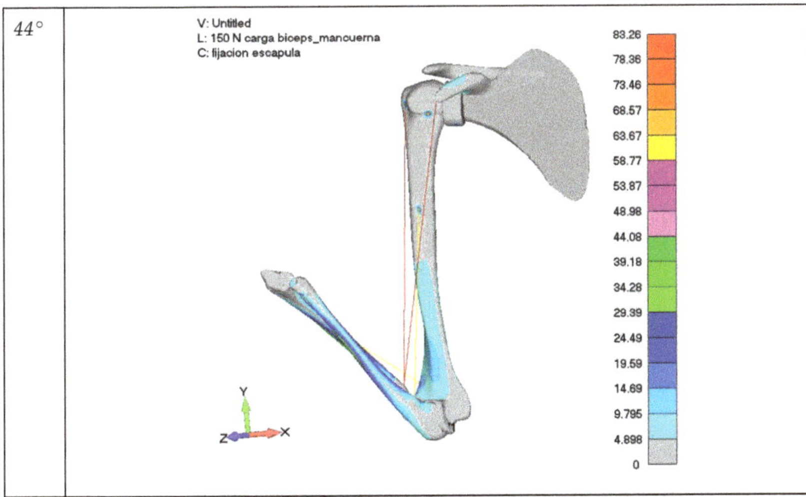

Figure A1. *Cont.*

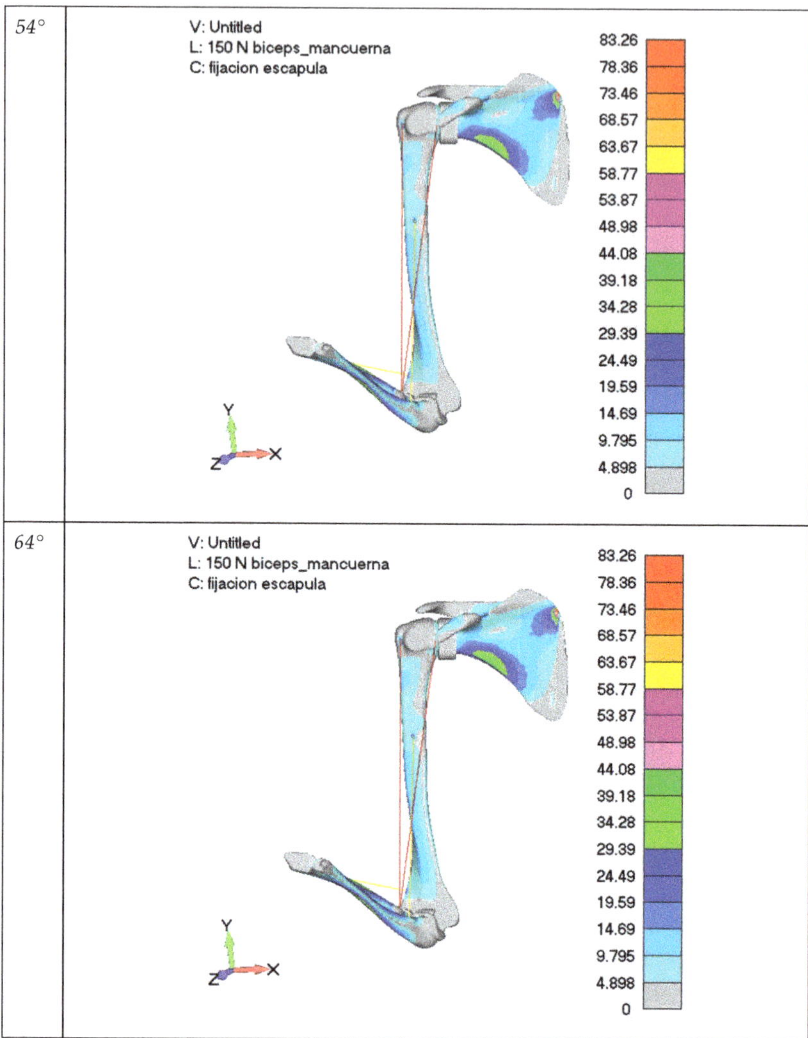

Figure A1. *Cont.*

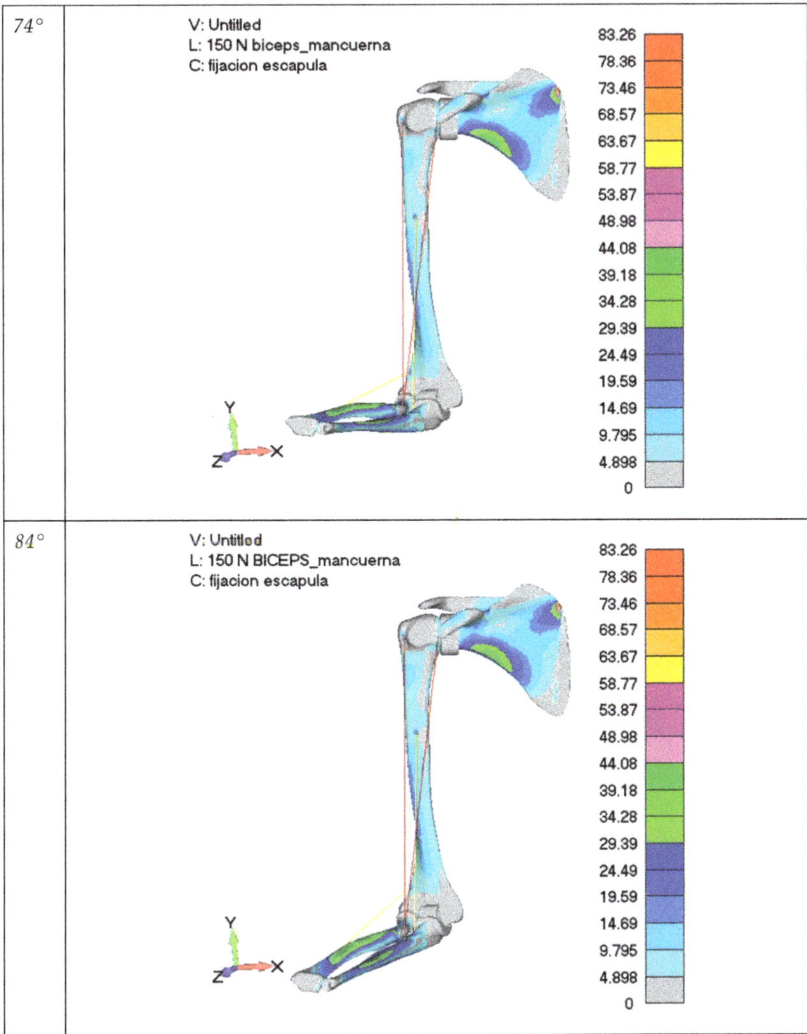

Figure A1. *Cont.*

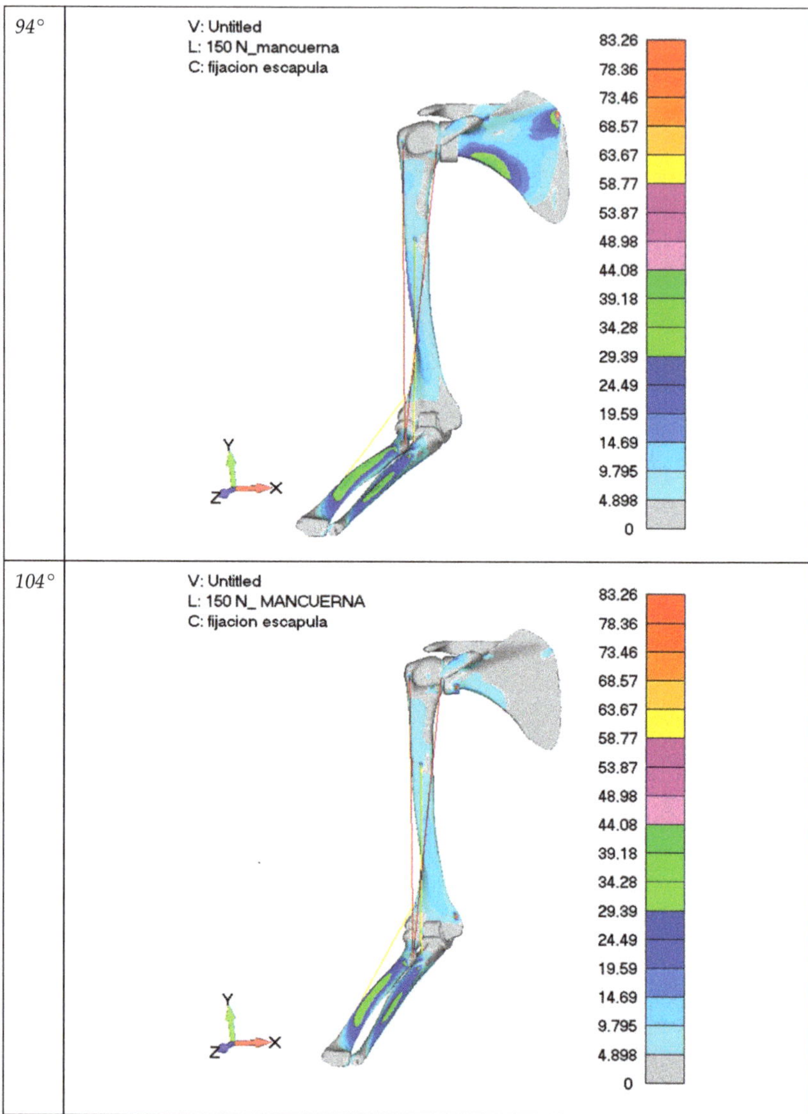

Figure A1. *Cont.*

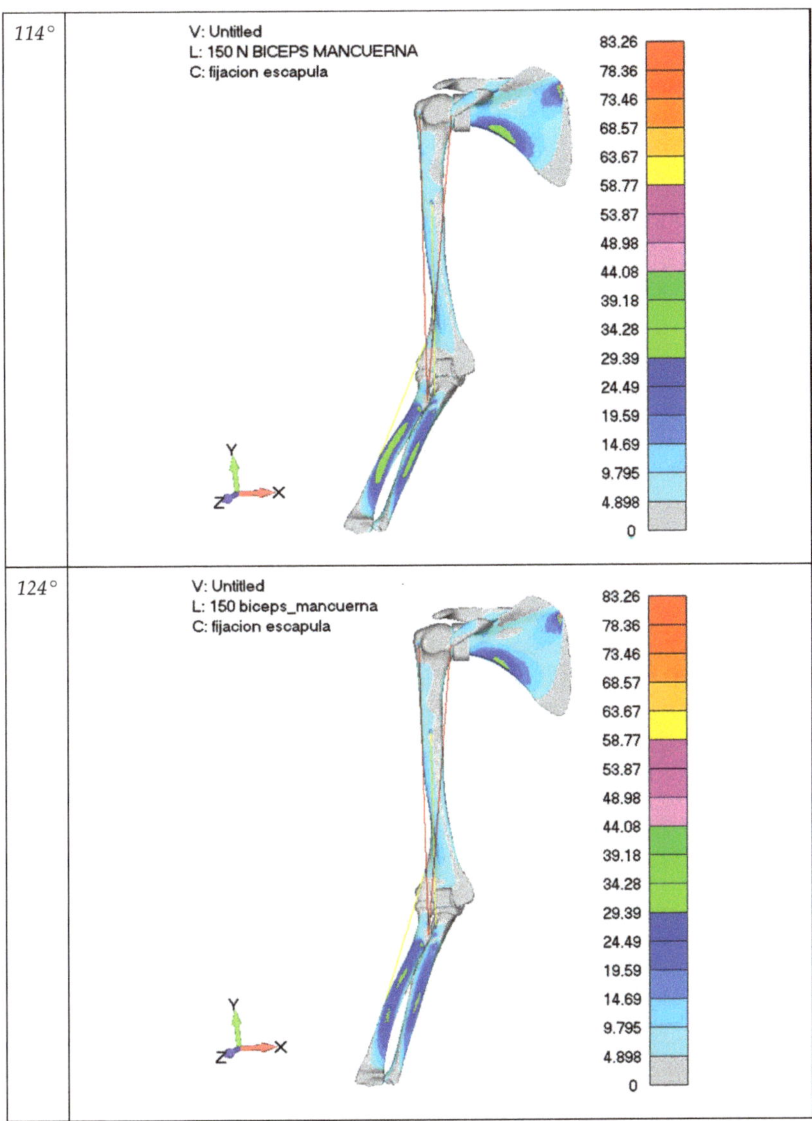

Figure A1. *Cont.*

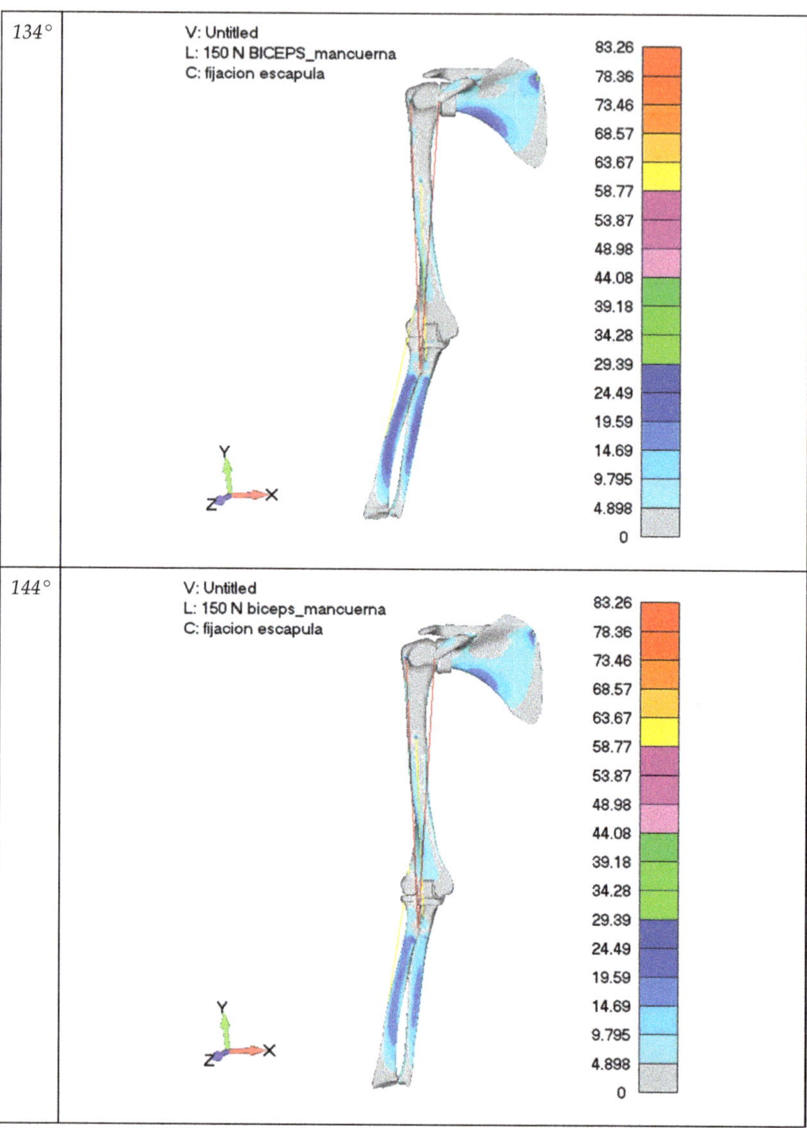

Figure A1. *Cont.*

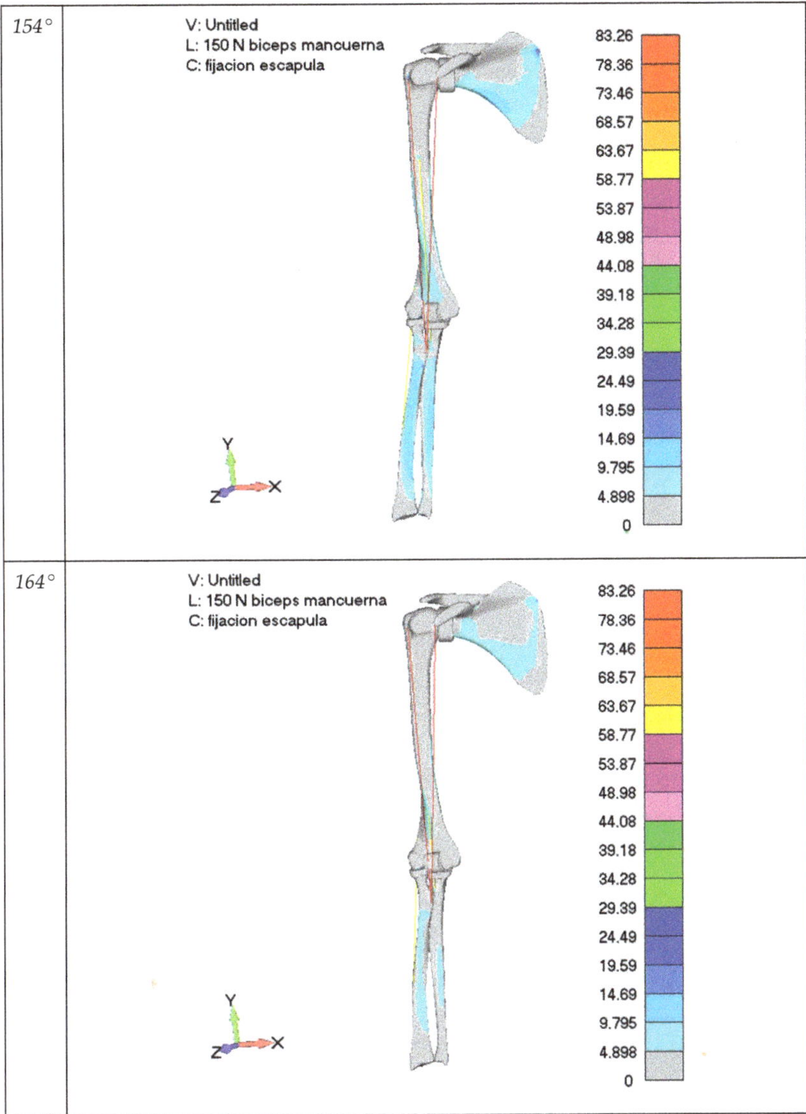

Figure A1. Different Elbow Angle Configuration: Bone Von Mises Stress Distribution Triceps Loading Case (150 N on wrist).

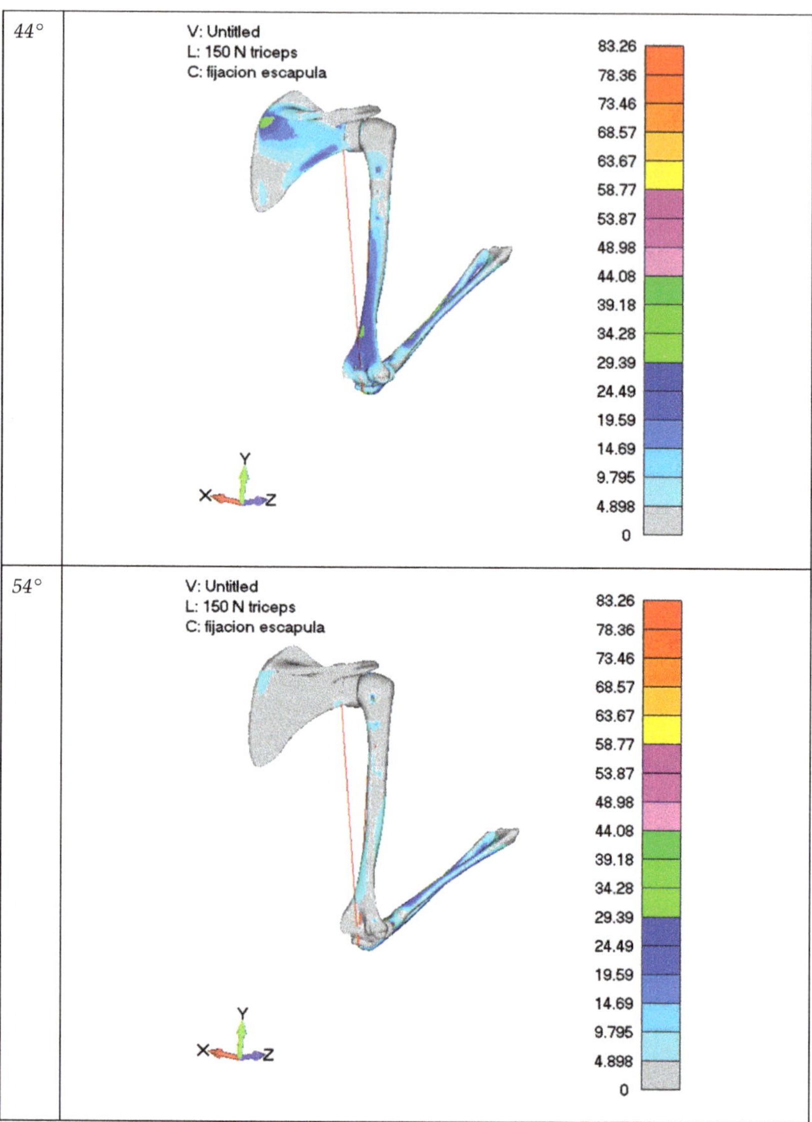

Figure A2. *Cont.*

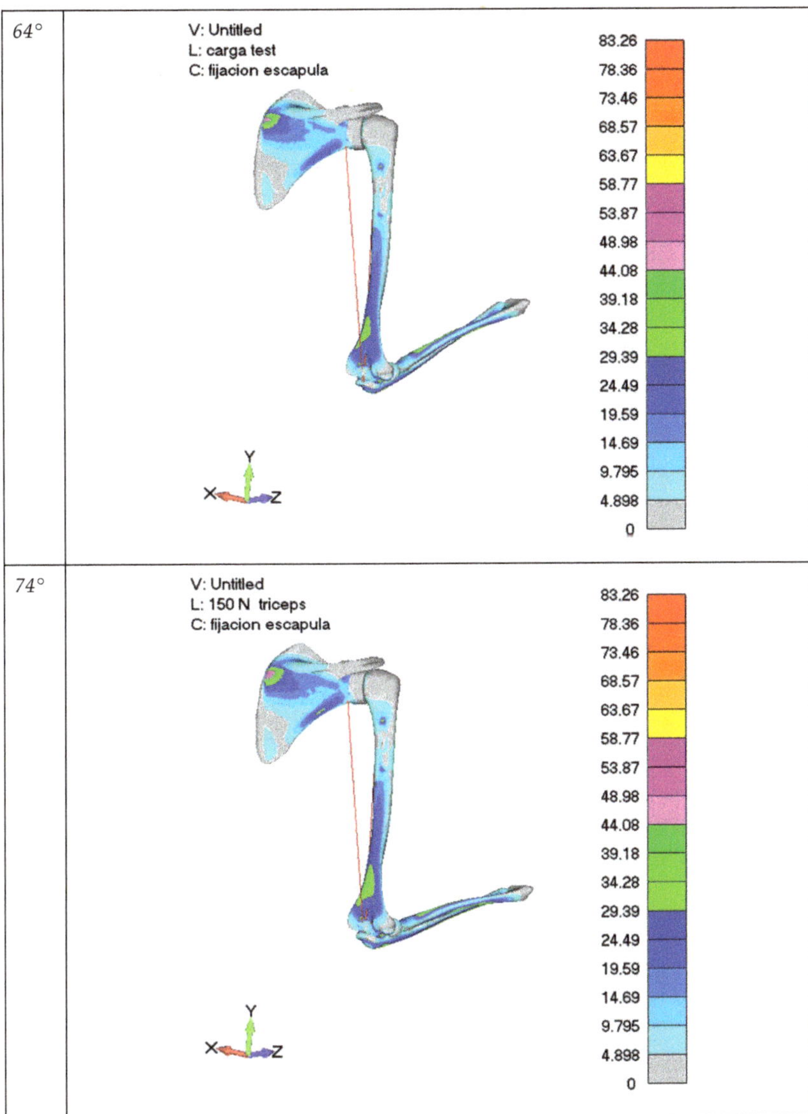

Figure A2. *Cont.*

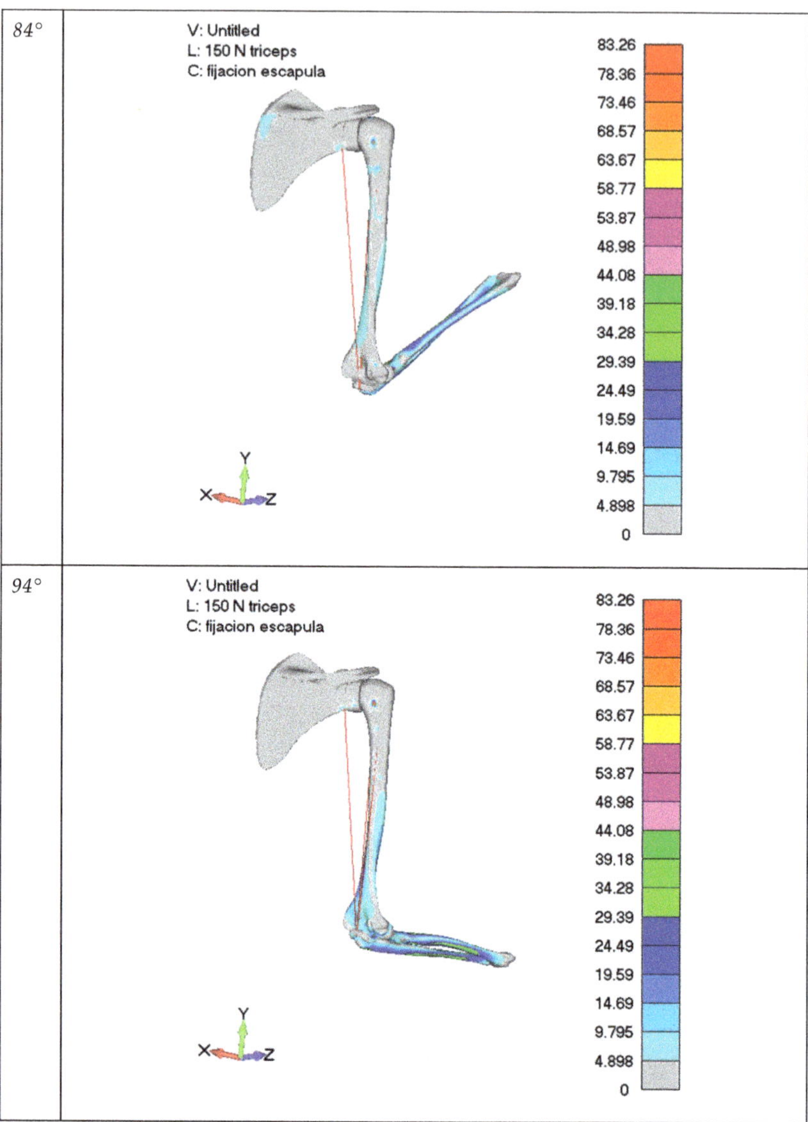

Figure A2. *Cont.*

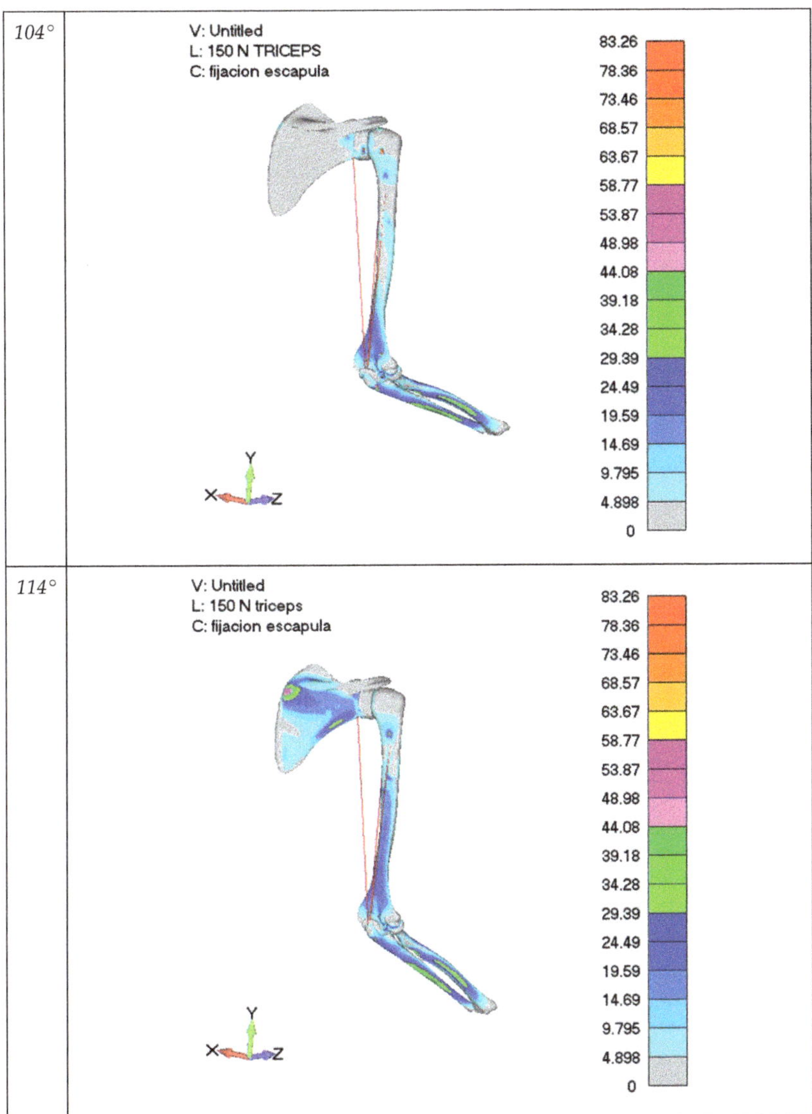

Figure A2. *Cont.*

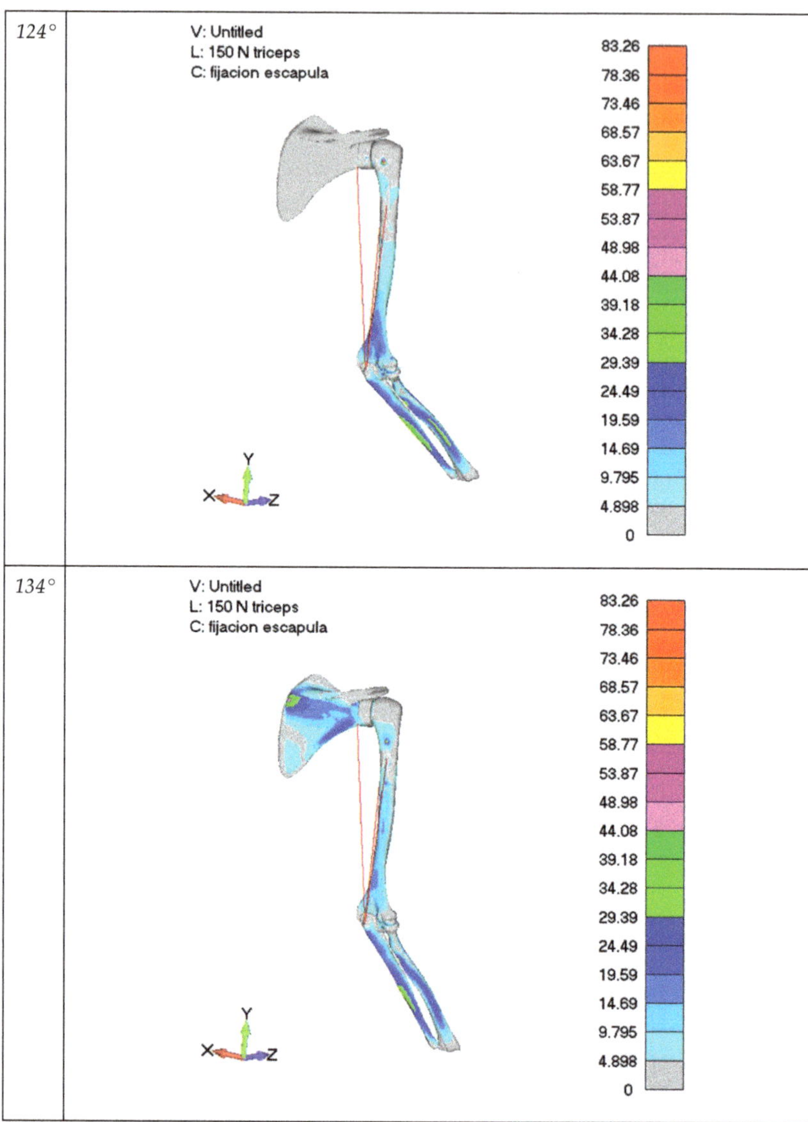

Figure A2. *Cont.*

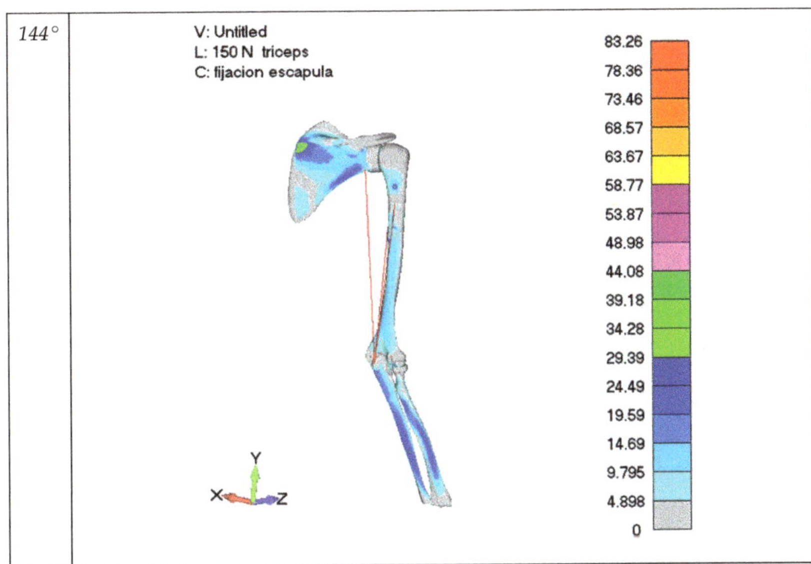

Figure A2. Different Elbow Angle Configuration: Bone Von Mises Stress Distribution Triceps Loading Case (150 N on wrist).

References

1. Strouboulis, T.; Babuška, I.; Whiteman, J.R. *The Finite Element Method and Its Reliability*; Clarendon Press: Oxford, UK, 2001; ISBN 978-0-19-850276-0.
2. Babuska, I.; Whiteman, J.; Strouboulis, T. *Finite Elements: An Introduction to the Method and Error Estimation*; OUP: Oxford, UK, 2010; ISBN 978-0-19-850669-0.
3. Pawełko, P.; Jastrzębski, D.; Parus, A.; Jastrzębska, J. A new measurement system to determine stiffness distribution in machine tool workspace. *Arch. Civ. Mech. Eng.* **2021**, *21*, 49. [CrossRef]
4. Wang, D.; Zhang, S.; Wang, L.; Liu, Y. Developing a Ball Screw Drive System of High-Speed Machine Tool Considering Dynamics. *IEEE Trans. Ind. Electron.* **2021**, *69*, 4966–4976. [CrossRef]
5. Li, Z.; Oger, G.; Le Touzé, D. A partitioned framework for coupling LBM and FEM through an implicit IBM allowing non-conforming time-steps: Application to fluid-structure interaction in biomechanics. *J. Comput. Phys.* **2021**, *449*, 110786. [CrossRef]
6. Della Rosa, N.; Bertozzi, N.; Adani, R. Biomechanics of external fixator of distal radius fracture, a new approach: Mutifix Wrist. *Musculoskelet. Surg.* **2020**, *106*, 89–97. [CrossRef] [PubMed]
7. Zhang, N.-Z.; Xiong, Q.-S.; Yao, J.; Liu, B.-L.; Zhang, M.; Cheng, C.-K. Biomechanical changes at the adjacent segments induced by a lordotic porous interbody fusion cage. *Comput. Biol. Med.* **2022**, *143*, 105320. [CrossRef] [PubMed]
8. Denozière, G.; Ku, D.N. Biomechanical Comparison between Fusion of Two Vertebrae and Implantation of an Artificial Intervertebral Disc. *J. Biomech.* **2006**, *39*, 766–775. [CrossRef] [PubMed]
9. Samani, A.; Bishop, J.; Yaffe, M.J.; Plewes, D.B. Biomechanical 3-D finite element modeling of the human breast using MRI data. *IEEE Trans. Med. Imaging* **2001**, *20*, 271–279. [CrossRef]
10. Jaecques, S.; Van Oosterwyck, H.; Muraru, L.; Van Cleynenbreugel, T.; De Smet, E.; Wevers, M.; Naert, I.; Sloten, J.V. Individualised, micro CT-based finite element modelling as a tool for biomechanical analysis related to tissue engineering of bone. *Biomaterials* **2003**, *25*, 1683–1696. [CrossRef]
11. Renner, S.M.; Natarajan, R.N.; Patwardhan, A.G.; Havey, R.M.; Voronov, L.I.; Guo, B.Y.; Andersson, G.B.; An, H.S. Novel model to analyze the effect of a large compressive follower pre-load on range of motions in a lumbar spine. *J. Biomech.* **2007**, *40*, 1326–1332. [CrossRef]
12. Martínez, A.M.R. Modelado y simulación del tejido músculo-esquelético. Validación Experimental con el Músculo Tibial Anterior de Rata. Ph.D. Thesis, Universidad de Zaragoza, Zaragoza, Spain, 2011. Available online: http://purl.org/dc/dcmitype/Text. (accessed on 5 November 2021).
13. Weiss, J.A.; Gardiner, J.C.; Ellis, B.J.; Lujan, T.J.; Phatak, N.S. Three-dimensional finite element modeling of ligaments: Technical aspects. *Med. Eng. Phys.* **2005**, *27*, 845–861. [CrossRef]
14. Islan, M.; Carvajal, J.; Pedro, P.S.; D'Amato, R.; Juanes, J.A.; Soriano, E. Linear Approximation of the Behavior of the Rotator Cuff under Fatigue Conditions. Violinist Case Study. In Proceedings of the ACM 5th International Conference on Technological Ecosystems for Enhancing Multiculturality, Cádiz, Spain, 18–20 October 2017; p. 58.

15. Sachenkov, O.A.; Hasanov, R.F.; Andreev, P.S.; Konoplev, Y.G. Numerical Study of Stress-Strain State of Pelvis at the Proximal Femur Rotation Osteotomy. *Russ. J. Biomech.* **2016**, *20*, 220–232. [CrossRef]
16. Martins, J.A.C.; Pato, M.P.M.; Pires, E.B. A finite element model of skeletal muscles. *Virtual Phys. Prototyp.* **2006**, *1*, 159–170. [CrossRef]
17. Tang, C.; Tsui, C.; Stojanovic, B.; Kojic, M. Finite element modelling of skeletal muscles coupled with fatigue. *Int. J. Mech. Sci.* **2007**, *49*, 1179–1191. [CrossRef]
18. Syomin, F.A.; Tsaturyan, A.K. Mechanical model of the left ventricle of the heart approximated by axisymmetric geometry. *Russ. J. Numer. Anal. Math. Model.* **2017**, *32*, 327–337. [CrossRef]
19. Perreault, E.J.; Sandercock, T.G.; Heckman, C.J. Hill Muscle Model Performance during Natural Activation and Electrical Stimulation. In Proceedings of the 23rd Annual International Conference of the IEEE Engineering in Medicine and Biology Society, Istanbul, Turkey, 25–28 October 2001; Volume 2, pp. 1248–1251.
20. Alonso, F.J.; Galán-Marín, G.; Salgado, D.R.; Pàmies Vilà, R.; Font Llagunes, J.M. Cálculo de Esfuerzos Musculares en la Marcha Humana Mediante Optimización Estática-Fisiológica. In Proceedings of the XVIII Congreso Nacional de Ingeniería Mecánica, Ciudad Real, Spain, 3–5 November 2010; pp. 1–9.
21. Holzbaur, K.R.S.; Murray, W.M.; Delp, S.L. A Model of the Upper Extremity for Simulating Musculoskeletal Surgery and Analyzing Neuromuscular Control. *Ann. Biomed. Eng.* **2005**, *33*, 829–840. [CrossRef]
22. Park, W.-I.; Lee, H.-D.; Kim, J. Estimation of isometric joint torque from muscle activation and length in intrinsic hand muscle. In Proceedings of the 2008 International Conference on Control, Automation and Systems, Seoul, Korea, 14–17 October 2008; pp. 2489–2493. [CrossRef]
23. Soechting, J.F.; Flanders, M. Evaluating an Integrated Musculoskeletal Model of the Human Arm. *J. Biomech. Eng.* **1997**, *119*, 93–102. [CrossRef]
24. Zajac, F.E. Muscle and Tendon: Properties, Models, Scaling, and Application to Biomechanics and Motor Control. *Crit. Rev. Biomed Eng.* **1989**, *17*, 359–411.
25. Lechosa Urquijo, E.; Blaya Haro, F.; D'Amato, R.; Juanes Méndez, J.A. Finite Element Model of an Elbow under Load, Muscle Effort Analysis When Modeled Using 1D Rod Element. In Proceedings of the Eighth International Conference on Technological Ecosystems for Enhancing Multiculturality, Salamanca, Spain, 21–23 October 2020; Association for Computing Machinery: New York, NY, USA, 21 October 2020; pp. 475–482.
26. Teo, E.C.; Zhang, Q.H.; Qiu, T.X. Finite Element Analysis of Head-Neck Kinematics Under Rear-End Impact Conditions. In Proceedings of the 2006 International Conference on Biomedical and Pharmaceutical Engineering, Singapore, 11–14 December 2006; pp. 206–209.
27. Donahue, T.L.H.; Hull, M.L.; Rashid, M.M.; Jacobs, C.R. A Finite Element Model of the Human Knee Joint for the Study of Tibio-Femoral Contact. *J. Biomech. Eng.* **2002**, *124*, 273–280. [CrossRef]
28. Abidin, N.A.Z.; Kadir, M.R.A.; Ramlee, M.H. Three Dimensional Finite Element Modelling and Analysis of Human Knee Joint-Model Verification. *J. Phys. Conf. Ser.* **2019**, *1372*, 012068. [CrossRef]
29. Sachenkov, O.A.; Hasanov, R.; Andreev, P.; Konoplev, Y. Determination of Muscle Effort at the Proximal Femur Rotation Osteotomy. *IOP Conf. Series: Mater. Sci. Eng.* **2016**, *158*, 012079. [CrossRef]
30. Jesal, N. Parekh Using Finite Element Methods to Study Anterior Cruciate Ligament Injuries: Understanding the Role of ACL Modulus and Tibial Surface Geometry on ACL Loading. Ph.D. Thesis, The University of Michigan, Ann Arbor, MI, USA, 2013.
31. CES EduPack Bulletin: January 2017. Available online: https://www.grantadesign.com/newsletters/ces-edupack-bulletin-ces-edupack-2017-new-products-database-symposia-deadlines-shared-resources-webinars-and-more/ (accessed on 20 June 2022).
32. Bruno, S.; José, M.; Filomena, S.; Vítor, C.; Demétrio, M.; Karolina, B. The Conceptual Design of a Mechatronic System to Handle Bedridden Elderly Individuals. *Sensors* **2016**, *16*, 725. [CrossRef] [PubMed]
33. Arcila Arango, J.C.; Cardona Nieto, D.; Giraldo, J.C. Abordaje Físico-Matemático Del Gesto Articular. Available online: https://www.efdeportes.com/efd171/abordaje-fisico-matematico-del-gesto-articular.htm (accessed on 5 November 2021).
34. Loss, J.F.; Candotti, C.T. Comparative Study between Two Elbow Flexion Exercises Using the Estimated Resultant Muscle Force. *Braz. J. Phys. Ther.* **2008**, *12*, 502–510. [CrossRef]
35. Murray, W.M.; Delp, S.L.; Buchanan, T.S. Variation of Muscle Moment Arms with Elbow and Forearm Position. *J. Biomech.* **1995**, *28*, 513–525. [CrossRef]

 *mathematics*

Article

Miura-Ori Inspired Smooth Sheet Attachments for Zipper-Coupled Tubes

Dylan C. Webb [1], Elissa Reynolds [1], Denise M. Halverson [1,*] and Larry L. Howell [2]

1 Department of Mathematics, Brigham Young University, Provo, UT 84602, USA; dylancw2@byu.edu (D.C.W.); elissar2@byu.edu (E.R.)
2 Department of Mechanical Engineering, Brigham Young University, Provo, UT 84602, USA; lhowell@byu.edu
* Correspondence: halverson@math.byu.edu

Abstract: Zipper-coupled tubes are a broadly applicable, deployable mechanism with an angular surface that can be smoothed by attaching an additional smooth sheet pattern. The existing design for the smooth sheet attachment, however, leaves small gaps that can only be covered by adding flaps that unfold separately, limiting applicability in situations requiring a seamless surface and simultaneous deployment. We provide a novel construction of the smooth sheet attachment that unfolds simultaneously with zipper-coupled tubes to cover the entire surface without requiring additional actuation and without inhibiting the tubes' motion up to an ideal, unfolded state of stability. Furthermore, we highlight the mathematics underlying the design and motion of the new smooth sheet pattern, thereby demonstrating its rigid-foldability and compatibility with asymmetric zipper-coupled tubes.

Keywords: zipper-coupled tubes; Miura-ori pattern; deployable mechanism; origami inspired design; smooth sheet attachment

MSC: 74-10; 51E24

Citation: Webb, D.C.; Reynolds, E.; Halverson, D.M.; Howell, L.L. Miura-Ori Inspired Smooth Sheet Attachments for Zipper-Coupled Tubes. *Mathematics* **2022**, *10*, 2643. https://doi.org/10.3390/math10152643

Academic Editors: Higinio Rubio Alonso, Alejandro Bustos, Jesus Meneses Alonso and Enrique Soriano-Heras

Received: 29 June 2022
Accepted: 24 July 2022
Published: 28 July 2022

Publisher's Note: MDPI stays neutral with regard to jurisdictional claims in published maps and institutional affiliations.

1. Introduction

Origami is the basis for many deployable mechanisms, including self-scaling, modular robots [1], satellite reflectarray antennas that pack efficiently [2], and multimodal biomedical devices that actuate electromagnetically [3]. Zipper-coupled tubes are multistable origami structures that fold up compactly and unfold bidirectionally to fill space and resist compression [4]. An asymmetric generalization of zipper-coupled tubes with smooth sheet attachments was introduced previously [5]. Together, these origami-based mechanisms form a deployable device with a smooth surface that is advantageous in applications, such as prefab architecture, when drivability and walkability are important, and in smooth medical devices, when sharp edges could harm the body. The smooth sheet attachment we presented previously, however, does not fully cover the surface of zipper-coupled tubes without additional flaps that actuate separately [5]. We rectify this problem by offering an alternative, Miura-ori inspired [6–8], construction of a smooth sheet attachment that fully covers the surface of, and deploys simultaneously with, asymmetric zipper-coupled tubes.

In this paper we briefly review the design of asymmetric zipper-coupled tubes and then demonstrate how to (1) construct a Miura-ori inspired smooth sheet attachment without gaps that attaches to the mountain folds of asymmetric zipper-coupled tubes and (2) handle design variations in the symmetric case. We present a mathematically robust design method, decomposing the attachment into pairs of compatible, tessellating cells inspired by the Miura-ori pattern and then defining the cells' vertices throughout the folding motion, thereby confirming rigid-foldability of the smooth sheet attachment and highlighting the mathematical processes involved in mechanism design. Our Miura-ori-based smooth sheet attachment expands the utility of asymmetric zipper-coupled tubes;

combined, these origami-inspired mechanisms are ideal for applications requiring a rigidly deployable structure with a smooth surface.

2. Zipper-Coupled Tubes Review

Zipper-coupled tubes are an origami pattern consisting of two or more deployable tubes coupled together in a zipper fashion that makes their motion compatible. Designed by Filipov et al., this structure is remarkable for its ability to deploy from a flat, stowable form into a stable, space-filling structure with only one degree of freedom [4]. In this section we review briefly the construction of asymmetric zipper-coupled tubes from two tube segments that are each, in turn, composed of two degree-four vertex cells; the interested reader may refer to our previous work for a thorough treatment of the construction [5].

Throughout this paper we use the notational convention that an arbitrary vector **x** has unit length direction vector $\hat{x}$ with length x, and hence

$$\mathbf{x} = x\hat{x}$$

Thus, the notation $\hat{x}$ always represents a unit length vector associated with a vector denoted **x**, where $\mathbf{x} = x\hat{x}$.

The building blocks for zipper-coupled tubes are the compatible degree-four vertex cells $\mathcal{C}_1$ and $\mathcal{C}_2$ illustrated in Figures 1 and 2. When combined, they form a single tube segment [5]. To couple correctly and satisfy rigid and flat-foldability [9], the design constraints of these degree-four vertex cells include:

$$\alpha_3 = \pi - \alpha_1 \qquad \alpha_1 + \alpha_3 = \alpha_2 + \alpha_4 \qquad \alpha_1 < \alpha_2$$
$$\alpha_4 = \pi - \alpha_2 \qquad d \sin a_4 = b \sin a_1 \qquad \alpha_1 + \alpha_2 \leq \pi$$

If $\alpha_1 + \alpha_2 < \pi$, the zipper-coupled tubes are called *asymmetric*, having a characteristic tilt and a customizable unfolding motion. In the special case that $\alpha_1 + \alpha_2 = \pi$ and $c = a$, the two cells in Figure 1 are congruent, symmetric Miura-ori cells and the resulting zipper-coupled tubes are those constructed by Filipov et al. [4].

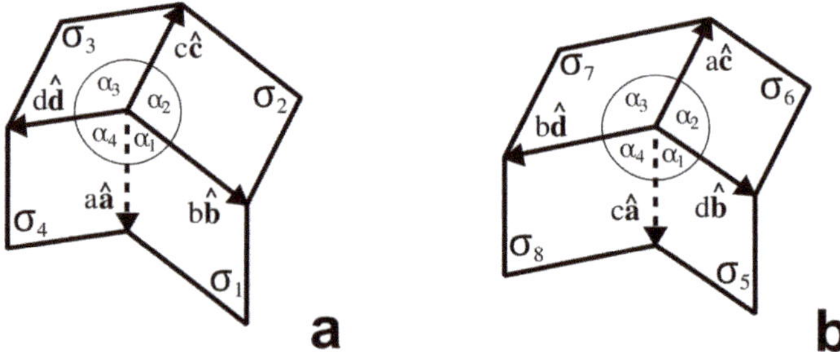

Figure 1. (**a**) Vectors defining the basic asymmetric degree-four vertex cell, $\mathcal{C}_1$. (**b**) Vectors defining the complementary degree-four vertex cell, $\mathcal{C}_2$. (Adapted from [5]).

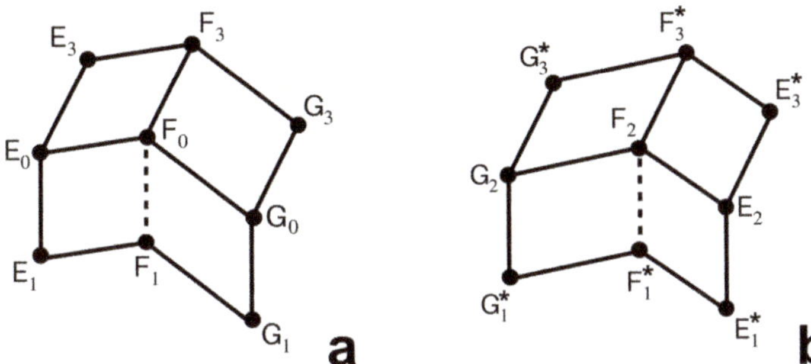

Figure 2. (**a**) Vertices of the basic asymmetric degree-four vertex cell, $\mathcal{C}_1$. (**b**) Vertices of the complementary degree-four vertex cell, $\mathcal{C}_2$. The points X_i^* identify with X_i. (Adapted from [5]).

Like a Miura-ori cell, the basic degree-four vertex cell has one degree of freedom in its motion. Let the basic and complementary cells lie flat in the xz-plane when unfolded, as depicted in Figures 1 and 2, and let the angles between the xz-plane and panels σ_1 and σ_4, respectively, be equal as σ_1 and σ_4 fold toward each other (see Figure 1a). Call this motion parameter γ. Then, placing F_0 at the origin and fixing F_1 on the negative z-axis in $\mathbb{R}^3$ (see Figure 2a), the motion of the basic cell is determined by the following vector paths:

$$\hat{\mathbf{a}}(\gamma) = [0,\, 0,\, -1]^T$$

$$\hat{\mathbf{b}}(\gamma) = [-\sin\alpha_1 \cos\gamma,\ \sin\alpha_1 \sin\gamma,\ -\cos\alpha_1]^T$$

$$\hat{\mathbf{c}}(\gamma) = \left[\frac{a_x c_3 + b_x}{k\cos\gamma},\ \frac{a_y c_3 + b_y}{k\sin\gamma},\ -\frac{a_x b_x \sin^2\gamma + a_y b_y \cos^2\gamma + k^2 \sin^2\gamma \cos^2\gamma}{a_x^2 \sin^2\gamma + a_y^2 \cos^2\gamma + k^2 \sin^2\gamma \cos^2\gamma} \right]^T$$

$$\hat{\mathbf{d}}(\gamma) = [\sin\alpha_4 \cos\gamma,\ \sin\alpha_4 \sin\gamma,\ -\cos\alpha_4]^T$$

where

$$a_x = \sin\alpha_1 \cos\alpha_4 - \sin\alpha_4 \cos\alpha_1 \qquad\qquad b_x = \sin\alpha_1 \cos\alpha_3 - \sin\alpha_4 \cos\alpha_2$$

$$a_y = \sin\alpha_1 \cos\alpha_4 + \sin\alpha_4 \cos\alpha_1 \qquad\qquad b_y = \sin\alpha_1 \cos\alpha_3 + \sin\alpha_4 \cos\alpha_2$$

$$k = 2\sin\alpha_1 \sin\alpha_4 \qquad\qquad\qquad\qquad \hat{\mathbf{c}} = \langle c_1, c_2, c_3 \rangle$$

The same vectors define the motion of both basic and complementary cells. By combining one basic cell ($\mathcal{C}_1$) and one complementary cell ($\mathcal{C}_2$), we obtain the first tube segment in a zipper-coupled pair, illustrated in Figure 3a. Its vertices, identified with their corresponding position vectors, are given by:

$$
\begin{aligned}
E_0(\gamma) &= \mathbf{d}(\gamma) & F_0(\gamma) &= \mathbf{0} & G_0(\gamma) &= \mathbf{b}(\gamma) \\
E_1(\gamma) &= \mathbf{a}(\gamma) + \mathbf{d}(\gamma) & F_1(\gamma) &= \mathbf{a}(\gamma) & G_1(\gamma) &= \mathbf{a}(\gamma) + \mathbf{b}(\gamma) \\
E_2(\gamma) &= \mathbf{a}(\gamma) + \mathbf{c}(\gamma) + \mathbf{d}(\gamma) & F_2(\gamma) &= \mathbf{a}(\gamma) + \mathbf{c}(\gamma) & G_2(\gamma) &= \mathbf{a}(\gamma) + \mathbf{b}(\gamma) + \mathbf{c}(\gamma) \\
E_3(\gamma) &= \mathbf{c}(\gamma) + \mathbf{d}(\gamma) & F_3(\gamma) &= \mathbf{c}(\gamma) & G_3(\gamma) &= \mathbf{b}(\gamma) + \mathbf{c}(\gamma)
\end{aligned}
$$

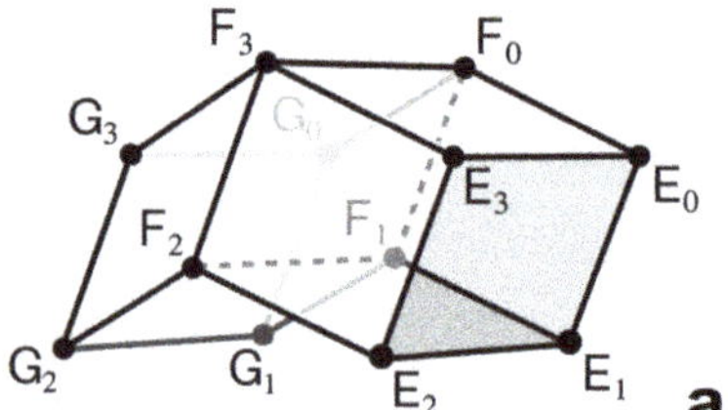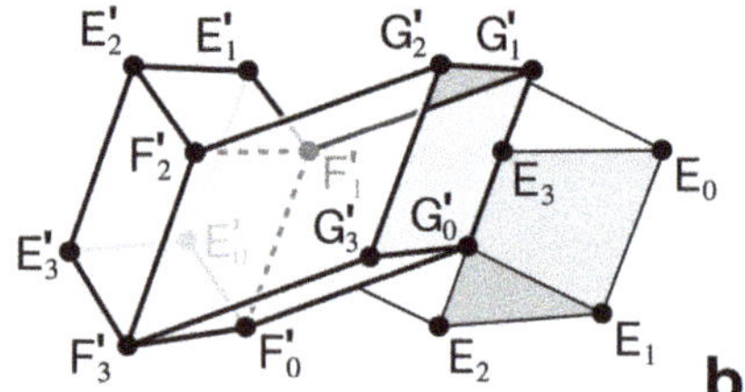

Figure 3. (**a**) The vertices in the first origami tube segment in the construction of an asymmetric zipper-coupled tube segments pair. (**b**) The vertices in the second origami tube segment in the construction of an asymmetric zipper-coupled tube segments pair. (Adapted from [5]).

The second tube segment is a copy of the first, but rotated 180° about the y-axis and then shifted by an offset vector so that it attaches to the first tube segment along the vertical creases (illustrated in Figure 3b). Let C_3 and C_4 denote the copies of C_1 and C_2, respectively, comprising the second tube segment. To define the motion of the second tube segment, let $\bar{\mathbf{x}}$ denote the 180° rotation of a vector $\mathbf{x}$ about the y-axis and define the offset vector:

$$\mathbf{s}(\gamma) = \left(1 + \frac{b\cos\alpha_1 + d\cos\alpha_4}{2a}\right)\mathbf{a}(\gamma) + \mathbf{c}(\gamma)$$

Then the vertices of the second tube segment are defined by:

$$E_0'(\gamma) = \bar{\mathbf{d}}(\gamma) + \mathbf{s}(\gamma) \qquad F_0'(\gamma) = \mathbf{s}(\gamma) \qquad G_0'(\gamma) = \bar{\mathbf{b}}(\gamma) + \mathbf{s}(\gamma)$$
$$E_1'(\gamma) = \bar{\mathbf{a}}(\gamma) + \bar{\mathbf{d}}(\gamma) + \mathbf{s}(\gamma) \qquad F_1'(\gamma) = \bar{\mathbf{a}}(\gamma) + \mathbf{s}(\gamma) \qquad G_1'(\gamma) = \bar{\mathbf{a}}(\gamma) + \bar{\mathbf{b}}(\gamma) + \mathbf{s}(\gamma)$$
$$E_2'(\gamma) = \bar{\mathbf{a}}(\gamma) + \bar{\mathbf{c}}(\gamma) + \bar{\mathbf{d}}(\gamma) + \mathbf{s}(\gamma) \qquad F_2'(\gamma) = \bar{\mathbf{a}}(\gamma) + \bar{\mathbf{c}}(\gamma) + \mathbf{s}(\gamma) \qquad G_2'(\gamma) = \bar{\mathbf{a}}(\gamma) + \bar{\mathbf{b}}(\gamma) + \bar{\mathbf{c}}(\gamma) + \mathbf{s}(\gamma)$$
$$E_3'(\gamma) = \bar{\mathbf{c}}(\gamma) + \bar{\mathbf{d}}(\gamma) + \mathbf{s}(\gamma) \qquad F_3'(\gamma) = \bar{\mathbf{c}}(\gamma) + \mathbf{s}(\gamma) \qquad G_3'(\gamma) = \bar{\mathbf{b}}(\gamma) + \bar{\mathbf{c}}(\gamma) + \mathbf{s}(\gamma)$$

The pair of tube segments, with vertices positioned as indicated above, form a single component in a pair of zipper-coupled tubes (Figure 3b), which will be denoted as $\mathcal{Z}_0$. The zipper-coupled tubes can be extended by taking multiple copies of $\mathcal{Z}_0$ and attaching them end-to-end. In particular, for $i = 1\dots n$, let

$$\mathcal{Z}_i = \mathcal{Z}_0 + i(\mathbf{d} - \mathbf{b})$$

Then

$$\mathcal{Z} = \bigcup_{i=0}^{n} \mathcal{Z}_i$$

denotes zipper-coupled tubes with $n+1$ components.

Of critical importance, at a certain point in the motion of $\mathcal{Z}$, the upper (or lower) creases simultaneously become coplanar, as bolded in Figure 4b. This state is the *ideal state*, and it occurs at a parameter value:

$$\gamma_0 = \cos^{-1}\left(\sqrt{\frac{a_y b_y - a_x^2 + k^2 + \sqrt{\left(a_y b_y + a_x^2 + k^2\right)^2 - 4a_y^2 a_x b_x}}{2k^2}}\right)$$

The value γ_0 will be the terminal value for the deployment of $\mathcal{Z}$ with a smooth sheet attachment. Thus, by construction, the smooth sheet attachment lies flat on the surface of the zipper-coupled tubes when $\gamma = \gamma_0$ and folds up with the zipper-coupled tubes until $\gamma = \pi/2$, at which point the entire structure lies in a plane and has no volume.

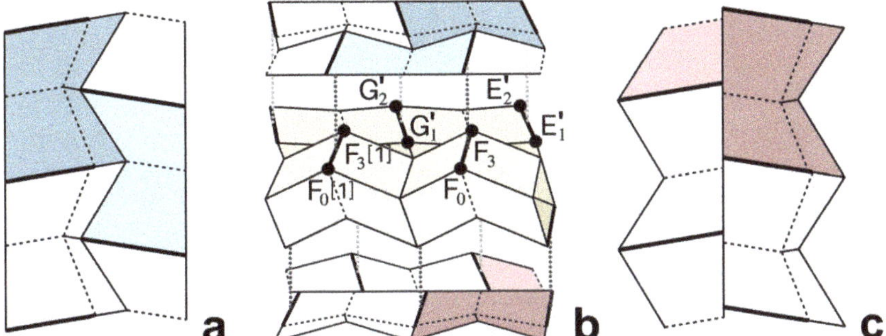

Figure 4. (**a**) The top smooth sheet attachment ($\mathcal{S}_1$ is lightly shaded, $\mathcal{S}_4$ is darkly shaded, and co-planar ridges are bolded). (**b**) Adding the top and bottom smooth sheet attachments to asymmetric zipper-coupled tubes (co-planar ridges are bolded). (**c**) The bottom smooth sheet attachment ($\mathcal{S}_3$ is lightly shaded, $\mathcal{S}_2$ is darkly shaded, co-planar ridges are bolded).

3. Asymmetric Smooth Sheet Attachment

In this section we design a Miura-ori inspired smooth sheet attachment that actuates simultaneously with the asymmetric zipper-coupled tubes pattern, folding up flat and unfolding into a rigid sheet without gaps in the ideal state. This additional pattern broadens the potential applications of the zipper-coupled tubes structure and expands on the design of the attachment described previously [5]. As with the zipper-coupled tubes, we decomposed the overall structure into its basic units: four distinct smooth sheet cells that tesselate. We define the vectors used to construct each cell to describe the motion of the cell's vertices and confirm its compatibility with the zipper-coupled tubes structure.

The seamless smooth sheet attachment consists of four distinct cells denoted $\mathcal{S}_i$, for $i = 1, 2, 3, 4$. In Figure 4a, the lightly shaded cell is $\mathcal{S}_1$ and the darkly shaded cell is $\mathcal{S}_4$, while in Figure 4c, the lightly shaded cell is half of $\mathcal{S}_3$ and the darkly shaded cell is $\mathcal{S}_2$. Cells $\mathcal{S}_1$ and $\mathcal{S}_3$ share the same configuration, but like the zipper-coupled tube segments, one is rotated $180°$ about the y-axis; the same is true of $\mathcal{S}_2$ and $\mathcal{S}_4$. The manner in which the smooth sheet cells attach to $\mathcal{Z}$ in the ideal state ($\gamma = \gamma_0$) is highlighted in Figure 4b. When the mechanism is fully deployed, cells on the top sheet meet along their jagged edges, as seen in Figure 4a. The cells on the bottom sheet, however, meet along their straight edges, as seen in Figure 4c—this is a key feature in defining a design that will cover zipper-coupled tubes without gaps and deploy without restricting the tubes' motion up to the ideal state.

Regarding the single component $\mathcal{Z}_0$, the cell $\mathcal{S}_i$ attaches to $\mathcal{C}_i$. However, whereas the cells $\mathcal{S}_2$ and $\mathcal{S}_4$ form a bridge between a pair of creases in $\mathcal{Z}_0$, the cells $\mathcal{S}_1$ and $\mathcal{S}_3$ form a bridge between a crease of $\mathcal{Z}_0$ and a crease of $\mathcal{Z}_1$ (bolded in Figure 4b). Upon careful examination of the zipper-coupled tubes, we observed that $\mathcal{S}_1$ can be designed as though it bridged two creases of $\mathcal{C}_4$ and then moved to bridge creases of the copies of $\mathcal{C}_1$ in $\mathcal{Z}_0$ and $\mathcal{Z}_1$; this strategy eliminates some complexity in defining $\mathcal{S}_1$. In contrast, $\mathcal{S}_4$ is designed directly, transversing the two creases of $\mathcal{C}_4$ where it attaches. Designing both cells atop $\mathcal{C}_4$ allows us to re-use the same vectors and enables the interested reader to easily compare the Miura-ori inspired cells and the original smooth sheet cells with gaps [5].

3.1. Design of $\mathcal{S}_1$

The first smooth sheet cell, $\mathcal{S}_1$, has two panels that fold toward each other as the zipper-coupled tubes fold up. We first discuss the design of $\mathcal{S}_1$ when built atop $\mathcal{C}_4$—the desired relation between $\mathcal{S}_1$ and $\mathcal{C}_4$ is shown in Figure 5a. The vectors that define $\mathcal{S}_1$ in this temporary configuration will be re-used to define $\mathcal{S}_4$, allowing the edges of these two cells to mesh when placed opposite each other. After construction, the smooth sheet cell $\mathcal{S}_1$ will be moved to attach to $\mathcal{C}_1$, as seen in Figure 5b.

In preparation of the design of $\mathcal{S}_1$, let the displacement between $G_2'(\gamma)$ and $E_2'(\gamma)$ be described by

$$\bar{\mathbf{q}}(\gamma) = \bar{\mathbf{d}}(\gamma) - \bar{\mathbf{b}}(\gamma)$$

and

$$\bar{\mathbf{q}}_\perp(\gamma) = \bar{\mathbf{q}}(\gamma) - (\bar{\mathbf{q}}(\gamma) \cdot \hat{\bar{\mathbf{c}}}(\gamma))\hat{\bar{\mathbf{c}}}(\gamma)$$

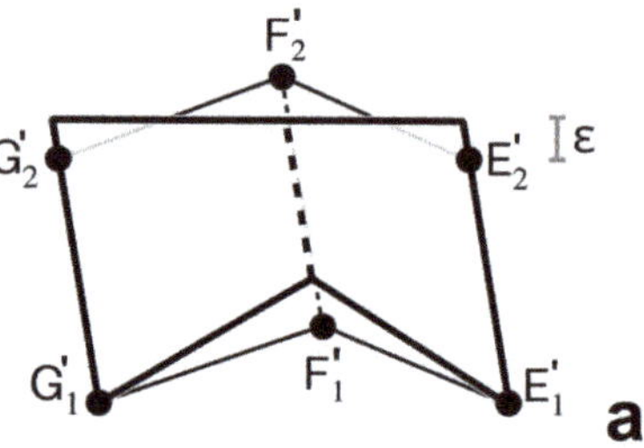
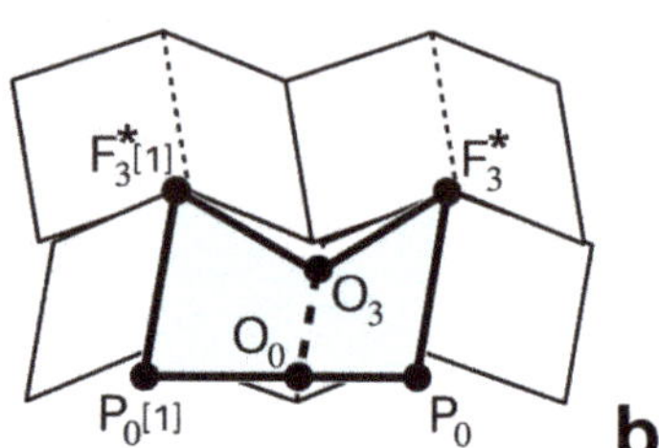

Figure 5. (a) Placing $\mathcal{S}_1$ on $\mathcal{C}_4$. (b) Attaching $\mathcal{S}_1$ to asymmetric zipper-coupled tubes.

As depicted in Figure 6a,b, $\bar{\mathbf{q}}_\perp(\gamma)$ is the component vector of $\bar{\mathbf{q}}(\gamma)$ orthogonal to $\bar{\mathbf{c}}(\gamma)$, the displacement between $E_1'(\gamma)$ and $E_2'(\gamma)$ (or equivalently, the displacement between $G_1'(\gamma)$ and $G_2'(\gamma)$). In the ideal state of the zipper-coupled tubes, the distance between the ridges of $\mathcal{Z}$ on which the smooth sheet will be attached is

$$\Delta = \|\bar{\mathbf{q}}_\perp(\gamma_0)\|$$

Figure 6. (a) Vectors in $\mathcal{S}_1$. (b) Vector components of $\bar{\mathbf{u}}$ and $\bar{\mathbf{v}}$. (c) Relation between h and w in the ideal state.

Note that, of necessity and by design, the top edges of both $\mathcal{S}_1$ and $\mathcal{S}_4$ are parallel with $\bar{\mathbf{q}}(\gamma_0)$. To remove gaps between zipper-coupled tubes stacked laterally to $\mathcal{Z}_0$ and in anticipation of the behavior depicted in Figure 4c, we extended the side edges of $\mathcal{S}_1$ in the direction of $\bar{\mathbf{c}}$ by the length ε, as shown in Figures 5a and 6a, where

$$\varepsilon = \frac{b \sin \alpha_1 \sin \gamma_0}{2|c_2(\gamma_0)|}$$

Toward this end, let

$$\bar{\boldsymbol{\varepsilon}}(\gamma) = \varepsilon \hat{\bar{\mathbf{c}}}$$

The smooth sheet cell is determined by the vectors $\bar{\mathbf{u}}$, $\bar{\mathbf{v}}$, and $\bar{\mathbf{w}}$ illustrated in Figure 6a. As highlighted in Figure 6b, let

$$u_1 = \frac{d \sin \alpha_1 - b \sin \alpha_4 + \Delta}{2} \tag{1}$$

$$v_1 = \frac{b \sin \alpha_4 - d \sin \alpha_1 + \Delta}{2} \tag{2}$$

Note that $\Delta = u_1 + v_1$. Examining the components of $\bar{\mathbf{u}}$ and $\bar{\mathbf{v}}$ in the ideal state, as shown in Figure 6a,b, we see that $u_2 + v_2 = ||\bar{\mathbf{q}}(\gamma_0) - \bar{\mathbf{q}}_\perp(\gamma_0)||$. Since

$$\frac{u_2}{u_2 + v_2} = \frac{u_1}{u_1 + v_1} = \frac{u_1}{\Delta} \quad \text{and} \quad \frac{v_2}{v_1} = \frac{u_2}{u_1}$$

it follows that

$$u_2 = \frac{u_1}{\Delta} ||\bar{\mathbf{q}}(\gamma_0) - \bar{\mathbf{q}}_\perp(\gamma_0)|| \tag{3}$$

$$v_2 = v_1 \frac{u_2}{u_1} \tag{4}$$

Rodrigues' rotation formula [10] rotates a vector $\mathbf{x}$ by angle θ in a counter-clockwise direction about a unit vector $\mathbf{k}$, and is given by:

$$R[\mathbf{k}, \theta](\mathbf{x}) = \mathbf{x} \cos \theta + (\mathbf{k} \times \mathbf{x}) \sin \theta + \mathbf{k} \langle \mathbf{k}, \mathbf{x} \rangle (1 - \cos \theta)$$

Let

$$\hat{\mathbf{u}}_1(\gamma) = R[-\hat{\bar{\mathbf{c}}}(\gamma), \lambda(\gamma)](\hat{\bar{\mathbf{q}}}_\perp(\gamma))$$
$$\hat{\mathbf{u}}_2(\gamma) = -\hat{\bar{\mathbf{c}}}(\gamma)$$

where

$$\lambda(\gamma) = \cos^{-1}\left(\frac{||\bar{\mathbf{q}}_\perp(\gamma)||^2 + u_1^2 - v_1^2}{2||\bar{\mathbf{q}}_\perp(\gamma)||u_1} \right)$$

Then we define $\bar{\mathbf{u}}(\gamma)$ and $\bar{\mathbf{v}}(\gamma)$ as

$$\bar{\mathbf{u}}(\gamma) = u_1 \hat{\mathbf{u}}_1(\gamma) + u_2 \hat{\mathbf{u}}_2(\gamma)$$
$$\bar{\mathbf{v}}(\gamma) = \bar{\mathbf{u}}(\gamma) - \bar{\mathbf{q}}(\gamma)$$

In order to satisfy flat-foldability, the direction of the vector $\bar{\mathbf{w}}$ must be the same as the direction of $-\bar{\mathbf{c}}$. Thus, for some positive constant w:

$$\bar{\mathbf{w}}(\gamma) = -w\hat{\bar{\mathbf{c}}}(\gamma)$$

The primary concern in choosing the length w was to avoid intersections with the zipper-coupled tubes during deployment. So that the smooth sheet cells $\mathcal{S}_1$ and $\mathcal{S}_4$ lined up correctly, we defined w and h to be the lengths necessary so that in the ideal state, the tips of the vectors $\bar{\mathbf{w}}$ $(-w\hat{\bar{\mathbf{c}}})$ and $h\hat{\mathbf{c}}$ meet, as shown in Figure 6c. In particular, we set

$$h\hat{\mathbf{c}}(\gamma_0) + F_3(\gamma_0) = -w\hat{\bar{\mathbf{c}}}(\gamma_0) + (E_2'(\gamma_0) + \bar{\boldsymbol{\varepsilon}}(\gamma_0) - \bar{\mathbf{u}}(\gamma_0))$$

Then, w (and h) can be obtained as follows:

$$[h, w]^T = [\text{proj}(\hat{\mathbf{c}}(\gamma_0)), \text{proj}(\hat{\bar{\mathbf{c}}}(\gamma_0))]^{-1} \text{proj}(\bar{\mathbf{a}} + \bar{\mathbf{c}} + \bar{\mathbf{d}} + \mathbf{s} + \bar{\boldsymbol{\varepsilon}} - \bar{\mathbf{u}} - \mathbf{c})(\gamma_0) \tag{5}$$

where the function proj : $\mathbb{R}^3 \rightarrow \mathbb{R}^2$ is defined such that

$$\text{proj}[x, y, z]^T = [x, y]^T$$

Remark 1. *Note that the matrix that is inverted in Equation (5) will be singular only in the symmetric case when $\bar{\mathbf{c}}$ has the same direction as $-\mathbf{c}$.*

We have now stipulated all three vectors—$\bar{\mathbf{u}}$, $\bar{\mathbf{v}}$, and $\bar{\mathbf{w}}$—which define $\mathcal{S}_1$ when it is attached to $\mathcal{C}_4$. What remains is to move $\mathcal{S}_1$ so that it attaches to $\mathcal{C}_1$. The desired placement is depicted in Figure 5b, where $\mathcal{S}_1$ is positioned on top of $\mathcal{C}_1$ and bridges $\mathcal{Z}_0$ and $\mathcal{Z}_1$; note that the zipper-coupled tubes have been rotated about the y-axis in this figure so that $\mathcal{S}_1$ lies parallel with the xy-plane (see [5] for further details). Let $X[i]$ denote the point in $\mathcal{Z}_i$ that is a copy of X in $\mathcal{Z}_0$, for $i \geq 1$. Then, the vertex $F_3{}^*$ in $\mathcal{S}_1$ attaches to F_3 in $\mathcal{Z}_0$ and the vertex $F_3{}^*[1]$ in $\mathcal{S}_1$ attaches to $F_3[1]$ in $\mathcal{Z}_1$. More particularly, the edge $\overline{P_0 F_3^*}$ attaches to $\overline{F_0 F_3}$ and $\overline{P_0[1] F_3^*[1]}$ attaches to $\overline{F_0[1] F_3[1]}$.

Recall that the $180°$ rotation of a vector $\bar{\mathbf{x}}$ about the y-axis is denoted $\mathbf{x}$. Thus, the vertices in $\mathcal{S}_1$ when attached to $\mathcal{C}_1$ as desired are:

$$F_3(\gamma) = \mathbf{c}(\gamma) \qquad O_0(\gamma) = -\boldsymbol{\varepsilon}(\gamma) + \mathbf{u}(\gamma) \qquad F_3[1](\gamma) = \mathbf{c}(\gamma) - \mathbf{b}(\gamma) + \mathbf{d}(\gamma)$$
$$P_0(\gamma) = -\boldsymbol{\varepsilon}(\gamma) \qquad O_3(\gamma) = -\boldsymbol{\varepsilon}(\gamma) + \mathbf{u}(\gamma) - \mathbf{w}(\gamma) \qquad P_0[1](\gamma) = -\boldsymbol{\varepsilon}(\gamma) - \mathbf{b}(\gamma) + \mathbf{d}(\gamma)$$

3.2. Design of $\mathcal{S}_3$

The cell $\mathcal{S}_3$ is a rotated copy of $\mathcal{S}_1$ attached to $\mathcal{C}_3$, so the vectors that define $\mathcal{S}_3$ are rotated copies of the vectors that define $\mathcal{S}_1$, shifted by $\mathbf{s}$. Let the vertices on smooth sheet cells attached to the rotated tube segment in $\mathcal{Z}_0$—i.e., the vertices in $\mathcal{S}_2$ and $\mathcal{S}_3$—be denoted with primes. Note that the smooth sheet cell $\mathcal{S}_3$ bridges $\mathcal{Z}_0$ and $\mathcal{Z}_{-1}$, so the smooth sheet cell contains the vertices F_3' and $F_3'[-1]$. Thus, the vertices of $\mathcal{S}_3$ when attached to $\mathcal{C}_3$ are:

$$F_3'(\gamma) = \bar{\mathbf{c}}(\gamma) + \mathbf{s}(\gamma) \quad O_0'(\gamma) = \mathbf{s}(\gamma) - \bar{\boldsymbol{\varepsilon}}(\gamma) + \bar{\mathbf{u}}(\gamma) \qquad F_3'[-1](\gamma) = \bar{\mathbf{c}}(\gamma) + \mathbf{s}(\gamma) - \bar{\mathbf{b}}(\gamma) + \bar{\mathbf{d}}(\gamma)$$
$$P_0'(\gamma) = \mathbf{s}(\gamma) - \bar{\boldsymbol{\varepsilon}}(\gamma) \quad O_3'(\gamma) = \mathbf{s}(\gamma) - \bar{\boldsymbol{\varepsilon}}(\gamma) + \bar{\mathbf{u}}(\gamma) - \bar{\mathbf{w}}(\gamma) \quad P_0'[-1](\gamma) = \mathbf{s}(\gamma) - \bar{\boldsymbol{\varepsilon}}(\gamma) - \bar{\mathbf{b}}(\gamma) + \bar{\mathbf{d}}(\gamma)$$

3.3. Design of $\mathcal{S}_4$

The smooth sheet cell $\mathcal{S}_4$ attaches on top of $\mathcal{C}_4$ and fits together with $\mathcal{S}_1$ in the ideal state, as illustrated in Figure 7. We make the edge of $\mathcal{S}_4$ opposite of $\mathcal{S}_1$ straight in the ideal state so that zipper-coupled tubes with smooth sheet attachments can be stacked laterally without gaps. From another point of view, the edge is made straight in preparation of the design of $\mathcal{S}_2$, a copy of $\mathcal{S}_4$ attached to the bottom of $\mathcal{Z}_0$—the straight edges of $\mathcal{S}_2$ and $\mathcal{S}_3$ meet in the ideal state, as illustrated in Figure 4c.

The smooth sheet cell $\mathcal{S}_4$ has a degree-four vertex folding pattern inspired by the Miura-ori cell, as shown in Figure 8a. This allows the cell to close the gap on the top of $\mathcal{Z}_0$ in the ideal state and fold up without intersecting the adjacent tube segment. For flat-foldability of the cell, we require the sum of opposite angles at the interior vertex to be $180°$ (see Kawasaki-Justin theorem [11,12]). The pattern in Figure 8a is described by the previously defined vectors $\bar{\mathbf{u}}$, $\bar{\mathbf{v}}$, and $\bar{\mathbf{w}}$ and the yet-to-be-defined vectors $\bar{\mathbf{r}}$, $\bar{\mathbf{t}}$, $\bar{\mathbf{h}}$, $\bar{\mathbf{f}}$, and $\bar{\mathbf{g}}$.

In the ideal state, $\bar{\mathbf{r}}$ and $\bar{\mathbf{t}}$ are the projections of $\mathbf{b}$ and $\mathbf{d}$, respectively, into the xy-plane; this is necessary to ensure flat foldability. Thus (see Figures 3 and 8a),

$$\beta_2 = \angle P_1' Q_1' Q_2' = \angle F_3 G_1' G_2' \quad \text{and} \quad \beta_3 = \angle P_1' O_1' O_2' = \angle F_3 E_1' E_2'$$

In particular,

$$\beta_2 = \cos^{-1}\left(-\hat{\bar{\mathbf{c}}}(\gamma_0) \cdot \frac{\mathbf{d}(\gamma_0) - s_0 \mathbf{a}(\gamma_0)}{\|\mathbf{d}(\gamma_0) - s_0 \mathbf{a}(\gamma_0)\|} \right)$$
$$\beta_3 = \cos^{-1}\left(-\hat{\bar{\mathbf{c}}}(\gamma_0) \cdot \frac{\mathbf{b}(\gamma_0) - s_0 \mathbf{a}(\gamma_0)}{\|\mathbf{b}(\gamma_0) - s_0 \mathbf{a}(\gamma_0)\|} \right)$$

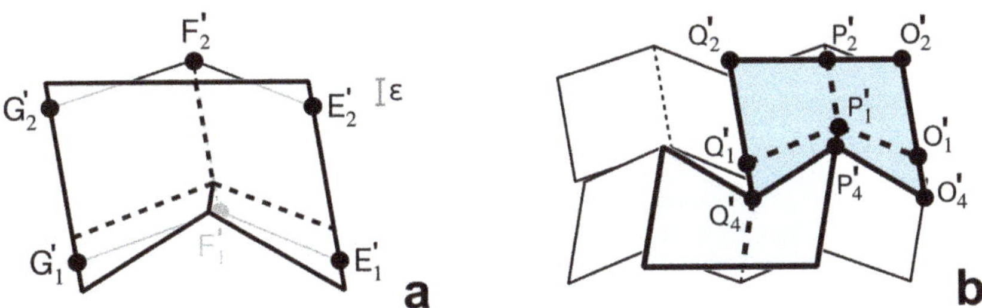

Figure 7. (a) Placing $\mathcal{S}_4$ on $\mathcal{C}_4$. (b) Attaching $\mathcal{S}_4$ to asymmetric zipper-coupled tubes.

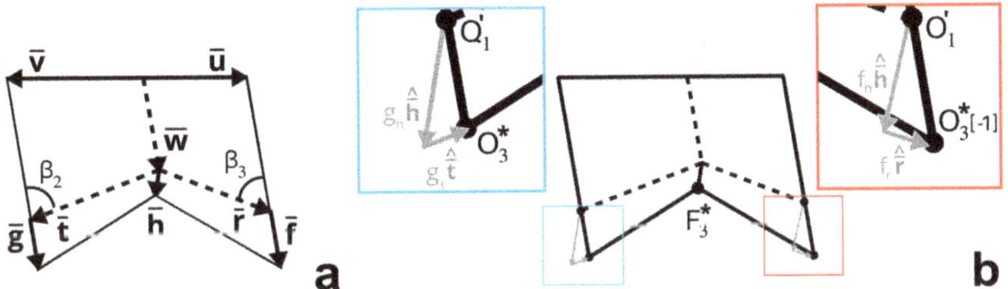

Figure 8. (a) Vectors in $\mathcal{S}_4$. (b) Vector components of $\bar{\mathbf{f}}$ and $\bar{\mathbf{g}}$.

Having derived these angles, we are now ready to define $\bar{\mathbf{r}}$ and $\bar{\mathbf{t}}$ as follows (see Equations (1)–(4)):

$$\bar{\mathbf{r}}(\gamma) = -\left(\frac{u_1}{\tan \beta_3} - u_2\right)\hat{\bar{\mathbf{c}}}(\gamma) + \bar{\mathbf{u}}(\gamma)$$

$$\bar{\mathbf{t}}(\gamma) = -\left(\frac{v_1}{\tan \beta_2} + v_2\right)\hat{\bar{\mathbf{c}}}(\gamma) + \bar{\mathbf{v}}(\gamma)$$

Observing Figure 8a, note that r and t can be expressed simply as:

$$r = \frac{u_1}{\sin \beta_3} \qquad t = \frac{v_1}{\sin \beta_2}$$

In the ideal state, the crease defined by $\bar{\mathbf{h}}$ must have the same direction as $-\mathbf{c}$ to satisfy flat-foldability. Moreover, so that there are no gaps when $\mathcal{S}_1$ and $\mathcal{S}_4$ come together in the ideal state, the length of $\bar{\mathbf{h}}$ should be the value h given by Equation (5), according to the premise upon which Equation (5) was derived (see also Figure 6c). Thus,

$$\bar{\mathbf{h}}(\gamma_0) = -h\hat{\bar{\mathbf{c}}}(\gamma_0)$$

For an arbitrary parameter value γ, the unit vectors adjacent to $\bar{\mathbf{h}}(\gamma)$ that emanate from the degree-four vertex in the interior of $\mathcal{S}_4$ are $\hat{\bar{\mathbf{r}}}(\gamma)$ and $\hat{\bar{\mathbf{t}}}(\gamma)$. Because opposite angles in a degree-four vertex sum to 180° [11,12] and we require a rigid folding, $\bar{\mathbf{h}}(\gamma)$ is determined by the following system of equations:

$$\hat{\bar{\mathbf{h}}}(\gamma) \cdot \hat{\bar{\mathbf{r}}}(\gamma) = \cos\left(\pi - \cos^{-1}\left(\hat{\bar{\mathbf{c}}}(\gamma) \cdot \hat{\bar{\mathbf{t}}}(\gamma)\right)\right)$$

$$\hat{\bar{\mathbf{h}}}(\gamma) \cdot \hat{\bar{\mathbf{t}}}(\gamma) = \cos\left(\pi - \cos^{-1}\left(\hat{\bar{\mathbf{c}}}(\gamma) \cdot \hat{\bar{\mathbf{r}}}(\gamma)\right)\right)$$

$$\hat{\bar{\mathbf{h}}}(\gamma) \cdot \hat{\bar{\mathbf{h}}}(\gamma) = 1$$

The first two equations are linear and the third is quadratic. Hence, there are precisely two solutions: one corresponding to a valley fold assignment and one corresponding to a mountain fold assignment. The solution corresponding to a mountain fold is the correct solution.

Because $\bar{\mathbf{g}}$ corresponds to an edge of the panel defined by $\bar{\mathbf{h}}$ and $\bar{\mathbf{t}}$, we can define it in terms of these vectors. We want the position of vertex Q_4' in $\mathcal{S}_4$ to equal that of O_3 in $\mathcal{S}_1$ when in the ideal state. This is equivalent to saying that $g_h\hat{\mathbf{h}}$ and $-g_t\hat{\mathbf{t}}$ define the same point when the former is extended from Q_1' and the latter is extended from O_3 in the ideal state, as shown in Figure 8b. The reader will recognize that this problem is formulated similarly to that in the end of Section 3.1, where the lengths h and w were computed using a system of equations based on two intersecting vectors. We employ the same technique, with the following system of equations:

$$g_h\hat{\mathbf{h}}(\gamma_0) + Q_1'(\gamma_0) = -g_t\hat{\mathbf{t}}(\gamma_0) + O_3(\gamma_0)$$

giving us

$$[g_h, g_t]^T = \left[\text{proj}\left(\hat{\mathbf{h}}(\gamma_0)\right), \text{proj}\left(\hat{\mathbf{t}}(\gamma_0)\right)\right]^{-1} \text{proj}(-\boldsymbol{\varepsilon} + \mathbf{u} - \mathbf{w} - \bar{\mathbf{a}} - \bar{\mathbf{c}} - \bar{\mathbf{d}} - \mathbf{s} - \bar{\boldsymbol{\varepsilon}} + \bar{\mathbf{u}} - \bar{\mathbf{w}} - \bar{\mathbf{t}})(\gamma_0)$$

We solve for g_h and g_t and use these components of projection to define $\bar{\mathbf{g}}$, using $\hat{\mathbf{h}}$ and $\hat{\mathbf{t}}$ as a basis:

$$\bar{\mathbf{g}}(\gamma) = g_h\hat{\mathbf{h}}(\gamma) + g_t\hat{\mathbf{t}}(\gamma)$$

We define $\bar{\mathbf{f}}$ similarly, solving for f_h and f_r via the same method and another system of equations:

$$f_h\hat{\mathbf{h}}(\gamma_0) + O_1'(\gamma_0) = -f_r\hat{\mathbf{r}}(\gamma_0) + O_3[-1](\gamma_0)$$

This gives us

$$[f_h, f_r]^T = \left[\text{proj}\left(\hat{\mathbf{h}}(\gamma_0)\right), \text{proj}\left(\hat{\mathbf{r}}(\gamma_0)\right)\right]^{-1} \text{proj}(-\boldsymbol{\varepsilon} + \mathbf{u} - \mathbf{w} - \bar{\mathbf{a}} - \bar{\mathbf{b}} - \bar{\mathbf{c}} - \mathbf{s} - \bar{\boldsymbol{\varepsilon}} + \bar{\mathbf{u}} - \bar{\mathbf{w}} - \bar{\mathbf{r}})(\gamma_0)$$

Thus, we have

$$\bar{\mathbf{f}}(\gamma) = f_h\hat{\mathbf{h}}(\gamma) + f_r\hat{\mathbf{r}}(\gamma)$$

Now that we have defined all the essential vectors in $\mathcal{S}_4$, we are ready to attach the degree-four vertex to $\mathcal{C}_4$ in the zipper-coupled tubes structure. In the ideal state, this smooth sheet cell matches exactly with the edges of $\mathcal{S}_1$, providing a smooth surface devoid of gaps on the top of asymmetric zipper-coupled tubes (Figure 4a). In particular, the edge $\overline{O_1'O_2'}$ attaches to $\overline{E_1'E_2'}$ and $\overline{Q_1'Q_2'}$ attaches to $\overline{G_1'G_2'}$. Likewise, the edges $\overline{O_1'O_4'}$ and $\overline{Q_1'Q_4'}$ align perfectly with adjacent cells in the ideal state and fold up at different rates to avoid intersections. The vertices in $\mathcal{S}_4$ identified with their corresponding position vectors are:

$$O_1'(\gamma) = \bar{\mathbf{a}}(\gamma) + \bar{\mathbf{c}}(\gamma) + \bar{\mathbf{d}}(\gamma) + \mathbf{s}(\gamma) + \bar{\boldsymbol{\varepsilon}}(\gamma) - \bar{\mathbf{u}}(\gamma) + \bar{\mathbf{w}}(\gamma) + \bar{\mathbf{r}}(\gamma)$$
$$O_2'(\gamma) = \bar{\mathbf{a}}(\gamma) + \bar{\mathbf{c}}(\gamma) + \bar{\mathbf{d}}(\gamma) + \mathbf{s}(\gamma) + \bar{\boldsymbol{\varepsilon}}(\gamma)$$
$$O_4'(\gamma) = \bar{\mathbf{a}}(\gamma) + \bar{\mathbf{c}}(\gamma) + \bar{\mathbf{d}}(\gamma) + \mathbf{s}(\gamma) + \bar{\boldsymbol{\varepsilon}}(\gamma) - \bar{\mathbf{u}}(\gamma) + \bar{\mathbf{w}}(\gamma) + \bar{\mathbf{r}}(\gamma) + \bar{\mathbf{f}}(\gamma)$$

$$P_1'(\gamma) = \bar{\mathbf{a}}(\gamma) + \bar{\mathbf{c}}(\gamma) + \bar{\mathbf{d}}(\gamma) + \mathbf{s}(\gamma) + \bar{\boldsymbol{\varepsilon}}(\gamma) - \bar{\mathbf{u}}(\gamma) + \bar{\mathbf{w}}(\gamma)$$
$$P_2'(\gamma) = \bar{\mathbf{a}}(\gamma) + \bar{\mathbf{c}}(\gamma) + \bar{\mathbf{d}}(\gamma) + \mathbf{s}(\gamma) + \bar{\boldsymbol{\varepsilon}}(\gamma) - \bar{\mathbf{u}}(\gamma)$$
$$P_4'(\gamma) = \bar{\mathbf{a}}(\gamma) + \bar{\mathbf{c}}(\gamma) + \bar{\mathbf{d}}(\gamma) + \mathbf{s}(\gamma) + \bar{\boldsymbol{\varepsilon}}(\gamma) - \bar{\mathbf{u}}(\gamma) + \bar{\mathbf{w}}(\gamma) + \bar{\mathbf{h}}(\gamma)$$

$$Q_1'(\gamma) = \bar{\mathbf{a}}(\gamma) + \bar{\mathbf{c}}(\gamma) + \bar{\mathbf{d}}(\gamma) + \mathbf{s}(\gamma) + \bar{\boldsymbol{\varepsilon}}(\gamma) - \bar{\mathbf{u}}(\gamma) + \bar{\mathbf{w}}(\gamma) + \bar{\mathbf{t}}(\gamma)$$
$$Q_2'(\gamma) = \bar{\mathbf{a}}(\gamma) + \bar{\mathbf{b}}(\gamma) + \bar{\mathbf{c}}(\gamma) + \mathbf{s}(\gamma) + \bar{\boldsymbol{\varepsilon}}(\gamma)$$
$$Q_4'(\gamma) = \bar{\mathbf{a}}(\gamma) + \bar{\mathbf{c}}(\gamma) + \bar{\mathbf{d}}(\gamma) + \mathbf{s}(\gamma) + \bar{\boldsymbol{\varepsilon}}(\gamma) - \bar{\mathbf{u}}(\gamma) + \bar{\mathbf{w}}(\gamma) + \bar{\mathbf{t}}(\gamma) + \bar{\mathbf{g}}(\gamma)$$

3.4. Design of $\mathcal{S}_2$

The smooth sheet cell $\mathcal{S}_2$ is a rotated copy of $\mathcal{S}_4$ that attaches to $\mathcal{C}_2$. The vertices in this smooth sheet cell are

$$O_1(\gamma) = \mathbf{a}(\gamma) + \mathbf{c}(\gamma) + \mathbf{d}(\gamma) + \boldsymbol{\varepsilon}(\gamma) - \mathbf{u}(\gamma) + \mathbf{w}(\gamma) + \mathbf{r}(\gamma)$$
$$O_2(\gamma) = \mathbf{a}(\gamma) + \mathbf{c}(\gamma) + \mathbf{d}(\gamma) + \boldsymbol{\varepsilon}(\gamma)$$
$$O_4(\gamma) = \mathbf{a}(\gamma) + \mathbf{c}(\gamma) + \mathbf{d}(\gamma) + \boldsymbol{\varepsilon}(\gamma) - \mathbf{u}(\gamma) + \mathbf{w}(\gamma) + \mathbf{r}(\gamma) + \mathbf{f}(\gamma)$$

$$P_1(\gamma) = \mathbf{a}(\gamma) + \mathbf{c}(\gamma) + \mathbf{d}(\gamma) + \boldsymbol{\varepsilon}(\gamma) - \mathbf{u}(\gamma) + \mathbf{w}(\gamma)$$
$$P_2(\gamma) = \mathbf{a}(\gamma) + \mathbf{c}(\gamma) + \mathbf{d}(\gamma) + \boldsymbol{\varepsilon}(\gamma) - \mathbf{u}(\gamma)$$
$$P_4(\gamma) = \mathbf{a}(\gamma) + \mathbf{c}(\gamma) + \mathbf{d}(\gamma) + \boldsymbol{\varepsilon}(\gamma) - \mathbf{u}(\gamma) + \mathbf{w}(\gamma) + \mathbf{h}(\gamma)$$

$$Q_1(\gamma) = \mathbf{a}(\gamma) + \mathbf{c}(\gamma) + \mathbf{d}(\gamma) + \boldsymbol{\varepsilon}(\gamma) - \mathbf{u}(\gamma) + \mathbf{w}(\gamma) + \mathbf{t}(\gamma)$$
$$Q_2(\gamma) = \mathbf{a}(\gamma) + \mathbf{b}(\gamma) + \mathbf{c}(\gamma) + \boldsymbol{\varepsilon}(\gamma)$$
$$Q_4(\gamma) = \mathbf{a}(\gamma) + \mathbf{c}(\gamma) + \mathbf{d}(\gamma) + \boldsymbol{\varepsilon}(\gamma) - \mathbf{u}(\gamma) + \mathbf{w}(\gamma) + \mathbf{t}(\gamma) + \mathbf{g}(\gamma)$$

We have now completed the details for the smooth sheet attachment in the asymmetric case; a summary of the edges and vertices in the attachment is given in Tables A1 and A2, suppressing γ for concision. This attachment folds up flat and actuates with the zipper-coupled tubes structure to form a smooth surface, leaving no gaps between the various asymmetric cells we have described. The symmetric case merits more discussion, however, because there are fewer constraints on the vectors in the $\mathcal{S}_i$, allowing for multiple rigidly foldable patterns given specific design parameters.

4. Symmetric Smooth Sheet Attachment

Unlike their asymmetric counterparts, symmetric zipper-coupled tubes have no tilt and fold parallel with the x-axis [5]. Furthermore, the center creases in $\mathcal{S}_1$ and $\mathcal{S}_4$ lie parallel with the y-axis in the ideal state, as illustrated in Figure 9. There are multiple valid lengths for these creases, therefore, in a design that covers all the gaps in the surface while maintaining rigid foldability without prohibitive intersections. In this section we comment on the diversity in symmetric, Miura-ori inspired smooth sheet cell construction and recommend values for certain lengths.

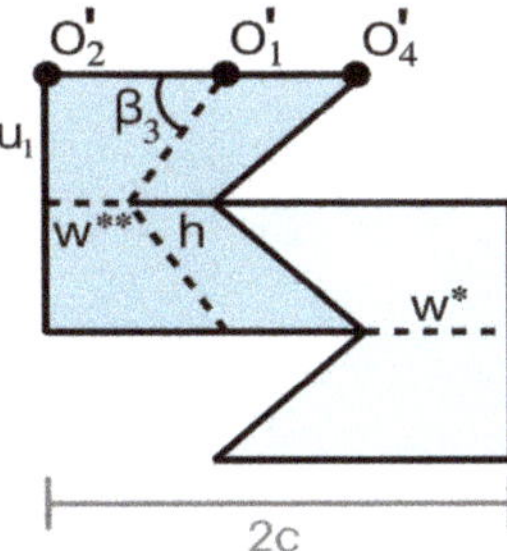

Figure 9. $\mathcal{S}_1$ and $\mathcal{S}_4$ in a symmetric smooth sheet attachment.

The design parameters for the smooth sheet cells $\mathcal{S}_i$ are uniquely determined for all cases where $\alpha_1 + \alpha_2 < \pi$. When $\alpha_1 + \alpha_2 = \pi$, however, there is no longer a unique solution to Equation (5). In particular, $\bar{\mathbf{c}}(\gamma_0)$ has the same direction as $-\mathbf{c}(\gamma_0)$ in the symmetric case, so the matrix inverted in Equation (5) is singular and the values h and w are not uniquely defined. Similarly, β_2 and β_3 are no longer constrained, and we may define these features of the design problem advantageously by choosing a solution that minimizes the amount by which the smooth sheet attachment protrudes from the structure when folded.

In the design of $\mathcal{S}_1$ and $\mathcal{S}_3$, let w^* replace the value of w. Likewise, in the design of $\mathcal{S}_4$ and $\mathcal{S}_2$, let w^{**} replace the value of w. As highlighted in Figure 9, for the symmetric case we no longer require that $w^* = w^{**}$. In selecting a value for w^*, we set it as large as possible to maximize the surface area of $\mathcal{S}_1$, thus minimizing the amount by which the edges of $\mathcal{S}_4$ can protrude from the zipper-coupled tubes structure. Applying the analysis given in [5] (see Section 7.1.3), the largest value for w^* can be shown to be

$$w^* = c - \frac{d \sin \alpha_1 + b \sin \alpha_4 - \Delta}{2 \tan (\alpha_2 - \alpha_1)} - \varepsilon$$

The only requirements for h, w^{**}, β_2, and β_3 in the symmetric case are

$$h + w^{**} = c - \varepsilon \quad \text{and} \quad \beta_2 = \beta_3$$

Adjustments to the values h, w^{**}, β_2, and β_3 can also assist in minimizing the protrusion of $\mathcal{S}_4$ from the zipper-coupled tubes. Optimal values can be determined by numerical methods according to the specific design application. However, care should be taken in making these adjustments to avoid intersections with the structure underneath.

On a final note, although $\bar{\mathbf{f}}$ and $\bar{\mathbf{g}}$ are determined after defining the previous quantities, a convenient simplification in their definition: because $\mathcal{S}_4$ is symmetric, the vectors $\bar{\mathbf{f}}$ and $\bar{\mathbf{g}}$ are parallel to $\bar{\mathbf{h}}$ and have equal lengths. Moreover, in the ideal state,

$$O_1' O_4' = O_2' O_4' - O_1' O_2'$$

Therefore,

$$g_t = f_r = 0 \quad \text{and} \quad g_h = f_h = (2c - w^*) - \left(w^{**} + \frac{u_1}{\tan \beta_3} \right)$$

5. Conclusions

We have successfully defined a smooth sheet attachment that folds up with the zipper-coupled tubes and unfolds to the ideal state without inhibiting their motion to form a flat surface without any gaps (Figure 10). This pattern is defined for both the asymmetric and symmetric cases, and we provide access to code which the reader may use to visualize the origami structures described and print out the corresponding fold patterns: https://github.com/dylanwebbc/azct (accessed on 23 July 2022).

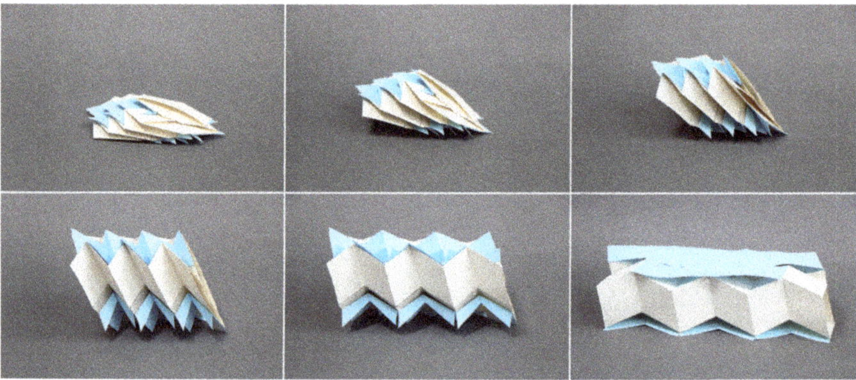

Figure 10. Model of asymmetric zipper-coupled tubes with a Miura-ori inspired smooth sheet attachment at different stages of unfolding; the nature of the deployment of the degree-four vertex cells is clearly observable on the bottom of the device. Parameters: $\alpha_1 = \frac{1}{3}\pi$, $\alpha_2 = \frac{5}{9}\pi$, $a = b = c$. Dimensions (inches): 4.9 by 3.3 when folded, 7.4 by 3.5 by 1.5 when unfolded.

Mathematics **2022**, 10, 2643

Note that the smooth sheet cells protrude from the zipper-coupled tubes structure when folded. When gaps are tolerable and the folded state must be minimized for transportation, constraining the cells to fold up within the zipper-coupled tubes while maximizing surface area in the ideal state results in the smooth sheet attachment described previously [5]. Thus, the smooth sheet design can be tailored to the situation, much like zipper-coupled tubes themselves. To inform future applications of these structures, we suggest dynamic and quasi-static analyses. Constructing a device for architectural applications will likely require the use of thick origami and compliant hinges, and remote self-actuation via magnetism or heat could be useful in space or medical applications [13–15].

Miura-ori-inspired smooth sheet attachments enhance the utility of zipper-coupled tubes in various situations. The tubes are useful in architecture because they pack tight and deploy to a rigid state [4]; our gapless smooth sheet attachments improve existing designs by increasing drivability and walkability. If a local bridge collapses, for example, a prefabricated bridge based on zipper-coupled tubes with smooth sheet attachments can easily be transported on a single vehicle and swiftly deployed on-site to provide smooth, emergency transit. Space structures are another popular application of origami-inspired mechanisms—the Miura-ori pattern that the smooth sheet is based on is common in deployable space array design. Accommodating for material thickness, however, makes Miura-ori sheets challenging to deploy [16]. In contrast, a thin solar array constructed from Miura-ori inspired smooth sheet cells can deploy rigidly because it is supported by zipper-coupled tubes.

The design of smooth sheet attachments without gaps is key to the development of more versatile zipper-coupled tubes. We have communicated a clear design method for the origami-based structure, examining the mathematics of its motion in detail. By elucidating the possibility for further enhancements on the zipper-coupled tubes structure, we hope to spur many novel and exciting applications beyond those mentioned.

Author Contributions: Conceptualization, D.C.W.; Methodology, D.M.H. and L.L.H.; Software, D.C.W.; Supervision, D.M.H. and L.L.H.; Validation, D.C.W.; Visualization, D.C.W. and E.R.; Writing—original draft, D.C.W. and D.M.H.; Writing—review & editing, D.C.W., E.R. and D.M.H. All authors have read and agreed to the published version of the manuscript.

Funding: This research was funded by a grant from the College of Physical and Mathematical Sciences at Brigham Young University.

Conflicts of Interest: The authors declare no conflict of interest.

Appendix A

Table A1. Defining the edges in Miura-ori based, smooth sheet attachments for a pair of asymmetric zipper-coupled tubes.

$\mathcal{S}_1$		$\mathcal{S}_2$			$\mathcal{S}_3$		$\mathcal{S}_4$		
$\overline{O_0P_0}$	$\overline{O_0P_0[1]}$	$\overline{O_1P_1}$	$\overline{O_2P_2}$	$\overline{O_4P_4}$	$\overline{O_0'P_0'}$	$\overline{O_0'P_0'[-1]}$	$\overline{O_1'P_1'}$	$\overline{O_2'P_2'}$	$\overline{O_4'P_4'}$
$\overline{O_3F_3}$	$\overline{O_3F_3[1]}$	$\overline{P_1Q_1}$	$\overline{P_2Q_2}$	$\overline{P_4Q_4}$	$\overline{O_3'F_3'}$	$\overline{O_3'F_3'[-1]}$	$\overline{P_1'Q_1'}$	$\overline{P_2'Q_2'}$	$\overline{P_4'Q_4'}$
$\overline{O_0O_3}$	$\overline{P_0F_3}$	$\overline{O_1O_2}$	$\overline{O_1O_4}$	$\overline{P_1P_2}$	$\overline{O_0'O_3'}$	$\overline{P_0'F_3'}$	$\overline{O_1'O_2'}$	$\overline{O_1'O_4'}$	$\overline{P_1'P_2'}$
$\overline{P_0[1]F_3[1]}$		$\overline{P_1P_4}$	$\overline{Q_1Q_2}$	$\overline{Q_1Q_4}$	$\overline{P_0'[-1]F_3'[-1]}$		$\overline{P_1'P_4'}$	$\overline{Q_1'Q_2'}$	$\overline{Q_1'Q_4'}$

Table A2. Defining the vertices in Miura-ori inspired smooth sheet attachments for a pair of asymmetric zipper-coupled tubes.

Vertex	Position	Vertex	Position
O_0	$-\boldsymbol{\varepsilon} + \mathbf{u}$	O_0'	$\mathbf{s} - \bar{\boldsymbol{\varepsilon}} + \bar{\mathbf{u}}$
O_1	$\mathbf{a} + \mathbf{c} + \mathbf{d} + \boldsymbol{\varepsilon} - \mathbf{u} + \mathbf{w} + \mathbf{r}$	O_1'	$\bar{\mathbf{a}} + \bar{\mathbf{c}} + \bar{\mathbf{d}} + \mathbf{s} + \bar{\boldsymbol{\varepsilon}} - \bar{\mathbf{u}} + \bar{\mathbf{w}} + \bar{\mathbf{r}}$
O_2	$\mathbf{a} + \mathbf{c} + \mathbf{d} + \boldsymbol{\varepsilon}$	O_2'	$\bar{\mathbf{a}} + \bar{\mathbf{c}} + \bar{\mathbf{d}} + \mathbf{s} + \bar{\boldsymbol{\varepsilon}}$
O_3	$-\boldsymbol{\varepsilon} + \mathbf{u} - \mathbf{w}$	O_3'	$\mathbf{s} - \bar{\boldsymbol{\varepsilon}} + \bar{\mathbf{u}} - \bar{\mathbf{w}}$
O_4	$\mathbf{a} + \mathbf{c} + \mathbf{d} + \boldsymbol{\varepsilon} - \mathbf{u} + \mathbf{w} + \mathbf{r} + \mathbf{f}$	O_4'	$\bar{\mathbf{a}} + \bar{\mathbf{c}} + \bar{\mathbf{d}} + \mathbf{s} + \bar{\boldsymbol{\varepsilon}} - \bar{\mathbf{u}} + \bar{\mathbf{w}} + \bar{\mathbf{r}} + \bar{\mathbf{f}}$
P_0	$-\boldsymbol{\varepsilon}$	P_0'	$\mathbf{s} - \bar{\boldsymbol{\varepsilon}}$
P_1	$\mathbf{a} + \mathbf{c} + \mathbf{d} + \boldsymbol{\varepsilon} - \mathbf{u} + \mathbf{w}$	P_1'	$\bar{\mathbf{a}} + \bar{\mathbf{c}} + \bar{\mathbf{d}} + \mathbf{s} + \bar{\boldsymbol{\varepsilon}} - \bar{\mathbf{u}} + \bar{\mathbf{w}}$
P_2	$\mathbf{a} + \mathbf{c} + \mathbf{d} + \boldsymbol{\varepsilon} - \mathbf{u}$	P_2'	$\bar{\mathbf{a}} + \bar{\mathbf{c}} + \bar{\mathbf{d}} + \mathbf{s} + \bar{\boldsymbol{\varepsilon}} - \bar{\mathbf{u}}$
P_4	$\mathbf{a} + \mathbf{c} + \mathbf{d} + \boldsymbol{\varepsilon} - \mathbf{u} + \mathbf{w} + \mathbf{h}$	P_4'	$\bar{\mathbf{a}} + \bar{\mathbf{c}} + \bar{\mathbf{d}} + \mathbf{s} + \bar{\boldsymbol{\varepsilon}} - \bar{\mathbf{u}} + \bar{\mathbf{w}} + \bar{\mathbf{h}}$
Q_1	$\mathbf{a} + \mathbf{c} + \mathbf{d} + \boldsymbol{\varepsilon} - \mathbf{u} + \mathbf{w} + \mathbf{t}$	Q_1'	$\bar{\mathbf{a}} + \bar{\mathbf{c}} + \bar{\mathbf{d}} + \mathbf{s} + \bar{\boldsymbol{\varepsilon}} - \bar{\mathbf{u}} + \bar{\mathbf{w}} + \bar{\mathbf{t}}$
Q_2	$\mathbf{a} + \mathbf{b} + \mathbf{c} + \boldsymbol{\varepsilon}$	Q_2'	$\bar{\mathbf{a}} + \bar{\mathbf{b}} + \bar{\mathbf{c}} + \mathbf{s} + \bar{\boldsymbol{\varepsilon}}$
Q_4	$\mathbf{a} + \mathbf{c} + \mathbf{d} + \boldsymbol{\varepsilon} - \mathbf{u} + \mathbf{w} + \mathbf{t} + \mathbf{g}$	Q_4'	$\bar{\mathbf{a}} + \bar{\mathbf{c}} + \bar{\mathbf{d}} + \mathbf{s} + \bar{\boldsymbol{\varepsilon}} - \bar{\mathbf{u}} + \bar{\mathbf{w}} + \bar{\mathbf{t}} + \bar{\mathbf{g}}$
F_3	$\mathbf{c}$	F_3'	$\bar{\mathbf{c}} + \mathbf{s}$
$F_3[1]$	$\mathbf{c} - \mathbf{b} + \mathbf{d}$	$F_3'[-1]$	$\bar{\mathbf{c}} + \mathbf{s} - \bar{\mathbf{b}} + \bar{\mathbf{d}}$
$P_0[1]$	$-\boldsymbol{\varepsilon} - \mathbf{b} + \mathbf{d}$	$P_0'[-1]$	$\mathbf{s} - \bar{\boldsymbol{\varepsilon}} - \bar{\mathbf{b}} + \bar{\mathbf{d}}$

References

1. Mena, L.; Muñoz, J.; Monje, C.A.; Balaguer, C. Modular and Self-Scalable Origami Robot: A First Approach. *Mathematics* **2021**, *9*, 1324. [CrossRef]
2. Rubio, A.J.; Kaddour, A.S.; Georgakopoulos, S.V.; Ynchausti, C.; Magleby, S.; Howell, L.L. A Deployable Hexagonal Reflectarray Antenna for Space Applications. In Proceedings of the 2021 United States National Committee of URSI National Radio Science Meeting (USNC-URSI NRSM), Boulder, CO, USA, 4–9 January 2021; pp. 136–137. [CrossRef]
3. Zhang, F.; Li, S.; Shen, Z.; Cheng, X.; Xue, Z.; Zhang, H.; Song, H.; Bai, K.; Yan, D.; Wang, H.; et al. Rapidly deployable and morphable 3D mesostructures with applications in multimodal biomedical devices. *Proc. Natl. Acad. Sci. USA* **2021**, *118*, e2026414118. [CrossRef] [PubMed]
4. Filipov, E.T.; Tachi, T.; Paulino, G.H. Origami tubes assembled into stiff, yet reconfigurable structures and metamaterials. *Proc. Natl. Acad. Sci. USA* **2015**, *112*, e1509465112. [CrossRef] [PubMed]
5. Webb, D.C.; Elissa, R.; Halverson, D.M.; Howell, L.L. Deployable Space-Filling Mechanisms: Asymmetric Zipper-coupled Tubes and Smooth Sheet Attachments. In Proceedings of the ASME 2022 International Design Engineering Technical Conferences and Computers and Information in Engineering Conference, St. Louis, MI, USA, 14–17 August 2022; Forthcoming.
6. Miura, K. The Science of Miura-Ori: A Review. In *Origami 4*; Lang, R.J., Ed.; A K Peters/CRC Press: Boca Raton, FL, USA, 2009; Chapter 4, pp. 87–99.
7. Miura, K. Method of Packaging and Deployment of Large Membranes in Space. *Inst. Space Astronaut. Sci. Rep.* **1985**, *618*, 1–9.
8. Miura, K.; Natori, M. 2-D Array Experiment on Board a Space Flyer Unit. *Space Sol. Power Rev.* **1985**, *5*, 345–356.
9. Zimmermann, L.; Stanković, T. Rigid and Flat Foldability of a Degree-Four Vertex in Origami. *J. Mech. Robot.* **2020**, *12*, 011004. [CrossRef]
10. Rodrigues, O. Des lois géométriques qui régissent les déplacements d'un système solide dans l'espace, et de la variation des coordonnées provenant de ces déplacements considérés indépendamment des causes qui peuvent les produire. *J. Mathématiques Pures Appliquées* **1840**, *5*, 380–440. Available online: http://xxx.lanl.gov/abs/http://sites.mathdoc.fr/JMPA/PDF/JMPA_1840_1_5_A39_0.pdf (accessed on 1 January 2022).
11. Kawasaki, T. On the Relation between Mountain-Creases and Valley-Creases of a Flat Origami. In *Proceedings of the 1st International Meeting on Origami Science and Technology*; Huzita, H., Ed.; Universita di Padova: Padova, Italy, 1989; pp. 229–237.
12. Justin, J. Mathematics of origami, part 9. *Br. Origami* **1986**, *118*, 28–30.
13. Yellowhorse, A.; Tolman, K.; Howell, L.L. Optimization of Origami-Based Tubes for Lightweight Deployable Structures. In Proceedings of the ASME 2017 International Design Engineering Technical Conferences and Computers and Information in Engineering Conference, Cleveland, OH, USA, 6–9 August 2017. [CrossRef]

14. Kim, Y.; Yuk, H.; Zhao, R.; Chester, S.A.; Zhao, X. Printing ferromagnetic domains for untethered fast-transforming soft materials. *Nature* **2018**, *558*, 274–279. [CrossRef] [PubMed]
15. Kuribayashi, K.; Tsuchiya, K.; You, Z.; Tomus, D.; Umemoto, M.; Ito, T.; Sasaki, M. Self-deployable origami stent grafts as a biomedical application of Ni-rich TiNi shape memory alloy foil. *Mater. Sci. Eng. A* **2006**, *419*, 131–137. [CrossRef]
16. Bolanos, D.; Ynchausti, C.; Brown, N.; Pruett, H.; Hunter, J.; Clark, B.; Bateman, T.; Howell, L.L.; Magleby, S.P. Considering thickness-accommodation, nesting, grounding and deployment in design of Miura-ori based space arrays. *Mech. Mach. Theory* **2022**, *174*, 104904. [CrossRef]

Article

Jump and Initial-Sensitive Excessive Motion of a Class of Relative Rotation Systems and Their Control via Delayed Feedback

Ziyin Cui and Huilin Shang *

School of Mechanical Engineering, Shanghai Institute of Technology, Shanghai 201418, China; 206091149@mail.sit.edu.cn
* Correspondence: shanghuilin@sit.edu.cn

Abstract: Jump and excessive motion are undesirable phenomena in relative rotation systems, causing a loss of global integrity and reliability of the systems. In this work, a typical relative rotation system is considered in which jump, excessive motion, and their suppression via delayed feedback are investigated. The Method of Multiple Scales and the Melnikov method are applied to analyze critical conditions for bi-stability and initial-sensitive excessive motion, respectively. By introducing the fractal of basins of attraction and the erosion of the safe basin to depict jump and initial-sensitive excessive motion, respectively, the point mapping approach is used to present numerical simulations which are in agreement with the theoretical prediction, showing the validity of the analysis. It is found that jump between bistable attractors can be due to saddle–node bifurcation, while initial-sensitive excessive motion can be due to heteroclinic bifurcation. Under a positive coefficient of the gain, the types of delayed feedback can both be effective in reducing jump and initial-sensitive excessive motion. The results may provide some reference for the performance improvement of rotors and main bearings.

Keywords: relative rotation; jump; safe basin; fractal; heteroclinic bifurcation; delayed feedback

MSC: 58Z05; 37G35; 37G15

Citation: Cui, Z.; Shang, H. Jump and Initial-Sensitive Excessive Motion of a Class of Relative Rotation Systems and Their Control via Delayed Feedback. *Mathematics* **2022**, *10*, 2676. https://doi.org/10.3390/math10152676

Academic Editors: Higinio Rubio Alonso, Alejandro Bustos, Jesus Meneses Alonso and Enrique Soriano-Heras

Received: 28 June 2022
Accepted: 27 July 2022
Published: 29 July 2022

Publisher's Note: MDPI stays neutral with regard to jurisdictional claims in published maps and institutional affiliations.

1. Introduction

Rotating machinery, widely used in engineering fields such as transmission systems of vehicles [1] and main bearings of aeroengines [2], plays an important role in the theory of nonlinear dynamics. Due to the nonlinearities of its dynamical systems, it usually undergoes complex dynamics, which have a negative impact on its operational stability and reliability [3,4]. Among these, relative rotation systems have been given much attention, as nonlinear vibration is undesirable. For instance, main shafts can be fatigue damaged by a jump [5,6] among multiple attractors, or by a chaotic response [7], and can even be fractured due to excessive motion [8]. Thus, comprehensive study of the global dynamics of relative rotation systems has become a practical issue.

So far, many researchers have studied the responses of relative rotation systems. Wang et al. [9] obtained the approximate periodic solution of a class of nonlinear systems with relative rotation by applying the Method of Multiple Scales and proved the uniqueness of the periodic response. Xiao et al. [10] verified the existence and uniqueness of periodic response in another type of relative rotation system with a time-varying stiffness. Li et al. [11] proposed a relative rotation system with nonlinear elastic force and nonlinear generalized damping force, and discussed its periodic solution problem theoretically. Shi et al. [12] numerically studied the chaotic behaviors of a relative rotation system under parametric excitation by Poincaré map and maximal Lyapunov exponent. By simulating relative torsional vibrations of an imbalanced shaft under a limited power supply, Verichev [13] found that the interaction of shaft and power supply may result in strange attractors such as classical Lorenz and Feigenbaum attractors. In a nonlinear relative rotation system with a triple-well Mathieu-Duffing oscillator, Liu et al. [14] obtained the threshold of chaos regarding Smale horseshoe commutation and exhibited the erosion process of safe basins. A class of nonlinear relative rotational systems containing two rotors was also built whose chaotic response was presented [15]. By introducing the erosion of the safe basin to describe the safe performance of the spur gear pair, Zhu et al. [16] classified the multiple meshing states and presented the transition process from safe to unsafe. To control the complex dynamics of the main transmission system of a scraper conveyor, Ju et al. [17] analyzed its local bifurcations and proposed a nonlinear state feedback controller whose effect was studied numerically. Considering the effectiveness of time-delay feedback on controlling fractal erosion of the safe basin and chaos in nonlinear dynamical systems [18,19], Zhao et al. [20] applied delayed displacement feedback in a relative torsional vibration system for reducing its response amplitude. On this basis, Shang et al. [21] discussed the effect of delayed position feedback on controlling the erosion of the safe basin and chaos. Although there has been meaningful research on the nonlinear vibration characteristics of the relative rotation system, the study of excessive motion and jump among multiple attractors is mainly carried out by numerical simulation and the mechanism of these complex dynamical behaviors and their control is still not yet clear.

To this end, we select a class of nonlinear relative rotation systems composed of two rotors and study the mechanisms behind its jump and initial-sensitive excessive motion, as well as the effect of delayed feedback on suppressing these phenomena. The paper is arranged as follows. In Section 2, the dynamical model of a relative rotation system is constructed and made dimensionless. In Section 3, the mechanism of jump and excessive motion is analyzed. In Section 4, two control strategies, namely, delay position feedback and delay velocity feedback, are applied to the original system, respectively, whose control mechanism is then discussed. Section 5 contains the discussion.

2. Dynamical Model and Unperturbed Dynamics

We consider a typical torsional vibrating system whose simplified diagram is presented in Figure 1. θ_1 and θ_2 are the absolute rotating angles of two rotors, respectively, and J_1 and J_2 represent their polar moments of inertia. Considering nonlinear torsional stiffness and nonlinear damping, the governing equation of this vibrating system can be derived by the momentum theorem as

$$
\begin{aligned}
J_1\ddot{\theta}_1 + c_{12}(\dot{\theta}_1 - \dot{\theta}_2) + K_1(\theta_1 - \theta_2) + K_3(\theta_1 - \theta_2)^3 + f_{12} &= T_{e1}, \\
J_2\ddot{\theta}_2 + c_{12}(\dot{\theta}_2 - \dot{\theta}_1) + K_1(\theta_2 - \theta_1) + K_3(\theta_2 - \theta_1)^3 - f_{12} &= T_{e2},
\end{aligned}
\tag{1}
$$

where K_1 is the coefficient of linear stiffness between the two rotors; K_3 is the coefficient of nonlinear stiffness satisfying $K_3 < 0$; c_{12} is the coefficient of linear damping; T_{e1} and T_{e2} are external rotational torques loaded on the two rotors, respectively; f_{12} is a class of nonlinear stick-slip frictions [20,21] given by

$$
f_{12} = c_0 + c_1(\dot{\theta}_1 - \dot{\theta}_2) + c_2(\dot{\theta}_1 - \dot{\theta}_2)^3,
\tag{2}
$$

in which c_0, c_1 and c_2 are the constant of static friction, the coefficient of linear damping and the coefficient of cubic nonlinear damping, respectively. By denoting $x = \theta_1 - \theta_2$ in Equation (1), the relative rotation system can be obtained as below:

$$
\ddot{x} + \frac{c_{12}J_2 + c_{12}J_1}{J_1 J_2}\dot{x} + \frac{K_1(J_1 + J_2)}{J_1 J_2}x + \frac{K_3(J_1 + J_2)}{J_1 J_2}x^3 + \frac{J_1 + J_2}{J_1 J_2}(c_1\dot{x} + c_2\dot{x}^3) = \frac{J_2 T_{e1} - J_1 T_{e2} - (J_1 + J_2)c_0}{J_1 J_2}.
\tag{3}
$$

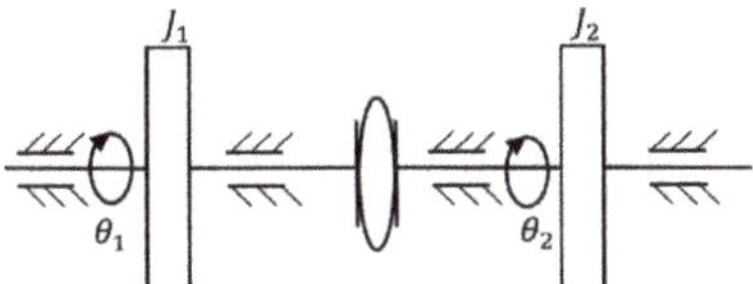

Figure 1. Simplified diagram of a two-rotor torsional vibrating system.

The torques T_{e1} and T_{e2} are usually harmonic excitations, thus can be represented as

$$\frac{J_2 T_{e1} - J_1 T_{e2} - (J_1 + J_2)c_0}{J_1 J_2} = F\cos(\Omega t) \tag{4}$$

where F and Ω are the amplitude and the frequency of the excitation, respectively. Introducing the following variables,

$$\omega_0{}^2 = \frac{K_1(J_1 + J_2)}{J_1 J_2}, b = -\frac{K_3}{K_1}, \mu = \frac{(c_{12} + c_1)(J_1 + J_2)}{J_1 J_2 \omega_0}, g = \frac{c_2(J_1 + J_2)\omega_0}{J_1 J_2}, T = \omega_0 t, \omega = \frac{\Omega}{\omega_0}, f = \frac{F}{\omega_0{}^2}, \tag{5}$$

the dimensionless form of the relative rotation system (3) is obtained as

$$y = \frac{dx}{dT}, \frac{dy}{dT} = -\mu y - x + bx^3 - gy^3 + f\cos(\omega T). \tag{6}$$

in which the parameters μ, b, g, and ω are positive, and the position $x(T)$ and the velocity $y(T)$ represent relative rotational angular and relative angular velocity at the moment T, respectively.

Since the parameters c_{12}, c_0, c_1, and c_2 in the original system (1) are very small, the relevant terms concerned, μ, g and f in Equation (6), can be considered as perturbed. The unperturbed system can be written as

$$\frac{dx}{dT} = y, \frac{dy}{dT} = -x + bx^3. \tag{7}$$

There will be three equilibria in the unperturbed system (7), namely, the center $S_1(0,0)$ and two saddle points $S_2(\frac{\sqrt{b}}{b},0)$ and $S_3(-\frac{\sqrt{b}}{b},0)$. The Hamiltonian of Equation (7) is

$$H(x,y) = \frac{1}{2}y^2 + \frac{1}{2}x^2 - \frac{1}{4}bx^4. \tag{8}$$

Accordingly, the heteroclinic orbits surrounding the center $S_1(0,0)$ can be given by

$$x_{\pm}(t) = \pm\frac{\sqrt{b}}{b}\tanh(\frac{\sqrt{2}}{2}t), y_{\pm}(t) = \pm\frac{\sqrt{2b}}{2b}\operatorname{sech}^2(\frac{\sqrt{2}}{2}t). \tag{9}$$

Fixing $b = 0.3$, the unperturbed orbits are depicted in Figure 2.

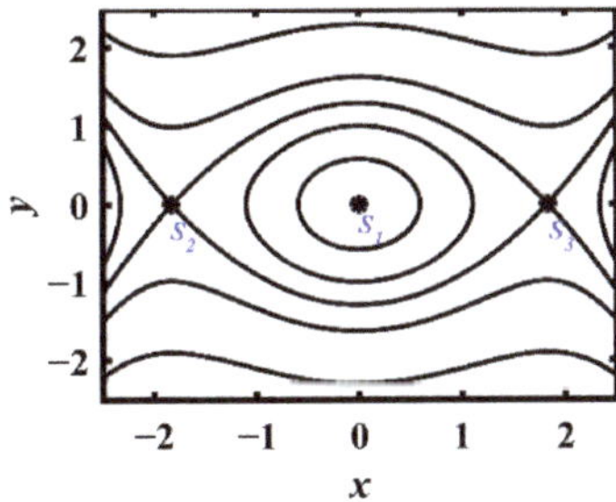

Figure 2. Orbits of the unperturbed system (7) for $b = 0.3$.

3. Complex Dynamics

3.1. Multistability and Jump

The periodic solution near the equilibrium $S_1(0,0)$ is considered. The Method of Multiple Scales (MMS) [22] is employed to obtain the approximate theoretical solution of Equation (3). Supposing a small parameter ε ($0 < \varepsilon \ll 1$) to rescale the nondimensional parameters as follows:

$$\mu = \varepsilon\widetilde{\mu},\ b = \varepsilon\widetilde{b},\ g = \varepsilon\widetilde{g},\ f = \varepsilon\widetilde{f}, \tag{10}$$

Equation (6) becomes

$$\frac{d^2x}{dT^2} + x = \varepsilon\left(-\frac{\widetilde{\mu}dx}{dT} + \widetilde{b}x^3 - \widetilde{g}\left(\frac{dx}{dT}\right)^3 + \widetilde{f}\cos(\omega T)\right). \tag{11}$$

Further rescaling parameters and operators in Equation (11) yields

$$\omega = 1 + \varepsilon\sigma,\ T_i = \varepsilon^i T,\ x(T) = x_0(T_0, T_1, \cdots) + \varepsilon x_1(T_0, T_1, \cdots) + \cdots,\ D_i = \frac{\partial}{\partial T_i},\ \frac{d}{dT} = \sum_{i=0}^{n}\varepsilon^i D_i\ (i = 0, 1, 2\cdots), \tag{12}$$

where σ is the detuning parameter, and separating the coefficients of ε^0 and ε^1 of the system (11) one finds

$$\varepsilon^0 :\ D_0{}^2x_0 + x_0 = 0, \tag{13}$$

and

$$\varepsilon^1 :\ D_0{}^2x_1 + x_1 = -2D_1D_0x_0 - \widetilde{\mu}D_0x_0 + \widetilde{b}x_0{}^3 - \widetilde{g}(D_0x_0)^3 + \widetilde{f}\cos(T_0 + \sigma T_1). \tag{14}$$

The solution of Equation (13) is

$$x_0(T_0, T_1) = A(T_1)e^{iT_0} + cc \tag{15}$$

where cc denotes the complex conjugate of the preceding terms. Substituting Equation (15) into Equation (14) and separating the secular terms yields

$$D_1A + \frac{\widetilde{\mu}A}{2} + \frac{3(\widetilde{g} + i\widetilde{b})A^2\overline{A}}{2} + \frac{i\widetilde{f}e^{i\sigma T_1}}{4} = 0. \tag{16}$$

In addition, by defining

$$A(T_1) = \frac{a(T_1)}{2}e^{i\beta(T_1)},\ \psi = \sigma T_1 - \beta, \tag{17}$$

separating the real and imaginary parts of Equation (16), and returning the parameters to the nondimensional parameters of the system (3), the modulation equation can be written as

$$\frac{da}{dT} = -\frac{\mu a}{2} - \frac{3ga^3}{8} + \frac{f\sin\psi}{2},\ a\frac{d\psi}{dT} = a(\omega - 1) + \frac{3ba^3}{8} + \frac{f\cos\psi}{2}. \tag{18}$$

The steady-state form of Equation (18) is obtained by setting $D_1a = D_1\psi = 0$ and returning the parameters of (18) to the nondimensional parameters as:

$$\mu a + \frac{3ga^3}{4} = f\sin\psi,\ -2(\omega - 1)a - \frac{3ba^3}{4} = f\cos\psi. \tag{19}$$

Accordingly, the approximate periodic solution can be expressed as $x_0 = a\cos(\omega T - \psi)$. By eliminating ψ from the relationships in Equation (19), its amplitude a can be solved from the following equation

$$\left(\mu + \frac{3ga^2}{4}\right)^2 a^2 + \left(2\omega - 2 + \frac{3}{4}ba^2\right)^2 a^2 = f^2. \tag{20}$$

The corresponding characteristic equation is

$$(\lambda + \frac{\mu}{2} + \frac{9ga^2}{8})(\lambda + \frac{\mu}{2} + \frac{3ga^2}{8}) + (\omega - 1 + \frac{9}{8}ba^2)(\omega - 1 + \frac{3}{8}ba^2) = 0, \tag{21}$$

meaning that for

$$\frac{27(g^2 + b^2)}{16}a^4 + 3(g\mu + b(2\omega - 2))a^2 + 4(\omega - 1)^2 + \mu^2 > 0, \tag{22}$$

the periodic solution is unstable and becomes a saddle. Namely, when the inequation (22) becomes an equation, saddle-node bifurcation will occur in the system (6). For

$$\mu = 0.03, b = 0.3, g = 0.03, \tag{23}$$

the response curve of Equation (20) is plotted in Figure 3 where the coexistence of two stable periodic branches in the same range of the excitation parameters implies bi-stability. For example, for f = 0.13 and ω within the range (0.65, 0.86), bistable periodic attractors coexist (see Figure 3a). Fixing ω = 0.80, the increase of the amplitude f can also lead to saddle-node bifurcation, and thus bi-stability. The numerical results totally match the theoretical prediction.

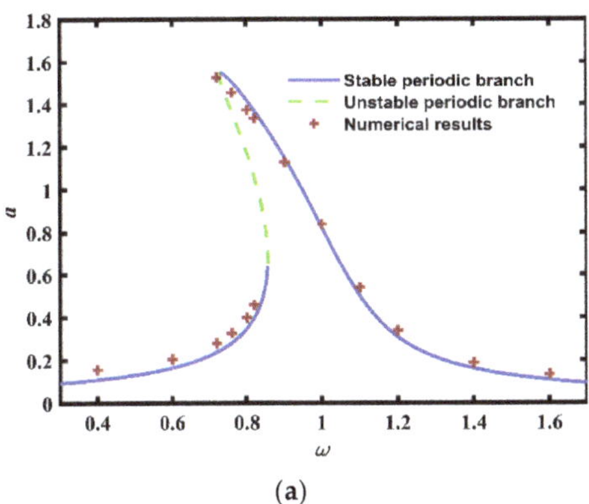

(a)

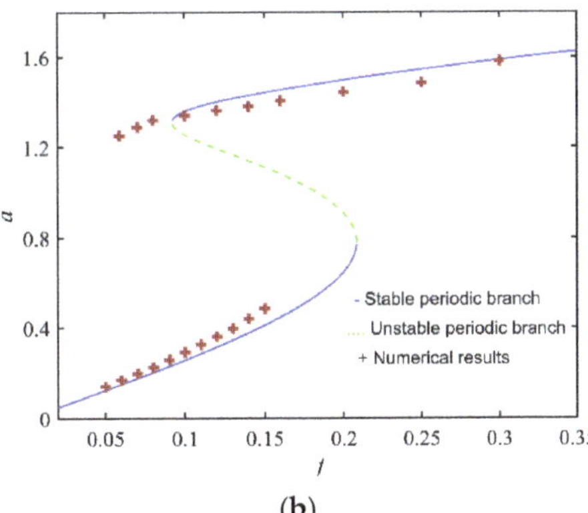

(b)

Figure 3. Response curves of system (2). (**a**) Amplitude a vs. ω when f = 0.13; (**b**) Amplitude a vs. f when ω = 0.80.

3.2. Initial-Sensitive Excessive Motion

Excessive motion of the system (6) can be understood as an escape or unbounded solution [21], for instance, a relative rotating angle in system (6) more than the material of the structure can bear. This may be induced by the variation of the system parameters or the initial conditions of the system. In the latter case, excessive motion coexists with bounded, meaning that a tiny disturbance of initial conditions leads to a change of dynamical behavior, such as from bounded vibration to unbounded. This is a typical initial-sensitive phenomenon, initial-sensitive excessive motion, indicating the loss of global integrity of dynamic systems [3,8]. Since it is often due to global bifurcation [15,21], we apply the Melnikov method [23] to detect the conditions for it. Substituting the heteroclinic orbits (9) into the Melnikov function yields

$$M_{\pm}(t_0) = \int_{-\infty}^{+\infty}(-\mu y(T) - gy^3(T) + f\cos(\omega(T + t_0)))y(T)dT = -\frac{2\sqrt{2}\mu}{3b} - \frac{8\sqrt{2}g}{35b^2} + \sqrt{\frac{2}{b}}\pi\omega f\,\mathrm{csch}(\frac{\sqrt{2}\pi\omega}{2})\cos(\omega t_0). \tag{24}$$

For

$$\sqrt{\frac{2}{b}}\,\pi\omega f\,\mathrm{csch}(\frac{\sqrt{2}\pi\omega}{2}) \geq \frac{2\sqrt{2}\mu}{3b} + \frac{8\sqrt{2}g}{35b^2}, \tag{25}$$

Namely,

$$f \geq f_0 = \frac{2(35\mu b + 12g)\sinh(\frac{\pi\omega}{\sqrt{2}})}{105\pi\omega b^{\frac{3}{2}}}, \tag{26}$$

there will be a value of t_0 satisfying $M_{\pm}(t_0) = 0$ and $M_{\pm}{}'(t_0) \neq 0$, meaning that the roots of Equation $M_{\pm}(t_0) = 0$ are simple, enabling the existence of the transverse heteroclinic orbits, and the system (6) may undergo initial-sensitive motion. According to Equation (26), f_0 is the amplitude threshold of heteroclinic bifurcation. For parameter values given by Equation (23), the critical value for heteroclinic bifurcation is $f_0 = 0.09$. In contrast, it follows from Inequation (25) that the threshold of the excitation frequency ω for heteroclinic bifurcation cannot be expressed as an explicit or monotonic increasing function of other parameters. For example, for $f = 0.13$ and the values of parameters in Equation (3), it can be calculated from Equation (25) that for $\omega < 1.22$, heteroclinic bifurcation will occur in the system (6).

In order to verify the criterion obtained in this subsection, numerical simulations are carried out by fixing parameter values as in Equation (23). The 4th Runge-Kutta approach and the point-mapping method [24] are employed to describe the phenomenon's initial-sensitive excessive motion. First, some terms such as basin of attraction and safe basin are introduced briefly. A basin of attraction is defined as the set of initial conditions that can lead to the same attractor [25]. If the boundary of basin of attraction of an attractor is fractal and mixed with another, jump among multiple attractors may occur [25]. Safe basin is defined as the union of basins of attraction for all bounded attractors [26]. Fractal of the safe basin of the system (6) induces the occurrence of initial-sensitive excessive motion. In this paper, the basin of attraction is drawn in the sufficiently large space region defined as $-3.0 \leq x(0) \leq 3.0$, $-3.0 \leq y(0) \leq 3.0$ by generating a 600×600 array of starting conditions, for each of these starting points. The escaping set for infinite time is approximated with good accuracy by a study with 1000 excited circles. The time step is taken as 0.01. The white region represents the numerical approximation to the basins of attraction of excessive motion. The red and blue regions are the basins of attraction for attractors from lower and higher stable periodic brunches of Figure 3, respectively. Thus, the union of the red region and the blue is the so-called safe basin.

The evolution of basins of attraction with the frequency ω for $f = 0.13$ and with the amplitude f for $\omega = 0.80$ can be observed in Figures 4 and 5. In Figure 4, each the safe basins, i.e., the union of red region and blue, is fractal-eroded when ω ranges from 0.62 to 0.87 satisfying $\omega < 1.22$ i.e., the condition of heteroclinic bifurcation, which shows the occurrence of initial-sensitive excessive motion, in agreement with the analytical prediction. In Figure 4a, there is only a red region, meaning that there is only one periodic attractor from the lower stable branch. When $\omega = 0.69$ (see Figure 4b), the red region and the blue one coexists, showing bistability. Specifically, on the left side of the origin, the red region is fractal and mixed with the white region and the blue, indicating that both jump and initial-sensitive excessive motion may occur there. As ω increases to 0.81 (see Figure 4c), most areas of the red region will be eroded by the blue one, showing that most of the safe initial conditions lead to the higher-amplitude attractor. For $\omega = 0.87$ (see Figure 4d), the red region disappears, and there is only the blue, which means that the response shifts from the lower-amplitude branch to the higher-amplitude, in agreement with the theoretical results of Figure 3a.

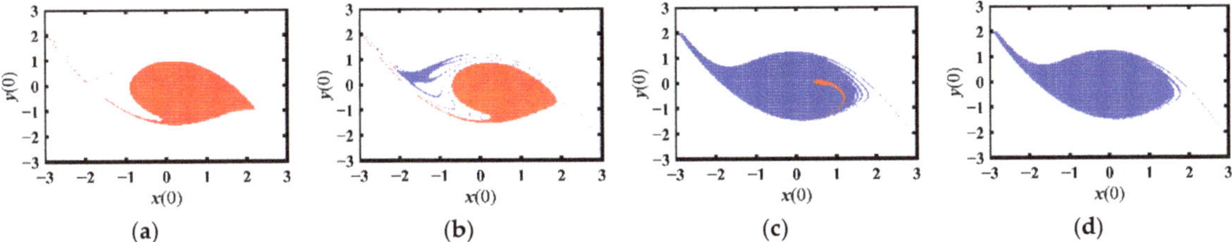

Figure 4. Evolution of basins of attraction with the increase of ω for $f = 0.13$. (**a**) $\omega = 0.62$; (**b**) $\omega = 0.69$; (**c**) $\omega = 0.81$; (**d**) $\omega = 0.87$.

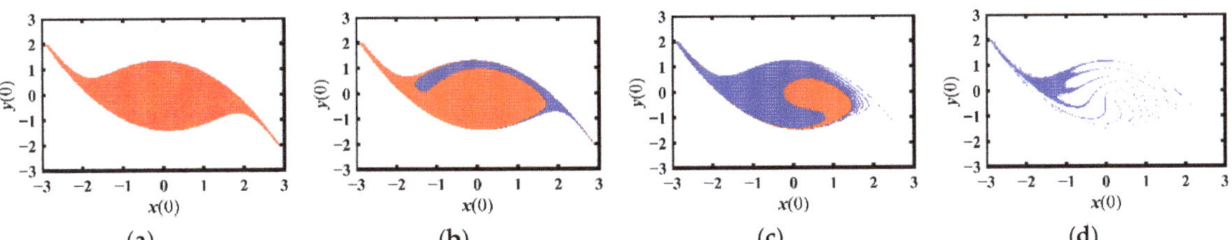

Figure 5. Evolution of basins of attraction with the increase of f for $\omega = 0.80$. (**a**) $f = 0.03$; (**b**) $f = 0.07$; (**c**) $f = 0.12$; (**d**) $f = 0.20$.

In Figure 5, the safe basin (the union of red and blue regions) is steadily eroded by the white region with the increase of the amplitude f, implying that initial-sensitive excessive motion can be induced by the increase of f. In Figure 5a,b, the boundary of safe basins is smooth. When f is more than 0.09, i.e., the critical threshold solved from Equation (26), the boundary of safe basin will become fractal (see Figure 5c,d). In Figure 5a–c, the red region and the blue one are mixed, indicating bi-stability as well as jump. Note that for $f = 0.12$, the red and blue regions are evenly mixed in the neighborhood of the origin (see Figure 5c). For $f = 0.20$, only the basin of attraction of the attractors in the higher stable branch of Figure 3b exists, which is seriously eroded by the white region, showing that the excessive motion is easy to trigger.

4. Time-Delay Feedback to Control Complex Dynamics

In this section, two types of linear delay feedback controllers, i.e., delayed position feedback and delayed velocity feedback, are applied to the system (1), and their effect on suppressing the phenomena jump and initial-sensitive excessive motion is discussed. The corresponding control diagram is shown in Figure 6 where Q is the coefficient of the gain, and τ time delay. $P(\theta_1, \theta_2, \dot{\theta}_1, \dot{\theta}_2, \theta_{1\tau}, \theta_{2\tau}, \dot{\theta}_{1\tau}, \dot{\theta}_{2\tau})$ represents delayed relative-angle feedback $(\theta_1(t - \tau) - \theta_2(t - \tau)) - (\theta_1(t) - \theta_2(t))$ or delayed relative-angular-velocity feedback $(\dot{\theta}_1(t - \tau) - \dot{\theta}_2(t - \tau)) - (\dot{\theta}_1(t) - \dot{\theta}_2(t))$, which is an active control transferring the motion state/velocity of the dynamical system [27]. The corresponding governing equation is

$$J_1\ddot{\theta}_1 + c_{12}(\dot{\theta}_1 - \dot{\theta}_2) + K_1(\theta_1 - \theta_2) + K_3(\theta_1 - \theta_2)^3 + f_{12} - QP(\theta_1, \theta_2, \dot{\theta}_1, \dot{\theta}_2, \theta_{1\tau}, \theta_{2\tau}, \dot{\theta}_{1\tau}, \dot{\theta}_{2\tau}) = T_{e1},$$
$$J_2\ddot{\theta}_2 + c_{12}(\dot{\theta}_1 - \dot{\theta}_2) + K_1(\theta_1 - \theta_2) + K_3(\theta_1 - \theta_2)^3 + f_{12} + QP(\theta_1, \theta_2, \dot{\theta}_1, \dot{\theta}_2, \theta_{1\tau}, \theta_{2\tau}, \dot{\theta}_{1\tau}, \dot{\theta}_{2\tau}) = T_{e2}.$$

(27)

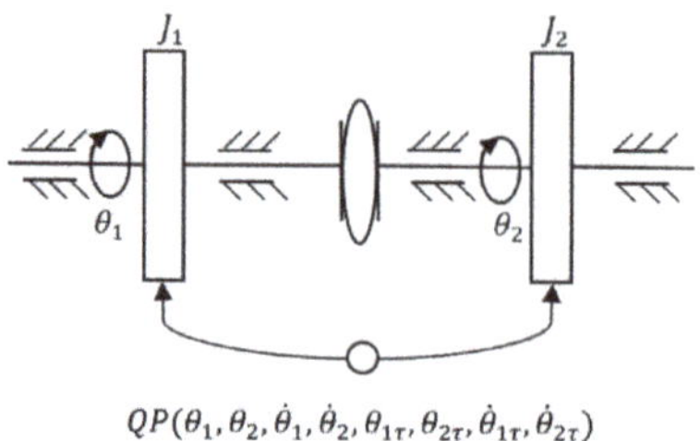

Figure 6. Diagram of the relative rotating system under delayed feedback control.

Letting

$$G_p = \frac{J_1 + J_2}{J_1 J_2} Q, \ G_v = \frac{J_1 + J_2}{J_1 J_2} Q \omega_0, \ \tilde{\tau} = \omega_0 \tau, \tag{28}$$

and based on Equations (2)–(6), we obtain the relative rotation system under delayed position feedback control, i.e.,

$$\frac{dx}{dT} = y, \ \frac{dy}{dT} = -\mu y - x + bx^3 - gy^3 + f\cos(\omega T) + G_p(x(T - \tilde{\tau}) - x(T)), \tag{29}$$

and the following delayed-velocity-feedback controlled system

$$y = \frac{dx}{dT}, \ \frac{dy}{dT} = -\mu y - x + bx^3 - gy^3 + f\cos(\omega T) + G_v(y(T - \tilde{\tau}) - y). \tag{30}$$

In the delayed-feedback-controlled systems (29) and (30), G_p, G_v and $\tilde{\tau}$ are independent parameters. For $G = 0$ or $\tilde{\tau} = 0$, the feedback terms in Equations (29) and (30) become 0, and the delayed systems (29) and (30) become the uncontrolled non-dimensional system (6). In this paper, considering the engineering application, we do not consider the periodic characteristics of $\tilde{\tau}$ but restrict that $0 \leq \tilde{\tau} \leq 2\pi$. Since there is no signal returned to the controlled systems (29) and (30) before $T = 0$, it is supposed that the initial states of the delayed system when $-\tilde{\tau} \leq T < 0$ satisfy $x(T) = y(T) = 0$. Then, safe basins of the delayed-feedback controlled systems can also be projected into the initial-state plane $x(0) - y(0)$, similar to the uncontrolled system (6).

4.1. Delayed Position Feedback Control

4.1.1. Primary Resonant Response and Stability of Solutions

As discussed in Section 3.1, the Method of Multiple Scales is applied to obtain the periodic response of the delayed system. Rescaling the gain as $G_p = \varepsilon \tilde{G}_p$, and following the similar procedure as in Section 3.1, one can finally obtain the slow flow equation:

$$\frac{da}{dT} = -\frac{\mu a}{2} - \frac{3ga^3}{8} + \frac{f\sin\psi}{2} - \frac{aG_p\sin\tilde{\tau}}{2}, \ a\frac{d\psi}{dT} = (\omega - 1)a + \frac{3ba^3}{8} + \frac{f\cos\psi}{2} - \frac{aG_p(1 - \cos\tilde{\tau})}{2}. \tag{31}$$

One can get the amplitude and the phases of the response by equating the right sides of Equation (31) to zero, i.e.,

$$\mu a + \frac{3ga^3}{4} + aG_p\sin\tilde{\tau} = f\sin\psi, \ -2(\omega - 1)a - \frac{3ba^3}{4} + aG_p(1 - \cos\tilde{\tau}) = f\cos\psi. \tag{32}$$

Accordingly, the approximate periodic solution can be expressed as $x_0 = a\cos(\omega T - \psi)$. Its amplitude a can be solved from the following equation

$$\left(\mu + G_p\sin\tilde{\tau} + \frac{3ga^2}{4}\right)^2 a^2 + \left(2\omega - 2 + \frac{3}{4}ba^2 + G_p\cos\tilde{\tau} - G_p\right)^2 a^2 = f^2. \tag{33}$$

The stability of the solutions, thus obtained, can be ascertained by computing the eigenvalues from the corresponding characteristic equation:

$$
\begin{aligned}
&\lambda^2 + \lambda\left(\mu + G_p \sin\tilde{\tau} + \tfrac{3ga^2}{2}\right) \\
&+ \tfrac{27(g^2+b^2)a^4}{64} + \tfrac{(\mu+G_p \sin\tilde{\tau})^2}{4} + \left(\omega - 1 + \tfrac{G_p \cos\tilde{\tau} - G_p}{2}\right)^2 + \tfrac{3a^2}{4}\left(gG_p \sin\tilde{\tau} + bG_p \cos\tilde{\tau} + b(2\omega - 2 - G_p) + g\mu\right) = 0.
\end{aligned}
\tag{34}
$$

Assuming $\lambda = iv$ in Equation (34) and separating the real and imaginary parts, we have

$$
v\left(\mu + G_p \sin\tilde{\tau} + \frac{3ga^2}{2}\right) = 0
\tag{35}
$$

and

$$
-v^2 + \tfrac{27(g^2+b^2)a^4}{64} + \tfrac{(\mu+G_p \sin\tilde{\tau})^2}{4} + \left(\omega - 1 + \tfrac{G_p \cos\tilde{\tau} - G_p}{2}\right)^2 + \tfrac{3a^2}{4}\left(gG_p \sin\tilde{\tau} + bG_p \cos\tilde{\tau} + b(2\omega - 2 - G_p) + g\mu\right) = 0.
\tag{36}
$$

As we know, when the delay is small, the delayed feedback can be expanded in a Taylor series so that the controlled system (29) will become

$$
\frac{dx}{dT} = y, \quad \frac{dy}{dT} = -(\mu + G_p\tilde{\tau})y - x + bx^3 - gy^3 + f\cos(\omega T).
\tag{37}
$$

For $G_p > 0$, it is much like increasing the damping of the system (6), thus reducing the periodic vibration [20,27]. Letting $G_p > 0$, one has $\mu + G_p \sin\tilde{\tau} + \tfrac{3ga^2}{2} > 0$ for $0 \leq \tilde{\tau} \leq \pi$. Thus, $v = 0$ in Equation (35); if

$$
\tfrac{27(g^2+b^2)a^4}{64} + \tfrac{3a^2}{4}\left(gG_p \sin\tilde{\tau} + bG_p \cos\tilde{\tau} + b(2\omega - 2 - G_p) + g\mu\right) + \left(\omega - 1 + \tfrac{G_p \cos\tilde{\tau} - G_p}{2}\right)^2 + \tfrac{(\mu+G_p \sin\tilde{\tau})^2}{4} = 0,
\tag{38}
$$

saddle-node bifurcation will occur in the periodic branch. For $G_p > 0$, when $\tilde{\tau}$ is long enough to satisfy

$$
\left(gG_p \sin\tilde{\tau} + bG_p \cos\tilde{\tau} + b(2\omega - 2 - G_p) + g\mu\right)^2 - 3(g^2 + b^2)\left(\left(\omega - 1 - \frac{G_p}{2} + \frac{G_p \cos\tilde{\tau}}{2}\right)^2 + \frac{(\mu + G_p \sin\tilde{\tau})^2}{4}\right) < 0,
\tag{39}
$$

there will be no positive roots for Equation (38), meaning under a positive gain, the delayed position feedback can be used to reduce saddle-node bifurcation and thus bi-stability. For example, given $G_p = 0.3$ and $\omega = 0.8$, one can calculate from Equation (39) that there is no saddle-node bifurcation for $\tilde{\tau} > 0.63$. In contrast, it is hard to express the range of $\tilde{\tau}$ for saddle-node bifurcation as a function of f. Fixing the value of f and combining Equations (33), (38) and (39), one can obtain the condition for $\tilde{\tau}$ to reduce saddle-node bifurcation. For instance, given f = 0.13, one can calculate that the condition of $\tilde{\tau}$ to reduce saddle-node bifurcation is $\tilde{\tau} > 0.27$.

Based on the above theoretical analysis, the results are verified by numerical simulation as shown in Figure 7. Figure 7a shows the frequency-response for $f = 0.13$, $G_p = 0.3$ and different values of time delay satisfying $\tilde{\tau} \ll \pi$. Obviously, in Figure 7a, the amplitude of the response reduces with the increase in time delay. Note that for $\tilde{\tau} = 0.35$ (longer than the theoretical values 0.27), there is no saddle-node bifurcation on the periodic branches, showing that bi-stability and jump will not occur. It can be seen from Figure 7a that the increase of time delay in the position feedback can reduce saddle-node bifurcation, and thus bi-stability and jump. This can also be verified by the evolution of basins of attraction with the increase of time delay (see Figure 8). It follows from Figure 8 that as $\tilde{\tau}$ increases, the basin of attraction of the higher-amplitude periodic attractor (the blue region) will be eroded by the basin of attraction of the lower-amplitude one (see the red region). Meanwhile, the safe basin (red and bule regions) becomes smooth. For $\omega = 0.80$, one can see from Figure 7b that even though the saddle-node bifurcation still exists, increase of time delay can help to reduce the range of f where bistable attractors coexist. The corresponding sequences of basin of attraction with the increase of f and $\widetilde{\tau}$ will be discussed in detail in the next subsection.

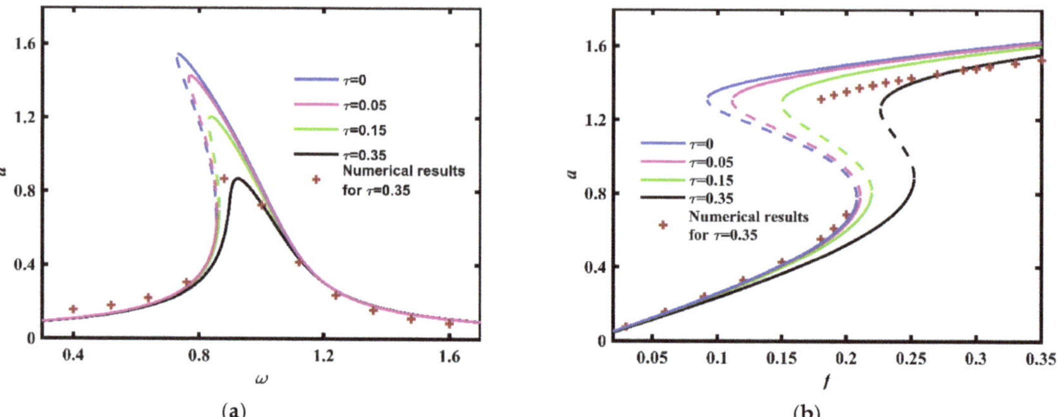

Figure 7. Variation of the response amplitude of the system (29) with the excitation for $G_p = 0.3$. (**a**) $f = 0.13$; (**b**) $\omega = 0.80$.

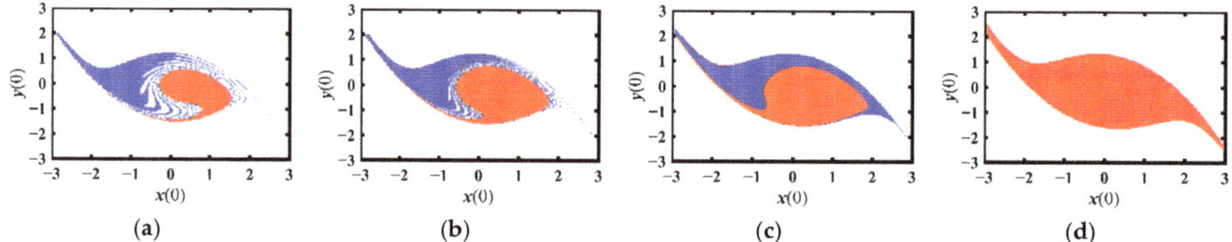

Figure 8. Evolution of basins of attraction with the increase of τ when $f = 0.13$ and $\omega = 0.78$. (**a**) $\tau = 0$; (**b**) $\tau = 0.05$; (**c**) $\tau = 0.15$; (**d**) $\tau = 0.35$.

4.1.2. Heteroclinic Bifurcation

To employ the Melnikov method to analyze the critical condition for heteroclinic bifurcation in a delayed-feedback-controlled system, it is necessary to consider the delayed position feedback as a perturbed term, namely, the values of the delay and the gain should satisfy that the stability of the equilibrium of the uncontrolled linear system cannot be changed by the delayed feedback. Since the equilibrium $S_1(0,0)$ is stable for $\mu > 0$, the delay should not exceed the first value of the stability switch of the following linear delayed system

$$\frac{d^2x}{dT^2} + \frac{\mu dx}{dT} + x = G_p(x(T - \widetilde{\tau}) - x(T)). \tag{40}$$

Its characteristic equation is given by

$$\lambda^2 + \mu\lambda + 1 + G_p - G_p e^{-\lambda\widetilde{\tau}} = 0. \tag{41}$$

Substituting $\lambda = iv$ into (41) and separating the imaginary and real parts yields

$$-\Omega_p{}^2 + 1 + G_p - G_p \cos(\Omega_p\widetilde{\tau}) = 0, \ \mu\Omega_p + G_p \sin(\Omega_p\widetilde{\tau}) = 0. \tag{42}$$

By eliminating the trigonometric functions, Equation (42) becomes

$$\Omega_p^4 - (2 + 2G_p - \mu^2)\Omega_p^2 + 1 + 2G_p = 0. \tag{43}$$

For

$$G_p > \max\{\frac{\mu^2}{2} - 1, \frac{\mu^2}{2} + \mu\}, \tag{44}$$

there are two different positive solutions of Equation (43) expressed as Ω_{p1} and Ω_{p2} ($\Omega_{p1} > \Omega_{p2} > 0$); the critical value of the delay for stability switch of $S_1(0,0)$ is

$$\tau_0 = \frac{1}{\Omega_{p1}}(2\pi - \arccos(1 + \frac{1 - \Omega_{p1}^2}{G_p})). \tag{45}$$

For $\tilde{\tau} < \tau_0$, the delay position feedback can be regarded as a disturbed term. Substituting the heteroclinic orbits (9) into the Melnikov function of the system (29) yields

$$M_{\pm}(t_0) = -\frac{2\sqrt{2}\mu}{3b} - \frac{8\sqrt{2}g}{35b^2} + \sqrt{\frac{2}{b}}\pi\omega f \operatorname{csch}(\frac{\pi\omega}{\sqrt{2}})\cos(\omega t_0) - \frac{G_p}{b}l_1(\tilde{\tau}), \tag{46}$$

where $l_1(\tilde{\tau}) = \operatorname{csch}^2(\frac{\sqrt{2}}{2}\tilde{\tau})(\sinh(\sqrt{2}\tilde{\tau}) - \sqrt{2}\tilde{\tau}) > 0$. The critical condition for heteroclinic bifurcation can be expressed as

$$f > f^P(\tilde{\tau}) = f_0 + \frac{G_p \operatorname{sech}(\frac{\pi\omega}{\sqrt{2}})l_1(\tilde{\tau})}{\pi\sqrt{2b\omega}}, \tag{47}$$

Accordingly, for $G_p > 0$, the threshold of heteroclinic bifurcation $f^P(\tilde{\tau})$ will increase with time delay and be higher than f_0. Fixing $\mu = 0.02$ and $G_p = 0.3$, it can be calculated that $\tau_0 = 2.55$.

Figure 9 shows the variation of $f^P(\tilde{\tau})$ with the increase of time delay $\tilde{\tau}$ ($\tilde{\tau} \ll \tau_0$). Then the numerical values of $f^P(\tilde{\tau})$ are obtained at which the boundary of the safe basin begin to be unsmooth. Each numerical critical value of $f^P(\tilde{\tau})$ is kept at two decimal places. We make sure that if f is less than the numerical results $f^P(\tilde{\tau})$, the boundary of safe basin will be smooth. In Figure 9, the numerical results for the critical values of f are in agreement with the analytical values, demonstrating that the threshold of the amplitude f for heteroclinic bifurcation will increase monotonically with the delay for $G_p > 0$ and $\tilde{\tau} < \tau_0$.

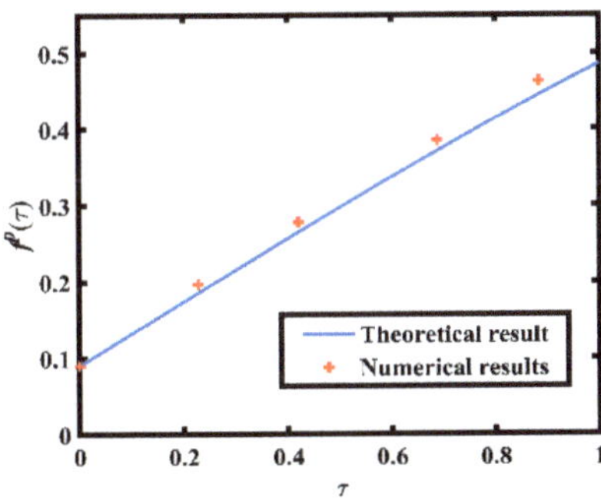

Figure 9. $f^P(\tau)$ of the controlled system (29) versus time delay when $G_p = 0.3$ and $\omega = 0.80$.

Figure 10 depicts the evolution of the safe basin of the controlled system (29) with the delay τ. In Figure 10a,e,i for $\tilde{\tau} = 0$, safe basins as well as basins of attraction can be considered as those of the uncontrolled system (6), illustrating that as the amplitude f increases, safe basin will be eroded. Specifically, in Figure 10e for $f = 0.30$, only the basin of attraction of the higher-amplitude attractor is left, which is seriously eroded by the white region, showing the high possibility of excessive motion. For $G_p = 0.3$ and $\tilde{\tau} = 0.11$, the safe basin is obviously expanded. Besides, comparing Figure 10a–d, one can observe that as $\tilde{\tau}$ increases, the safe basin changes from the union of red region and blue to the red region itself, meaning that jump and initial-sensitive excessive motion are reduced by the basin expansion of the lower-amplitude attractor. In Figure 10e–h, one can also draw this conclusion. Note that Figure 10e,f, three color regions are mixed in the neighborhood of the origin, which indicates the high probability of jump and initial-sensitive excessive motion. As $\tilde{\tau}$ increases, they can be controlled. For $f = 0.30$ (see Figure 10i–l), even though the safe basin is still fractal, its area becomes much larger with the increase of $\tilde{\tau}$. This shows that the possibility of excessive motion is reduced. It follows from Figure 9 that the delayed position feedback can suppress the erosion of the safe basin effectively when G_p is positive.

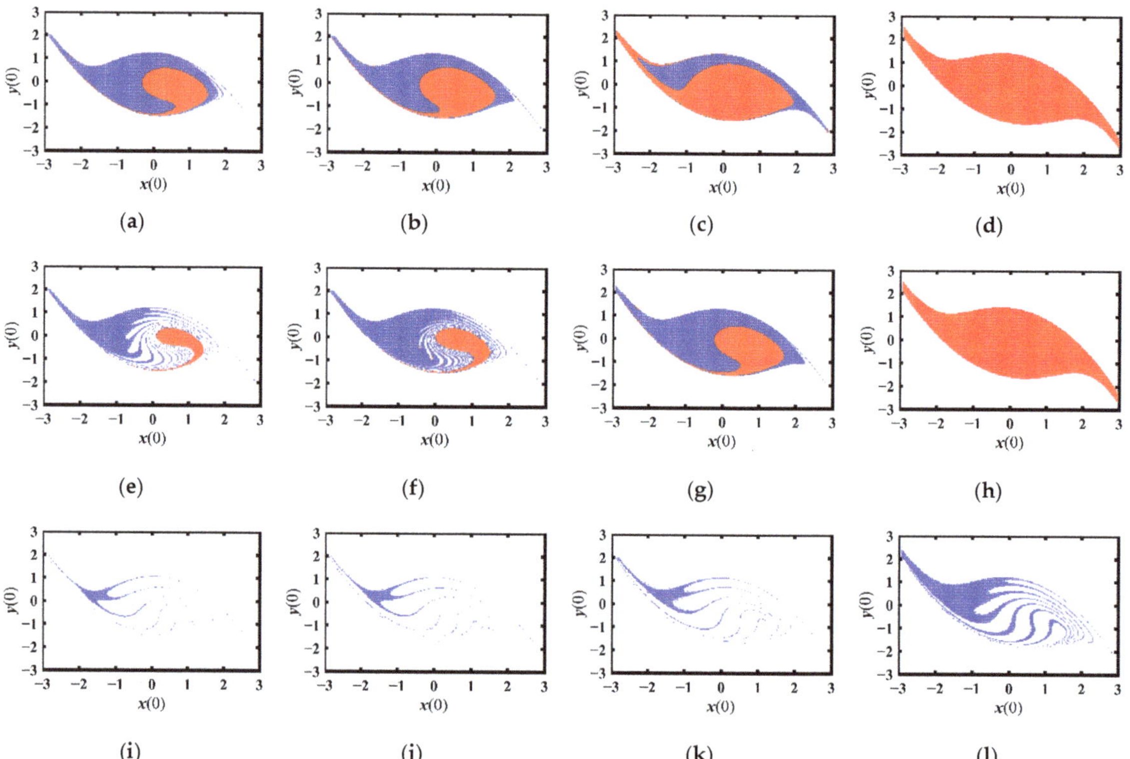

Figure 10. Sequences of safe basin of the system (29) with the increase of f and $\tilde{\tau}$ when $G_p = 0.3$. (**a**) $f = 0.11$, $\tilde{\tau} = 0$; (**b**) $f = 0.11$, $\tilde{\tau} = 0.05$; (**c**) $f = 0.11$, $\tilde{\tau} = 0.15$; (**d**) $f = 0.11$, $\tilde{\tau} = 0.35$; (**e**) $f = 0.15$, $\tilde{\tau} = 0$; (**f**) $f = 0.15$, $\tilde{\tau} = 0.05$; (**g**) $f = 0.15$, $\tilde{\tau} = 0.15$; (**h**) $f = 0.15$, $\tilde{\tau} = 0.35$; (**i**) $f = 0.30$, $\tilde{\tau} = 0$; (**j**) $f = 0.30$, $\tilde{\tau} = 0.05$; (**k**) $f = 0.30$, $\tilde{\tau} = 0.15$; (**l**) $f = 0.30$, $\tilde{\tau} = 0.35$.

4.2. Delayed Velocity Feedback

4.2.1. Primary Resonant Response and Stability of Solutions

As in Section 4.1.1, by applying the Method of Multiple Scales, one can get the slow flow equation of the delayed-velocity-feedback controlled system (30) as below:

$$\frac{da}{dT} = -\frac{\mu}{2}a - \frac{3ga^3}{8} + \frac{f \sin \psi}{2} - \frac{G_v(1 - \cos \tilde{\tau})}{2}a, \quad a\frac{d\psi}{dT} = a(\omega - 1) + \frac{3ba^3}{8} + \frac{f \cos \psi}{2} + \frac{G_v \sin \tilde{\tau}}{2}a. \tag{48}$$

The amplitude and phases of the system can be obtained via letting the right side of Equation (48) be zero, given by

$$f \sin \psi = \mu a + \frac{3g}{4} a^3 + G_v(1 - \cos \tilde{\tau})a, \quad f \cos \psi = (2 - 2\omega - \frac{3b}{4} a^2 - G_v \sin \tilde{\tau})a. \tag{49}$$

Thus, the amplitude of the approximated periodic solution $x_0 = a \cos(\omega T - \psi)$ can be solved from the following equation:

$$\left(\mu + \frac{3g}{4} a^2 + G_v(1 - \cos \tilde{\tau})\right)^2 a^2 + \left(2 - 2\omega - \frac{3b}{4} a^2 - G_v \sin \tilde{\tau}\right)^2 a^2 = f^2. \tag{50}$$

According to its characteristic equation, the stability switch will occur if there exists a real number $\tilde{v}$ satisfying

$$v(\mu + G_v(1 - \cos \tilde{\tau}) + \frac{3ga^2}{2}) = 0,$$
$$-v^2 + \frac{(\mu + G_v(1 - \cos \tilde{\tau}))^2}{4} + \frac{(2 - 2\omega - G_v \sin \tilde{\tau})^2}{4} + \frac{\mu}{2} G_v(1 - \cos \tilde{\tau}) + \frac{3a^2}{4}(\mu g + b(2\omega - 2 + G_v \sin \tilde{\tau})) + \frac{27(g^2 + b^2)a^4}{64} = 0. \tag{51}$$

For $G_v > 0$, in Equation (51), one has $\mu + \frac{3g}{2} a^2 + G_v(1 - \cos \tilde{\tau}) > 0$, $\tilde{v} = 0$, and

$$\frac{(\mu + G_v(1 - \cos \tilde{\tau}))^2}{4} + \frac{(2 - 2\omega - G_v \sin \tilde{\tau})^2}{4} + \frac{\mu}{2} G_v(1 - \cos \tilde{\tau}) + \frac{3a^2}{4}(\mu g + b(2\omega - 2 + G_v \sin \tilde{\tau})) + \frac{27(g^2 + b^2)a^4}{64} = 0, \tag{52}$$

implying that saddle-node bifurcation will occur in the delayed-velocity-feedback controlled system. For $G_v > 0$, when $\tilde{\tau}$ is long enough to satisfy

$$4(\mu g + b(2\omega - 2) + bG_v \sin \tilde{\tau})^2 - 3(g^2 + b^2)((\mu + G_v)^2 + 2\mu G_v + G_v^2 + 4(1 - \omega)^2 - 2G_v(2\mu + G_v) \cos \tilde{\tau} + 4(\omega - 1)G_v \sin \tilde{\tau}) < 0, \tag{53}$$

there will be no positive roots for Equation (52), meaning no saddle-node bifurcation or bi-stability. For example, given $G_v = 0.3$ and $\omega = 0.8$, one can calculate from Equation (53) that saddle-node bifurcation will disappear for $\tilde{\tau} > 0.68$. Similarly to Section 4.1.1, although delayed velocity feedback can be used to reduce the occurrence of saddle-node bifurcation, it is hard to express the range of $\tilde{\tau}$ for SN bifurcation as a function of f. Fixing the value of f and combining Equations (50), (52) and (53), one can obtain the condition for $\tilde{\tau}$ to reduce saddle-node bifurcation. For instance, given $f = 0.13$, one can calculate that the condition of $\tilde{\tau}$ to reduce saddle-node bifurcation is $\tilde{\tau} > 0.77$.

The similar as in Section 4.1.1, for $G_v = 0.3$ and different values of time delay in the velocity feedback, Figure 11 shows the variation of the response amplitude with the excitation. It follows from the theoretical and numerical results depicted in Figure 11 that, even though the response amplitude is not necessary to be suppressed with the increase of time delay, the velocity feedback can reduce saddle-node bifurcation, and thus bi-stability and jump. The controlling effect of velocity feedback on jump can also be verified by the evolution of basins of attraction with the increase of time delay for $f = 0.13$, $\omega = 0.78$ (see Figure 12). To be different from the case of the delayed position feedback, as the delay increases, the basin of attraction of the higher-amplitude attractor is expanded, whose boundary becomes smoother and smoother. Finally, it becomes the only bounded attractor (see Figure 12d). For $\omega = 0.80$, if can be concluded from Figure 11b that the delayed velocity feedback can suppress bi-stability. For the fixed value of ω, sequences of the basin of attraction with the increase of f and $\tilde{\tau}$ will be discussed in detail in the next subsection.

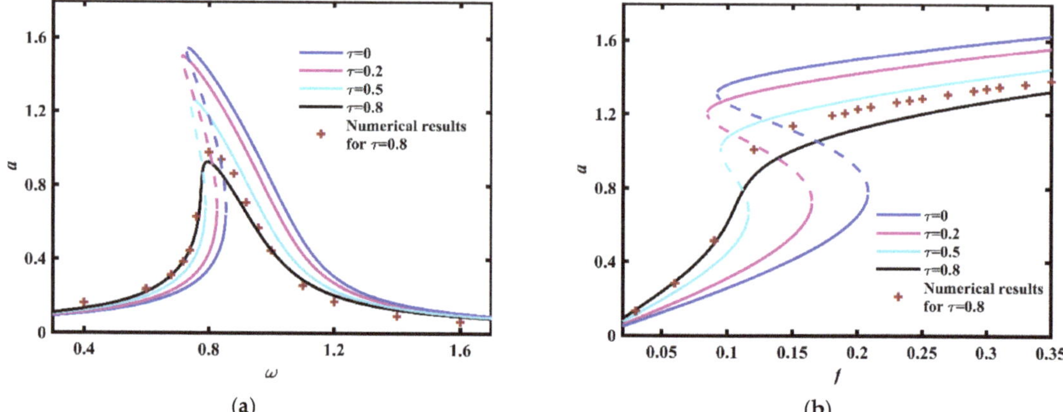

Figure 11. Variation of the response amplitude of the system (30) with the excitation for $G_v = 0.3$. (a) $f = 0.13$; (b) $\omega = 0.80$.

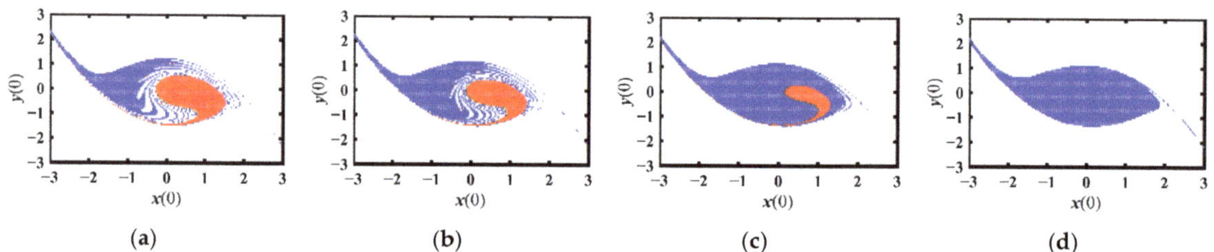

Figure 12. Evolution of basins of attraction with the increase of τ for $f = 0.13$ and $\omega = 0.78$. (a) $\widetilde{\tau} = 0$; (b) $\widetilde{\tau} = 0.2$; (c) $\widetilde{\tau} = 0.5$; (d) $\widetilde{\tau} = 0.8$.

4.2.2. Heteroclinic Bifurcation under Delayed Velocity Feedback

As in Section 4.1.2, to apply the Melnikov method in the delayed-velocity-feedback controlled system (30), we also need to treat the delayed feedback as a perturbed term of this system. To this end, the delay $\widetilde{\tau}$ should not exceed the first value of stability switch of $S_1(0,0)$ in the linear delayed system whose characteristic equation can expressed as

$$\lambda^2 + \mu\lambda + 1 + G_v\lambda - G_v\lambda e^{-\lambda\widetilde{\tau}} = 0 \tag{54}$$

The first value of time delay for stability switch of $S_1(0,0)$ will occur when $\lambda = \pm\Omega_v i$. Substituting it into Equation (54) and separating the imaginary and the real parts of Equation (54) yields

$$-\Omega_v{}^2 + 1 = G_v\Omega_v \sin(\Omega_v\widetilde{\tau}), \ \mu\Omega_v = G_v\Omega_v(\cos(\Omega_v\widetilde{\tau}) - 1). \tag{55}$$

By eliminating the trigonometric functions, the equation above becomes

$$(\Omega_v^2 - 1)^2 + (\mu^2 + 2\mu G_v)\Omega_v{}^2 = 0. \tag{56}$$

Since $\mu > 0$, for $G_v > 0$, there will be no real roots of Equation (56), meaning that the stability of the origin will not be changed. Then the delay velocity feedback of the delayed system (30) can be regarded as a disturbed term. The Melnikov function of the controlled system (30) can be expressed as

$$M_{\pm}(t_0) = \int_{-\infty}^{+\infty}(-\mu y(T) - gy^3(T) + f\cos(\omega(T+t_0)) - G_v y(T) + G_v y(T-\tilde{\tau}))y(T)dT$$
$$= -\frac{2\sqrt{2}\mu}{3b} - \frac{8\sqrt{2}g}{35b^2} + \sqrt{\frac{2}{b}}\pi\omega f\,\mathrm{csch}(\frac{\sqrt{2}\pi\omega}{2})\cos(\omega t_0) - \frac{2G_v}{b}l_2, \tag{57}$$

where $l_2(\tilde{\tau}) = \mathrm{csch}^2(\frac{\sqrt{2}\tilde{\tau}}{2})(\sqrt{2} - \tilde{\tau}\coth(\frac{\sqrt{2}}{2}\tilde{\tau})) > 0$. Accordingly, for

$$f > f^V(\tilde{\tau}) = f_0 + \frac{\sqrt{2b}G_v}{b\pi\omega}\sinh(\frac{\sqrt{2}\pi\omega}{2})l_2(\tilde{\tau}), \tag{58}$$

And heteroclinic bifurcation may occur. In Equation (58), for $G_v > 0$, the threshold of heteroclinic bifurcation $f^V(\tilde{\tau})$ increases monopoly with the delay $\tilde{\tau}$. Given $G_v = 0.3$, the variation of $f^V(\tilde{\tau})$ with the increase of $\tilde{\tau}$ is presented in Figure 13. Numerical and analytical results both illustrate that, for $G_v > 0$, the delayed velocity feedback can be used to increase the threshold of the amplitude f for heteroclinic bifurcation, thus being effective in reducing the probability of initial-sensitive excessive motion.

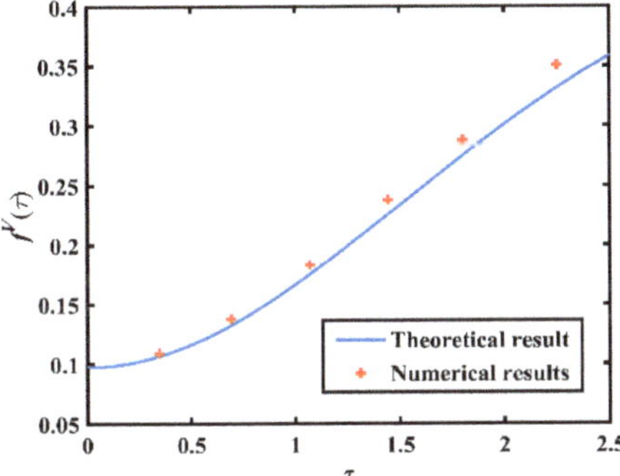

Figure 13. $f^V(\tau)$ of the controlled system (30) versus time delay when $G_v = 0.3$.

Figure 14 shows the evolution of the safe basin of the controlled system (30) with the increase of f and $\tilde{\tau}$. For $\tilde{\tau} = 0$, Figure 14a,e,i are the same as Figure 10a,e,i, depicting the safe basin of the uncontrolled system (6). Under a positive gain $G_v = 0.3$, when $\tilde{\tau}$ increases, the area of safe basin of the delayed-velocity-feedback-controlled system will expand obviously. To be different from the results of Section 4.1.2, for $f = 0.11$ and 0.15 where the two periodic attractors coexisting in the uncontrolled system, as $\tilde{\tau}$ increases, the basin of attraction of the higher-amplitude attractor is expanded and finally becomes the safe basin (see the blue regions in Figure 14a–h). This illustrates that, although the delayed velocity feedback does not necessary reduce the response amplitude, it reduces jump and excessive motion.

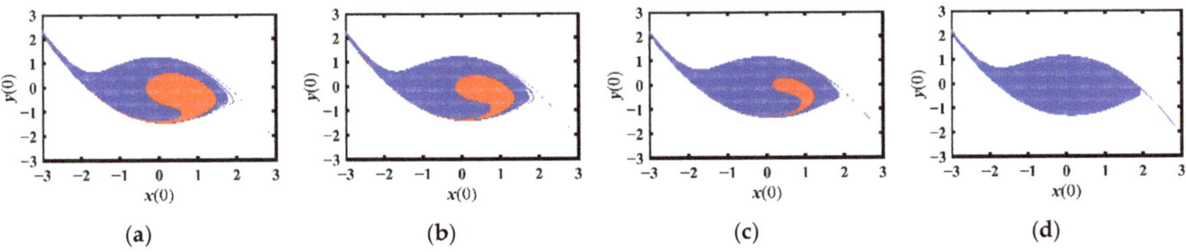

Figure 14. *Cont.*

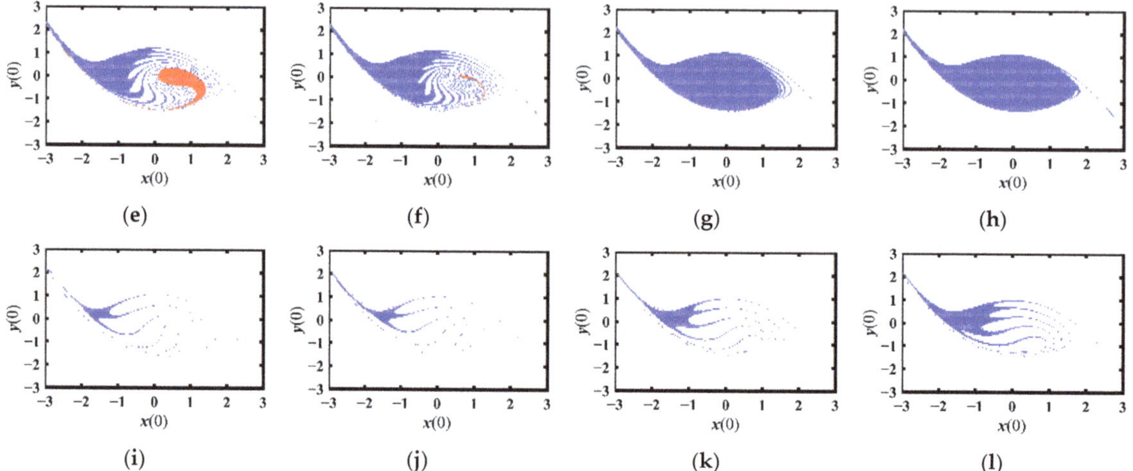

Figure 14. Sequences of safe basin of the system (30) with the increase of f and $\widetilde{\tau}$ when $G_v = 0.3$. (**a**) $f = 0.11$, $\widetilde{\tau} = 0$; (**b**) $f = 0.11$, $\widetilde{\tau} = 0.20$; (**c**) $f = 0.11$, $\widetilde{\tau} = 0.50$; (**d**) $f = 0.11$, $\widetilde{\tau} = 0.80$; (**e**) $f = 0.15$, $\widetilde{\tau} = 0$; (**f**) $f = 0.15$, $\widetilde{\tau} = 0.20$; (**g**) $f = 0.15$, $\widetilde{\tau} = 0.50$; (**h**) $f = 0.15$, $\widetilde{\tau} = 0.80$; (**i**) $f = 0.30$, $\widetilde{\tau} = 0$; (**j**) $f = 0.30$, $\widetilde{\tau} = 0.20$; (**k**) $f = 0.30$, $\widetilde{\tau} = 0.50$; (**l**) $f = 0.30$, $\widetilde{\tau} = 0.80$.

5. Discussion

In this paper, a typical relative rotation system is considered, and the phenomena of jump and initial-sensitive excessive motion as well as their suppression via delayed feedback is investigated. The Method of Multiple Scales and the Melnikov method are applied to analyze the conditions for bi-stability and initial-sensitive excessive motion, respectively. By introducing the basins of attraction and safe basin to describe the extent of jump and initial-sensitive excessive motion, respectively, the point mapping approach is employed to present the numerical results which matches the theoretical approach, verifying the validity of the analysis. Some significant results are presented as follows.

(1) The variation of excitation may induce the coexistence of bistable periodic attractors, which can be ascribed to saddle-node bifurcation.
(2) The increase of the excitation amplitude may cause initial-sensitive excessive motion, which can be due to heteroclinic bifurcation.
(3) Under positive coefficients of the feedback gain, the delayed position feedback and the delayed velocity feedback can reduce saddle-node bifurcation and heteroclinic bifurcation so as to suppress jump and initial-sensitive excessive motion. Comparatively, the former can also reduce the amplitude of the response, while the latter may not; the former works well if time delay does not exceed the first stability switch of the trivial equilibrium, while the latter does not have that restriction.

This work presents a detailed analysis of jump and initial-sensitive excessive motion of a typically relative rotation system, which may be beneficial for the performance improvement of rotors and main bearings. The relevant experimental investigations will be included in our future work.

Author Contributions: Conceptualization, H.S.; formal analysis, H.S.; funding acquisition, H.S.; Investigation, Z.C.; methodology, H.S.; project administration, H.S.; software, Z.C.; supervision, H.S.; validation, H.S.; visualization, Z.C.; writing—original draft, Z.C. and H.S.; writing—review & editing, H.S. All authors have read and agreed to the published version of the manuscript.

Funding: This research was funded by the National Natural Science Foundation of China, grant number 11472176.

Institutional Review Board Statement: Not applicable.

Informed Consent Statement: Not applicable.

Data Availability Statement: Not applicable.

Conflicts of Interest: The authors declare no conflict of interest.

References

1. Liang, Y.; Li, N. Optimal vibration control for nonlinear systems of tracked vehicle half-car suspensions. *Int. J. Control. Autom. Syst.* **2017**, *15*, 1675–1683. [CrossRef]
2. Renson, L.; Noël, J.P. Complex dynamics of a nonlinear aerospace structure: Numerical continuation and normal modes. *Nonlinear Dyn.* **2015**, *79*, 1293–1309. [CrossRef]
3. Shi, J.; Gou, X. Bifurcation and Erosion of Safe Basin for a Spur Gear System. *Int. J. Bifurc. Chaos* **2018**, *28*, 1830048. [CrossRef]
4. Li, J.; Wu, H. Bifurcation, chaos, and their control in a wheelset model. *Math. Methods Appl. Sci.* **2020**, *43*, 7152–7174. [CrossRef]
5. Luo, Z.; Wang, J. Research on vibration performance of the nonlinear combined support-flexible rotor system. *Nonlinear Dyn.* **2019**, *98*, 113–128. [CrossRef]
6. Li, Y.; Luo, Z. Dynamic modeling and stability analysis of a rotor-bearing system with bolted-disk joint. *Mech. Syst. Signal Process.* **2021**, *158*, 107778. [CrossRef]
7. Shi, P.; Liu, B. Chaotic motion of some relative rotation nonlinear dynamic system. *Acta Phys. Sin.* **2008**, *57*, 1321–1328.
8. Li, Z.; Gou, X. Erosion and bifurcation of safety-attraction basin for multi-state meshing gear transmission system under tooth safety condition. *J. Vbration Shock* **2021**, *40*, 63–74.
9. Wang, K.; Guan, X. Precise periodic solutions and uniqueness of periodic solutions of some relative rotation nonlinear dynamic system. *Acta Phy. Sin.* **2010**, *59*, 3648–3653. [CrossRef]
10. Xiao, L.; Xuan, C. The periodic solution problem of a relative rotation nonlinear dynamic system with time-varying stiffness. *Acta Phy. Sin.* **2013**, *62*, 21–25.
11. Li, X.; Yan, J. The periodic solution problem of a relative rotation nonlinear system with nonlinear elastic force and generalized damping force. *Acta Phy. Sin.* **2014**, *63*, 36–41.
12. Shi, P.; Han, D. Chaos and chaotic control in a relative rotation nonlinear dynamical system under parametric excitation. *Chin. Phys. B* **2010**, *19*, 116–121.
13. Verichev, N. Chaotic torsional vibration of imbalanced shaft driven by a limited power supply. *J. Sound Vib.* **2012**, *331*, 384–393. [CrossRef]
14. Liu, B.; Zhao, H. Bifurcation and chaos of some relative rotation system with triple-well Mathieu-Duffing oscillator. *Acta Phy. Sin.* **2014**, *63*, 174502.
15. Liu, S.; Tian, S. Chaos of a kind of nonlinear relative rotation system based on the effect of Coulomb friction. *Acta Phy. Sin.* **2015**, *64*, 247–254.
16. Zhu, L.; Li, Z. Evolutionary mechanism of safety performance for spur gear pair based on meshing safety domain. *Nonlinear Dyn.* **2021**, *104*, 215–239. [CrossRef]
17. Ju, J.; Wei, L. Dynamics and nonlinear feedback control for torsional vibration bifurcation in main transmission system of scraper conveyor direct-driven by high-power PMSM. *Nonlinear Dyn.* **2018**, *93*, 307–321. [CrossRef]
18. Shang, H. Pull-in instability of a typical electrostatic MEMS resonator and its control by delayed feedback. *Nonlinear Dyn.* **2017**, *90*, 171–183. [CrossRef]
19. Wang, Q.; Wu, H. The effect of fractional damping and time-delayed feedback on the stochastic resonance of asymmetric SD oscillator. *Nonlinear Dyn.* **2022**, *107*, 2099–2114. [CrossRef]
20. Zhao, Y.; Li, C. The delayed feedback control to suppress the vibration in a torsional vibrating system. *Acta Phy. Sin.* **2011**, *60*, 417–425.
21. Shang, H.; Han, Y. Suppression of chaos and basin erosion in a nonlinear relative rotation system by delayed position feedback. *Acta Phy. Sin.* **2014**, *63*, 88–95.
22. Rezaei, M.; Khadem, E.S.; Friswell, I.M. Energy harvesting from the secondary resonances of a nonlinear piezoelectric beam under hard harmonic excitation. *Meccanica* **2020**, *55*, 1463–1479. [CrossRef]
23. Siewe, S.M.; Hegazy, H.U. Homoclinic bifurcation and chaos control in MEMS resonators. *Appl. Math. Model.* **2011**, *35*, 5533–5552. [CrossRef]
24. Danico, K.; Tanmoy, C.; Milan, C.; Sondipon, A.; Karličić, D.; Chatterjee, T.; Cajić, M.; Adhikari, S. Parametrically amplified Mathieu-Duffing nonlinear energy harvesters. *J. Sound Vib.* **2020**, *488*, 115677.
25. Zhu, Y.; Shang, H. Multistability of the vibrating system of a micro resonator. *Fractal Fract.* **2022**, *6*, 141. [CrossRef]
26. Rega, G.; Lenci, S. Dynamical integrity and control of nonlinear mechanical oscillators. *J. Vib. Control* **2008**, *14*, 159–179. [CrossRef]
27. Liu, C.; Yan, Y.; Wang, W. Resonances and chaos of electrostatically actuated arch micro/nanoresonators with time delay velocity feedback. *Chaos Solitons Fractals* **2019**, *10*, 109512. [CrossRef]

 *mathematics*

Article

The General Dispersion Relation for the Vibration Modes of Helical Springs

Leopoldo Prieto *, Alejandro Quesada, Ana María Gómez Amador and Vicente Díaz

Department of Mechanical Engineering, Universidad Carlos III de Madrid, 28911 Leganes, Spain; alejandro@ing.uc3m.es (A.Q.); amgomez@ing.uc3m.es (A.M.G.A.); vdiaz@ing.uc3m.es (V.D.)
* Correspondence: leprieto@ing.uc3m.es

Abstract: A system of mathematical equations was developed for the calculation of the natural frequencies of helical springs, its predictions being compared with finite element simulation with ANSYS®. Authors derive the general equations governing the helical spring vibration relative to the Frenet trihedral representing the normal, binormal and tangent unit vectors to the spring medium line. The dispersion relation $\omega = f(k)$ has been obtained to model a wave traveling along the axis of the wire.

Keywords: helical spring; vibration; Frenet trihedral; dispersion relation; natural frequency

MSC: 70B15

Citation: Prieto, L.; Quesada, A.; Gómez Amador, A.M.; Díaz, V. The General Dispersion Relation for the Vibration Modes of Helical Springs. *Mathematics* **2022**, *10*, 2698. https://doi.org/10.3390/math10152698

Academic Editor: Dan B. Marghitu

Received: 24 June 2022
Accepted: 28 July 2022
Published: 30 July 2022

Publisher's Note: MDPI stays neutral with regard to jurisdictional claims in published maps and institutional affiliations.

1. Introduction

Helical springs are one of the most frequently used elastic elements in mechanical engineering. They are used in the most diverse structures as elastic energy accumulators supplementing damping devices, e.g., automobile or railroad suspensions, and in advance and return devices, e.g., camshafts and valves of internal combustion engines (Wahl [1], Shigley [2]; Kobelev [3]). One of the spring failure modes is caused by resonance vibration that occurs when the spring is excited with a periodic signal of frequency equal to its natural frequency.

Love [4] develops equations to study the static response of helical springs subjected to large deformations. Stokes [5] investigates vibrations in the case of a spring subjected to shock loads. Gironnet and Louradour [6] determine more precise expressions for the calculation of the natural frequencies of the helical spring. There are several numerical methods to determine the natural frequencies, the main ones being: the transfer matrix method (Yildirim [7]), the dynamic stiffness formulation (Lee [8]), and the pseudo-spectral method (Lee [9]). Becker [10] determines with the transfer matrix method the resonant frequencies of a spring subjected to an axial compression load. Jiang [11] studies forced vibrations and wave propagation in springs using the Laplace transform.

Wahl [1] already proposed in the middle of the 20th century the equations that are widely used today for the design of springs. Among these equations, Equation (1), which provides the stiffness constant of the spring subjected to an axial force, and Equation (2), which provides the natural frequency of the spring placed between two parallel flat plates, stand out.

$$k_{Wahl} = \frac{Gd^4}{8D^3N} \tag{1}$$

where k_{Wahl} is the spring axial stiffness, G is the shear modulus of elasticity, d is the wire diameter, D is the helix diameter and N is the number of active coils.

Where f is the natural harmonic frequencies of a spring in Hz, and W is the mass of the spring in kg. The fundamental frequency is determined for $m = 1$ and is usually the most important frequency in practice. Replacing in Equation (2) the value of the axial stiffness we obtain Equation (3) also called Harignx's equation [12], who experimentally verified its validity.

$$f = \frac{m}{2}\sqrt{\frac{k_{Wahl}}{W}} \qquad m = 1,2,3\ldots \tag{2}$$

$$f = \frac{d}{2\pi R^2 N}\sqrt{\frac{G}{32\rho}} \tag{3}$$

R being the helix radius and ρ the density of the material.

Equation (3) indicates that for a given material, the fundamental frequency of a helical spring is proportional to the wire diameter and inversely proportional to the product of the helix diameter and the number of active coils, i.e., the resonance frequency of the spring depends on all the spring design parameters. If a higher frequency is desired for the same spring diameter, it is sufficient to increase the wire diameter (this increases the spring stiffness), and vice versa. If you want to vary the resonance frequency while keeping the wire diameter constant, just change the spring diameter or the number of coils of the spring. All the expressions shown above are raised with respect to an inertial reference system whose main X axis coincides with the symmetry axis of the helical spring.

By design of the machine itself, the spring is always inserted between two masses in relative oscillatory motion and its function is to keep them away from each other ensuring at all times that they do not come into contact. The design of the helical spring depends directly on the inertia between the masses at any given moment. In addition to the diameter of the wire and the spring itself, the number of coils and the pitch are responsible for achieving non-interaction between the mechanical elements in relative motion.

It is interesting to note that the behavior of the spring changes radically when its operation is associated with a mass much greater than that of the spring itself. For example, in the damping systems of a railway bogie such as the one in Figure 1. In that case, the mass of the spring is much lower than the mass of the railway car it is supporting and Equations (2) and (3) are no longer valid. Den Hartog [13] shows that in that configuration the natural frequency follows Equation (4).

$$f = \frac{1}{2\pi}\sqrt{\frac{g}{\delta_{est}}} \tag{4}$$

where g is the gravitational constant and δ_{est} is the static deflection of the spring under the weight of the suspended load.

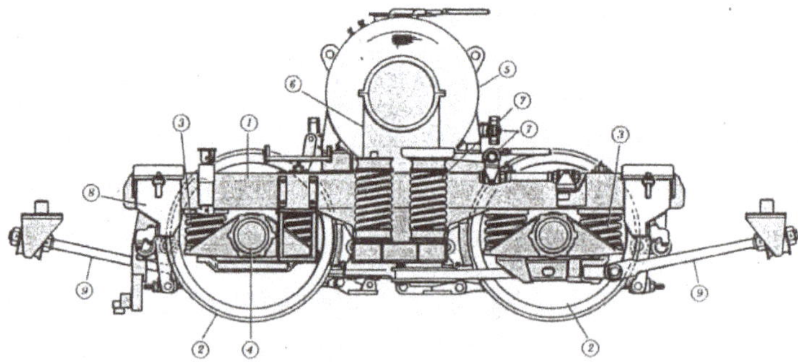

Figure 1. Damping springs of a railway bogie (Díaz [14]).

Sauvage [15] and Campedelli [16] studied in detail the behavior of the springs of a bogie and simulated in finite elements the anomaly presented by the springs during axial compression called "effort de chasse". This anomaly consists of a rotation and a lateral displacement appearing on the bearing surface of the spring. The "effort de chasse", also called transverse flexibility of the spring, is caused by the forces and moments that appear at the ends of the spring during its axial compression. Campedelli [16] determines that the transverse flexibility is determined by both the geometry of the spring end and the way in which the gap between the coils closes during axial compression. The "effort de chasse" is an interesting anomaly in railway suspension springs since in their displacement the springs could interfere with other mechanical elements of the bogie generating interferences and breakages. To avoid this, the railway standard [17] requires a test for the springs of the highest category according to the scheme in Figure 2a. From the test, both the direction and the transverse bowing force (Φc) of the spring subjected to a defined axial load are obtained. The direction of the transverse direction is marked on the spring by a permanent system to take it into account in its mounting on the bogie. Figure 2b shows an example of a bogie spring assembly compensating for the spring bowing forces.

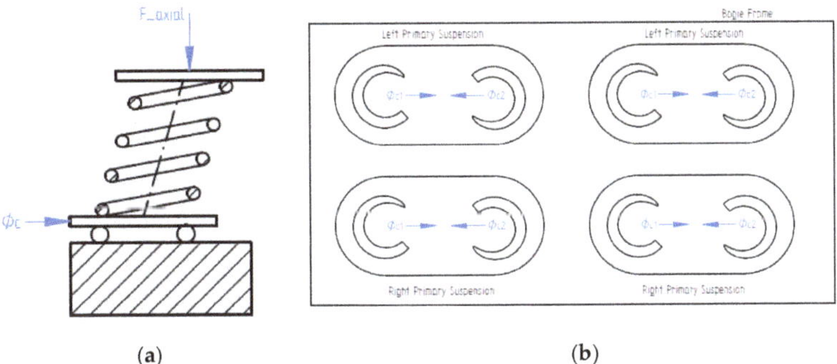

(a)   (b)

Figure 2. "Effort de chasse" or transverse flexibility: (**a**) schematic of the test device; (**b**) mounting on a spring bogie compensating the "effort de chasse".

Kobelev [18] also studies another type of anomaly in the behavior of springs such as the transverse vibration of the spring once it enters into oscillations close to resonant frequencies. Kobelev [18] found that the fundamental natural frequency of transverse oscillations turns to be to zero when the lateral buckling of the spring occurs.

Yildirim [19] demonstrates that Equation (1), which is usually used to determine the spring stiffness, is only valid for spring angles $\alpha \leq 10°$. Figure 3a shows the spring geometry indicating the position of the angle α.

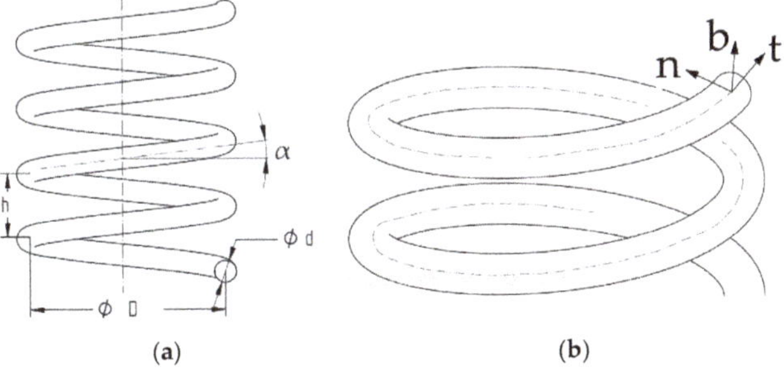

(a)   (b)

Figure 3. Helical spring: (**a**) geometrical parameters; (**b**) Frenet trihedral.

Yildirim [19] indicates that Equations (1) and (2) only take into account the effect of the torsion of the spring section when compressing it. However, as Yildirim [19] points out when the axial compression force acts, both stresses due to torsion and bending moment and forces normal and shear to the section appear. These last three effects are negligible for spring angles $\alpha \leq 10°$ where the torsional effect dominates. Yildirim [19] proposes a global equation to determine the deformation of the spring subjected to an axial force. Additionally, Kato [20] proposes recently an equation outside the elastic range to determine large spring deformations.

In general, the previous authors focus the study of the spring behavior using a conventional Cartesian system in which the static or dynamic displacements are referred to as the first axis of symmetry of the helical spring, X. With this methodology the global behavior of a spring as a whole can be known with precision, i.e., constituted by a series of coils of a certain diameter and a certain thickness of the wire, contemplating in some of the occasions the ending effect.

This type of methodology, therefore, allows us to analyze the elastic effect of the spring as a whole, but it is not able to explain some mechanical behaviors intrinsic to its own constitution, i.e., it does not study the phenomenon occurring in the coil itself. In this way, anomalous spring behaviors such as those described above cannot be fully explained.

In order to achieve a reference system that meets these needs, the Frenet trihedral is used in this article, since it is intrinsic to the wire of the spring itself, configuring a reference system positioned according to the neutral line of the helix of the coil as shown in Figure 3b.

In the present study, the equations of motion of the coil of a helical spring are obtained in order to determine its modes of vibration. This is an original and novel approach since the usual practice is to obtain the modes of vibration of the complete spring referred to as a reference system according to its joint motion. Additionally, a sensitivity analysis has been performed for the main geometrical parameters of the coil: i.e., the helix diameter, the wire diameter and pitch. The obtained vibration modes have been validated satisfactorily with finite element calculations with ANSYS® R17 (release 17).

2. Model of Coupled Bending and Torsional Oscillation in a Spring

2.1. Frenet's Trihedral

For the approach of the oscillation model of the coil, the coordinate system formed by the Frenet trihedral, intrinsic to the neutral line of the spring, has been considered.

Figure 4 shows a differential element of the length of the curve defined by the midline of the coil of a helical spring. The Frenet trihedral or intrinsic trihedral is positioned on it. It is an orthonormal trihedron ($\vec{t}, \vec{n}, \vec{b}$) whose components are named tangent, normal and binormal. The tangent vector is oriented to the neutral line of the loop. The normal is directed towards the bending center of curvature of the loop. The binormal vector follows the orientation defined by the other two. The vectors t and n form a tangent plane to the loop, containing the element ds, and perpendicular to the binormal vector which is directed towards the bending curvature of this element.

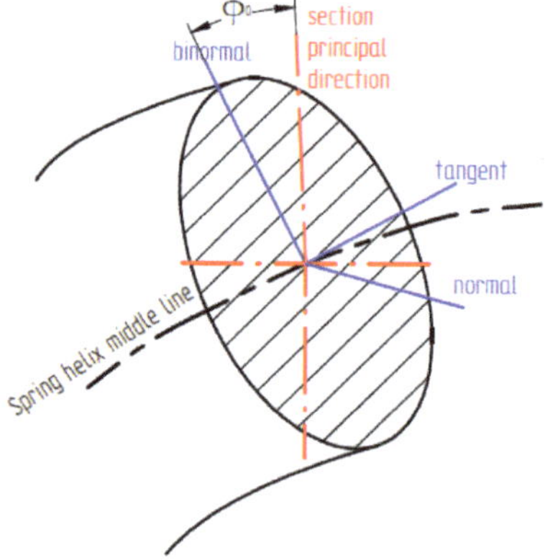

Figure 4. Intrinsic trihedral and principal directions of inertia of the section.

Considering as a scalar variable the arc length of the midline of the loop(s), we obtain the well-known Frenet–Serret equations, Equation (5), which give us the derivative with respect to s of the components of the trihedral as a function of the curvature of the midline of the coil.

$$\begin{pmatrix} \frac{d\vec{t}}{ds} \\ \frac{d\vec{n}}{ds} \\ \frac{d\vec{b}}{ds} \end{pmatrix} = \begin{pmatrix} \vec{t}\prime \\ \vec{n}\prime \\ \vec{b}\prime \end{pmatrix} = \begin{pmatrix} 0 & \Omega_3 & -\Omega_2 \\ -\Omega_3 & 0 & \Omega_1 \\ \Omega_2 & -\Omega_1 & 0 \end{pmatrix} \begin{pmatrix} \vec{t} \\ \vec{n} \\ \vec{b} \end{pmatrix} \tag{5}$$

where $\Omega_1 = \tau_0$, $\Omega_2 = -k_0 \sin \phi_0$ and $\Omega_3 = k_0 \cos \phi_0$. Where k_0 is the bending curvature, τ_0 the torsional curvature and ϕ_0 the angle formed by the principal directions of inertia with the Frenet trihedron.

Equation (5) can be expressed as a function of the vector product of the Darboux vector $\vec{\Omega}(s)$ by the components of the intrinsic trihedron obtaining Equation (6).

$$\begin{pmatrix} \frac{d\vec{t}}{ds} \\ \frac{d\vec{n}}{ds} \\ \frac{d\vec{b}}{ds} \end{pmatrix} = \vec{\Omega} \times \begin{pmatrix} \vec{t} \\ \vec{n} \\ \vec{b} \end{pmatrix} \tag{6}$$

where the Darboux vector is defined according to Equation (7):

$$\vec{\Omega}(s) = \Omega_1 \vec{t} + \Omega_2 \vec{n} + \Omega_3 \vec{b} \tag{7}$$

The variation of any vector function, for example, the rotation and translation of the section, with respect to the length s of the arc follows the expression of Equation (8a) that we will simplify using the notation of Equation (8b) where we will call $[R]$ the tensor associated with the operation $\vec{\Omega}\times$.

$$\frac{d\vec{V}}{ds} = \frac{d\vec{V}}{ds}\bigg|_s + \vec{\Omega} \times \vec{V} \tag{8a}$$

$$\left\{\vec{V}\right\}\prime = \left\{\vec{V}\prime\right\} + [R]\,\vec{V} \tag{8b}$$

$$[R] = \begin{pmatrix} 0 & \Omega_3 & -\Omega_2 \\ -\Omega_3 & 0 & \Omega_1 \\ \Omega_2 & -\Omega_1 & 0 \end{pmatrix} \tag{8c}$$

where $\frac{d\vec{V}}{ds}\big|_s$ is the derivative calculated as if the Frenet trihedron were fixed along the curve. That is, the variation of the function along the curve has two terms, the first represents the variation of the vector within the basis itself and the second represents the variation of the basis itself along the curve.

2.2. Rotation of the Spring Cross-Section

The spring in its natural geometrical configuration undergoes a deformation caused by external forces and moments. The plane cross-section remains plane and we neglect its small deformation. As shown in Figure 5, the cross-section of the deformed part is obtained by a translation $\vec{\eta}(s)$ of the center and a rotation $\vec{\psi}(s)$ of the section. Therefore, the point P the intersection of the cross-section with the midline, will be transformed into the point P_1 (where $\vec{PP_1} = \vec{\eta}(s)$) and the perpendicular axes (j,k), coinciding with the normal and binormal of the intrinsic trihedron before the deformation, will become the axes $\vec{j_1}$ and $\vec{k_1}$ which we also assume to be perpendicular to each other. The axis $\vec{i_1}$ is obtained as the vector product of the previous ones, $\vec{i_1} = \vec{j_1} \times \vec{k_1}$.

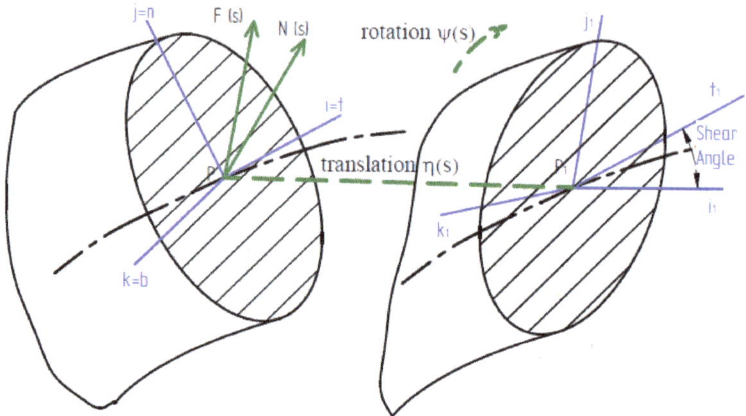

Figure 5. Schematic of the oscillations of the spring section.

Due to the shear produced by the shear forces, the i-axis is not transformed into the i-axis$_1$ by rotation and translation. Consequently, the tangent vector to a point on the midline of the part will no longer coincide with the first axis of the trihedron as it did in the initial equilibrium state but will form an angle $\vec{\beta}$, called the shear angle, proportional to the shear stress. If we do not consider the deformation due to the shear stress, the shear angle is canceled, so that the tangent to the midline of the part will be perpendicular to the section, or what is the same, will coincide with the first axis of the trihedron.

The derivative of $\vec{\psi}$ with respect to s is given to first order by Equation (9a,b).

$$\{\vec{\psi}\}\prime = [M_N]\,\vec{N} + [R]\,\vec{\psi} \tag{9a}$$

$$[M_N] = \begin{pmatrix} \frac{1}{k_1 G I_1} & 0 & 0 \\ 0 & \frac{1}{E I_2} & 0 \\ 0 & 0 & \frac{1}{E I_3} \end{pmatrix} = \begin{pmatrix} \frac{1}{I_1^*} & 0 & 0 \\ 0 & \frac{1}{I_2^*} & 0 \\ 0 & 0 & \frac{1}{I_3^*} \end{pmatrix}. \tag{9b}$$

where M_N is the tensor associated with the torsional-flector torque increment, E is the modulus of elasticity, G is the shear modulus of elasticity, I_i is the geometric moment of inertia and k_1 is the torsion constant. $\vec{N}$ is the vector of internal torque increase. Its first component is the torsional moment, and the other two components are the bending moments in the normal and binormal directions. I_1^* is the torsional stiffness, I_2^* is the bending stiffness along the normal axis, and I_3^* is the bending stiffness along the binormal axis.

2.3. Translation of the Spring Cross-Section

The derivative of the displacement vector $\vec{\eta}$ with respect to s is given to first order by Equation (10a–c).

$$\{\vec{\eta}\}\prime = [M_F]\,\vec{F} + [S]\,\vec{\psi} + [R]\,\vec{\eta} \tag{10a}$$

$$[M_F] = \begin{pmatrix} \frac{1}{EA} & 0 & 0 \\ 0 & \frac{1}{k_2 G A} & 0 \\ 0 & 0 & \frac{1}{k_3 G A} \end{pmatrix} = \begin{pmatrix} E^* & 0 & 0 \\ 0 & G_1^* & 0 \\ 0 & 0 & G_2^* \end{pmatrix} \tag{10b}$$

$$[S] = \begin{pmatrix} 0 & 0 & 0 \\ 0 & 0 & 1 \\ 0 & -1 & 0 \end{pmatrix} \tag{10c}$$

where M_F is the tensor associated with the forces and k_i the shear constants, or the so-called Timoshenko's k-factors [21]. Being for symmetric sections $k_2 = k_3$. $\vec{F}$ is the vector of internal force increment. Its first component is the axial force, and the other two components are the shear forces according to the normal and binormal. EA is the axial stiffness and $k_2 GA$ is the shear stiffness.

Equations (9a,b) and (10a–c) constitute a system of six scalar equations for the twelve components of $\vec{\psi}$, $\vec{\eta}$, $\vec{F}$, $\vec{N}$. The remaining six equations are obtained from the dynamic study of the system for which the equations of the linear momentum and angular momentum of the part element comprised between s and $s + ds$ are posed.

2.4. Linear Momentum

Considering the linear momentum, the derivative of $\vec{F}$ with respect to s is given by Equation (11).

$$\left\{\vec{F}\right\}' = \lambda\,\ddot{\vec{\eta}} + [R]\,\vec{F} \tag{11}$$

where $\lambda = \rho \cdot A$ is the linear density per unit length.

2.5. Angular Momentum

Considering the angular momentum, the derivative of $\vec{N}$ with respect to s is given by Equation (12a,b), which takes into account the effects of rotational inertia introduced by Rayleigh [22].

$$\left\{\vec{N}\right\}' = \frac{\lambda}{A}[I]\,\ddot{\vec{\psi}} + [S]\,\vec{F} + [R]\,\vec{N} \tag{12a}$$

$$[I] = \begin{pmatrix} I_1 & 0 & 0 \\ 0 & I_2 & 0 \\ 0 & 0 & I_3 \end{pmatrix} \tag{12b}$$

Table 1 summarizes the complete system of equations obtained and which describes the behavior of the spring coil subjected to bending and torsion. It consists of four vector equations in linear partial derivatives, non-homogeneous and of second order.

Table 1. System of Vectorial Equations to model spring behavior.

Vectorial Equations
$\left\{\vec{\psi}\right\}' = [M_N]\,\vec{N} + [R]\,\vec{\psi}$
$\left\{\vec{\eta}\right\}' = [M_F]\,\vec{F} + [S]\,\vec{\psi} + [R]\,\vec{\eta}$
$\left\{\vec{F}\right\}' = \lambda\,\ddot{\vec{\eta}} + [R]\,\vec{F}$
$\left\{\vec{N}\right\}' = \frac{\lambda}{A}[I]\,\ddot{\vec{\psi}} + [S]\,\vec{F} + [R]\,\vec{N}$

2.6. Dispersion Relation

If in the Table 1 system of equations we change the variable using Equation (13a–d), we can simulate the movements in time of the spring and then perform a vibration analysis of its oscillatory behavior in the vector variables $\vec{\psi}$, $\vec{\eta}$, $\vec{F}$, $\vec{N}$. We will assume a wave traveling along the axis of the wire. Thus

$$\vec{\psi}(s,t) = Re\left(\vec{\psi}_0(s) \cdot e^{i(ks-\omega t)}\right) \tag{13a}$$

$$\vec{\eta}(s,t) = Re\left(\vec{\eta}_0(s) \cdot e^{i(ks-\omega t)}\right) \tag{13b}$$

$$\vec{F}(s,t) = Re\left(\vec{F_0}(s)\cdot e^{i(ks-\omega t)}\right) \tag{13c}$$

$$\vec{N}(s,t) = Re\left(\vec{N_0}(s)\cdot e^{i(ks-\omega t)}\right) \tag{13d}$$

The new parameters introduced are the wavenumber, k, and the angular velocity of the wave, ω (rad/s). Where the wavenumber k, is related to the length λ of the vibrating wave, with Equation (14).

$$k = \frac{2\pi}{\lambda} \tag{14}$$

Substituting the system of equations in Table 1 we arrive at the following system in Equation (15a–d).

$$[\Lambda]\,\vec{F} - \lambda\omega^2\,\vec{\eta} = 0 \tag{15a}$$

$$[\Lambda]\,\vec{\eta} + [M_F]\,\vec{F} + [S]\,\vec{\psi} = 0 \tag{15b}$$

$$[\Lambda]\,\vec{\psi} + [M_N]\,\vec{N} = 0 \tag{15c}$$

$$[\Lambda]\,\vec{N} - \frac{\lambda\omega^2}{A}[I]\,\vec{\psi} + [S]\,\vec{F} = 0 \tag{15d}$$

where the matrix $[\Lambda]$ follows the expression of Equation (16).

$$[\Lambda] = \begin{pmatrix} -ik & k_0 & 0 \\ -k_0 & -ik & \tau_0 \\ 0 & -\tau_0 & -ik \end{pmatrix} \tag{16}$$

Solving for $\vec{\eta}$ in Equation (15a) and replacing it in Equation (15b), and similarly solving for $\vec{N}$ in Equation (15c) and replacing in Equation (15d) we obtain the following vector Equation (17a,b):

$$[A]\,\vec{F} + [S]\,\vec{\psi} = 0 \tag{17a}$$

$$[B]\,\vec{\psi} + [S]\,\vec{F} = 0 \tag{17b}$$

where the matrices $[A]$, $[B]$ are calculated with Equations (18) and (19).

$$[A] = \frac{1}{\lambda\omega^2}[\Lambda]^2 + [M_F] \tag{18}$$

$$[B]\,\vec{\psi} + [S]\,\vec{F} = 0 \tag{19}$$

Equation (17a,b) constitutes a system of six equations with six unknowns $(F_1, F_2, F_3,\ \psi_1,\ \psi_2, \psi_3)$ and with null independent terms. It is therefore a homogeneous system and for it to have a solution other than the trivial, null solution, it must be verified that its determinant is null, i.e., Equation (20).

$$\begin{vmatrix} \lambda\omega^2 E^* - (k^2 + \beta_9) & \beta_4 - 2ik\Omega_3 & \beta_5 + 2ik\Omega_2 & 0 & 0 & 0 \\ 2ik\Omega_3 + \beta_4 & \lambda\omega^2 G_1^* - (k^2 + \beta_8) & \beta_6 - 2ik\Omega_1 & 0 & 0 & \lambda\omega^2 \\ \beta_5 - 2ik\Omega_2 & \beta_6 + 2ik\Omega_1 & \lambda\omega^2 G_2^* - (k^2 + \beta_7) & 0 & -\lambda\omega^2 & 0 \\ 0 & 0 & 0 & k^2 I_1^* + \alpha_4 - \omega^2\lambda_1 & -\beta_4 I_3^* + ik\alpha_1 & -\beta_5 I_2^* - ik\alpha_2 \\ 0 & 0 & 1 & -\beta_4 I_3^* - ik\alpha_1 & k^2 I_2^* + \alpha_5 - \lambda_2\omega^2 & -\beta_6 I_1^* + ik\alpha_3 \\ 0 & 1 & 0 & \beta_5 I_2^* - ik\alpha_2 & \beta_6 I_1^* + ik\alpha_3 & -k^2 I_3^* + \omega^2\lambda_3 - \alpha_6 \end{vmatrix} = 0 \tag{20}$$

Operating the determinant and simplifying we obtain the alphanumerical Equation (21), which would be the dispersion relation, $\omega = f(k)$, which relates the length of a wave with its oscillation frequency in the spring coil.

$$m_1\omega^{12} + m_2\omega^{10} + m_3\omega^8 + m_4\omega^6 + m_5\omega^4 + m_6\omega^2 + m_7 = 0 \tag{21}$$

where the coefficients of the polynomial are calculated according to the expressions given in Equation (A1) of Appendix A.

The real roots in ω^2 of Equation (21) will give us the number of branches of the dispersion equation for each value of the wavenumber k.

If in the dispersion relation obtained for the spring we cancel the independent term, m_7 (which is equivalent to making $\omega \to 0$), we will have the solutions of the wave number shown in Equation (22a,b).

$$k^* = 0 \tag{22a}$$

$$k^{**} = \sqrt{\Omega_1^2 + \Omega_3^2} \tag{22b}$$

That is to say, there exists a value of the wave number k different from zero (to which corresponds, therefore, a finite wavelength) for which one of the real and positive roots of the dispersion relation (degree 12) cancels, so that the branch corresponding to this solution, for any value of k, is tangent to the x-axis in $k = k^{**}$.

If we do $k \to 0$ we will obtain the number of branches of the dispersion relation (which will be equivalent to the number of real roots of the equation). For this case, the real and positive branches are determined by Equation (23a–c):

$$\omega = 0 \tag{23a}$$

$$\rho^2 I_1 I_2 \omega^4 - \omega^2[kG\{A\rho I_1 + (\beta_1 + \beta_3)I_2\} + EI_2\rho(I_1\beta_1 + I_2\beta_3)] \\ + kGEI_2[(\beta_1 + \beta_3)(I_1\beta_1 + I_2\beta_3) + A\beta_3] = 0 \tag{23b}$$

$$\rho^2 I_2 \omega^4 - \omega^2\rho[EI_2(\beta_1 + \beta_3) + kG[(\beta_1 I_2 + A) + \beta_3 kGI_1]] \\ + (\beta_3 kGI_1 + \beta_1 EI_1)(\beta_3 E + \beta_1 kG) + \beta_3 kGEA = 0 \tag{23c}$$

The wave numbers k obtained from Equation (23a–c) are indicators of the spring oscillation modes. Equation (23b) represents the bending oscillations of a differential element of spring length in plane 1, 2 determined by the principal directions of inertia of a cross-section (Figure 6). Equation (23c) represents the torsional oscillations of the differential element around the principal direction 1 of Figure 6. These oscillation modes will be explained also later in the finite elements section.

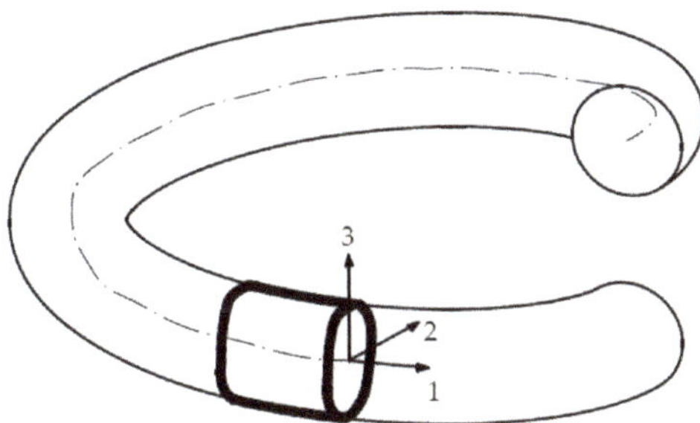

Figure 6. Spring Oscillation Modes.

The result of the oscillation is bending accompanied by torsion which tends to make the ends of the coil open and close in time, oscillating according to the movement. This is composed of two modes: opening and closing of the ends, the coil remaining at all times inscribed in the original helix of the spring (of diameter D) and the other mode consists of the ends of the coil entering and leaving the original helix of diameter D, forming an ellipse. This last mode responds to the reason why the springs, under certain dynamic conditions, pull the coils out of their original helix, and may impact other elements of the machine, thus affecting its operation. Authors consider this can explain part of the "Effort de Chasse" anomaly described in Section 1.

In this article, we are in the case of a freely oscillating coil. In the case of several coils joined together, it would be necessary to consider the ligatures of their ends as shown in Figure 7 for the end of a squared and ground spring. That is to say, with a helix angle of zero degrees and a flat seat, which allows a better load transfer. The assembly thus defined becomes an oscillatory mode in which the spring as a whole adopts the barrel configuration. Pearson [23] studies the important contribution of the ends to the calculation of the spring natural frequencies.

Figure 7. The end of squared and ground helical spring.

The dispersion relation allows us to obtain the normal modes of oscillation by applying the appropriate boundary conditions. For example, in the case of a helical spring with one end clamped and the other free, the boundary conditions would be as follows: the rotation vector $\vec{\psi}$ and the translation vector $\vec{\eta}$ cancel out at the clamped end and the internal torque $\vec{N}$ and the internal stress $\vec{F}$ cancel at the free end. Assuming the clamped end for s = 0 and the free end for $s = L$, Equation (24) is satisfied which implies that the wave number k takes the values of Equation (25).

$$\vec{\psi}(0,t) = \vec{\psi}\prime(L,t) = 0 \tag{24}$$

$$k = \frac{n\cdot\pi}{2L} = \frac{n\cdot\pi}{2N\sqrt{h^2 + \pi^2 D^2}} \qquad n = 1,3,\ldots \tag{25}$$

Entering the value of k calculated according to Equation (25) in the dispersion Equation (21), the natural oscillation frequencies for the helical spring are obtained.

In order to validate the proposed theoretical model, a comparison with finite elements and with the experimental tests of a helical spring is made below.

3. Application of Finite Elements

The previous calculation performed to obtain the natural frequencies of the system has been compared with the results of the resonance frequencies obtained by numerical simulation. In addition, a complete modal analysis is performed. In order to compare the results, the oscillation modes of the one spring turn are calculated using standard computational methods. The simulations have been carried out by modeling the spring geometry by Finite Elements using ANSYS® software.

A three-dimensional geometry of the helicoid representing the neutral fiber of a helical spring has been faithfully reproduced. The three-dimensional geometry has been generated from SOLID EDGE software, and it has been imported by the FEM software in the *.iges format. The SOLID EDGE software provides flexibility in the modelization and facilitates the actual geometrical definition compared to the CAD tool integrated in ANSYS. The spring is thus reproduced volumetrically.

We proceed to calculate in both cases, i.e., the theoretical model proposed in the previous section with one solved in finite elements, in order to compare the results.

In finite elements, a meshing has been performed using extruded hexahedral elements of 2 mm characteristic size, which represents the circumference of the wire cross-section (10 mm diameter) with 16 nodes in its contour. This results in 6000 nodes per turn, 54,000 nodes for the complete spring and 162,000 degrees of freedom.

The meshing strategy used consists of an extrusion method from the two-dimensional meshing (with quadrilaterals) of the cross-section, which guarantees a regular meshing that avoids geometric distortions of the elements and achieves a low bandwidth for the stiffness matrix, resulting in a better performance of the iterative methods of resolution.

The SOLID185 element has been selected in its homogeneous form, which is frequently used to model three-dimensional structures in ANSYS. The element consists of eight nodes with three degrees of freedom per node: translations in the three directions of space. SOLID185 uses the reduced selective integration method.

The material has been modeled linearly, with steel properties considering Young's modulus and Poisson's ratio.

The problem has been plausibly solved using the Lanczos method for the determination of eigenfrequencies and eigenmodes.

In free-free solid conditions, the different natural frequencies of one spring turn belonging to a helical spring, with different helix diameters and wire thickness have been calculated. In all cases, the pitch of the helix has always been the same.

In all cases, the corresponding oscillation modes are studied. Table 2 shows the results. Columns 4 to 9 show the frequency and the mode of vibration. Essentially, the first mode corresponds to an opening between ends, such that the original helix is transformed into an ellipsoidal helix (Figure 8a). In this case, the predominant deformation of the helix is bending. The second mode corresponds to the relative motion between these ends (Figure 8b) but preserves the circumference of the helix, which corresponds to a mode in which the torsional effect of the helix is more predominant than its bending. In this second case, the pitch of the coil would increase significantly.

Table 2. FEM calculation of the first natural frequencies (Hz) of helical spring.

Natural Frequencies (Hz) Calculated by ANSYS								
D (mm)	d (mm)	Pitch (mm)	Bending 1	Torsional 1	Bending 2	Torsional 2	Bending 3	Torsional 3
80	18	39.8	960	1218	2111	3184	4543	6322
80	22	39.8	1168	1468	2556	3733	5428	7414
80	28	39.8	1475	1823	3199	4435	6623	8254
108	18	39.8	539	664	1183	1815	2595	3721
108	22	25	667	792	1451	2155	3178	4339
108	22	39.8	657	806	1439	2162	3132	4349
108	22	50	648	817	1426	2169	3091	4355
108	28	39.8	832	1012	1815	2630	3899	5127
220	18	39.8	133	158	290	455	648	968
220	22	39.8	162	193	354	553	790	1169
220	28	39.8	207	245	450	695	1000	1454

$N = 1$ turn, $E = 210$ GPa, $G = 84$ GPa, $\rho = 7800$ kg/m^3, $\nu = 0.24407$.

From these two basic modes, the deformation profiles are reproduced with increasing inflection points. The third (Figure 8c) and fifth (Figure 8e) modes constitute double bending and triple bending, respectively. Always without leaving the imaginary plane that would contain the spiral if it were a ring in a plane and not helical, but distorting the circumference that defines it. The fourth (Figure 8d) and sixth (Figure 8f) modes constitute double torsion (with one inflection point) and triple torsion (with two inflection points), respectively.

If we continue to search for natural modes at increasing frequencies, increasingly complex deformation profiles with small displacements and numerous inflection points appear.

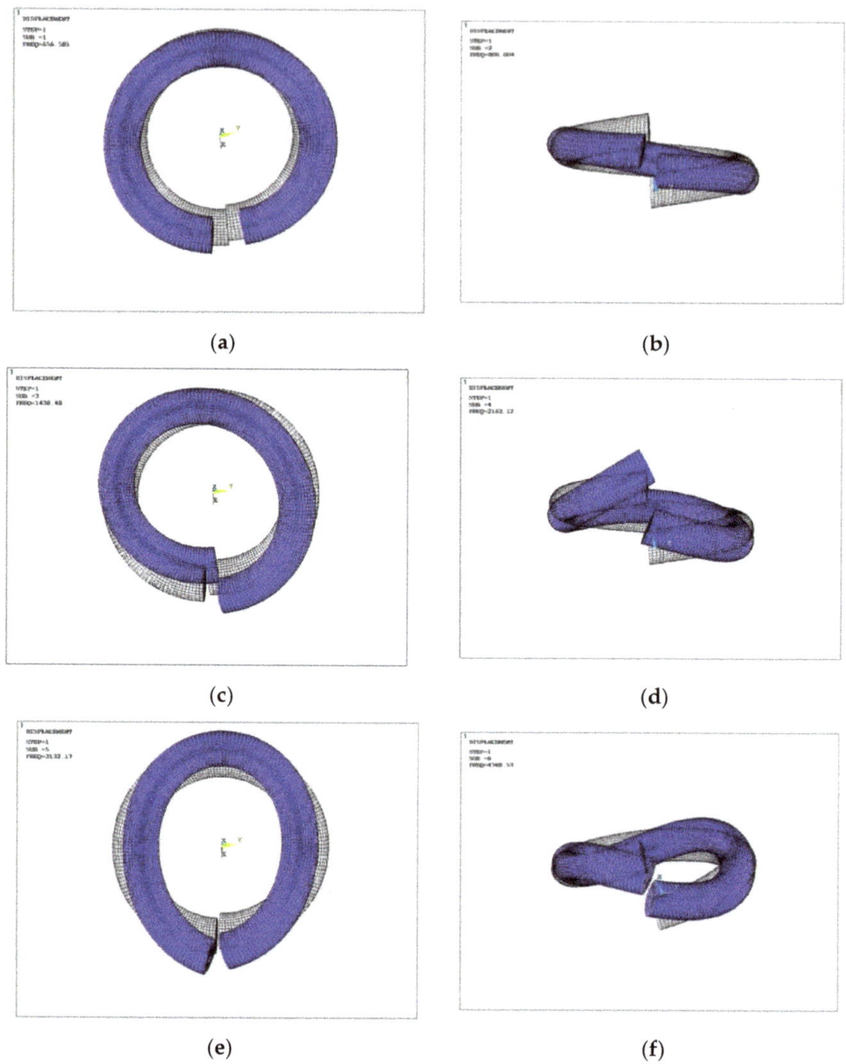

Figure 8. Modes of vibration of a spring: (**a**) Bending 1; (**b**) Torsional 1; (**c**) Bending 2; (**d**) Torsional 2; (**e**) Bending 3; (**f**) Torsional 3.

With respect to the results shown in Table 2, it can be deduced that, in the case of the first resonance mode (bending), keeping the helix diameter constant and varying the wire thickness, the resonance frequency increases with increasing thickness. This is logical because the stiffness of the spring increases. Figure 9 shows this trend. Moreover, as shown in Figure 9, as the diameter of the helix increases, the resonance frequency decreases. Despite maintaining the stiffness, the loop flexes with lower frequency values. A similar situation occurs with the first torsional mode.

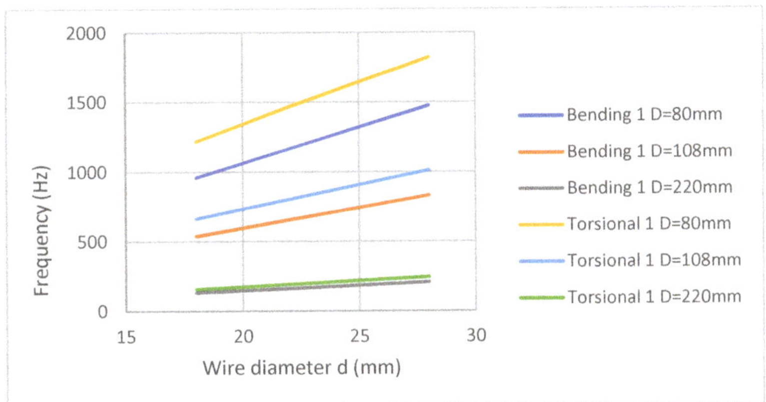

Figure 9. Sensitivity of the first resonance modes to the wire diameter, d.

Proceeding analogously for the first bending mode, the wire thickness is now kept constant, and the average diameter of the helix is varied (Figure 10). The frequencies decrease as the helix diameter increases. Moreover, as the wire diameter increases, the resonant frequency increases. The situation is repeated for the first torsional mode of vibration. It is a situation analogous to that of the previous paragraph.

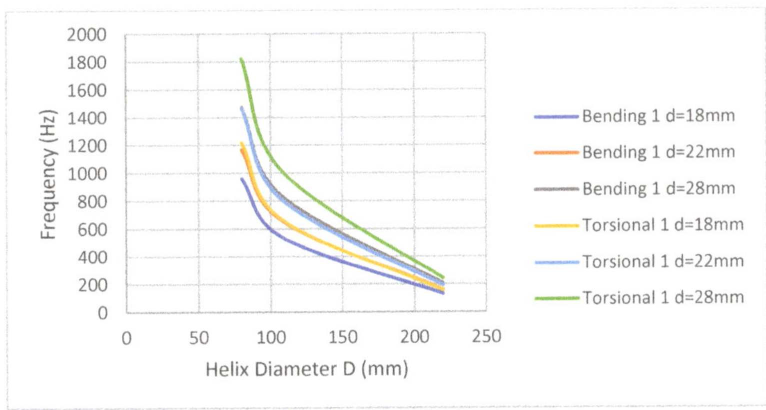

Figure 10. Sensitivity of the first resonance modes to the helix diameter, D.

Analyzing the variation of the frequency as a function of the variation of the average diameter of the helix, it is observed (Figure 10) that the first bending and torsional mode of the spring decrease as the mass increases due to a higher helix diameter. This has to do with the wire diameter/helix diameter ratio and it can be seen that there is a certain characteristic value of this ratio that linearizes the resonant behavior.

As shown in Figure 9 the relationship between frequencies and wire diameter (d) is linear. This result is consistent with Equation (3).

It has also been considered whether the pitch of the spring can be an influential design parameter when calculating resonances. Figure 11 shows that the pitch variation does not affect the frequencies. Figure 11 shows the different modes obtained and commented.

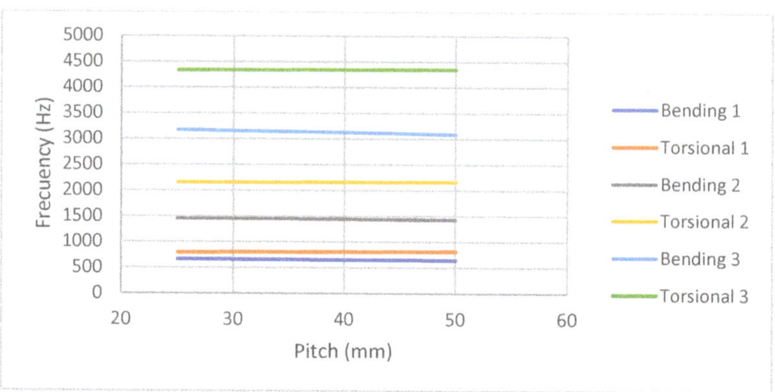

Figure 11. Sensitivity of the first resonance modes to the pitch of the helix.

Table 3 includes a comparison for the first harmonic of the results obtained with the finite element simulation and the numerical model presented in this paper. The results of the numerical model are obtained by solving ω in Equation (21) with the wavenumber k calculated with Equation (22b) or Equation (25). In general, it is observed that the numerical model overestimates the frequency values with an average difference of 3%.

Table 3. Comparison of the numerical model with FEM for the first harmonic (Hz).

D (mm)	d (mm)	Pitch (mm)	Wavenumber k (m^{-1})	Bending 1 FEM	Bending 1 Numerical Model Equation (21)
			Fundamental Frecuency (Hz)		
80	18	39.8	24.69	960	989
80	22	39.8	24.69	1168	1190
80	28	39.8	24.69	1475	1495
108	18	39.8	18.39	539	555
108	22	25	18.47	667	680
108	22	39.8	18.39	657	670
108	22	50	18.32	648	665
108	28	39.8	18.39	832	840
220	18	39.8	9.08	133	145
220	22	39.8	9.08	162	170
220	28	39.8	9.08	207	211

$N = 1$ loop, $E = 210$ GPa, $G = 84$ GPa, $\rho = 7800$ kg/m^3, $v = 0.24407$.

4. Conclusions

The study of a single coil has been proposed since, in general, the authors who propose equations to model the vibration behavior of a spring consider the spring as a whole, including its ends, but do not analyze the behavior of the coil itself. From the analysis of a single coil, the singular behavior of the spring known as "effort de chasse" described in the introduction could be justified.

For the study of the vibratory behavior of a spring, an analytical model has been developed based on a reference system consisting of a Frenet trihedron whose tangent vector is the midline of the coil. One of the results obtained from the model is the dispersion relation that relates the wavelength to its oscillation frequency, $\omega = f(k)$. Applying the appropriate boundary conditions the dispersion relation allows us to calculate the normal modes of oscillation.

Another of the results obtained is that there is a value of the wavenumber $k = \sqrt{\Omega_1^2 + \Omega_3^2}$ for which the roots of the dispersion equation cancel out. Being Ω_1 and Ω_3 the torsional curvature and the bending curvature, respectively.

Two clear modes of oscillation have been characterized, one in which the coil moves according to the torsion of the cross-section increasing the pitch of the coil, and another bending mode in which it moves by making an opening of its ends so that the original helix is transformed into an ellipsoidal helix.

A spring has been modeled by finite elements with ANSYS® software in order to contrast results with the numerical model, obtaining an average difference of 3% in the fundamental oscillation frequency.

Author Contributions: Individual contributions are as follows: original concept: L.P., A.Q., A.M.G.A. and V.D.; mathematical development: L.P. and A.M.G.A.; implementation, simulation and draft preparation: L.P., A.M.G.A., A.Q. and V.D.; review, editing, validation, and formal analysis: L.P., A.Q., A.M.G.A. and V.D. All authors have read and agreed to the published version of the manuscript.

Funding: This research received no external funding.

Institutional Review Board Statement: Not applicable.

Informed Consent Statement: Not applicable.

Data Availability Statement: Not applicable.

Conflicts of Interest: The authors declare no conflict of interest.

Appendix A

In this appendix, the expressions for the calculation of the coefficients of Equation (21) are indicated. Equation (A1a–g) contains the expressions for the calculation of m_i

$$m_1 = \phi_1 \cdot \phi_5 \tag{A1a}$$

$$m_2 = \phi_1 \cdot \phi_6 + \phi_2 \cdot \phi_5 + \phi_9 \tag{A1b}$$

$$m_3 = \phi_1 \cdot \phi_7 + \phi_2 \cdot \phi_6 + \phi_3 \cdot \phi_5 + \phi_{10} \tag{A1c}$$

$$m_4 = \phi_1 \cdot \phi_8 + \phi_2 \cdot \phi_7 + \phi_3 \cdot \phi_6 + \phi_4 \cdot \phi_5 + \phi_{11} \tag{A1d}$$

$$m_5 = \phi_2 \cdot \phi_8 + \phi_3 \cdot \phi_7 + \phi_4 \cdot \phi_6 + \phi_{12} \tag{A1e}$$

$$m_6 = \phi_3 \cdot \phi_8 + \phi_4 \cdot \phi_7 + \phi_{13} \tag{A1f}$$

$$m_7 = \phi_4 \cdot \phi_8 \tag{A1g}$$

With Equation (A2a–m) for the calculation of the parameters of the ϕ_i

$$\phi_1 = \lambda_1 \lambda_2 \lambda_3 \tag{A2a}$$

$$\phi_2 = -[\lambda_2 \lambda_3 a_4 + \lambda_1 \lambda_2 a_6 + \lambda_1 \lambda_3 a_5] \tag{A2b}$$

$$\phi_3 = \{\lambda_1(a_5 a_6 - c_3) + \lambda_2(a_4 a_6 - c_5) + \lambda_3(a_4 a_5 - c_4)\} \tag{A2c}$$

$$\phi_4 = c_4 a_6 + c_5 a_5 - a_4(a_5 a_6 - c_3) + b_5 \tag{A2d}$$

$$\phi_5 = \lambda^3 G_1^* G_2^* E^* \tag{A2e}$$

$$\phi_6 = -\lambda^2 \{G_1^* G_2^* a_3 + E^*(G_1^* a_1 + G_2^* a_2)\} \tag{A2f}$$

$$\phi_7 = \lambda\{a_3[G_1^* a_1 + G_2^* a_2] + E^*[a_2 a_1 - c_6] + G_2^* c_2 - G_1^* c_1\} \tag{A2g}$$

$$\phi_8 = -a_3(a_1 a_2 - c_6) + c_1 a_2 - c_2 a_1 + b_6 \tag{A2h}$$

$$\phi_9 = -\lambda^3 E^* \lambda_1 (\lambda_3 G_1^* + \lambda_2 G_2^*) \tag{A2i}$$

$$\phi_{10} = \lambda^2 \{\lambda_1 \lambda_3 (E^* a_2 + G_1^* a_3) + \lambda E^* G_1^* (\lambda_1 a_6 + \lambda_3 a_4) + \lambda_1 \lambda_2 (E^* a_1 + G_2^* a_3) \\ + \lambda E^* G_2^* (\lambda_2 a_4 + \lambda_1 a_5) + \lambda \lambda_1 E^*\} \tag{A2j}$$

$$\phi_{11} = \lambda\{\lambda^2 E^*\left[G_1^*(c_5 - a_4 a_6) + G_2^*(c_4 - a_4 a_5)\right]$$
$$-\lambda\left[(E^* a_2 + G_1^* a_3)(\lambda_1 a_6 + \lambda_3 a_4)\right.$$
$$+(E^* a_1 + G_2^* a_3)(\lambda_2 a_4 + \lambda_1 a_5)]$$
$$-\lambda_1[\lambda_2(a_1 a_3 - c_1) + \lambda_3(a_2 a_3 + c_2)]$$
$$+\lambda[\lambda_1(E^* b_1 - a_3) - E^* a_4]\}$$

(A2k)

$$\phi_{12} = \lambda\{(a_2 a_3 + c_2)(\lambda_1 a_6 + \lambda_3 a_4) + (\lambda_2 a_4 + \lambda_1 a_5)(a_1 a_3 - c_1)$$
$$-\lambda_1(b_3 + a_3 b_1)$$
$$-\lambda[(E^* a_1 + G_2^* a_3)(c_4 - a_4 a_5) + (E^* a_2 + G_1^* a_3)(c_5 - a_4 a_6)$$
$$-\lambda a_3 a_4 + \lambda E^*(a_4 b_1 + b_2)]\}$$

(A2l)

$$\phi_{13} = \lambda[(a_2 a_3 + c_2)(c_5 - a_4 a_6) + (a_1 a_3 - c_1)(c_4 - a_4 a_5) + a_3(b_1 a_4 + b_2) \ + a_4 b_3 + b_4]$$

(A2m)

With Equation (A3a–f) for the calculation of the parameters of the a_i

$$a_1 = k^2 + \beta_7 \tag{A3a}$$

$$a_2 = k^2 + \beta_8 \tag{A3b}$$

$$a_3 = k^2 + \beta_9 \tag{A3c}$$

$$a_4 = k^2 I_1^* + \alpha_4 \tag{A3d}$$

$$a_5 = k^2 I_2^* + \alpha_5 \tag{A3e}$$

$$a_6 = k^2 I_3^* + \alpha_6 \tag{A3f}$$

With Equation (A4a–f) for the calculation of the parameters of the b_i

$$b_1 = 2\left[\beta_6^2 I_1^* - 2k^2 \beta_1(I_2^* + I_3^*)\right] \tag{A4a}$$

$$b_2 = 2\beta_5 I_2^*\left[\beta_4 \beta_6 I_3^* + 2k^2 \beta_5(I_1^* + I_2^*)\right] - 2k^2\left[\alpha_1 \alpha_2 \beta_6 - 2\beta_4^2 I_3^*(I_1^* + I_3^*)\right] \tag{A4b}$$

$$b_3 = 2\beta_5\left[\beta_4 \beta_6 I_1^* + 2k^2 \beta_5(I_2^* + I_3^*)\right] - 4k^2\left[2\beta_6^2 I_1^* - \beta_4^2(I_2^* + I_3^*)\right] \tag{A4c}$$

$$b_4 = 2(\beta_4 \beta_5 I_2^* I_3^* - k^2 \alpha_1 \alpha_2)(\beta_4 \beta_5 - 4k^2 \beta_6)$$
$$-2k^2[\beta_5 I_2^*(I_1^* + I_2^*)(\beta_5 \beta_3 + \beta_6 \beta_4)$$
$$+\beta_4 I_3^*(I_1^* + I_3^*)(\beta_5 \beta_6 + \beta_2 \beta_4)] \tag{A4d}$$

$$b_5 = 2\beta_5 I_2^*\left(\beta_6 \beta_4 I_1^* I_3^* - k^2 \alpha_1 \alpha_3\right) - 2k^2 \alpha_2(\alpha_1 \beta_6 I_1^* + \alpha_3 \beta_4 I_3^*) \tag{A4e}$$

$$b_6 = 2\beta_5 \beta_4 \beta_6 - 8k^2\left(\beta_4^2 + \beta_5^2 + \beta_6^2\right) \tag{A4f}$$

With Equation (A5a–f) for the calculation of the parameters of the c_i

$$c_1 = \beta_1 \beta_3 + 4k^2 \beta_2 \tag{A5a}$$

$$c_2 = -\left(4k^2 \beta_3 + \beta_4^2\right) \tag{A5b}$$

$$c_3 = \beta_6^2 I_1^{*2} + k^2 \alpha_3^2 \tag{A5c}$$

$$c_4 = \beta_4^2 I_3^{*2} + k^2 \alpha_1^2 \tag{A5d}$$

$$c_5 = \beta_5^2 I_2^{*2} + k^2 \alpha_2^2 \tag{A5e}$$

$$c_6 = \beta_6^2 + 4k^2 \beta_1 \tag{A5f}$$

where Equation (A6a–i) indicates the calculation of the parameters β_i

$$\beta_1 = \Omega_1^2 \tag{A6a}$$

$$\beta_2 = \Omega_2^2 \tag{A6b}$$

$$\beta_3 = \Omega_3^2 \tag{A6c}$$

$$\beta_4 = \Omega_1 \Omega_2 \tag{A6d}$$

$$\beta_5 = \Omega_1 \Omega_3 \tag{A6e}$$

$$\beta_6 = \Omega_2 \Omega_3 \tag{A6f}$$

$$\beta_7 = \beta_1 + \beta_2 \tag{A6g}$$

$$\beta_8 = \beta_1 + \beta_3 \tag{A6h}$$

$$\beta_9 = \beta_2 + \beta_3 \tag{A6i}$$

where Equation (A7a–f) indicates the calculation of the parameters α_i

$$\alpha_1 = \Omega_3(I_1^* + I_2^*) \tag{A7a}$$

$$\alpha_2 = \Omega_2(I_1^* + I_3^*) \tag{A7b}$$

$$\alpha_3 = \Omega_1(I_2^* + I_3^*) \tag{A7c}$$

$$\alpha_4 = \beta_3 I_2^* + \beta_2 I_3^* \tag{A7d}$$

$$\alpha_5 = \beta_3 I_1^* + \beta_1 I_3^* \tag{A7e}$$

$$\alpha_6 = \beta_1 I_2^* + \beta_2 I_1^* \tag{A7f}$$

where Equation (A8a–c) indicates the calculation of the parameters λ_i

$$\lambda_1 = \frac{\lambda I_1}{A} \tag{A8a}$$

$$\lambda_2 = \frac{\lambda I_2}{A} \tag{A8b}$$

$$\lambda_3 = \frac{\lambda I_3}{A} \tag{A8c}$$

References

1. Wahl, A.M. *Mechanical Spring*, 2nd ed.; McGraw-Hill: New York, NY, USA, 1963.
2. Shigley, J.E. *Mechanical Engineering Design*, 10th ed.; McGraw-Hill: New York, NY, USA, 2015.
3. Kobelev, V. *Durability of Springs*; Springer: Berlin/Heidelberg, Germany, 2018.
4. Love, A.E.H. *A Treatise on the Mathematical Theory of Elasticity*, 4th ed.; Dover Publications: Mineola, NY, USA, 1927.
5. Stokes, V.K. On the dynamic radial expansion of helical springs due to longitudinal impact. *J. Sound Vib.* **1974**, *35*, 77–99. [CrossRef]
6. Gironnet, B.; Louradour, G. *Comportement Dynamique des Resorts*; Techniques de l'Ingénieur: Paris, France, 1983.
7. Yldirim, V. An efficient numerical method for predicting the natural frequencies of cylindrical helical springs. *Int. J. Mech. Sci.* **1999**, *41*, 919–939. [CrossRef]
8. Lee, J.; Thompson, D.J. Dynamic stiffness formulation, free vibration and wave motion of helical springs. *J. Sound Vib.* **2001**, *239*, 297–320. [CrossRef]
9. Lee, L. Free vibration analysis of cylindrical helical springs by the pseudo spectral method. *J. Sound Vib.* **2007**, *302*, 185–196. [CrossRef]
10. Becker, L.E.; Chassie, G.G.; Cleghorn, W.L. On the natural frequencies of helical compression springs. *Int. J. Mech. Sci.* **2002**, *44*, 825–841. [CrossRef]
11. Jiang, W.; Wang, T.L.; Jones, W.K. The forced vibration of helical spring. *Int. J. Mech. Sci.* **1992**, *34*, 549–562. [CrossRef]
12. Haringx, J.A. On highly compressible helical springs and rubber rods, and their application for vibration-free mountings. *Philips Res. Rep.* **1949**, *4*, 49–80.
13. Den Hartog, J.P. *Mechanical Vibrations*; Dover Civil and Mechanical Engineering; Courier Corporation: Chelmsford, MA, USA, 1956.
14. Díaz, V. *Automóviles y Ferrocarriles*; Universidad Nacional de Educación a Distancia: Madrid, Spain, 2012; p. 287, ISBN 978-84-362-6568-2.

15. Sauvage, G. Determining the Characteristics of Helican Springs: A simplification for Application in Railway Suspensions. *Veh. Syst. Dyn. Int. J. Veh. Mech. Mobil.* **1984**, *13*, 43–59.
16. Campedelli, J. Modelisation Globale Statique des Systemes Mecaniques Hyperstatiques Pre-Charges Application a un Bogie Moteur. Ph.D. Thesis, INSA Lyon, Lyon, France, 2002.
17. *UNE-EN 13298*; Railway Applications. Suspension Components. Helical Suspension Springs, Steel. The European Committee for Standardisation: Brussels, Belgium, 2003.
18. Kovelev, V. Effect of static axial compression on the natural frequencies of helical springs. *Multidiscip. Model. Mater. Struct.* **2014**, *10*, 379. [CrossRef]
19. Yildirim, V. Axial Static Load Dependence Free Vibration Analysis of Helical Springs Based on the Theory of Spatially Curved Barx. *Lat. Am. J. Solids Struct.* **2016**, *13*, 2852–2875. [CrossRef]
20. Kato, H.; Suzuki, H. Nonlinear deflection analysis of helical spring in elastic-perfect plastic material: Application to the plastic extension of piano wire spring. *Mech. Mater.* **2021**, *160*, 103971. [CrossRef]
21. Timoshenko, S.P. On the correction for shear of the differential equation for transverse vibrations of prismatic bars. *Philos. Mag.* **1921**, *41*, 744–746. [CrossRef]
22. Rayleigh, J.W.S. *The Theory of Sound*; Dover Publications: Mineola, NY, USA, 1945.
23. Pearson, D. Modelling the ends of compression helical springs for vibration calculations. *Proc. Inst. Mech. Eng.* **1986**, *200*, 3–11. [CrossRef]

 *mathematics*

Article

Fault Classification in a Reciprocating Compressor and a Centrifugal Pump Using Non-Linear Entropy Features

Ruben Medina [1,*], Mariela Cerrada [2], Shuai Yang [3], Diego Cabrera [2], Edgar Estupiñan [4] and René-Vinicio Sánchez [2,*]

1   CIBYTEL-Engineering School, University of Los Andes, Mérida 5101, Venezuela
2   GIDTEC, Universidad Politécnica Salesiana, Cuenca 010105, Ecuador
3   National Research Base of Intelligent Manufacturing Service, Chongqing Technology and Business University, 19# Xuefu Avenue, Nan'an District, Chongqing 400067, China
4   Mechanical Engineering Department, University of Tarapaca, Arica 1130000, Chile
*   Correspondence: ruben.djmedina@ieee.org (R.M.); rsanchezl@ups.edu.ec (R.-V.S.)

Abstract: This paper describes a comparison of three types of feature sets. The feature sets were intended to classify 13 faults in a centrifugal pump (CP) and 17 valve faults in a reciprocating compressor (RC). The first set comprised 14 non-linear entropy-based features, the second comprised 15 information-based entropy features, and the third comprised 12 statistical features. The classification was performed using random forest (RF) models and support vector machines (SVM). The experimental work showed that the combination of information-based features with non-linear entropy-based features provides a statistically significant accuracy higher than the accuracy provided by the Statistical Features set. Results for classifying the 13 conditions in the CP using non-linear entropy features showed accuracies of up to 99.50%. The same feature set provided a classification accuracy of 97.50% for the classification of the 17 conditions in the RC.

Keywords: approximate entropy; non-linear systems; phase space reconstruction; fault classification; random forest; support vector machines

MSC: 28D20

Citation: Medina, R.; Cerrada, M.; Yang, S.; Cabrera, D.; Estupiñan, E.; Sánchez, R.-V. Fault Classification in a Reciprocating Compressor and a Centrifugal Pump Using Non-Linear Entropy Features. *Mathematics* **2022**, *10*, 3033. https://doi.org/10.3390/math10173033

Academic Editor: Andrey Jivkov

Received: 30 June 2022
Accepted: 17 August 2022
Published: 23 August 2022

Publisher's Note: MDPI stays neutral with regard to jurisdictional claims in published maps and institutional affiliations.

1. Introduction

In industrial applications, centrifugal pumps are equipment for transferring energy to fluids to enable pipeline transportation. Similarly, reciprocating compressors are essential in petrochemical plants and refineries for gas and other fluid transportation. Reliable functioning of these types of equipment is essential in industry to avoid unexpected halting of processes and economic losses. Consequently, several condition monitoring (CM) methods have been proposed for diagnosing faults in centrifugal pumps and reciprocating compressors [1–3].

The CM methods are based on sensing several variables from the target equipment, such as electrical current, vibration, sound, temperature, or acoustic emission. Some of the most common variables are vibration signals recorded using accelerometer sensors. Such a set of signals is processed for extracting features useful for fault detection and classification. Traditional techniques for vibration signal analysis are divided into time, frequency, and time-frequency methods. In general, signal processing techniques assume that signals represent linear systems that comply with the stationary and periodicity condition. Unfortunately, this assumption is not always met. In particular, in most rotatory machines, the vibration signals represent a non-linear system [4]. These constraints can be efficiently handled using chaos theory and non-linear dynamics techniques.

In this work, we propose to use the approximate entropy (AppEn) [5,6] and variants, such as the sample entropy (SampEn) [7,8] and fuzzy entropy (FuzzyEn) [9,10] for fault classification. The emphasis of this research is feature extraction. The classification stage can be performed using either classical or deep learning-based classifiers. However, as the feature extraction is computationally expensive, selecting deep learning models would impose an unnecessary cost. In consequence, we have selected two efficient classical models corresponding to RF and the multi-class SVM [11,12]. The original contributions of this research are the following:

- Extraction of a non-linear entropy-based feature set that provides high classification accuracy using RF and SVM models. The accuracy attained by the SVM model trained with the non-linear *Entropy Features* set was higher than the accuracy attained by a CNN model trained with 2D spectrogram images for both the CP and the RC.
- Detailed comparison of three feature sets useful for classifying a large number of faults in a CP and an RC. These are the non-linear *Entropy Features*, the *Information Entropy*, and the *Statistical Features* sets. The first set is composed of the approximate entropy and several variants, and the second set is composed of the combination of the wavelet packet transform (WPT)-based features and the power spectrum entropy (PSE)-based features. Finally, the third set is composed of classical time series statistical features.
- The non-linear *Entropy Features* set and the *All Features* set corresponding to the fusion of the three feature sets when compared provided a classification accuracy of up to 99.59% for the CP and up to 97.90% for the RC. For the CP, 13 different conditions were classified, and 17 valve conditions were classified for the RC.

The paper is organized as follows: in Section 3 the theoretical background is presented. In Section 4, the test bed for acquiring the vibration signals from the CP and RP is presented. In Sections 5 and 6, the methodology for extraction of the different sets of features is described. The methodology for classifying faults using RF and SVM models based on the extracted features is presented in Section 7. The results are described in Section 8, and finally, the conclusion and future work are presented in Section 10.

2. Related Research

The related research is discussed in the following subsections. A compendium with several publications is presented in Table A1 in terms of their main features. A brief revision concerning these publications is presented in Sections 2.1–2.3.

2.1. Rotating Machinery

Although non-linear entropy methods have been used for classifying a small set of faults in centrifugal pumps and reciprocating compressors, attaining high accuracy, their application to other rotatory machines (RM) has also been reported. Research concerning the correlation dimension (CD) is presented in [13]. The approach analyzes the vibration signal measured from a roller element bearing. Results allow them to conclude that CD features show significant statistical differences between a roller element close to failure and a new one. The largest Lyapunov exponent (LLE) [14] has been used to show that faults in non-linear rotatory systems can be detected early when the onset of the fault occurs. This feature evolves as the fault progresses. The research reported in [4] combines the CD, the approximate entropy, the LLE, and visualization of the phase space for detecting damage in bearings and gearboxes. In [15], a method for classifying seven different conditions in roller bearings and ten types of conditions on a gearbox is reported. The approach is based on features extraction from the Poincaré plot and their classification using SVM. The accuracy attained was 100% for the roller bearings and 99.3% for the gearbox dataset. A combination of the spectral graph wavelet transform (SGWT) with detrended fluctuation analysis (DFA) was used for denoising the vibration signals in RM [16]. The first stage is converting the vibration signal into a graph signal. The SGWT is used for decomposing the graph signal into scaling function coefficients and spectral graph wavelet coefficients. DFA is used for selecting the level of decomposition of SGWT. A bi-dimensional representation of vibration signals denoted as the complexity-entropy causality plane (CECP) useful for fault classification in RM is reported in [17]. This 2D representation was obtained with the normalized permutation entropy and the Jensen–Shannon complexity of time series. The authors showed the utility of this representation for improving the classification accuracy of faults in roller bearing and gear applications using SVM models. A review of entropy algorithms and its variants in RM is reported in [18,19]. The reviews briefly introduced several entropy methods and their application in fault detection in RM.

2.2. Reciprocating Compressor

An application of faults diagnosis in an RC using multi-scale entropy is reported in [20]. After performing the signals' local mean decomposition (LMD), vibration signals' multi-scale entropy (MSE) features were calculated. Four conditions were classified using SVM, obtaining a classification accuracy of 95.0%. In a previous study [2], a method for feature extraction from vibration signals recorded from an RC is proposed to classify 13 different conditions of combined faults in valves and roller bearings. The feature extraction method was based on symbolic dynamics and complex correlation measures. The extracted features were used for fault classification using several machine learning models, attaining accuracy of up to 99%. Research concerning the fault diagnosis and severity of valve leakage in an RC is reported in [21]. The diagnosis was performed using features extracted from the $p - V$ diagram using principal component analysis (PCA) and linear discriminant analysis (LDA). The method showed that the $p - V$ diagram's features help classify two types of faults with several levels of severity, attaining classification accuracies of up to 99.62%. The research reported in [22] concerns the classification of faults based on images representing the $p - V$ diagram. A set of features is extracted from the images for classifying eight health conditions in valves, piston rings, and valve springs. Accuracies of up to 97.9% were obtained using artificial neural network (ANN) models. Research reported in [23] combines the binary moth flame optimization (BMFO) algorithm with the K-nearest neighbor (KNN) algorithm for classifying three types of valve faults in an RC, attaining accuracies higher than 99%. Research concerning convolutional neural networks (CNNs) for classifying faults in an RC is reported in [24]. The vibration signal was re-arranged into a 2D format by bringing its time samples into lines that moved 45 degrees counterclockwise to reinforce their features. This array was fed to the CNN that incorporated the attention mechanism through squeeze-and-excitation (SE) modules. The maximum accuracy attained was 99.4%. Classification of four different valve conditions using 1D and 2D CNNs is reported in [25]. The classification was performed using vibration signals in the case of the 1D CNN and considering the fusion of seven channels (four channels with vibration signals and three channels with pressure) in the case of 2D CNN. The classification was performed for different SNR values of the signals, and the highest classification accuracy was 100% for signals with more than 20 dB of SNR. The CNN used has a complex architecture.

2.3. Centrifugal Pump

The permutation entropy features are used in a fault detection method reported in [26]. The method was applied to hydraulic pump (HYP) fault detection. The vibration signal was processed using resonance-based sparse signal decomposition (RSDD). A Multi-scale hierarchical amplitude-aware permutation entropy (MHAAPE) method was applied for feature extraction. Four fault conditions were classified using a probabilistic neural network, SVM models, and extreme learning machines (ELM). The approach attained an accuracy of up to 100%. A method for faults diagnosis in a CP is reported in [27]. The method is based on applying complementary ensemble empirical mode decomposition (CEEMD) to vibration signals. The decomposed signals or intrinsic mode functions are processed for extracting the SampEn. The set of features is used for classification; RF attained classification accuracies of 94.58% for classifying five conditions. A feature extraction approach from vibration signals is reported [28] to classify faults in a CP. The discriminant feature extraction method includes three stages. The first stage is the healthy baseline signal selection. The second stage is devoted to cross-correlating the healthy baseline signals with the vibration signals of different fault classes. A set of features is extracted from the resultant correlation sequence. In the third stage, time, frequency, and time–frequency domain raw hybrid features are extracted from baseline signal and vibration signals from faulty classes. A new set of discriminant features is composed of the correlation coefficient between raw hybrid feature pools. The combination of all extracted features is fed to an SVM model for classification. The method is used to classify four conditions in a CP, attaining accuracy of up to 98.4%. A method for fault classification in a CP based on vibration signals is reported in [29]. A feature extraction pre-processing is used to obtain a time-frequency representation using the continuous wavelet transform (CWT). The CWT is converted to gray images fed to a CNN model with an adaptive learning rate. Accuracies of 100% are obtained for classifying four different fault conditions. A fault classification method of a multi-stage CP reported in [30] starts by selecting the fault-specific frequency band and continues by extracting statistical features in time, frequency, and wavelet domains from this band. The next step is to select a low dimensional features vector using the informative ratio PCA. The classification of four conditions in a multi-stage CP using KNN models. The accuracy attained was 100%. Fault detection and classification for water pump bearings based on features extracted from the instantaneous power spectrum is reported in [31]. The instantaneous power spectrum (IPS) was obtained with the voltage and current measured, and several features were extracted from the IPS for classifying three different conditions at different load levels. The classification was performed using an extreme gradient boosting (XBG) model [32]. The application of CNN for classifying faults in a CP based on acoustic images is reported in [33]. In this application, the sound signals were acquired from a CP test rig where five different health conditions were implemented. The sound signals were converted into acoustic images using the analytical wavelet transform (AWT). The acoustic images were fed to the CNN for classification. The attained accuracy was 100%.

2.4. Anomaly Detection

Anomaly detection corresponds to identifying events within a stream of data features that deviate significantly from most available data. Anomaly detection has applications in many fields. This topic is relevant because the non-linear entropy-based features have found applications in the context of anomaly detection. Examples of applications have been reported in [34–39]. Several authors have proposed anomaly detection applications centered around deep learning. Approaches such as CNN, autoencoders, and generative adversarial networks [40–42] can be used for anomaly detection implementation. Research concerning anomaly detection in RM is active. An example is reported in [41] for anomaly detection in gearboxes where a prototype selection algorithm is used for improving the training of the feature extractor implemented with a CNN. The research reported in [40] uses an autoencoder for learning to model the healthy state of the rotating machine. Such a learning process is performed using vibration signals representing the healthy state. The anomaly is detected using a threshold-based approach to the reconstruction error of data acquired from an unknown state. The algorithm was validated with the IMS bearings dataset. Other recent applications of anomaly detection in RM and also in other fields are reported in [41,43–46].

3. Theoretical Background

3.1. Phase Space Reconstruction

The phase space reconstruction method is based on the embedding theorem of Taken, reported in [47]. The theorem postulates that we can recover the equivalent dynamics of a non-linear system by using time delays from a recorded time-domain signal. The approximate dynamics correspond to a 1D projection of the system trajectory. The theorem is mainly applied to univariate time series [47]; however, the multivariate version has also been reported in [48]. In this research, we are mainly concerned with univariate embedding.

The average mutual information (AMI) enables the selection of the time delay or lag denoted τ [49]. The AMI is plotted versus τ. According to [50], the selection of τ could be the location of the first local minimum of the AMI. The system's embedding dimension, denoted as m, could be selected by applying the false nearest neighbors (FNN) method, as reported in [51].

3.2. Approximate Entropy

Given a signal denoted $\mathbf{x}$ including N samples, the AppEn [5,6] measures the regularity of the time-series. The time-series regularity and dynamics are estimated considering a higher order dimensional space generated by a series of vectors represented by time-delayed components of $\mathbf{x}$. The AppEn is calculated for a time-series of length N, representing a system with embedding dimension m, lag τ, and a distance threshold between vectors denoted r. In general, the calculation of the AppEn should consider a window of the signal with length N large enough to reduce the bias in the estimation [6]. However, a limitation is that extending N can significantly increase the computational cost. AppEn has the advantage of being robust to noise and outliers (see [52] for a detailed review concerning entropy calculation methods).

3.3. Sample Entropy

The SampEn is obtained through modification of the AppEn algorithm [7]. The modification consists of the elimination of self-matches in the evaluation of the distance between lagged vectors. The SampEn avoids the bias introduced in ApEn by considering self-matches. The resultant features are more consistent and unbiased, even when their calculation uses short-length signal windows. In addition, the calculation of SampEn is more straightforward than the calculation of AppEn.

3.4. Fuzzy Entropy

The FuzzyEn measure [10] is a variant of the AppEn that modifies the function used for estimating the similarity between lagged vectors. The algorithm's modification is based on changing the Heaviside function for a fuzzy exponential membership function. The calculation of FuzzyEn also incorporates the modifications introduced for the calculation of the SampEn.

3.5. Shannon Entropy and Other Measures of Signal Complexity

The entropy measure enables quantification of the complexity of a non-linear dynamical system. The Shannon entropy (ShannEn) quantifies the amount of information in a system. In addition, the conditional entropy quantifies the rate of information generation [53,54]. Other entropy measures, such as corrected conditional entropy, can be estimated based on these measures. A detailed review of ShannEn estimation is presented in [55].

3.6. Permutation Entropy

The permutation entropy (PermEn) is based in partitions [56]. When the time series $\mathbf{x}$ takes many values, their range can be replaced with a symbol sequence $\{s_i\}$. This is done by defining the partition of the signal range $\mathbf{x} = P_1 \cup, \ldots, \cup P_\xi$. Based on this partition, the symbol $s_i = j$ is defined if x_i is in P_j. The PermEn is calculated for different embedding dimensions m; however, for practical purposes $m = 3, 4, \ldots, 7$. Denote by π the permutation of order m with probability given as $p(\pi) = \#\{Cardinality\ of\ permutation\ \pi\}/N - m + 1$; then, the PermEn is calculated using Equation (1):

$$PermEn = -\sum p(\pi)logp(\pi) \tag{1}$$

This research used the method reported in [57] for computing the PermEn. The method is based on pre-calculating values for successive ordinal patterns of order m. This approach allows the calculation of PermEn in large time series using overlapped successive time windows.

3.7. The Correlation Dimension

The system's phase space enables the calculation of complexity [58] that is quantified using several non-linear features such as SampEn. In addition, another feature is the CD which quantifies the number of independent lagged vectors required for defining a system [59]. Details concerning the estimation of this feature are presented in [60].

3.8. Detrended Fluctuation Analysis

The fractal scaling properties quantified in short intervals of time-series [61] can be performed using DFA. This approach enables quantitative analysis of non-stationary time-series [62]. Their calculation requires performing the integration of the signal, subdividing such a signal into several segments, finding the trend of the signal using least-square approximations, and calculating the average fluctuation of the signal around the trend. Details of the estimation of this feature are presented in [63].

3.9. Largest Lyapunov Exponent

The Lyapunov exponent λ quantifies the relative motion of trajectories in the phase-space of a system [64]. Its value allows for determining if a signal is chaotic or periodic. A system can be represented by several Lyapunov exponents determined by the embedded dimension. However, the largest one, denoted *LLE*, is typically used. There are several algorithms for calculating the Lyapunov exponent [65]. The method proposed by Rosenstein et al. [66] was implemented in this research.

3.10. Information Entropy and Chaotic Time Series

Information entropy was proposed by Claude Shannon [67] in the context of theoretical communication modeling. A collection of messages sent through a communication channel can be considered as independent random events with probabilities $p_i \geq 0$, where $\sum_{i=1}^{n} p_i = 1$. Information entropy is defined as:

$$H = -\sum_{i=1}^{n} p_i log\, p_i \qquad (2)$$

The entropy concept is also used in classical thermodynamics in the context of heat transfer [68]. Gibbs and Boltzmann attained a definition of entropy with an expression similar to Equation (2) [69]. When dealing with discrete-time signals represented as a vector, the entropy concept has also been applied to select the best basis for orthogonal wavelet packets [70]. The authors showed that the assumption that the Karhunen–Loeve basis is the best, even for a single vector which enables the calculation of several entropy features. Consider the time-series **x** where the vector can be arranged in decreasing order such that $x_i = 0$ for large i. Then, $p_i = |x_i|^2 \|\mathbf{x}\|^{-2}$. Without loss of generality, the time-series can be normalized so $\|\mathbf{x}\|^{-2} = 1$. The following features can be estimated: the log energy entropy, the Shannon information entropy, the norm entropy, the threshold entropy, and the "sure" entropy.

3.10.1. Tsallis and Rényi Entropy

In the context of systems with complex behavior (including, for instance, chaotic motion or fractal evolution), Tsallis [71] has proposed an alternative definition of entropy as:

$$HT_q = \frac{K}{q-1}\left(1 - \sum_{i=1}^{N} p_i^q\right) \qquad (3)$$

where K is a positive constant and q denotes any real number. Tsallis showed that when $q \to 1$, HT_q tends to the classical Boltzmann–Gibbs–Shannon entropy represented by Equation (2). Similarly, q is used in the Rényi entropy [72] to amplify probabilities as:

$$HR_q = \frac{K}{q-1} log \left(\sum_{i=1}^{N} p_i^q \right) \tag{4}$$

Both entropies are parameterized representations of the Boltzmann–Gibbs–Shannon entropy, where one parameter is used for weighting the probabilities to evaluate different fractal properties of the system.

3.10.2. Power Spectral Entropy

The complexity can also be quantified using the power spectral density of time-series [73,74]. This feature has been used for fault detection in gearboxes and roller bearings. The normalized power spectral density of the time series $\mathbf{x}$ is denoted $P(\omega_i)$, and it can be expressed as:

$$p_i = \frac{P(\omega_i)}{\sum_i P(\omega_i)}. \tag{5}$$

This equation enables the calculation of the power spectral entropy (PSE) as:

$$S = -\sum_{i=1}^{n} p_i log \, p_i \tag{6}$$

The PSE is calculated for each time frame and is a vector denoted as $\mathbf{S}$. When the signal has high complexity, the spectrum of the time series tends to be uniform, and the PSE is high. In contrast, when the spectrum has a narrow peak, the PSE feature is lower, and the signal is less complex.

3.11. Machine Learning Techniques

3.11.1. Random Forest

The RF algorithm is a machine learning algorithm that combines a set of weak learners using ensemble methods to attain a high-quality classification or regression model [75]. The most common weak learners in the RF model are tree-type classifiers used for growing forests of randomized trees using bootstraps resampling techniques [76].

3.11.2. Support Vector Machines

The machine learning model proposed by Vapnik [77] is the SVM. In its more basic form, the SVM is used for solving problems of binary classification where the data in each of the classes can be separated using a hyperplane that can be constructed based on the support vectors, as detailed in [78]. In practical applications, the SVM incorporates higher-dimensional transformation represented as kernels for transforming the feature space to increase class separability. Similarly, multi-class support vector machines are also available, as reported in [11].

4. Experimental Test-Bed

In this research, two different vibration signal datasets were considered. The first was acquired from a CP and the second from an RC.

4.1. Centrifugal Pump Test-Bed

In this research, a multi-stage vertical CP is considered. The CP model 3SV10GE4F20 was a 2 HP pump operated at 3500 rpm. The pump had ten stages for a nominal flow of 15 GPM. The vibration signals were recorded using model 603C01 accelerometers. The module NI 9234 connected to a laptop computer was used to acquire the vibration signal using a sampling frequency of 50 kHz. The physical description and locations of sensors are shown in Figure 1. The signal acquisition was performed under controlled operating conditions concerning the room temperature, relative humidity, and environmental noise.

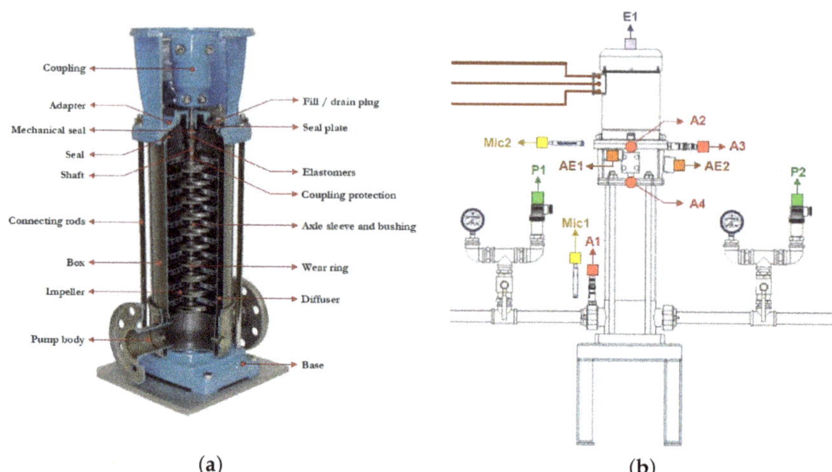

Figure 1. CP and sensors. (**a**) Multi-stage vertical CP. (**b**) Locations of sensors. The vibration signals were recorded using A1, A2, A3, and A4 accelerometers.

Six experimental conditions were considered related to the discharge pressure. The conditions are denoted as Ci with i ranging from 1 to 6. The pressure started at 5.5 bar, increasing by 1 bar for each condition. $C1$ corresponds to a discharge pressure of 5.5 bar.

4.1.1. Faults of the CP

A set of 13 impeller faults were configured in the CP. One condition corresponded to a CP where all stages were healthy. Three conditions involved pitting at the entrance of the impeller blades (PEB), three conditions involved pitting at the output of the impeller blades (POB), three conditions involved impeller channel blockade (ICB), and finally, three conditions involved an imbalanced impeller (IB). The pitting and impeller blockade can affect several stages of a pump with different levels of severity. For instance, in fault P3, the severity level for stage 6 was 1, and the severity increased with the stage number, and it was 5 for stage 10. The pitting severity levels were attained by changing the number and diameter of holes in the blades. The ICB severity levels were attained by varying the percent of blocked channels in the impeller. The impeller imbalance severity was obtained by increasing the surface of the cut portion of the impeller front cover. Table 1 presents the list of configured faults.

Table 1. Faults conditions in the CP.

Fault Label	Impeller Fault	Stages Status	Severity
P1	Healthy	All stages healthy	healthy
P2	PEB	Faulty stages 9-10	1-2
P3	PEB	Faulty stages 6-10	1-5
P4	PEB	Faulty stages 5-10	3-8
P5	POB	Faulty stages 9-10	1-2
P6	POB	Faulty stages 6-10	1-5

Table 1. *Cont.*

Fault Label	Impeller Fault	Stages Status	Severity
P7	POB	Faulty stages 5-10	3-8
P8	ICB	Faulty stages 10	1
P9	ICB	Faulty stages 7-10	1-4
P10	ICB	Faulty stages 5-10	1-6
P11	IB	Faulty stages 10	1
P12	IB	Faulty stages 7-10	1-4
P13	IB	Faulty stages 5-10	1-6

4.1.2. CP Vibration Signal Dataset

The vibration signal acquisition was performed by maintaining the environmental condition within a controlled range. The motor rotation was set to 3600 rpm. A pressure condition Ci was set, and the vibration signal was acquired during 10 s. Ten repetitions were performed for each of the faults conditions P1 to P13. Consequently, 60 vibration signals were recorded for each condition. A total of 780 vibration signals are included in the dataset, including all fault conditions.

4.2. Reciprocating Compressor Test-Bed

The RC test bed is centered around a compressor model EGB-250 with two stages. The compressor is driven by an induction motor with a power of 5.5 HP that operates at 3470 rpm.

A detailed description of the test bed is described in [2]. The test bed incorporates several accelerometer sensors for acquiring the vibration signals. The accelerometers are denoted as A1, A2, A3, and A4 and were located as shown in Figure 2. Module NI9234 was used to acquire the vibration signals from model 603C01 accelerometers. The sampling frequency was 50 kHz. The acquired vibration signals were transferred to a laptop computer.

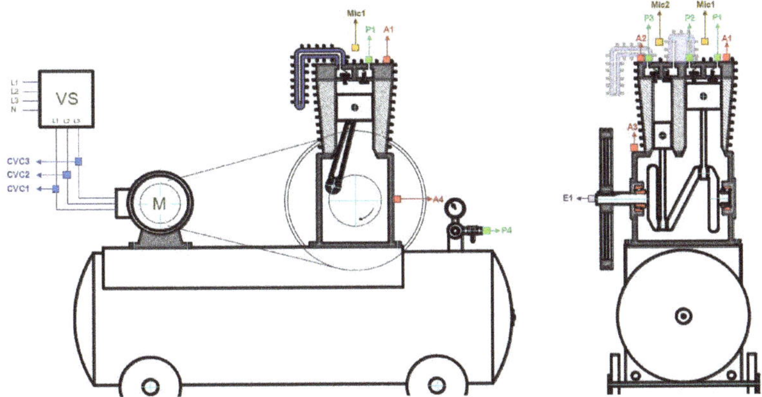

Figure 2. Sensors' locations in the RC. The accelerometers are denoted as A1, A2, A3, and A4.

4.2.1. Faults of the Reciprocal Compressor

The test bed enables the configuration of a set of 17 fault conditions concerning the valves. Four faults were configured for compressor stages and the discharge and inlet valves. The faults configured were (a) valve seat wear, (b) corrosion of the valve plate, (c) fracture of the valve plate, and broken spring. Table 2 presents the completed list of configured faults.

Table 2. Valve faults in the RC.

Fault	Label Stage and Valve Type	Fault Type
P1	All stages	Healthy
P2	Second stage, discharge	Valve seat wear
P3	Second stage, discharge	Corrosion of valve plate
P4	Second stage, discharge	Fracture of valve plate
P5	Second stage, discharge	Broken Spring
P6	Second stage, inlet valve	Valve seat wear
P7	Second stage, inlet valve	Corrosion of valve plate
P8	Second stage, inlet valve	Fracture of valve plate
P9	Second stage, inlet valve	Broken Spring
P10	First stage, discharge	Valve seat wear
P11	First stage, discharge	Corrosion of valve plate
P12	First stage, discharge	Fracture of valve plate
P13	First stage, discharge	Broken Spring
P14	First stage, inlet valve b	Valve seat wear
P15	First stage, inlet valve b	Corrosion of valve plate
P16	First stage, inlet valve b	Fracture of valve plate
P17	First stage, inlet valve b	Broken Spring

4.2.2. RC Vibration Signal Dataset

Each vibration signal has a duration of 10 s. The motor rotated at 3462 rpm, representing a crankshaft rotation of 768 rpm. The operating conditions of the RC were kept constant during the vibration signal acquisition. In particular, the tank pressure was kept at 3 bar, the inlet pressure and discharge pressure were kept within the nominal range, and the surface temperature of the valve cover was maintained. Similarly, the environmental conditions corresponding to the room temperature and relative humidity were maintained within a predefined range. A total of 15 vibration signals were acquired for each fault condition, and a total of 255 vibration signals were acquired for each sensor.

5. Feature Extraction from the CP

5.1. Non-Linear Entropy Based Features

Each vibration signal in the CP dataset was processed for extracting a set of 29 features. The set of features was composed of a subset of 14 features corresponding to non-linear entropy features (Table A2) defined in Equations (A1)–(A14), a subset of 15 features corresponding to information-based features (Table A3) defined in Equations (A15)–(A29). The first step consisted of performing the phase space reconstruction [48] for estimating the spatial dimensions m and the lag τ of the system. The entropy-based features were estimated using these parameters. Each feature set estimated in a widow was considered a feature sample. Five non-contiguous segments of size 12,000 samples from each vibration signal were considered for calculating features. As the number of vibration signals files was 780, the total number of feature samples was $780 \times 5 = 3900$.

5.2. Statistical Features

Additionally, a set of 12 statistical features (Table A4), defined in Equations (A30)–(A41), were extracted to compare to the entropy features set and be combined with other feature sets to evaluate their accuracy. The procedure for extracting the statistical features was similar to extracting the non-linear features. Three-thousand nine-hundred (3900) samples for the CP dataset were extracted and considered for classification.

6. Feature Extraction from the RC

6.1. Entropy Based Features

The procedure for extracting the features from the RC was similar to extracting the features from the CP. However, the number of segments with 12,000 samples was increased to ten. This increment was justified because this dataset had fewer files (255). A set of 29 entropy-based features was extracted from each signal segment. In total, 2550 feature samples were obtained for this dataset.

6.2. Statistical Features

A set of 2550 feature samples was obtained from the RC vibration signal dataset. The procedure was similar to the case of the CP. Each feature sample comprised 12 statistical features extracted from each of the windows considered, and each signal was subdivided into ten fragments.

7. Classification of Faults Using RF and SVM

Each feature dataset was used as input to both classical machine learning algorithms. Each dataset was randomly split into a training set (85%) and a test set (15%). The training set was used for 10-fold cross-validation, and the accuracy of the partitioned trained model was calculated with the test set. This entire process (including the splitting) was repeated ten times, and the accuracy is the average obtained during the repetitions.

The accuracy and the area under the Receiver Operator Curve (ROC) are two relevant metrics for classifier performance. Several metrics [79] were used for evaluating the classification results. The confusion matrix elements were used for defining the metrics [80].

8. Results

8.1. Results for the CP Dataset

The set of fault conditions for the CP included several severity levels in the impeller (see Table 1). Examples of the vibration signals for the CP dataset are shown in Figure 3. Four conditions are shown, and in this subset, only subtle differences in signal amplitude can be visually detected. Any fault signature could be affected by noise.

In Figure 4, the AppEn for four vibration signals of the CP dataset are shown. Each of the vibration signals represents a fault condition. The utility of non-linear features for detecting fault conditions in vibration signals is shown in the following paragraphs. The feature estimation in each signal starts by performing the phase space reconstruction [47] aimed at estimating the system dimension m and the lag τ. These parameters are used for calculating the AppEn for several values of r. The AppEn for the CP signals increases with r until attaining a maximum and decreases with the increase in r. In this example, the differences are evident between each impeller's conditions.

The LLE is calculated using four vibration signals extracted from the CP dataset. The calculation procedure was explained in Section 3.9. Results concerning the LLE are shown in Figure 5. The LLE is negative for the healthy class, class P6, and class P13. However, it is positive for class P10. The sign of the Lyapunov exponent could be related to the presence of chaos when the LLE is positive. When the LLE is negative, the time series could represent a system with periodic dynamics, as suggested in [65].

A non-integer dimension usually characterizes non-linear systems that could have a chaotic nature. Generally, when dealing with chaotic systems, the Rényi dimension is used as a feature of strange attractors that enables the estimation of several feature dimensions. The D_0 is the fractal dimension, D_1 is the information dimension, and D_2 is the CD [81]. Like AppEn estimation, the first step corresponds to the phase space reconstruction that provides the embedding dimension m. Parameter m is the upper bound of the CD [82]. The CD feature was estimated for several vibration signals of the CP dataset. The results are shown in Figure 6. This set of signals has estimated embedding dimensions of 5 for P1 and P6, and 4 for P10 and P13. The CD is lower than 5 and higher than 4 for the signals from classes P1 and P6. Similarly, for classes P10, and P13, the CD values are higher than 3 and lower than 4.

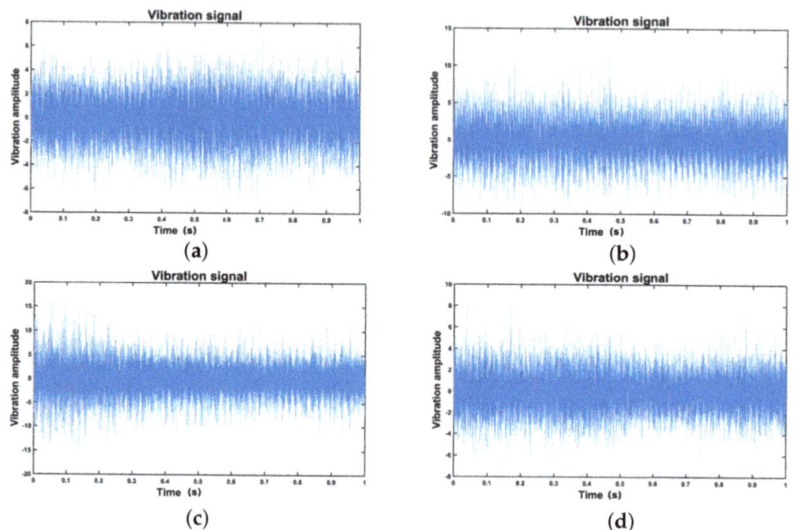

Figure 3. Vibration signals from the CP dataset. (**a**) Signal from class P1. (**b**) Signal from class P6. (**c**) Signal from class P10, (**d**) Signal from class P13.

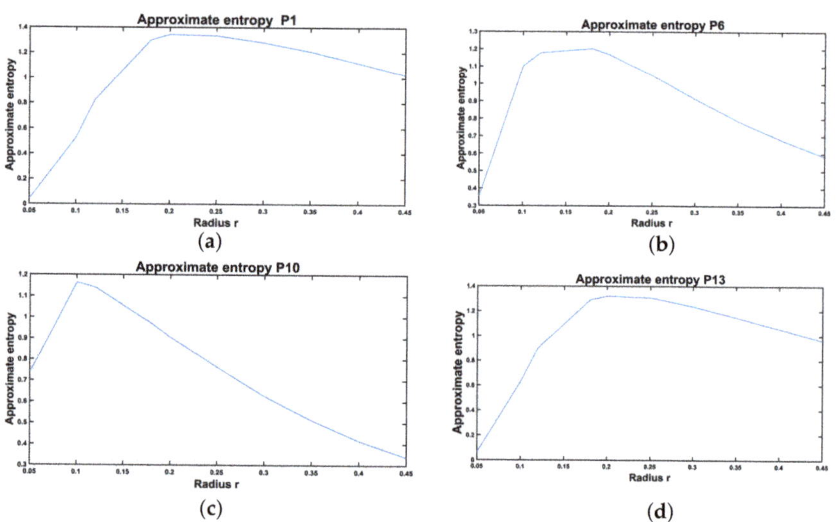

Figure 4. AppEn for several signals from the pump dataset. (**a**) ApEn for a signal from class P1, (**b**) ApEn for a signal from class P6, (**c**) ApEn for a signal from class P10, (**d**) ApEn for a signal from class P13.

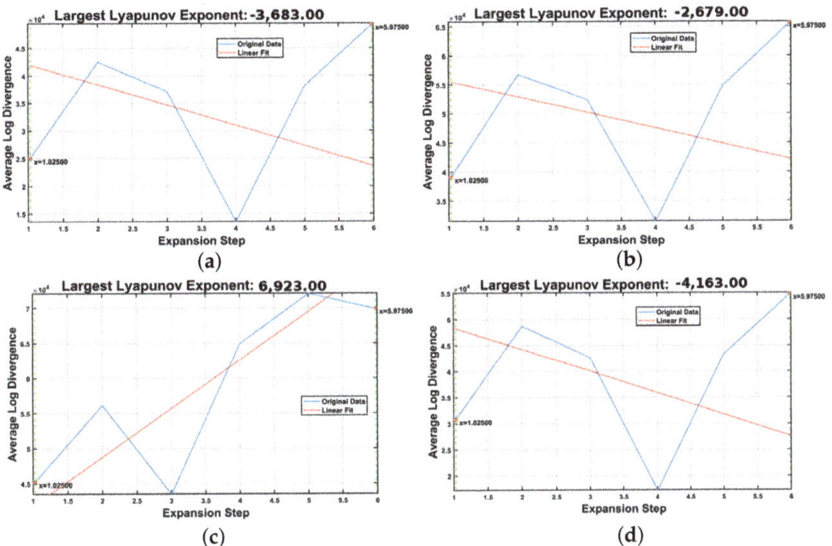

Figure 5. LLE for several signals from the CP dataset. (**a**) LLE for class P1, (**b**) LLE for class P6, (**c**) LLE for class P10, (**d**) LLE for class P13.

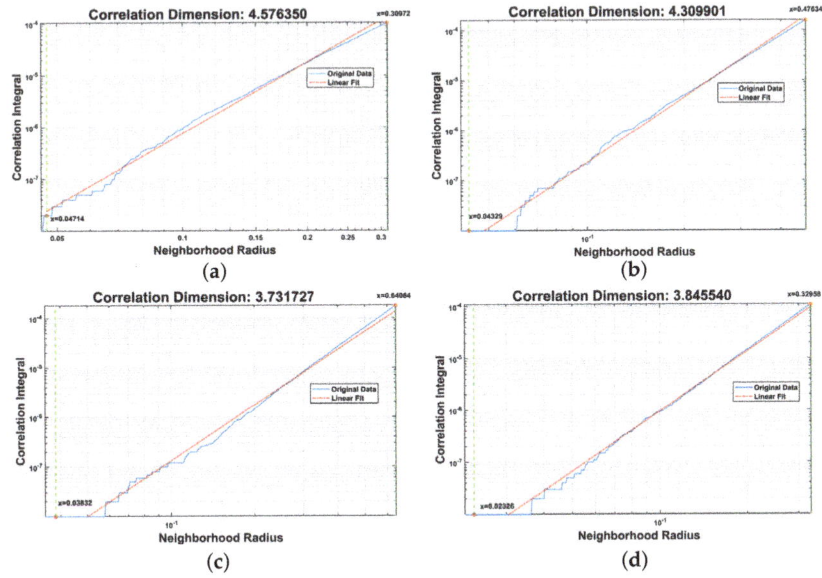

Figure 6. CD for several signals from the CP dataset. (**a**) CD for a signal from class P1, (**b**) CD for a signal from class P6, (**c**) CD for a signal from class P10, (**d**) CD for a signal from class P13.

The percentages of accuracy for classifying the fault considering the CP dataset using RF and SVM are presented in Table 3. The accuracy is presented for each of the models and feature sets considered. The vibration signal channels A2, A3, and A4 were considered for feature extraction. Vibration signal channel A1 was not considered for two reasons. Firstly, the computational time of the feature extraction is high, and to keep the total feature extraction time feasible with the available hardware, we only considered three channels.

Secondly, we selected the channels close to the mechanical connection with the motor, where a roller bearing is located, because the pump stages in this neighborhood have a higher probability of failing.

Table 3. Percentages of accuracy for fault classification using the vibration signals of the CP dataset.

Features	Model	A2	A3	A4
Statistical	RF	67.76	68.34	68.50
	SVM	68.50	70.51	71.03
Entropy-based	RF	98.09	97.83	94.70
	SVM	99.50	98.96	97.30
Info-Statistical	RF	84.60	84.80	93.81
	SVM	86.26	86.96	96.32
All Features	RF	97.76	99.38	99.37
	SVM	99.59	99.69	99.81

The lowest classification accuracy was attained using RF trained with the *Statistical Features* set. In this case, the accuracy attained from channel A2 was 67.76%, for channel A3 was 68.34%, and for channel A4 was 68.50%. The accuracy attained using SVM with this set of features was 68.5% for channel A2, 70.51% for channel A3, and 71.03% for channel A4. The highest accuracy was 99.59% for channel A2, 99.69% for channel A3, and 99.81% for channel A4. This accuracy was attained with the *All Features* set and SVM. With this set of features, using the RF model, the accuracy was 97.76% for channel A2, 99.38% for channel A3, and 99.37% for channel A4.

A comparison of accuracy considering both machine learning models and feature sets is presented in Figure 7. In the box plot, a red line represents the median of the data that divides the box or inter-quartile range in two parts. This box represents 50% of the data. The horizontal line over the box represents the upper quartile that indicates that 75% of the data have values below this quartile. The horizontal line below the box represents the lower quartile that indicates that 25% of the data have values below this quartile. The red cross below or over the box represents the outliers. The accuracy is presented for channel A2 in box plots calculated based on the ten repetitions performed for obtaining ten different cross-validated models. The highest accuracy (99.50% and 99.59% in Table 3) was attained based on the *Entropy Features* and the complete set of features, *All Features*, with the SVM model. The lower accuracies (67.76% and 68.50%) for the RF and SVM models were attained by considering the *Statistical Features* set. The combination of information entropy and statistical features (*InfoStat Features*) provided an intermediate accuracy value (84.60% and 86.96%) for both considered models.

Table 4 presents several performance metrics [79,80,83] results expressed in percentages. The included metrics are the sensitivity, the specificity, the error, the false positive rate, and the area under the curve (AUC). The metrics were calculated during the cross-validation results of the SVM model trained with the *All Features* set extracted from vibration signals recorded in the channel A2 from the CP. The highest sensitivity of 100.00% was attained by classes P1, P6, P12, and P13. In contrast, the lowest sensitivity of 98.67% was attained by class P9. The highest specificity of 100% was attained by classes P1, P3, P5, and P12. The lowest specificity of 99.87% was attained by classes P4 and P8. Concerning the AUC [84], the highest value of 100% was attained by classes P1 and P12, and the lowest AUC value was 99.19%.

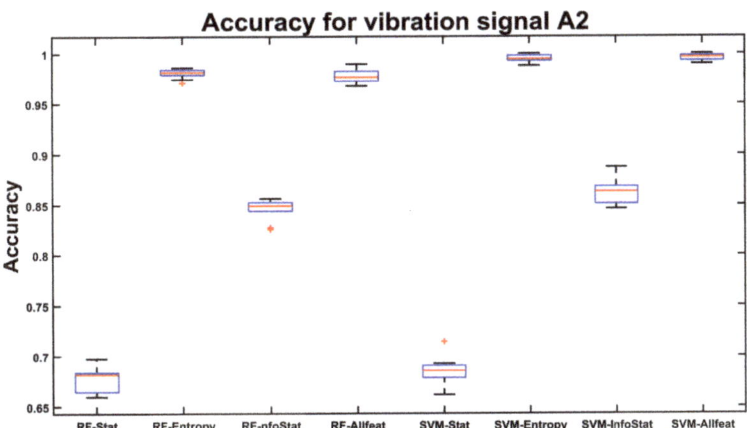

Figure 7. Comparison of accuracy for the machine learning models tested, considering several combinations of features from channel A2 acquired from the CP.

Table 4. Performance metrics (in percent) obtained with the SVM model and the *All Features* set. The model was trained using vibration signals in channel A2 from the CP dataset.

Class	Sensitivity	Specificity	Error	FPR	AUC
P1	100.00	100.00	0.00	0.00	100.00
P2	99.12	99.96	0.88	0.00	99.74
P3	99.78	100.00	0.22	0.00	99.99
P4	99.11	99.87	0.89	0.13	99.19
P5	98.68	100.00	1.32	0.00	99.94
P6	100.00	99.98	0.00	0.02	99.89
P7	98.90	99.96	1.10	0.04	99.73
P8	99.55	99.87	0.45	0.13	99.20
P9	98.67	99.93	1.33	0.07	99.50
P10	99.56	99.98	0.44	0.02	99.87
P11	99.33	99.91	0.67	0.09	99.42
P12	100.00	100.00	0.00	0.00	100.00
P13	100.00	99.93	0.00	0.07	99.56

8.2. Results for the RC Dataset

A set of vibration signals for the RC is presented in Figure 8. The vibration signals presented are for the healthy class P1, and for class P6 representing the valve seat wear (in the inlet valve) of the second stage. The bottom row shows the vibration signal for class P11 corresponding to corrosion of the valve plate located in the discharge valve of the first stage. Finally, the vibration signal for the class P16 represents the fracture of the valve plate for inlet valve b of the first stage. The appearance of this vibration signal set differs from that of the CP dataset's signals. In this dataset, the signals are more peaked, and each prominent peak indicating an event is surrounded by signals with lower amplitudes.

The AppEn estimated for each of the signals shown in Figure 8 is shown in Figure 9. The curve shape of AppEn is different from that of the CP. The AppEn in the RC is higher for low amplitudes and decreases as the threshold of amplitudes represented by the radius r increases.

The calculation of the LLE for the set of vibration signals selected from the RC dataset is shown in Figure 10. Although the LLE value is different for each signal presented, all the estimated values are positive, suggesting that the system is chaotic [65]. The CD is presented in Figure 11 for the set of signals of the RC. The CD for these signals has different values according to the fault type. The estimated values of CD are close to the embedded dimensions [82] (four, calculated using the phase space reconstruction method).

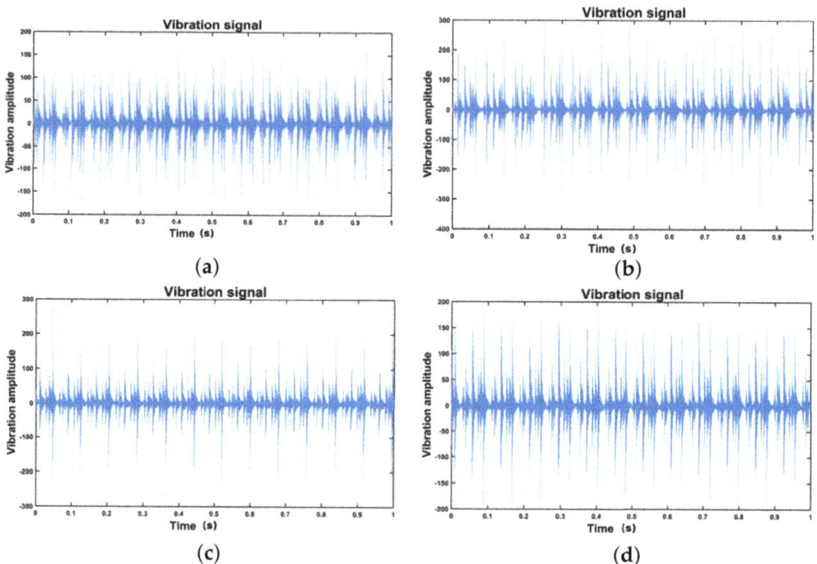

Figure 8. Vibration signals from the RC dataset. (**a**) Signal from the class P1, (**b**) signal from class P6, (**c**) signal from class P11, (**d**) signal from class P16.

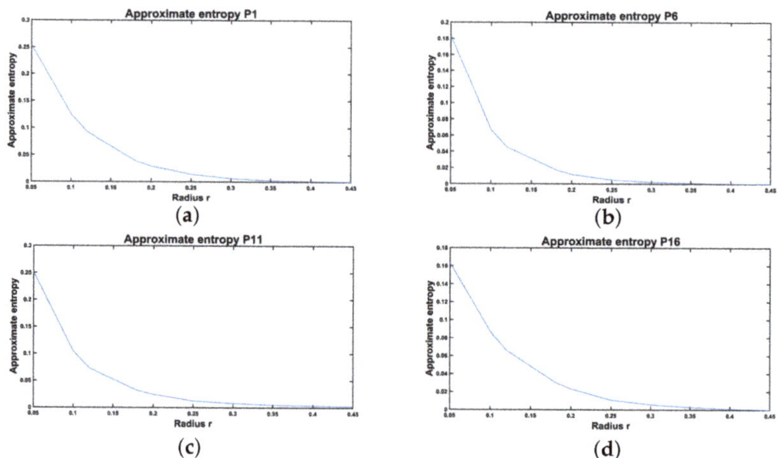

Figure 9. AppEn for several signals from the RC dataset. (**a**) ApEn for a signal from the class P1, (**b**) ApEn for a signal from the class P6, (**c**) ApEn for a signal from the class P11, (**d**) ApEn for a signal from the class P16.

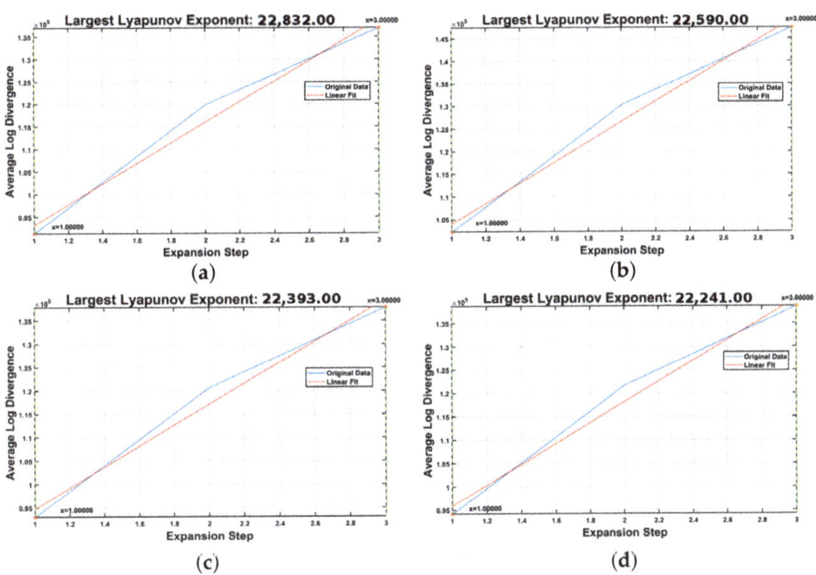

Figure 10. LLE for several signals from the RC dataset. (**a**) LLE for class P1, (**b**) LLE for class P6, (**c**) LLE for a signal from the class P11, (**d**) LLE for class P16.

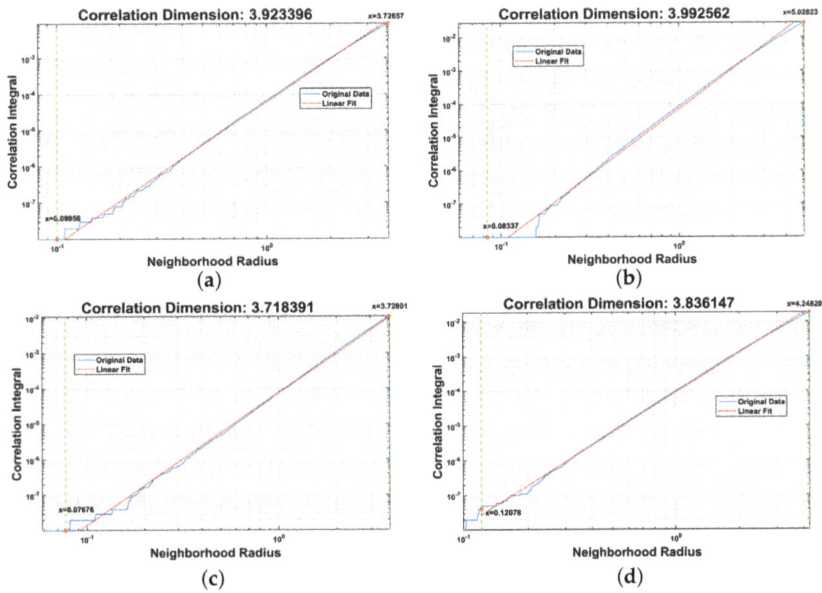

Figure 11. CD for several signals from the RC dataset. (**a**) CD for a signal from the class P1, (**b**) CD for a signal from the class P6, (**c**) CD for a signal from the class P11, (**d**) CD for a signal from the class P16.

Concerning the RC, we selected three vibration channels from the four available. The selected channels were A1, A2, and A3. We excluded A4 to save computational time because A4 was far from the valve's location. The percent accuracy attained using RF and SVM is presented in Table 5. Four feature sets are considered: the first is the *Statistical Features* set, and the second is the *Entropy Features*. The third is the *InfoStat Features* set.

Finally, we have the *All Features* set. The lowest accuracy was attained with the *Statistical Features* set. In this case, the accuracy attained by the RF model was 83.96% for vibration signal channel A1, 74.07% for A2; and 72.33% for A3. The accuracy attained by the SVM model was 86.91% for vibration signal channel A1, 76.91% for A2, and 75.55% for A3. The highest accuracy was attained using the *All Features* set. The accuracy attained by the RF model was 95.35% for vibration signal channel A1, 94.40% for A2, and 88.82% for A3. The SVM model attained a classification accuracy of 97.90% for signal channel A1, 95.47% for A2 and 93.63% for A3. The rest of the feature combinations attained intermediate accuracy close to the optimal accuracy.

Table 5. Percentage accuracy for fault classification using the vibration signals of the RC dataset.

Features	Model	A1	A2	A3
Statistical	RF	83.96	74.07	72.33
	SVM	86.91	76.55	75.55
Entropy based	RF	95.35	94.12	86.21
	SVM	97.57	94.55	91.87
Info-Statistical	RF	93.96	90.66	85.78
	SVM	96.83	94.27	90.38
All Features	RF	95.35	94.40	88.82
	SVM	97.90	95.47	93.63

A comparison of results obtained during the ten repetitions of the cross-validation is presented in Figure 12 for vibration signal A1. The comparison is presented using box plots. The accuracies attained by the *Statistical Features* set with RF and SVM were the lowest (83.96% and 86.91% in Table 5). The feature sets *Entropy Features*, *InfoStat Features*, and *All Features* attained higher accuracy with both machine learning models. In particular, the highest accuracies were attained by the *All Features* set with RF and SVM (95.35% and 97.90%). The accuracies attained by the *Entropy Features* were slightly lower, corresponding to 95.35% and 97.57% with the RD and SVM models, respectively. The accuracies attained by the *InfoStat Features* set were 93.96% and 96.83% with the RF and SVM models, respectively.

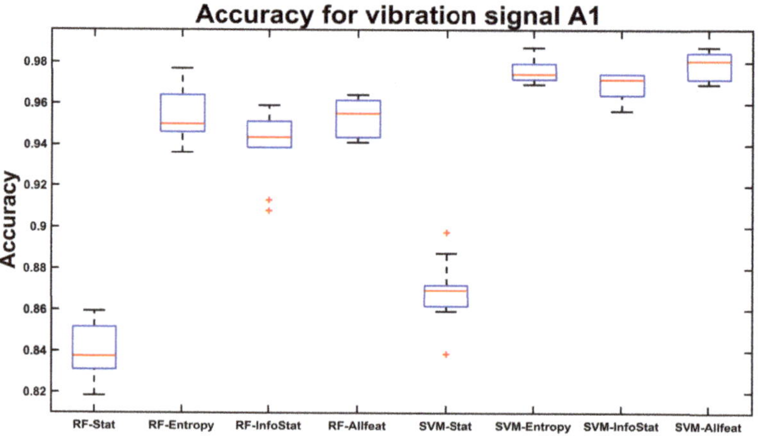

Figure 12. Comparison of accuracy for the machine learning models tested, considering several combinations of features extracted from channel A1 of the RC dataset. In the box plot, a red line represents the median of the data. The horizontal line over the box represents the upper quartile. The horizontal line below the box represents the lower quartile. The red cross, below or over the box, represents the outliers.

A set of performance metrics [79,80,83] expressed in percentages is presented in Table 6. Such metrics were obtained with the SVM model trained with the *All Features* set extracted from vibration signals recorded in channel A1 from the RC dataset. The highest sensitivity of 100.00% was attained by classes P1, P2, P6, P7, P9, P10, and P11; and 93.99% attained by class P3 was the lowest. The highest specificity of 100% was attained by classes P1, P2, P10, and P17. The highest AUC value of 100.00% was attained by classes P1, P2, and P10; and the lowest value of 97.42% was attained by class P5.

Table 6. Performance metrics (in percent) obtained with the SVM model and the *All Features* set. The model was trained using vibration signals in channel A1 from the RC dataset.

Class	Sensitivity	Specificity	Error	FPR	AUC
P1	100.00	100.00	0.00	0.00	100.00
P2	100.00	100.00	0.00	0.00	100.00
P3	96.07	99.73	3.93	0.27	97.70
P4	99.56	99.92	0.44	0.08	99.33
P5	93.99	99.70	6.01	0.30	97.42
P6	100.00	99.97	0.00	0.03	99.78
P7	100.00	99.78	0.00	0.22	98.26
P8	97.01	99.92	2.99	0.08	99.25
P9	100.00	99.95	0.00	0.05	99.57
P10	100.00	100.00	0.00	0.00	100.00
P11	100.00	99.86	0.00	0.14	98.91
P12	94.56	99.89	5.44	0.11	98.95
P13	99.13	99.92	0.87	0.08	99.32
P14	99.56	99.95	0.44	0.05	99.55
P15	99.56	99.92	0.44	0.08	99.33
P16	97.42	99.92	2.58	0.08	99.27
P17	98.29	100.00	1.71	0.00	99.95

9. Discussion

The proposed methodology was implemented on a laptop computer. The laptop was equipped with an intel(R) Core(TM) i7-6700HQ CPU, 12 GB of RAM memory, and a graphics card NVIDIA GeForce GTX 950M. The Matlab software was used to implement the feature extraction and classification. Concerning the computational time, the *Entropy Features* set required an average time of 311.85 s when extracted from the five windows (each with a length of 12,000-time samples) of the vibration signals from the CP dataset. In contrast, the *Information Entropy* features require 0.97 s, and the *Statistical Features* set requires only 0.156 s. In the RC, when the sampling rate is similar to that of the CP and the window length is similar, the computational time is similar to the time required in the CP.

In the case of the CP, multiple pairwise comparisons of machine learning models are presented in Figure A1. A one-way analysis of variance by ranks was performed using a Kruskal–Wallis approach [85]. The test showed no significant statistical differences between results attained by either machine learning models when trained with the non-linear *Entropy Features* set or *All Features* set. Results attained by the SVM machine learning models trained with the non-linear *Entropy Features* set or the *All Features* set are statistically significantly superior to than those of RF or SVM models trained with the *Statistical Features* or the combination *InfoStat Features*. The comparison also shows that the non-linear *Entropy Features* set provides higher accuracy without combining with the *Statistical Features* set.

The plot representing the Kruskal–Wallis with multiple one-way comparisons is presented in Figure A2 for the RC. The combination of features corresponding to *Entropy Features* and *All Features* sets attained higher accuracies using both machine learning models, and there were no significant statistical differences between them. In contrast, there were significant statistical differences in the results attained with the *Statistical Features* set using both machine learning models. Additionally, the results obtained by the features sets *Entropy Features*, and *All Features* with the SVM model were statistically significantly superior to the results obtained with SVM using the *InfoStat Features* set.

A comparison with a deep-learning method was performed. For this purpose, the spectrogram of the vibration signal was calculated and fed to a CNN for fault classification. The spectrogram was calculated considering a Blackman widow of size 81.92 ms that was moved along the time axis with an overlap of 81.44 ms. The feature array had a size of 96×128, for each signal, and it was standardized using the Z-score normalization. The feature array for each dataset was fed to a CNN for 10-fold cross-validation. Results of the comparison are presented in Table 7. Concerning the case of the CP, the highest accuracy attained by the CNN was 98.86% for vibration channel A2, whereas using the *Entropy Features* set with SVM provided a classification accuracy of 99.50%. The accuracy attained by the RF model was 98.09%, which was slightly lower than the accuracy provided by the CNN. Results for the RC show that the highest accuracy attained by the CNN models was 96.74% when trained with signals from the A1 vibration channel. In contrast, the accuracy provided by the SVM trained with the *Entropy Features* extracted from the same channel was 97.57%. In summary, the accuracy provided by the SVM trained with the *Entropy Features* was higher than the accuracy attained by the CNN trained with the spectrogram of the vibration signal.

Table 7. Comparison of performances of the tested models obtained during the 10-fold cross-validation between entropy features with classical models and the spectrogram with a CNN model.

Machine	Signal	Accuracy *Entropy Features* +SVM	Accuracy *Entropy Features* +RF	Accuracy Spectrogram +CNN
Centrifugal Pump	A2	99.50	98.09	98.86
	A3	98.96	97.83	97.36
	A4	97.30	94.70	98.61
Reciprocating Compressor	A1	97.57	95.35	96.74
	A2	94.55	94.12	95.55
	A3	91.87	86.21	93.98

Results can be compared to previous research that used non-linear entropy features with classical models. In [27], the average accuracy for classifying five different conditions in a CP was 94.58%. The work reported in [28] used an SVM for classifying four different conditions in a CP using vibration signals. The method attained an average accuracy of 98.4%. In [21] the $p - V$, was used with LDA to classify six different conditions, attaining an accuracy of 99.62% in an RC. Several papers reported CNN-based approaches for fault classification in CP and RC [24,25,29,33]. These methods can highly accurately classify a small set of faults in a CP or an RC. In contrast, our method was able to classify 13 types of faults in a CP and 17 different conditions in an RC with accuracies close to those of these deep learning-based methods.

10. Conclusions

The non-linear *Entropy Features* set enables fault classification in a CP or an RC with high accuracy. In this research, 13 impeller faults in a CP incorporating different degrees of severity were classified using RF and SVM models. Additionally, a set of 17 valve faults in an RC were also classified.

The research showed that the non-linear *Entropy Features* set provides high accuracy (99.50% for the CP and 97.57% for the RC) considering the SVM model. The accuracy attained using the *Entropy Features* set and the RF model was 98.09% for the CP and 95.35% for the RC. Results showed no significant statistical differences between *Entropy Features* and *All Features*. However, both feature sets provided accuracies statistically significantly higher than the accuracy provided by the *Statistical Features* set. The *Statistical Features* set provided the lowest accuracy in this application. In this case, the accuracy attained with the SVM was 71.03% for the CP and 76.55% for the RC. Concerning the comparison of machine learning models, when both models RF and SVM were trained with the *Entropy Features* set, there were no statistically significant differences in the accuracy of results. The results suggest that the *Entropy Features* set or the *All Features* set could be used for fault classification in centrifugal pumps and reciprocating compressors, where multiple fault types in impellers and valves should be diagnosed at early stages.

The novelty of the approach is to propose using the *entropy features* for classifying faults in centrifugal pumps and reciprocating compressors with high accuracy. In addition, we showed that the entropy-based feature sets provide high classification accuracy even for a large number of faults in both the CP and the RC. Concerning the comparison with respect to a deep learning model, the accuracy results obtained by the SVM trained with the *Entropy Features* set (99.50% for the CP and 97.57% for the RC) were superior to those obtained by a CNN trained with the spectrogram (98.86% for the CP and 96.74% for the RC).

The limitation of the proposed approach is the high computational cost necessary for extracting the entire set of entropy-based features.

Further research is oriented at reducing the computational time required for extracting the non-linear *Entropy Features* set and exploring their implementation using multi-scale approaches. We also plan to select some of the entropy-based features that would be useful for anomaly detection in RM, combined with deep learning approaches, such as generative adversarial networks and autoencoders.

Author Contributions: Conceptualization, R.-V.S., M.C. and R.M.; methodology, D.C. and M.C.; software, R.M.; validation, R.M., E.E. and R.-V.S.; formal analysis, S.Y.; investigation, E.E.; resources, S.Y.; data curation, D.C.; writing—original draft preparation, R.M.; writing—review and editing, M.C. and R.-V.S.; visualization, D.C.; supervision, R.-V.S.; project administration, R.-V.S.; funding acquisition, R.-V.S. All authors have read and agreed to the published version of the manuscript.

Funding: This research was funded by the MoST Science and Technology Partnership Program (KY201802006) and National Research Base of Intelligent Manufacturing Service, Chongqing Technology and Business University, and by the Universidad Politécnica Salesiana through the GIDTEC research group.

Institutional Review Board Statement: Not applicable.

Informed Consent Statement: Not applicable.

Data Availability Statement: The data used in this research are available upon reasonable request to the corresponding authors. We also plan to prepare a paper describing the datasets and submit it to the MDPI journal *Data* to release them for public use.

Conflicts of Interest: The authors declare no conflict of interest.

Abbreviations

The following abbreviations are used in this manuscript:

CP	Centrifugal Pump
RC	Reciprocating Compressor
CM	Condition Monitoring
CD	Correlation Dimension
LLE	Largest Lyapunov Exponent
LMD	Local Mean Decomposition
SVM	Support Vector Machine
CEEMD	Complementary Ensemble Empirical Mode Decomposition
SampEn	Sample Entropy
RF	Random Forest
ELM	Extreme Learning Machine
AMI	Average Mutual Information
FNN	False Nearest Neighbor
AppEn	Approximate Entropy
FuzzyEn	Fuzzy Entropy
ShannEn	Shannon Entropy
PermEn	Permutation Entropy
DFA	Detrended Fluctuation Analysis
LLE	Larger Lyapunov Exponent
S	Power Spectral Entropy
PEB	Pitting at the Entrance of Impeller Blades
POB	Pitting at the Output of the Impeller Blades
ICB	Impeller Channel Blockage
IB	Imbalance Impeller
ROC	Receiver Operator Curve
AUC	Area Under the Curve
HP	Horse Power
RMS	Root Mean Square
RPM	Revolutions per minute
GPM	Gallons per minute
RM	Rotating Machinery
CNN	Convolutional Neural Network
SGWT	Spectral Graph Wavelet Transform
CECP	Complexity-Entropy Causality Plane
MSE	Multi-Scale Entropy
PCA	Principal Component Analysis
LDA	Linear Discriminant Analysis
BMFO	Binary Moth Flame Optimization
SE	Squeeze and Excitation
HYP	Hydraulic Pump
RSDD	Resonance-based sparse signal decomposition
MHAAPE	Multi-scale Hierarchical Amplitude Aware Permutation Entropy
ELM	Extreme Learning Machine
CEEMD	Complementary Ensemble Empirical Mode Decomposition
CWT	Continous Wavelet Transform
XBG	Extreme Gradient Boosting
AWT	Analitical Wavelet Transform

Appendix A. Additional Tables

Appendix A.1. Related Research Compendium

Table A1. Related research and characteristics.

Contributors	Signal	Methods	Application	Advantages	Limitations
Li et al. [21]	Pressure vibration	PCA LDA	RC	Accuracy 99.62%	Six types of fault
Patil et al. [23]	Vibration Signal	BMFO KNN	RC	Accuracy >99%	Can get trapped in local minimum
Lv et al. [22]	Pressure	ANN	RC	Accuracy 97.9%	Several Structuring Elements
Zhao et al. [24]	Vibration	CNN	RC	Accuracy 99.4%	Complex CNN architecture
Xiao et al. [25]	Vibration Pressure Phase	CNN	RC	Acc = 100%	Complex CNN Architecture, 4 Conditions
Ahmad et al. [28]	Vibration	SVM	CP	Accuracy 98.4%	4 conditions
Hasan et al. [29]	Vibration	CNN	CP	Acc = 100%	4 conditions
Ahmad et al. [30]	Vibration	KNN	CP	Acc = 100%	4 conditions
Irfan et al. [31]	Voltage Current	XGB	Water pump bearings	Acc = 100%	Threshold Selection
Kumar et al. [33]	Sound	CNN	CP	Sound Sensor	Acoustic noise
Zhao et al. [20]	Vibration	SVM MSE	RC	Weak fault detection	LMD Mode mixing
Wang et al. [27]	Vibration	RF + CEEMD SampEn	CP	Accurate	High Comp. Cost
Zhou et al. [26]	Vibration	RSDD + RF MHAAPE	HYP	Accurate	Large set of parameters
Xin et al. [16]	Vibration	DFA + SGWT	RM	Retain fine signatures	Lacks more testing
Radhakrishnan et al. [17]	Vibration	CECP + SVM	RM	Robust and easy	Setting of parameters

Appendix A.2. Entropy-Based Features

Table A2. Entropy features extracted.

Feature	Equation	
ApEn [5,6]	$ApEn(m,k,\tau,N) = \phi^{m}(r) - \phi^{m+1}(r), \quad k = 0.06$	(A1)
ApEn [5,6]	$ApEn(m,k,\tau,N) = \phi^{m}(r) - \phi^{m+1}(r), \quad k = 0.21$	(A2)
FuzzyEn [10]	$FuzzyEn(m,n,k) = ln\phi^{m}(n,r) - ln\phi^{m+1}(n,r), \quad k = 0.06$	(A3)
FuzzyEn [10]	$FuzzyEn(m,n,k) = ln\phi^{m}(n,r) - ln\phi^{m+1}(n,r), \quad k = 0.21$	(A4)

Table A2. *Cont.*

Feature	Equation	
SampEn [7]	$SampEn(m, r) = -ln\left[\dfrac{A^m(r)}{B^m(r)}\right], \quad k = 0.06$	(A5)
SampEn [7]	$SampEn(m, r) = -ln\left[\dfrac{A^m(r)}{B^m(r)}\right], \quad k = 0.21$	(A6)
ShannonEn [54]	$ShanEn(m) = -\sum\limits_m p_m log p_m$	(A7)
CondEn [54]	$CondEn(m/m-1) = ShanEn(m) - ShanEn(m-1)$	(A8)
CCondEn [54]	$CCondEn(m) = CondEn(m/m-1) + perc(m) ShanEn(1)$	(A9)
PermEn [57]	$PermEn(m) = lnM - \dfrac{1}{M}\sum\limits_{j=0}^{(m+1)!-1} q_j ln q_j$	(A10)
CorDim [59]	$CD(m) = \lim\limits_{r \to 0} \dfrac{\partial ln C_m(r)}{\partial ln r}$	(A11)
LLE [66]	$LLE = \lim\limits_{t \to \infty} \dfrac{1}{t} ln \dfrac{\|\delta Z(t)\|}{\|\delta Z_0\|}$	(A12)
DFA1(α1) [66]	$\alpha 1 = \tau'(1), \quad \tau(1) = h(1) - 1$	(A13)
DFA2(α2) [66]	$\alpha 2 = \tau'(2), \quad \tau(2) = 2h(2) - 1$	(A14)

Appendix A.3. Information-Based Features

Table A3. Information entropy features.

Feature	Equation	
WPT-Shannon [70,86]	$WPT - Shannon(s) = -\sum\limits_i s_i^2 log(s_i^2)$	(A15)
WPT-Norm [70,86]	$WPT - Norm(s) = -\sum\limits_i |s_i|^p, \quad p \geq 1$	(A16)
WPT-LogEn [70,86]	$WPT - LogEn(s) = -\sum\limits_i log(s_i^2)$	(A17)
WPT-Thres [70,86]	$WPT - Thres(s) = \#\{i, / |s_i| > p\}, \quad |s_i| > p$	(A18)
WPT-Sure [70,86]	$WPT - Sure(s) = n + \sum\limits_i min(s_i^2, p^2) - \#\{i, / |s_i| \leq p\}$	(A19)
RényiEn [72]	$HR_q = \dfrac{K}{q-1} log\left(\sum\limits_{i=1}^N p_i^q\right)$	(A20)
TsallisEn [71]	$HT_q = \dfrac{K}{q-1}\left(1 - \sum\limits_{i=1}^N p_i^q\right)$	(A21)
PSE-mean [73,74]	$PSE - mean = mean(\mathbf{S})$	(A22)
PSE-std [73,74]	$PSE - std = std(\mathbf{S})$	(A23)
PSE-rms [73,74]	$PSE - rms = rms(\mathbf{S})$	(A24)

Table A3. *Cont.*

Feature	Equation	
PSE-shape [73,74]	$PSE - shape = \dfrac{rms(\mathbf{S})}{\frac{1}{N}\sum_i |s_i|}$	(A25)
PSE-MaxToRms [73,74]	$PSE - MaxToRms = max(\mathbf{S})/rms(\mathbf{S})$	(A26)
PSE-median [73,74]	$PSE - median = median(\mathbf{S})$	(A27)
PSE-Skew [73,74]	$PSE - Skew = skewness(\mathbf{S})$	(A28)
PSE-Kur [73,74]	$PSE - Kur = kurtosis(\mathbf{S})$	(A29)

Appendix A.4. Statistical Features

Table A4. Statistical features.

Feature	Equation	
Mean	$\mu = \dfrac{1}{N}\sum_{i=1}^{N} x_i$	(A30)
Root Mean Square (RMS)	$RMS = \sqrt{\dfrac{1}{N}\sum_{i=1}^{N}(x_i)^2}$	(A31)
Standard deviation	$\sigma = \sqrt{\dfrac{1}{N}\sum_{i=1}^{N}(x_i - \mu)^2}$	(A32)
Kurtosis	$Kurtosis = \dfrac{N\sum_{i=1}^{N}(x_i - \mu)^4}{\left[\sum_{i=1}^{N}(x_i - \mu)^2\right]^2}$	(A33)
Maximum value	$Max = max(\mathbf{x}_n)$	(A34)
Crest factor	$CrestFact = \dfrac{max(\mathbf{x}_n)}{rms(\mathbf{x}_n)}$	(A35)
Rectified mean value	$RMV = \dfrac{1}{N}\sum_{i=1}^{N}|x_i|$	(A36)
Shape factor	$ShapeFact = \dfrac{rms(\mathbf{x}_n)}{\frac{1}{N}\sum_{i=1}^{N}|x_i|}$	(A37)
Impulse factor	$ImpulseFact = \dfrac{max(\mathbf{x}_n)}{\frac{1}{N}\sum_{i=1}^{N}|x_i|}$	(A38)
Variance	$Var = \dfrac{1}{N}\sum_{i=1}^{N}(x_i - \mu)^2$	(A39)
Minimum value	$Min = min(\mathbf{x}_n)$	(A40)
Skewness	$Skewness = \dfrac{N\sum_{i=1}^{N}(x_i - \mu)^3}{\sigma^3}$	(A41)

Appendix B. Additional Figures

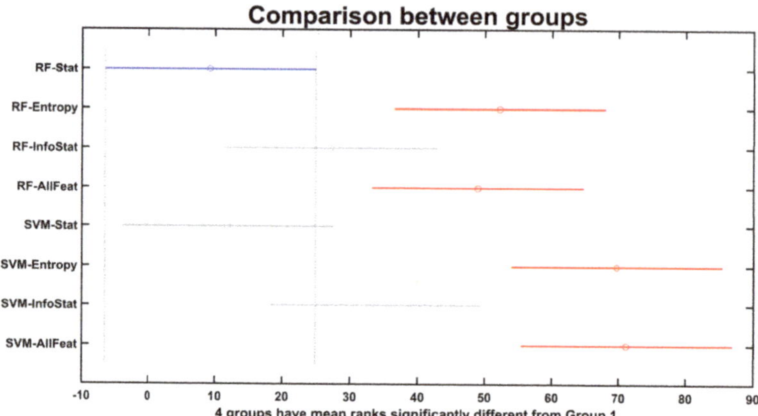

Figure A1. Kruskal–Wallis comparison between the machine learning models tested. The plot shows several combinations of features calculated from the channel A2 recorded from the CP. There are statistical significant differences between the groups labeled with red and the group labeled with blue, between the groups SVM-Entropy and SVM-AllFeat and the groups labeled with blue and gray. In general there are statistically significant differences, when the projection of the horizontal lines of a group over the horizontal axis does not have any interception with the projection of another group.

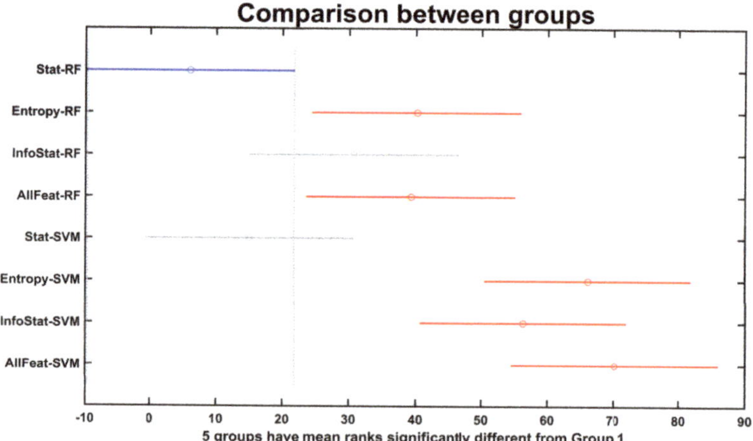

Figure A2. Kruskal–Wallis comparison between the machine learning models tested, considering several combinations of features extracted from the channel A1 of the RC dataset. There are statistical significant differences between the groups labeled with red and the group labeled with blue. In general there are statistically significant differences, when the projection of the horizontal lines of a group over the horizontal axis does not have any interception with the projection of another group.

References

1. Ali, S.M.; Hui, K.; Hee, L.; Leong, M.S. Automated valve fault detection based on acoustic emission parameters and support vector machine. *Alex. Eng. J.* **2018**, *57*, 491–498. [CrossRef]
2. Cerrada, M.; Macancela, J.C.; Cabrera, D.; Estupiñan, E.; Sánchez, R.V.; Medina, R. Reciprocating compressor multi-fault classification using symbolic dynamics and complex correlation measure. *Appl. Sci.* **2020**, *10*, 2512. [CrossRef]
3. Sharma, V.; Parey, A. Performance evaluation of decomposition methods to diagnose leakage in a reciprocating compressor under limited speed variation. *Mech. Syst. Signal Process.* **2019**, *125*, 275–287. [CrossRef]

4. Soleimani, A.; Khadem, S. Early fault detection of rotating machinery through chaotic vibration feature extraction of experimental data sets. *Chaos Solitons Fractals* **2015**, *78*, 61–75. [CrossRef]

5. Pincus, S.M. Approximate entropy as a measure of system complexity. *Proc. Natl. Acad. Sci. USA* **1991**, *88*, 2297–2301. [CrossRef] [PubMed]

6. Pincus, S.M.; Huang, W.M. Approximate entropy: Statistical properties and applications. *Commun. Stat. Theory Methods* **1992**, *21*, 3061–3077. [CrossRef]

7. Richman, J.S.; Moorman, J.R. Physiological time-series analysis using approximate entropy and sample entropy. *Am. J. Physiol. Heart Circ. Physiol.* **2000**, *278*, H2039–H2049. [CrossRef]

8. Weippert, M.; Behrens, M.; Rieger, A.; Behrens, K. Sample entropy and traditional measures of heart rate dynamics reveal different modes of cardiovascular control during low intensity exercise. *Entropy* **2014**, *16*, 5698–5711. [CrossRef]

9. Zhao, L.; Wei, S.; Zhang, C.; Zhang, Y.; Jiang, X.; Liu, F.; Liu, C. Determination of sample entropy and fuzzy measure entropy parameters for distinguishing congestive heart failure from normal sinus rhythm subjects. *Entropy* **2015**, *17*, 6270–6288. [CrossRef]

10. Chen, W.; Wang, Z.; Xie, H.; Yu, W. Characterization of surface EMG signal based on fuzzy entropy. *IEEE Trans. Neural Syst. Rehabil. Eng.* **2007**, *15*, 266–272. [CrossRef]

11. Wang, Z.; Xue, X. Multi-class support vector machine. In *Support Vector Machines Applications*; Springer: Berlin/Heidelberg, Germany, 2014; pp. 23–48. [CrossRef]

12. Escalera, S.; Pujol, O.; Radeva, P. On the decoding process in ternary error-correcting output codes. *IEEE Trans. Pattern Anal. Mach. Intell.* **2010**, *32*, 120–134. [CrossRef] [PubMed]

13. Janjarasjitt, S.; Ocak, H.; Loparo, K. Bearing condition diagnosis and prognosis using applied nonlinear dynamical analysis of machine vibration signal. *J. Sound Vib.* **2008**, *317*, 112–126. [CrossRef]

14. Sun, Y. Fault Detection in Dynamic Systems Using the Largest Lyapunov Exponent. Ph.D. Thesis, Texas A & M University, College Station, TX, USA, 2012.

15. Medina, R.; Macancela, J.C.; Lucero, P.; Cabrera, D.; Sánchez, R.V.; Cerrada, M. Gear and bearing fault classification under different load and speed by using Poincaré plot features and SVM. *J. Intell. Manuf.* **2020**, *33*, 1031–1055. [CrossRef]

16. Dong, X.; Li, G.; Jia, Y.; Li, B.; He, K. Non-iterative denoising algorithm for mechanical vibration signal using spectral graph wavelet transform and detrended fluctuation analysis. *Mech. Syst. Signal Process.* **2021**, *149*, 107202. [CrossRef]

17. Radhakrishnan, S.; Lee, Y.T.T.; Rachuri, S.; Kamarthi, S. Complexity and entropy representation for machine component diagnostics. *PLoS ONE* **2019**, *14*, e0217919. [CrossRef]

18. Li, Y.; Wang, X.; Liu, Z.; Liang, X.; Si, S. The entropy algorithm and its variants in the fault diagnosis of rotating machinery: A review. *IEEE Access* **2018**, *6*, 66723–66741. [CrossRef]

19. Huo, Z.; Martínez-García, M.; Zhang, Y.; Yan, R.; Shu, L. Entropy measures in machine fault diagnosis: Insights and applications. *IEEE Trans. Instrum. Meas.* **2020**, *69*, 2607–2620. [CrossRef]

20. Zhao, H.-Y.; Wang, J.-D.; Xing, J.-J.; Gao, Y.-Q. A feature extraction method based on LMD and MSE and its application for fault diagnosis of reciprocating compressor. *J. Vibroeng.* **2015**, *17*, 3515–3526.

21. Li, X.; Ren, P.; Zhang, Z.; Jia, X.; Peng, X. A p−V Diagram Based Fault Identification for Compressor Valve by Means of Linear Discrimination Analysis. *Machines* **2022**, *10*, 53. [CrossRef]

22. Lv, Q.; Cai, L.; Yu, X.; Ma, H.; Li, Y.; Shu, Y. An Automatic Fault Diagnosis Method for the Reciprocating Compressor Based on HMT and ANN. *Appl. Sci.* **2022**, *12*, 5182. [CrossRef]

23. Patil, A.; Soni, G.; Prakash, A. A BMFO-KNN based intelligent fault detection approach for reciprocating compressor. *Int. J. Syst. Assur. Eng. Manag.* **2021**, 1–13. [CrossRef]

24. Zhao, D.; Liu, S.; Zhang, H.; Sun, X.; Wang, L.; Wei, Y. Intelligent fault diagnosis of reciprocating compressor based on attention mechanism assisted convolutional neural network via vibration signal rearrangement. *Arab. J. Sci. Eng.* **2021**, *46*, 7827–7840. [CrossRef]

25. Xiao, S.; Nie, A.; Zhang, Z.; Liu, S.; Song, M.; Zhang, H. Fault Diagnosis of a Reciprocating Compressor Air Valve Based on Deep Learning. *Appl. Sci.* **2020**, *10*, 6596. [CrossRef]

26. Zhou, F.; Liu, W.; Yang, X.; Shen, J.; Gong, P. A new method of health condition detection for hydraulic pump using enhanced whale optimization-resonance-based sparse signal decomposition and modified hierarchical amplitude-aware permutation entropy. *Trans. Inst. Meas. Control* **2021**, *43*, 3360–3376. [CrossRef]

27. Wang, Y.; Lu, C.; Liu, H.; Wang, Y. Fault diagnosis for centrifugal pumps based on complementary ensemble empirical mode decomposition, sample entropy and random forest. In Proceedings of the 2016 12th World Congress on Intelligent Control and Automation (WCICA), Guilin, China, 12–15 June 2016; IEEE: Piscataway, NJ, USA, 2016; pp. 1317–1320. [CrossRef]

28. Ahmad, Z.; Rai, A.; Maliuk, A.S.; Kim, J.M. Discriminant feature extraction for centrifugal pump fault diagnosis. *IEEE Access* **2020**, *8*, 165512–165528. [CrossRef]

29. Hasan, M.J.; Rai, A.; Ahmad, Z.; Kim, J.M. A fault diagnosis framework for centrifugal pumps by scalogram-based imaging and deep learning. *IEEE Access* **2021**, *9*, 58052–58066. [CrossRef]

30. Ahmad, Z.; Nguyen, T.K.; Ahmad, S.; Nguyen, C.D.; Kim, J.M. Multistage centrifugal pump fault diagnosis using informative ratio principal component analysis. *Sensors* **2021**, *22*, 179. [CrossRef]

31. Irfan, M.; Alwadie, A.S.; Glowacz, A.; Awais, M.; Rahman, S.; Khan, M.K.A.; Jalalah, M.; Alshorman, O.; Caesarendra, W. A novel feature extraction and fault detection technique for the intelligent fault identification of water pump bearings. *Sensors* **2021**, *21*, 4225. [CrossRef]

32. Chen, T.; Guestrin, C. Xgboost: A scalable tree boosting system. In Proceedings of the 22nd ACM Sigkdd International Conference on Knowledge Discovery and Data Mining, San Francisco, CA, USA, 13–17 August 2016; pp. 785–794. [CrossRef]

33. Kumar, A.; Gandhi, C.; Zhou, Y.; Kumar, R.; Xiang, J. Improved deep convolution neural network (CNN) for the identification of defects in the centrifugal pump using acoustic images. *Appl. Acoust.* **2020**, *167*, 107399. [CrossRef]

34. Howedi, A.; Lotfi, A.; Pourabdollah, A. An entropy-based approach for anomaly detection in activities of daily living in the presence of a visitor. *Entropy* **2020**, *22*, 845. [CrossRef]

35. Callegari, C.; Giordano, S.; Pagano, M. Entropy-based network anomaly detection. In Proceedings of the 2017 International Conference on Computing, Networking and Communications (ICNC), Santa Clara, CA, USA, 26–29 January 2017; IEEE: Piscataway, NJ, USA, 2017; pp. 334–340. [CrossRef]

36. Palmieri, F. Network anomaly detection based on logistic regression of nonlinear chaotic invariants. *J. Netw. Comput. Appl.* **2019**, *148*, 102460. [CrossRef]

37. Wu, S.; Moore, B.E.; Shah, M. Chaotic invariants of lagrangian particle trajectories for anomaly detection in crowded scenes. In Proceedings of the 2010 IEEE Computer Society Conference on Computer Vision and Pattern Recognition, San Francisco, CA, USA, 13–18 June 2010; IEEE: Piscataway, NJ, USA, 2010; pp. 2054–2060. [CrossRef]

38. Garland, J.; Jones, T.R.; Neuder, M.; Morris, V.; White, J.W.; Bradley, E. Anomaly detection in paleoclimate records using permutation entropy. *Entropy* **2018**, *20*, 931. [CrossRef] [PubMed]

39. Wang, S.; Lu, M.; Kong, S.; Ai, J. A Dynamic Anomaly Detection Approach Based on Permutation Entropy for Predicting Aging-Related Failures. *Entropy* **2020**, *22*, 1225. [CrossRef] [PubMed]

40. Ahmad, S.; Styp-Rekowski, K.; Nedelkoski, S.; Kao, O. Autoencoder-based condition monitoring and anomaly detection method for rotating machines. In Proceedings of the 2020 IEEE International Conference on Big Data (Big Data), Atlanta, GA, USA, 10–13 December 2020; IEEE: Piscataway, NJ, USA, 2020; pp. 4093–4102. [CrossRef]

41. de Paula Monteiro, R.; Lozada, M.C.; Mendieta, D.R.C.; Loja, R.V.S.; Bastos Filho, C.J.A. A hybrid prototype selection-based deep learning approach for anomaly detection in industrial machines. *Expert Syst. Appl.* **2022**, *204*, 117528. [CrossRef]

42. Li, D.; Chen, D.; Jin, B.; Shi, L.; Goh, J.; Ng, S.K. MAD-GAN: Multivariate anomaly detection for time series data with generative adversarial networks. In Proceedings of the International Conference on Artificial Neural Networks, Munich, Germany, 17–19 September 2019; Springer: Berlin/Heidelberg, Germany, 2019; pp. 703–716. [CrossRef]

43. Kumar, S.R.; Iniyal, U.; Harshitha, V.; Abinaya, M.; Janani, J.; Jayaprasanth, D. Anomaly Detection in Centrifugal Pumps Using Model Based Approach. In Proceedings of the 2022 8th International Conference on Advanced Computing and Communication Systems (ICACCS), Coimbatore, India, 25–26 March 2022; IEEE: Piscataway, NJ, USA, 2022; Volume 1, pp. 427–433. [CrossRef]

44. Dutta, N.; Kaliannan, P.; Subramaniam, U. Application of machine learning algorithm for anomaly detection for industrial pumps. In *Machine Learning Algorithms for Industrial Applications*; Springer: Berlin/Heidelberg, Germany, 2021; pp. 237–263.[CrossRef]

45. Dutta, A.; Karimi, I.A.; Farooq, S. PROAD (Process Advisor): A Health Monitoring Framework for Centrifugal Pumps. *Comput. Chem. Eng.* **2022**, *163*, 107825. [CrossRef]

46. Charoenchitt, C.; Tangamchit, P. Anomaly Detection of a Reciprocating Compressor using Autoencoders. In Proceedings of the 2021 Second International Symposium on Instrumentation, Control, Artificial Intelligence, and Robotics (ICA-SYMP), Bangkok, Thailand, 20–22 January 2021; IEEE: Piscataway, NJ, USA, 2021; pp. 1–4. [CrossRef]

47. Rand, D.; Young, L.S. Dynamical Systems and Turbulence. In *Lecture Notes in Mathematics*; Springer: Cham, Switzerland, 1981; Volume 898. [CrossRef]

48. Vlachos, I.; Kugiumtzis, D. State space reconstruction for multivariate time series prediction. *arXiv* **2008**, arXiv:0809.2220. https://doi.org/10.48550/arXiv.0809.2220.

49. Abarbanel, H.D.; Brown, R.; Sidorowich, J.J.; Tsimring, L.S. The analysis of observed chaotic data in physical systems. *Rev. Mod. Phys.* **1993**, *65*, 1331. [CrossRef]

50. Fraser, A.M.; Swinney, H.L. Independent coordinates for strange attractors from mutual information. *Phys. Rev. A* **1986**, *33*, 1134. [CrossRef]

51. Kennel, M.B.; Brown, R.; Abarbanel, H.D. Determining embedding dimension for phase-space reconstruction using a geometrical construction. *Phys. Rev. A* **1992**, *45*, 3403. [CrossRef]

52. Ribeiro, M.; Henriques, T.; Castro, L.; Souto, A.; Antunes, L.; Costa-Santos, C.; Teixeira, A. The entropy universe. *Entropy* **2021**, *23*, 222. [CrossRef] [PubMed]

53. Azami, H.; Escudero, J. Amplitude-and fluctuation-based dispersion entropy. *Entropy* **2018**, *20*, 210. doi: 10.3390/e20030210. [CrossRef] [PubMed]

54. Porta, A.; Baselli, G.; Liberati, D.; Montano, N.; Cogliati, C.; Gnecchi-Ruscone, T.; Malliani, A.; Cerutti, S. Measuring regularity by means of a corrected conditional entropy in sympathetic outflow. *Biol. Cybern.* **1998**, *78*, 71–78. [CrossRef] [PubMed]

55. Feutrill, A.; Roughan, M. A Review of Shannon and Differential Entropy Rate Estimation. *Entropy* **2021**, *23*, 1046. [CrossRef]

56. Bandt, C.; Pompe, B. Permutation entropy: A natural complexity measure for time series. *Phys. Rev. Lett.* **2002**, *88*, 174102. [CrossRef]

57. Unakafova, V.A.; Keller, K. Efficiently measuring complexity on the basis of real-world data. *Entropy* **2013**, *15*, 4392–4415. [CrossRef]

58. Steven, V. Heart rate variability linear and nonlinear analysis with applications in human physiology. *Diss. Abstr. Int.* **2010**, *71*.

59. Lu, C.; Sun, Q.; Tao, L.; Liu, H.; Lu, C. Bearing health assessment based on chaotic characteristics. *Shock Vib.* **2013**, *20*, 519–530. [CrossRef]

60. Rolo-Naranjo, A.; Montesino-Otero, M.E. A method for the correlation dimension estimation for on-line condition monitoring of large rotating machinery. *Mech. Syst. Signal Process.* **2005**, *19*, 939–954. [CrossRef]

61. Ihlen, E.A. Introduction to multifractal detrended fluctuation analysis in Matlab. *Front. Physiol.* **2012**, *3*, 141. [CrossRef]

62. JiaQing, W.; Han, X.; Yong, L.; Tao, W.; Zengbing, X. Detrended Fluctuation Analysis and Hough Transform Based Self-Adaptation Double-Scale Feature Extraction of Gear Vibration Signals. *Shock Vib.* **2016**, *2016*, 3409897. [CrossRef]

63. Golińska, A.K. Detrended fluctuation analysis (DFA) in biomedical signal processing: Selected examples. *Stud. Log. Gramm. Rhetor.* **2012**, *29*, 107–115.

64. Henry, B.; Lovell, N.; Camacho, F. Nonlinear dynamics time series analysis. In *Nonlinear Biomedical Signal Processing: Dynamic Analysis and Modeling*; Wiley: Hoboken, NJ, USA, 2012; Volume 2, pp. 1–39. [CrossRef]

65. Wolf, A.; Swift, J.B.; Swinney, H.L.; Vastano, J.A. Determining Lyapunov exponents from a time series. *Phys. D Nonlinear Phenom.* **1985**, *16*, 285–317. [CrossRef]

66. Rosenstein, M.T.; Collins, J.J.; De Luca, C.J. A practical method for calculating largest Lyapunov exponents from small data sets. *Phys. D Nonlinear Phenom.* **1993**, *65*, 117–134. [CrossRef]

67. Shannon, C.E. A mathematical theory of communication. *ACM SIGMOBILE Mob. Comput. Commun. Rev.* **2001**, *5*, 3–55. [CrossRef]

68. Gao, J.; Liu, F.; Zhang, J.; Hu, J.; Cao, Y. Information entropy as a basic building block of complexity theory. *Entropy* **2013**, *15*, 3396–3418. [CrossRef]

69. Goldstein, S.; Lebowitz, J.L.; Tumulka, R.; Zanghì, N. Gibbs and Boltzmann entropy in classical and quantum mechanics. *arXiv* **2019**, arXiv:1903.11870.

70. Coifman, R.R.; Wickerhauser, M.V. Entropy-based algorithms for best basis selection. *IEEE Trans. Inf. Theory* **1992**, *38*, 713–718. [CrossRef]

71. Tsallis, C. Possible generalization of Boltzmann-Gibbs statistics. *J. Stat. Phys.* **1988**, *52*, 479–487. [CrossRef]

72. Rényi, A. On measures of entropy and information. In *Contributions to the Theory of Statistics, Proceedings of the Fourth Berkeley Symposium on Mathematical Statistics and Probability, Davis, CA, USA, 20–30 June 1960*; The Regents of the University of California: Davis, CA, USA, 1961; Volume 1.

73. Sharma, V.; Parey, A. A review of gear fault diagnosis using various condition indicators. *Procedia Eng.* **2016**, *144*, 253–263. [CrossRef]

74. Pan, Y.; Chen, J.; Li, X. Spectral entropy: A complementary index for rolling element bearing performance degradation assessment. *Proc. Inst. Mech. Eng. Part C J. Mech. Eng. Sci.* **2009**, *223*, 1223–1231. [CrossRef]

75. Breiman, L. Random forests. *Mach. Learn.* **2001**, *45*, 5–32. [CrossRef]

76. Efron, B.; Tibshirani, R.J. *An Introduction to the Bootstrap*; CRC Press: Boca Raton, FL, USA, 1994.

77. Vapnik, V.N. An overview of statistical learning theory. *IEEE Trans. Neural Netw.* **1999**, *10*, 988–999. [CrossRef]

78. Scholkopf, B. Support vector machines: A practical consequence of learning theory. *IEEE Intell. Syst.* **1998**, *13*, 4. [CrossRef]

79. Hossin, M.; Sulaiman, M. A review on evaluation metrics for data classification evaluations. *Int. J. Data Min. Knowl. Manag. Process* **2015**, *5*, 1. [CrossRef]

80. Sokolova, M.; Lapalme, G. A systematic analysis of performance measures for classification tasks. *Inf. Process. Manag.* **2009**, *45*, 427–437. [CrossRef]

81. Wang, W.; Chen, J.; Wu, Z. The application of a correlation dimension in large rotating machinery fault diagnosis. *Proc. Inst. Mech. Eng. Part C J. Mech. Eng. Sci.* **2000**, *214*, 921–930. [CrossRef]

82. Boon, M.Y.; Henry, B.I.; Suttle, C.M.; Dain, S.J. The correlation dimension: A useful objective measure of the transient visual evoked potential? *J. Vis.* **2008**, *8*, 6. [CrossRef] [PubMed]

83. Brown, J. Classifiers and their metrics quantified. *Mol. Inform.* **2018**, *37*, 1700127. [CrossRef] [PubMed]

84. Bradley, A.P. The use of the area under the ROC curve in the evaluation of machine learning algorithms. *Pattern Recognit.* **1997**, *30*, 1145–1159. [CrossRef]

85. Pohlert, T. The pairwise multiple comparison of mean ranks package (PMCMR). *R Package* **2014**, *27*, 9.

86. Leite, G.d.N.P.; Araújo, A.M.; Rosas, P.A.C.; Stosic, T.; Stosic, B. Entropy measures for early detection of bearing faults. *Phys. A Stat. Mech. Its Appl.* **2019**, *514*, 458–472. [CrossRef]

Article

Lateral Dynamic Simulation of a Bus under Variable Conditions of Camber and Curvature Radius

Ester Olmeda [1,*], Enrique Roberto Carrillo Li [2], Jorge Rodríguez Hernández [2] and Vicente Díaz [1]

[1] Mechanical Engineering Department, Research Institute of Vehicle Safety (ISVA), University Carlos III de Madrid, Avda. de la Universidad, 30, 28911 Leganes, Madrid, Spain
[2] Mechanical Engineering Section, Engineering Department, Pontificia Universidad Catolica del Perú, Lima 15088, Peru
* Correspondence: eolmeda@ing.uc3m.es

Abstract: The objective of this paper is to describe a model for the simulation of the lateral dynamics of a vehicle, specifically buses, under variable trajectory conditions, such as camber and radius of curvature; in addition, a variable speed is added as a simulation parameter. The objective of this study is the prevention of vehicle rollover and sideslip. An 8 degrees of freedom model was developed, considering a front and a rear section of the bus with its respective suspension system, and both sections have been connected by a torsion spring that emulates the torsional stiffness of the vehicle chassis. A Panhard bar is also added at the rear as an additional element to the suspension and the behavior of the bus when it is added is analyzed. This model also allows the evaluation of the force on each suspension component, which allows for future controllability of the active suspension components. The results show the dynamic behavior of the vehicle, and some indicators are introduced to show the possible sideslip or rollover. As a conclusion, the influence of the road parameters on the dynamic behavior of the bus and the effect of the Panhard bar on the dynamic behavior of the bus can be pointed out.

Keywords: lateral dynamics; track parameters; vehicle mathematical modeling; safety vehicle index

MSC: 70E50

Citation: Olmeda, E.; Carrillo Li, E.R.; Rodríguez Hernández, J.; Díaz, V. Lateral Dynamic Simulation of a Bus under Variable Conditions of Camber and Curvature Radius. *Mathematics* **2022**, *10*, 3081. https://doi.org/10.3390/math10173081

Academic Editor: Armin Fügenschuh

Received: 30 June 2022
Accepted: 18 August 2022
Published: 26 August 2022

Publisher's Note: MDPI stays neutral with regard to jurisdictional claims in published maps and institutional affiliations.

1. Introduction

Vehicle dynamics deals with the mechanical modeling of vehicles as well as the mathematical description and analysis of vehicles.

Vehicle research models have undergone major changes in recent decades, from the traditional grouped parameter model to the finite element model, the dynamic substructure model and the dynamic multibody model; and from linear models to nonlinear models with nonlinear stiffness or damping. The responses of these models can be obtained theoretically or numerically depending on the complexity of the resulting mathematical models. This complexity is closely linked to the number of degrees of freedom considered for the respective modeling [1].

According to [2–4], among others, there are two ways to perform modeling of mechanical systems: an empirical approach and an axiomatic or theoretical approach. In this work, the axiomatic modeling method has been used for a vehicular system. The mathematical or virtual model is obtained through the application of the Newton-D'Alembert principle, obtaining a system of ordinary differential equations.

In general, the analytical modeling of any vehicle involves analyzing the interrelationships between a large number of components, as well as their response to different external disturbances and the effects of these on the rest of the components. Because of this, different models must be proposed to predict the behavior of the vehicle in the presence of a set of external stimuli in order to determine a specific type of response. Such modeling implies leaving aside some components and disturbances with little or no relevance to the phenomenon we are trying to understand.

In the case of the present work, the objective is to analyze the lateral dynamics of a vehicle, and therefore to know the vertical and lateral response of it, as well as the vertical and lateral movements of its components due to vertical disturbances and transverse loads; therefore, longitudinal disturbances are left aside. In addition, the kinematics of the vehicle in the longitudinal axis and its rotation around the z-axis will not be determined.

In recent years, different vehicle models have been proposed for the analysis of lateral dynamics and accident consequences in case of rollover or sideslip. In [5], a restricted lateral dynamics model of articulated vehicles and an algorithm for the estimation of sideslip angle and cornering stiffness were proposed. The articulated vehicle was modeled by using the bicycle model, the linear tire model, and the modified Dug-off model. In [6] is presented an application of combined dynamic and finite element (FE) simulations to evaluate the rollover crash of a bus. A nonlinear mathematical model with six degrees of freedom is shown in [7] from which the effect of different factors on the stability of the vehicle on a curved trajectory (braking angle, vehicle speed and acceleration a.s.o.) is studied. In [8], a model that simulates the lateral rollover test including the effect of a variable position of the center of gravity and calculates the maximum speed at which the bus can travel in a curve without rolling over is proposed. The influence of several parameters (i.e., height of the center of gravity, weight distribution between axles, chassis torsional stiffness, etc.) on bus roll stability during a lane change maneuver, has been analyzed by means of six degrees of freedom beam finite element model in [9].

The aim of lateral dynamics is to describe the behavior of a vehicle when it is subjected to the action of accelerations and/or forces in its lateral direction in order to prevent a possible rollover or sideslip of the vehicle. These solicitations, whose magnitudes are a function of speed, commonly appear when the vehicle travels on a curved track. Different studies show that a high number of traffic accidents take place on horizontal curves, where the risk of accidents is significantly higher than on other road sections [10–12]. According to several recent studies, the radius of curvature represents the most important factor in the geometric design of horizontal curves, and the accident rate is directly related to it [13,14].

The most significant parameters of a curved path for this study are radius of curvature and camber since these geometrical characteristics of a curved path influence the lateral behavior. Due to these solicitations, vehicle driving can be difficult or even dangerous when driving on a curved path. The vehicle is subjected to a lateral force caused by centripetal acceleration acting on the center of gravity of each of the vehicle's components. Consequently, in order to achieve a balance, the adherence forces and the normal forces between the road and the tire create a balance that prevents the vehicle from rollover or sideslip. However, these adhesion forces have a limit given by Coulomb's dry friction law and another limit can also be considered when the normal force of any wheel reaches a value of zero, the moment in which the vehicle rolls over. These limits are directly related to the geometrical factors of the curve and to the environmental conditions that may cause these limits to vary.

Therefore, the study of track characteristics in lateral vehicle dynamics is also of great relevance. the safety issue of horizontal road curves was evaluated by [15] according to the American Association of Highway Transportation Officials (AASHTO) standard. Different parameters, such as vehicle weight, vehicle dimensions, longitudinal slopes, and vehicle velocity are evaluated in the geometric design of horizontal curves using a multi-body dynamic simulation process. A combination of simple circular and clothoid transition curves with several longitudinal upslopes and downslopes were designed.

Ref. [16] studied the radius of curvature of the real vehicle trajectory under driver's instantaneous emergency steering maneuvers. It was calculated based on the bicycle model. They analyzed the possibility of rollover taking into account the curve entry velocity and camber.

In this study, a trajectory composed of three paths (a straight path, a path of variable radius of curvature and a path of constant curvature) is modeled. For simulation purposes, only geometric variations are considered in order to obtain the values of the adherence and normal forces present, while a vehicle follows a simulation trajectory composed of these three paths with a different radius of curvature. This design fits to realistic transport routes since it has been calculated and parameterized with curves of agreement (trajectories that connect a straight and circular trajectory in such a way that it starts with an infinite radius of curvature and ends in a constant radius of curvature) used in road design.

A bus model is provided using rigid multibody systems with which a virtual or mathematical model (system of ordinary differential equations) has been generated. The passenger bus is then modeled in two sections, front and rear. The front part contemplates several subsystems, such as the unsprung masses, the unsprung masses are connected by a stability or torsion bar, the flexible wheels, etc. These parts are represented by a set of inertia, stiffness, and damping elements. The rear part has been modeled in an analogous way although, unlike the front, in this there is only one unsprung mass and a Panhard type bar is used to improve its vertical and lateral dynamic behavior. For this model, the suspension system and tire stiffness as well as the damping of both bodies have been modeled as linear components.

The Panhard bar is a suspension element that prevents lateral displacements of an axle and also restricts the relative vertical movement between the suspended and unsuspended mass. Independent rear suspension systems do not require a Panhard bar. The bar is attached to two points that allow only the up and down movement of its ends, so that it can only move in the vertical plane. The forces that restrict the movement between the connected elements take part in the dynamics of these, since the forces and the consequent moments that originate, both in the non-suspended mass and in the suspended, intervene in their dynamics. Some previous studies have incorporated a Panhard bar in their vehicle models. In [17], a roll plane model of a road vehicle is provided incorporating the kinematic constraint provided by a Panhard bar. The results show that the location and orientation of the Panhard bar significantly influences the kinetostatic roll properties of the suspension when the vehicle is subjected to vertical and lateral forces. In [18], a common front axle suspension is designed for four different tractor models, in which a Panhard bar is used to improve safety and driving stability. Related with the Panhard bar [19], it compares the structural durability behavior of commercial vehicles Panhard bars of a ferritic cast nodular iron with the behavior of rods of an austempered.

Both sections of the bus are connected by a torsional spring (defined by its torsional stiffness) to emulate the dynamic influence of the vehicle chassis. Due to the different longitudinal position of the two sections and, consequently, the different roll angle, the chassis exerts a torsional moment on both sections, which influences the lateral stability.

This bus model travels along the simulated trajectory and allows the simulation of the lateral dynamics of a bus when it travels on a cambered curve.

2. Geometric Path Description

In Figure 1, a schematic model of the path considered for this work is shown.

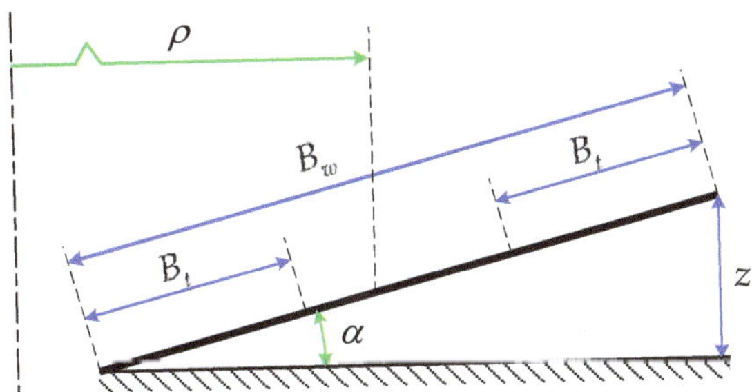

Figure 1. Path's geometric parameters.

These dimensions are subsequently applied to modeling of the road, with the radius of curvature ρ and camber angle α as geometric parameters defining the dynamic behavior of the vehicle on a curved path. The track's width B_t and roadway width B_w define the maximum value of the camber height and the vehicle width, respectively; the track width also defines the number of lanes on the road. In order to simplify the simulation, it has been assumed that the vehicle is driving in the middle roadway lane.

2.1. Kinematic Effects

The radius of curvature of a track section measures the magnitude of the curvature in that section. According to the formulation deduced by dynamics, the normal acceleration a_y, caused by said curvature, is defined by the following equation:

$$a_y = v^2/\rho \tag{1}$$

This acceleration, mainly dependent on the speed (v) and inversely proportional to the radius of curvature, is translated into an inertial force, opposite in sense to the normal acceleration and oriented on the horizontal of the plane in which it is located. This inertial force is commonly known as centrifugal force, F_i:

$$F_i = \frac{v^2}{\rho} m_T \tag{2}$$

where m_T is the total mass of the vehicle.

The camber is defined as a relative elevation between the ends of the track, such that a component of the vehicle's weight compensates the inertial force caused by normal acceleration; its length is measured between the ends of the track according to the vertical direction (Figure 1).

Figure 2 shows a resultant force R as the vectorial sum of the weight and inertial force (centrifugal force). As an effect of the inclination, there is a component of this resultant force, which is perpendicularly pointed to the path's plane R_N, which keeps an adherence force between the wheel and the path, and a perpendicular oriented component R_T of the resultant force R, which is responsible for the skidding and rollover of the vehicle; when the resultant force R is completely oriented in the perpendicular direction of the path's plane, it is said that the vehicle is traveling at design velocity v_D.

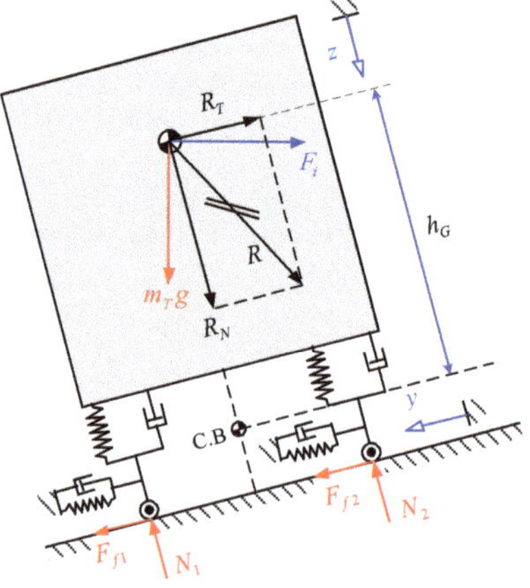

Figure 2. Force Schema for a vehicle traveling over a curved path.

For a skidding situation, it is said that the parallel component of R force is higher than the adherence force determined by the Coulomb's dry friction force; on the other hand, when a rollover is produced, the vehicle will turn into the external side of the curve, it means that it will turn over one or more wheel from the external side of the vehicle. Consequently, we can affirm that when the normal force of one or more wheels in the inner side are zero the rollover is produced, or is at least imminent. To analyze the vehicles rollover, a fixed point is required thus the concept of roll center is introduced as a reference point, which despite the strains that suspension suffers, it remains static on the vehicle frontal plane. From this point, all moments are calculated so it is important to highlight that the higher the vehicle's mass center, the higher the possibility of rollover.

2.2. Kinds of Path

Three path types with different geometric characteristics (curvature radius and camber) have been defined as shown in Table 1.

Table 1. Geometrical characteristic variations for type of path.

Type of Path	Camber Angle	Curvature Radius
Straight path	None	Infinite
Transition path	Variable	Variable
Circular path	Constant	Constant

The transition curves enhance the horizontal curve safety as they help to gradually apply camber and centrifugal force. Since the transition path is the only one which varies its parameters, it is the only one that defines parameter variation equations. Such paths connect a straight and a circular path so that begins with an infinite curvature radius and ends at a constant curvature radius; furthermore, over those paths, the camber angle starts to increase or decrease. To describe those paths a variety of mathematical curves are applied to determine their points in the space. Among those curves, the most known are Bernoulli's lemniscate and the clothoid, which is the most widely applied [15]. The use of a clothoid transition curve between tangent and a circular curve has been suggested to increase the safety margin of safety against sideslip, especially when the camber exceeds 12% [20]. So, the clothoid has been used in this study.

A clothoid is defined by the following equation:

$$\rho(s) \cdot s = A^2 \tag{3}$$

where s is the traveled distance over the curve and $\rho(s)$ is the curvature radius as a function of the traveled distance, it means that there is a linear correspondence through a constant. This could be interpreted as while the vehicle is moving forward over the path, the magnitude of the curvature radius increases or decreases. A clothoid curve is depicted in Figure 3.

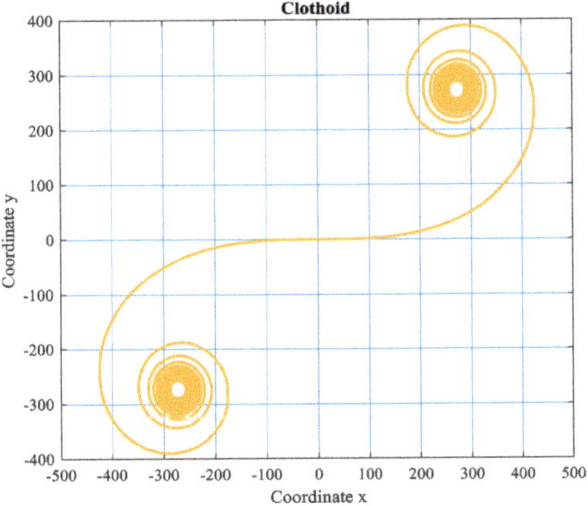

Figure 3. Clothoid curve.

The position vector is described by the following equation which is expressed in terms of Fresnel functions:

$$\vec{r}(s) = [x(s); y(s)] = A\sqrt{\pi}\left[\text{FresnelC}\left(\frac{s}{A\sqrt{\pi}}\right); \text{FresnelS}\left(\frac{s}{A\sqrt{\pi}}\right)\right] \tag{4}$$

One of the most important functions is avoid the appearance of sudden uncompensated acceleration, thus it is required to evaluate the magnitude of the over acceleration, which is the rate of variation of the acceleration. According to [21], for railway vehicles, it is recommended that this value must not overcome 0.2 m/s^3; nevertheless, for a road vehicle, this value should be controlled by the maneuvering of the car's rudder. For this simulation, this value must be settled as an input parameter, hence the maneuvering of the vehicles is not being simulated as it is a random parameter.

2.3. Simulation Path Parameterization

Once all geometric characteristics of the path are known, it is necessary to build, in a parametric way, the simulation path over which the vehicle will travel during the simulation. For that reason, the following equations are developed to describe the camber and curvature radiuses over the clothoid:

$$\rho(t) = \frac{A^2}{v(t)t} \tag{5}$$

$$z(t) = v(t)t\frac{z_b}{s_b} \tag{6}$$

In the previous expressions, z_b is the constant camber value in the circular path and s_b is the total length of the clothoid. The velocity is going to be considered as a constant value for simulation path generation purpose. Due to the vehicle's behavior is studied just a transition and circular path are required, hence in a straight path there is no risk of rollover because lateral acceleration is not present.

From Equation (6) an equivalence to determine $s_{b,\min}$ clothoid curve minimum length can be calculated. According with [21], the maximum noncompensated lateral acceleration should be 1 m/s^2 for comfort and safety reasons. From the given conditions, the following expression can be developed:

$$\frac{\left(\dfrac{9.81\,z_b}{B_w} - \dfrac{v^2\sqrt{B_w^2 - z_b^2}}{\rho_{\min}\,B_w}\right)v_D}{0.2} < s_{b,\min} \tag{7}$$

From this expression, the clothoid's constant can be determined:

$$A = \sqrt{s_{b,\min}\rho_b} \tag{8}$$

Once the transition path is defined, a circular path can be added giving as a result the whole simulation path.

For this work, a simulation path was developed considering the following parameters:

- Camber over the circular path (Final value): 0.25 m
- Curvature radius over the circular path (Final value): 400 m
- Track's width: 0.9 m
- Number of tracks: 3
- Maximum over acceleration: 0.15 m/s^3
- Path's total length: 1500 m
- Average Speed: 120 km/h

Figure 4 shows a plane view of the track and the center of the circular path:

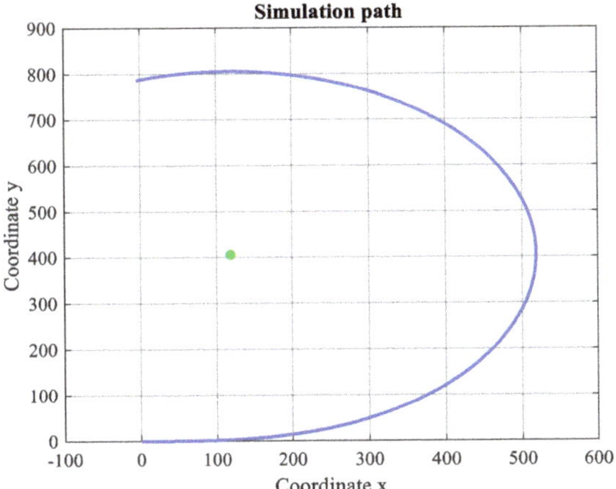

Figure 4. Simulation path—plane view.

Taking as a reference the average speed, function for camber angle and curvature radius can be developed in terms of time (Figure 5).

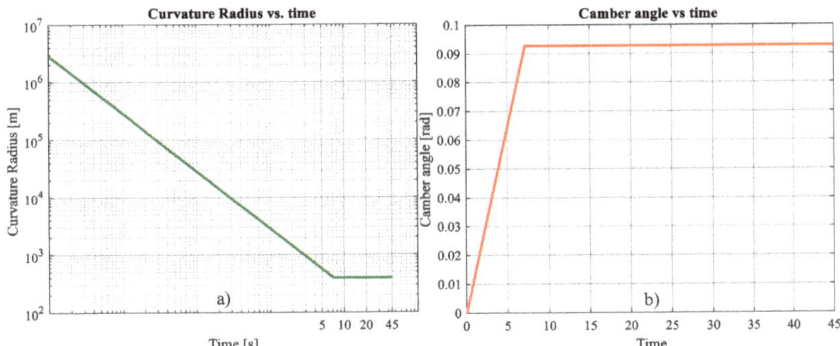

Figure 5. (**a**) Curvature radius vs. time and (**b**) camber angle vs. time for simulation.

3. Vehicle Modeling

As it has been described, to simplify the vehicle analysis, the bus has been divided into two sections, front and rear, both are joined by a torsional spring, which emulates the stiffness belonging to the bus structure.

3.1. Frontal Section

For the front section, the unsprung mass includes elements, such as the mass of the tires and other components of the steering system, all of them are modelled as punctual masses, which have only one degree of freedom on the perpendicular direction of path's plane. The Figure 6 shows the theoretical representation of the front section of the bus:

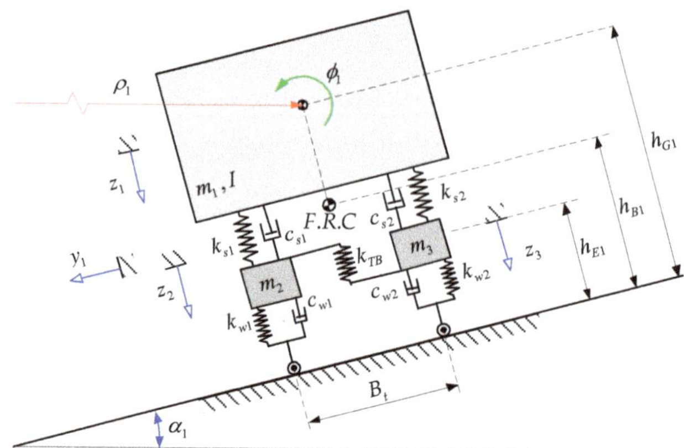

Figure 6. Bus front section—mathematical model.

The model proposed is based on [22,23]. This model considers not only the stiffness of the wheels but also the damping of that element so that it is possible to get a more realistic bus mechanical behavior. Additionally, a sway bar is considered as passive element, this component has the functionality of link the vertical displacement of the unsprung masses. It is performed taking advantage of the lever arms generated by the displacement giving as a result a torsional moment along the bar so that restrict the movement of both unsprung masses, it is desirable to avoid excessive vibrations, as well as a large displacement of the roll center.

3.2. Rear Section

In the rear section, the unsprung mass is modeled as a unique body, which is linked with the bus chassis through the suspension system so that the unsprung mass has two degrees of freedom, the vertical displacement, and a rotation around the longitudinal axis. Most of the components are quite like the front section except for a rigid element known as a Panhard bar which, as mentioned above, is used to avoid lateral displacement, and restrict relative vertical displacement and rotation between the unsprung and sprung masses. In terms of modeling, the Panhard bar is considered to be embedded in the unsprung mass and articulated to the sprung mass.

A model of the rear section is depicted in Figure 7.

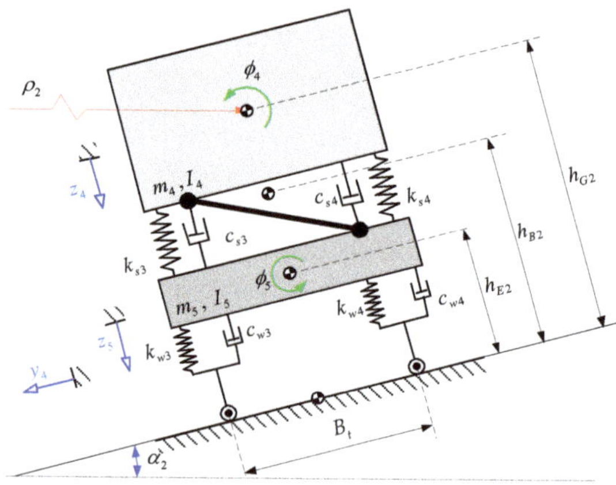

Figure 7. Bus rear section—mathematical model.

Regarding the camber angle and curvature radius, the values of both parameters are different from the front section, hence they are in different positions over the path; for that reason, a time interval delay is introduced to simulate that effect.

3.3. Complete Model

From the aforementioned description, a torsional spring, with a torsional stiffness k_E that links both sections to emulate the stiffness of the bus structure chassis. Altogether, an 8 degree of freedom model is composed by the two sections. The whole model is shown in Figure 8.

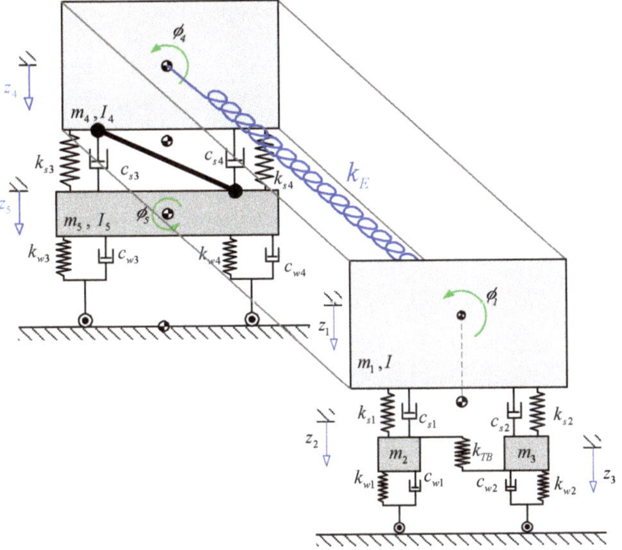

Figure 8. Complete bus model.

4. Kinetic Analysis of the Vehicle

4.1. Frontal Section

Hence, the forces on some vehicle components will be analyzed and a few parameters will be modified to study the influence it has on behavior, the motion equations are going to be determined by applying the D'Alembert principle. Firstly, equations belonging to the suspended mass are obtained and then the equations corresponding to the unsprung masses.

From the geometric parameters of the road, which has been previously described, it is possible to notice that the horizontal acceleration is the known centripetal acceleration produced by the vehicle's travel over a curved road, therefore the only unknown acceleration is vertical acceleration. Both accelerations can be decomposed into two orthogonal components, one parallel and the other normal to the road plane (Figure 9). Transposing those accelerations to the equivalent system and analyzing the so-called dynamic equilibrium, motion equations are going to be obtained.

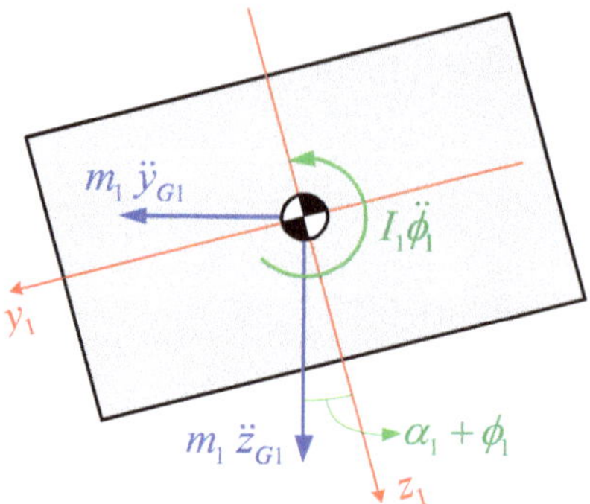

Figure 9. Acceleration decomposition.

Suspended mass rotation, its angular speed and its vertical displacement are responsible for the forces on the suspension's system components and each of them is linked to the displacement of the correspondent unsprung mass connected to them. From those forces and considering the center of gravity location, the moments around its roll center can be calculated. In Figure 10, the front of the body diagram for dynamic equilibrium is shown.

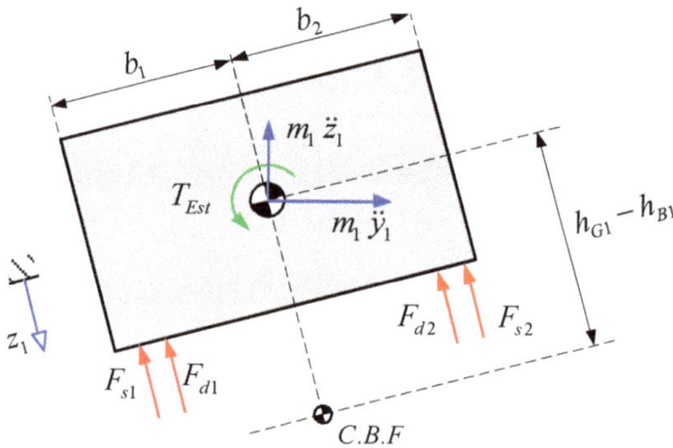

Figure 10. Front cabin's free body diagram for dynamic equilibrium.

From the free body diagram, the following differential equations are settled. For the vertical displacement z_1:

$$m_1\ddot{z}_1 = -\left[k_{s1}(z_1 + b_1\,\phi_1 - z_2) + c_1(\dot{z}_1 + b_1\,\dot{\phi}_1 - \dot{z}_2)\right]$$
$$-\left[k_{s2}(z_1 - b_2\,\phi_1 - z_3) + c_2(\dot{z}_1 - b_2\,\dot{\phi}_1 - \dot{z}_3)\right] \tag{9}$$
$$+\frac{m_1 v^2}{\rho_1(t)}\sin[\alpha_1(t) + \phi_1(t)]$$

For rotation ϕ_1:

$$\left[I_1 + m_1(h_{G1} - h_{B1})^2\right]\ddot{\phi}_1 = b_2\left[k_{s2}(z_1 - b_2\phi_1 - z_3) + c_2(\dot{z}_1 - b_2\dot{\phi}_1 - \dot{z}_3)\right]$$
$$-b_1\left[k_{s1}(z_1 + b_1\phi_1 - z_2) + c_1(\dot{z}_1 + b_1\dot{\phi}_1 - \dot{z}_2)\right] \tag{10}$$
$$-\left[\left(m_1\frac{v^2}{\rho_1(t)}\cos(\alpha_1(t) + \phi_1(t))\right)(h_{G1} - h_{B1})\right]$$
$$+[(m_1 g\sin\alpha_1(\alpha_1(t) + \phi_1(t)))(h_{G1} - h_{B1})] + T_{est}$$

This equation considers a resistant torque, which is caused by the torsional stiffness of the vehicle's structure; said moment is denoted by T_{est}, which will be determined later. For the unsprung masses, the analysis is quite similar as that carried out for the suspended mass, with the difference of the intervention of the torsion bar, which links the movements of both suspended masses. Figure 11 shows the free body diagram and equivalent system of both unsprung masses.

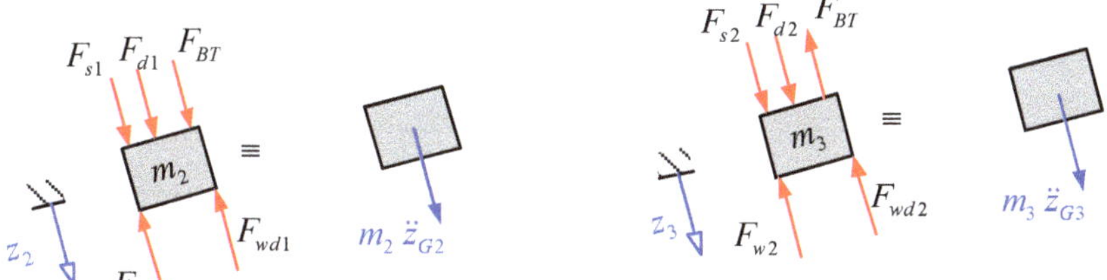

Figure 11. Unsprung frontal masses free body diagram.

In Figure 11, it is possible to note that there is a force produced by the sway bar F_{TB}, which is opposite at each unsprung mass. This element takes advantage of the lever arms generated by each unsprung mass, which due to the displacement difference between the ends of the bar generates a torsion moment.

The displacement can be calculated as follow:

$$z_2 = \frac{F_1 l_b^3}{3 E_b I_b} \tag{11}$$

where F_1 is the force applied by the left unsprung mass, l_b is the arm bar's length, E_b Young Modulus of the bar material and I_b is the bar's inertia moment. The force applied at each end of the bar can be determined by the following equation:

$$F_1(z_2) = \frac{3 z_2 E_b I_b}{l_b^3} \tag{12}$$

From Equation (12), a correlation between unsprung mass displacements and force in the bar can be obtained. Figure 12 shows the free body diagram of the sway bar.

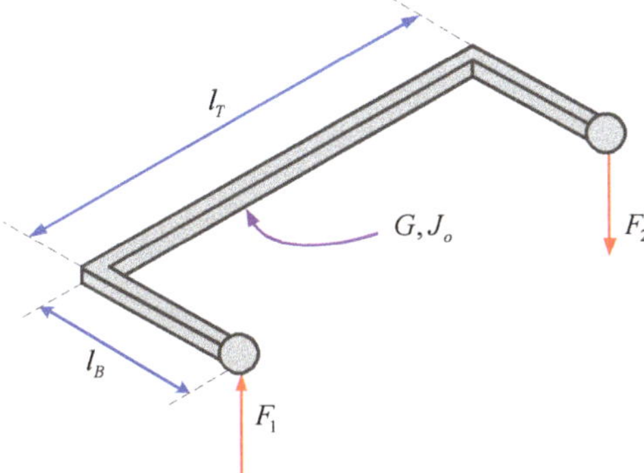

Figure 12. Free body diagram—sway bar.

From the force calculated in (12), the torsion moment over the bar is calculated:

$$M_{t1} = F_1 l_b = \frac{3 z_2\ E_b I_b}{l_b^2} \tag{13}$$

In a similar way, it is possible calculate the moment generated by the unsprung right mass as follow:

$$M_{t2} = F_2 l_b = \frac{3 z_3\ E_b I_b}{l_b^2} \tag{14}$$

Due to the moment exerted by both arms being opposite, the resultant torsion moment is calculated from the difference between them:

$$M_{TB} = M_{t1} - M_{t2} = \frac{3\,E_b I_b}{l_b^2}(z_2 - z_3) \tag{15}$$

The resultant torsion moment generates a rotation angle on the bar:

$$\Phi_B = \frac{M_{TB}\, l_T}{G\, J_0} \tag{16}$$

This angle can be related to the displacement arm difference as follows:

$$z_\Delta = z_\Delta = \Phi\, l_b = \frac{3\,E_b I_b\, l_T}{l_b\, G\, J_0}(z_2 - z_3) \tag{17}$$

From Equation (17) and due to the linear correlation between the displacement and the unsprung mass displacements, an elastic constant is determined:

$$k_{tb} = \frac{9\,E_b^2 I_b^2\, l_T}{l_b^4\, G\, J_0} \tag{18}$$

Considering the influence of the sway bar and applying the acceleration decomposition and transposing them to the first member of the equivalence (Dynamic equilibrium), the following motion equations are developed:

Unsprung left mass:

$$m_2\,\ddot{z}_2 - m_2\frac{v^2}{\rho_1(t)}\sin[\alpha_1(t) + \phi_1(t)] = k_{s1}(z_1 + b_1\,\phi_1 - z_2) + c_1(\dot{z}_1 + b_1\,\dot{\phi}_1 - \dot{z}_2) \\ -k_{TB}(z_2 - z_3) - k_{w1}\,z_2 - c_{w1}\,\dot{z}_2 \tag{19}$$

Unsprung right mass:

$$m_3\,\ddot{z}_3 = k_{s2}(z_1 - b_1\,\phi_1 - z_3) + c_2(\dot{z}_1 - b_1\,\dot{\phi}_1 - \dot{z}_3) \\ + k_{TB}(z_2 - z_3) - k_{w2}\,z_3 - c_{w2}\,\dot{z}_3 \\ + m_3\frac{v^2}{\rho_1(t)}\sin[\alpha_1(t) + \phi_1(t)] \tag{20}$$

4.2. Rear Section

In a similar way as the front section, the free body diagram of the suspended mass for dynamic equilibrium is depicted in Figure 13.

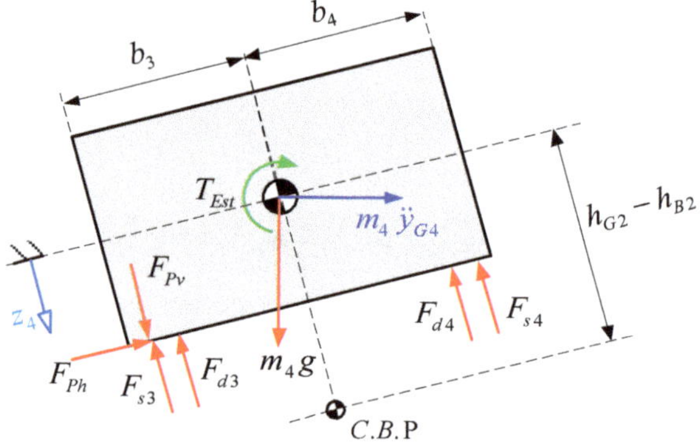

Figure 13. Rear cabin's free body diagram for dynamic equilibrium.

In comparison with the front section, here there is a noticeable effect of the Panhard bar, which is represented by the forces F_{Ph} and F_{Pv}, which are the horizontal and vertical component of the force along the bar. Additionally, for the acceleration decomposition, the influence of the rotation angle of both masses, unsprung and suspended, must be taken into account, including the torsional torque produced by the structure. From this and the free body diagram the following motion equations are obtained:

For the displacement z_4

$$
m_4\left(\ddot{z}_4 - \frac{v^2}{\rho_2(t)}\sin[\alpha_2(t) + \phi_4(t) + \phi_5(t)]\right) = -k_{s3}(z_4 + b_3\phi_4 - z_5 - b_5\phi_5)
$$
$$
-k_{s4}(z_4 - b_4\phi_4 - z_5 + b_6\phi_5)
$$
$$
-c_{s3}(\dot{z}_4 + b_3\dot{\phi}_4 - \dot{z}_5 - b_5\dot{\phi}_5)
$$
$$
-c_{s4}(\dot{z}_4 - b_4\dot{\phi}_4 - \dot{z}_5 + b_6\dot{\phi}_5) + F_{Pv}
$$

(21)

For the rotation ϕ_4:

$$
\left[I_4 + m_4(h_{G2} - h_{B2})^2\right]\ddot{\phi}_4 = b_4\left[k_{s4}(z_4 - b_4\phi_4 - z_5 + b_6\phi_5) + c_{s4}(\dot{z}_4 - b_4\dot{\phi}_4 - \dot{z}_5 + b_6\dot{\phi}_5)\right]
$$
$$
-b_3\left[k_{s3}(z_4 + b_3\phi_4 - z_5 - b_5\phi_5) + c_{s3}(\dot{z}_4 + b_3\dot{\phi}_4 - \dot{z}_5 - b_5\dot{\phi}_5) - F_{Pv}\right]
$$
$$
-\left[\left(m_4\frac{v^2}{\rho_2(t)}\cos[\alpha_2(t) + \phi_4(t) + \phi_5(t)]\right)(h_{G2} - h_{B2})\right] - T_{est}
$$
$$
+[(m_4 g \sin[\alpha_2(t) + \phi_4(t) + \phi_5(t)])(h_{G2} - h_{B2})]
$$
$$
-[F_{Ph}(h_{G2} - h_{S2} - h_{B2})]
$$

(22)

For the unsprung mass, the analysis is similar obtaining the free body diagram for the dynamic equilibrium showed in Figure 14.

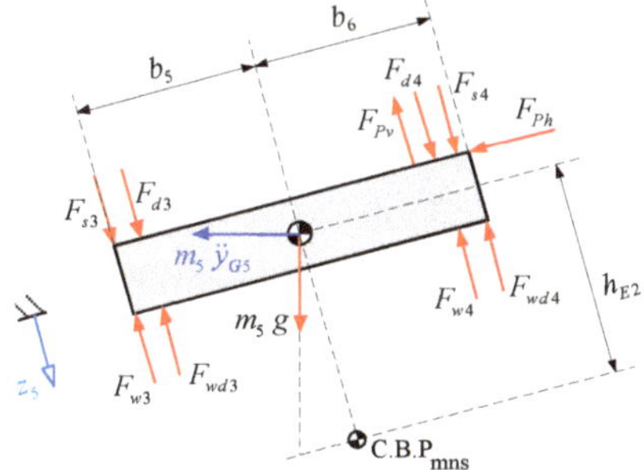

Figure 14. Rear unsprung mass free body diagram for dynamic equilibrium.

From the free body diagram, the following motion equations are obtained:
For the displacement z_5

$$
m_5\ddot{z}_5 = -k_{w3}(z_5 + b_5\phi_5) - c_{w3}(\dot{z}_5 + b_5\dot{\phi}_5) - k_{w4}(z_5 - b_6\phi_5)
$$
$$
-c_{w4}(\dot{z}_5 - b_6\dot{\phi}_5) + k_{s3}(z_4 + b_3\phi_4 - z_5 - b_5\phi_5)
$$
$$
+k_{s4}(z_4 - b_4\phi_4 - z_5 + b_6\phi_5) + c_{s3}(\dot{z}_4 + b_3\dot{\phi}_4 - \dot{z}_5 - b_5\dot{\phi}_5)
$$
$$
+c_{s4}(\dot{z}_4 - b_4\dot{\phi}_4 - \dot{z}_5 + b_6\dot{\phi}_5) - F_{Pv} + \frac{m_5 v^2}{\rho_2(t)}\sin(\alpha_2(t) + \phi_5(t))
$$

(23)

For the rotation ϕ_5:

$$\left[I_5 + m_5(h_{E2})^2\right]\ddot{\phi}_5 = b_6 \begin{bmatrix} k_{w4}(z_5 - b_6\phi_5) + c_{w4}(\dot{z}_5 - b_6\dot{\phi}_5) \\ -k_{s4}(z_4 - b_4\phi_4 - z_5 + b_6\phi_5) \\ -c_{s4}(\dot{z}_4 - b_4\dot{\phi}_4 - \dot{z}_5 + b_6\dot{\phi}_5) + F_{PB} \end{bmatrix}$$
$$-b_5 \begin{bmatrix} k_{s3}(z_4 + b_3\phi_4 - z_5 - b_5\phi_5) \\ +c_{s3}(\dot{z}_4 + b_3\dot{\phi}_4 - \dot{z}_5 - b_5\dot{\phi}_5) \\ -k_{w3}(z_5 + b_5\phi_5) - c_{w3}(\dot{z}_5 + b_5\dot{\phi}_5) \end{bmatrix}$$
$$+h_{E2} \begin{bmatrix} (m_5 g \sin[\alpha_2(t) + \phi_5(t)]) - \\ \left(m_5 \frac{v^2}{\rho_2(t)} \cos[\alpha_2(t) + \phi_5(t)]\right) + F_{Ph} \end{bmatrix} \tag{24}$$

The forces produced by the Panhard bar require a kinematic analysis of both ends of it. Such a component could be modeled as a bar element on which its inner force acts along itself, therefore the force's direction is already known.

Considering only the vertical displacement at the ends of the bar, Figure 15 shows the displacement of each point connected to the Panhard bar:

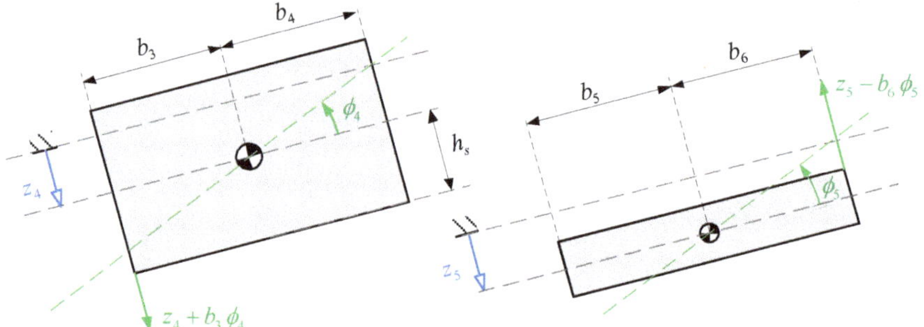

Figure 15. Rear masses displacements at the ends connected to the Panhard Bar.

From that displacement, the bar vertical displacement can be analyzed as it is shown in Figure 16.

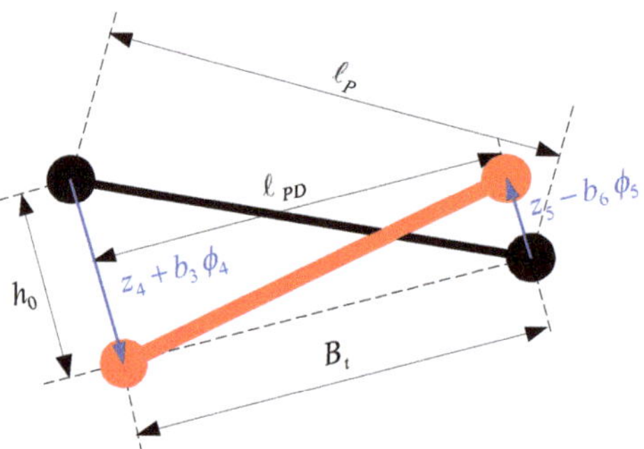

Figure 16. Panhard Bar vertical displacement.

From this analysis, the vertical dimension of the bar can be calculated as follow:

$$h = h_0 + z_5 - b_6\phi_5 - z_4 - b_3\phi_4 \tag{25}$$

With the vertical dimension determined, the bar's length can be calculated directly:

$$\updownarrow_{PD} = \sqrt{h^2 + B_t^2} \tag{26}$$

Additionally, its orientation:

$$\varphi_B = \arctan\left(\frac{B_t}{h_0 + z_5 - b_6\phi_5 - z_4 - b_3\,\phi_4}\right) \tag{27}$$

Later, the bar stiffness can be calculated instantly as follow:

$$k_P = \frac{EA}{\updownarrow_{PD}} \tag{28}$$

where E is the material's Young Modulus and A its transversal area.

As a final step the influence of the torsional stiffness of the structure is seen on the motion equations, so the torsional moment originated this element is equivalent to the following expression:

$$T_{est} = k_E(\phi_4 - \phi_1) \tag{29}$$

This component links both bus sections, so all differential equations developed are coupled.

5. Simulation and Results

Due to the complexity of the model, a state space model must be implemented, so that the motion equations settled previously can be arranged as a differential equation system, which can be divided into two matrices, one as a state matrix **A** which is entirely lineal, and an input matrix **B**, which due to the Panhard bar and its dependency to the variable states is highly non-lineal.

From the state space equation:

$$\begin{aligned} d\mathbf{x} &= \mathbf{Ax} + \mathbf{Bu} \\ \mathbf{y} &= \mathbf{Cx} + \mathbf{Du} \end{aligned} \tag{30}$$

The state vector is defined as follow:

$$x = \begin{bmatrix} z_1 & \dot{z}_1 & z_2 & \dot{z}_2 & z_3 & \dot{z}_3 & \phi_1 & \dot{\phi}_1 & z_4 & \dot{z}_4 & z_5 & \dot{z}_5 & \phi_4 & \dot{\phi}_4 & \phi_5 & \dot{\phi}_5 \end{bmatrix} \tag{31}$$

Due to its dimensions and simplicity, the state matrix which is a 16×16 square matrix will not be shown on this document, but the input matrix will be presented for its non-linear formulation:

$$\begin{pmatrix} 0 \\ \frac{v^2}{\rho(t)}\sin(\alpha(t) + \phi_1) \\ 0 \\ \frac{v^2}{\rho(t)}\sin\alpha(\alpha(t) + \phi_1) \\ 0 \\ \frac{v^2}{\rho(t)}\sin\alpha(\alpha(t) + \phi_1) \\ 0 \\ \frac{-\left(m_1\frac{v^2}{\rho(t)}\cos(\alpha(t)+\phi_1)\right)(h_{G1}-h_{B1})+(m_1 g\sin(\alpha(t)+\phi_1))(h_{G1}-h_{B1})}{I_{1B}} \\ 0 \\ \frac{v^2}{\rho(t-1)}\sin(\alpha(t-1) + \phi_4 + \phi_5) + \frac{F_{Pv}}{m_4} \\ 0 \\ \frac{v^2}{\rho(t-1)}\sin(\alpha(t-1) + \phi_5) - \frac{F_{Pv}}{m_5} \\ 0 \\ \frac{\left(-m_4\frac{v^2}{\rho(t-1)}\cos(\alpha(t-1)+\phi_4+\phi_5) + m_4 g\sin(\alpha(t-1)+\phi_4+\phi_5)\right)(h_{G2}-h_{B2})-F_{Ph}(h_{S2})+F_{Pv}b_3}{I_{4D}} \\ 0 \\ \frac{\left[(m_5 g\sin(\alpha(t-1)+\phi_5))-\left(m_5\frac{v^2}{\rho(t-1)}\cos(\alpha(t-1)+\phi_5)\right)+F_{Ph}\right]h_{E2}+F_{Pv}b_6}{I_{5B}} \end{pmatrix} \tag{32}$$

From the equations given in this matrix, it is highly dependent on the state matrix so that matrix **B** is a function of **x**. Although there is a control vector **u** and matrix **C**, output, and matrix **D**, feedforward, this matrix will be replaced by an identity matrix and a row vector with ones, hence the main purpose of the settled system is simulation. In further works, this matrix can be replaced to obtain a control system. All equations for the path and other elements, such as a Panhard bar, are modeled in a Simulink® algorithm.

Once the differential equations are expressed into the state space equation, it is necessary to numerically fix all the mechanic parameters which have been taken from [24] are expressed in the following list:

$m_1 = 7000$ kg—Front suspended mass

$m_2 = 500$ kg—Front left unspring mass

$m_3 = 500$ kg—Front right unspring mass

$I_1 = 125,000$ kg $\cdot$ m^2—Front suspended mass inertia respect to its mass center

$c_1 = 600$ N $\cdot$ s/m—Damping coefficient of the left front suspension element

$c_2 = 600$ N $\cdot$ s/m—Damping coefficient of the right front suspension element

$b_1 = 0.8$ m—Distance from suspended mass center to the left front suspension

$b_2 = 0.8$ m—Distance from suspended mass center to the right front suspension

$k_{s1} = 200,000$ N/m—Stiffness coefficient of the left front suspension element

$k_{s2} = 200,000$ N/m—Stiffness coefficient of the right front suspension element

$h_{B1} = 0.75$ m—Roll center front suspended mass height respect to the path's plane.

$h_{G1} = 1.55$ m—Mass center height of the suspended front mass

$k_{TB} = 1,000,000$ N/m—Stiffness coefficient of the sway bar

$k_{w1} = 921,607$ N/m—Stiffness coefficient of the front left tire

$k_{w2} = 921,607$ N/m—Stiffness coefficient of the front right tire

$c_{w1} = 800$ N $\cdot$ s/m—Damping coefficient of the front left tire

$c_{w2} = 800$ N $\cdot$ s/m—Damping coefficient of the front right tire

$k_E = 600,000,000$ N $\cdot$ m/rad—Torsional stiffness coefficient of the vehicle structure

$m_4 = 7000$ kg—Read suspended mass

$m_5 = 1000$ kg—Rear left unspring mass

$I_4 = 125,000$ kg $\cdot$ m^2—Rear suspended mass inertia respect to its mass center

$I_5 = 1000$ kg $\cdot$ m^2—Rear unspring mass inertia respect to its mass center

$c_3 = 600$ N $\cdot$ s/m—Damping coefficient of the left rear suspension element

$c_4 = 600$ N $\cdot$ s/m—Damping coefficient of the right rear suspension element

$b_3 = 0.8$ m—Distance from suspended mass center to the left rear suspension

$b_4 = 0.8$ m—Distance from suspended mass center to the right rear suspension

$b_5 = 0.8$ m—Distance from unspring mass center to the left rear tire

$b_6 = 0.8$ m—Distance from unspring mass center to the right rear tire

$k_{s3} = 200,000$ N/m—Stiffness coefficient of the left rear suspension element

$k_{s4} = 200,000$ N/m—Stiffness coefficient of the right rear suspension element

$h_{B2} = 0.75$ m—Roll center rear suspended mass height respect to the path's plane.

$h_{G2} = 1.55$ m—Mass center height of the suspended rear mass

$h_{E2} = 0.5$ m—Mass center height of the unspring rear mass

$k_{w3} = 921,607$ N/m—Stiffness coefficient of the rear left tire

$k_{w4} = 921,607$ N/m—Stiffness coefficient of the rear right tire

$c_{w3} = 800$ N $\cdot$ s/m—Damping coefficient of the rear left tire

$c_{w4} = 800$ N $\cdot$ s/m—Damping coefficient of the rear right tire

Also, for the Panhard bar, the following parameters are taken:

$E = 2.1 \times 10^5$ N/mm^2—Panhard bar material's Young modulus

$d_P = 30$ mm—Panhard bar diameter

From the constants given above, the following results are obtained from the simulation of the vehicle's travel over a transition and circular path, as previously given (Figure 17).

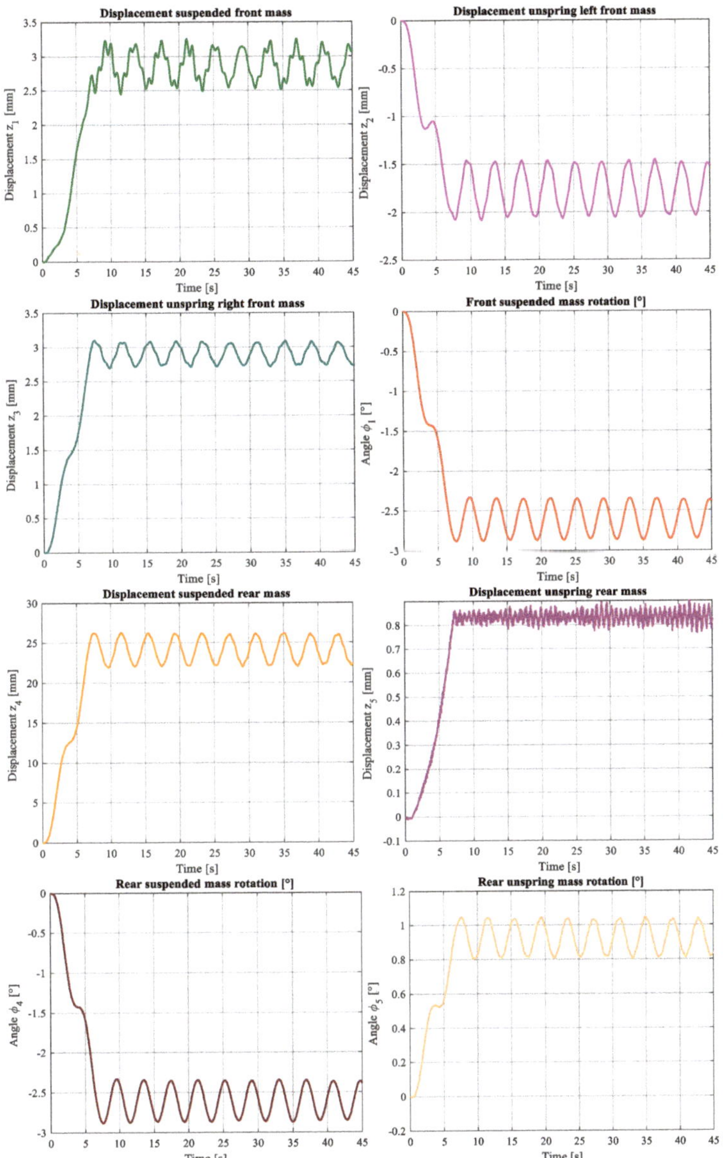

Figure 17. Simulation results—Displacement and rotation angles.

Furthermore, the instantaneous Panhard's bar force and angle are also obtained from the simulation (Figure 18).

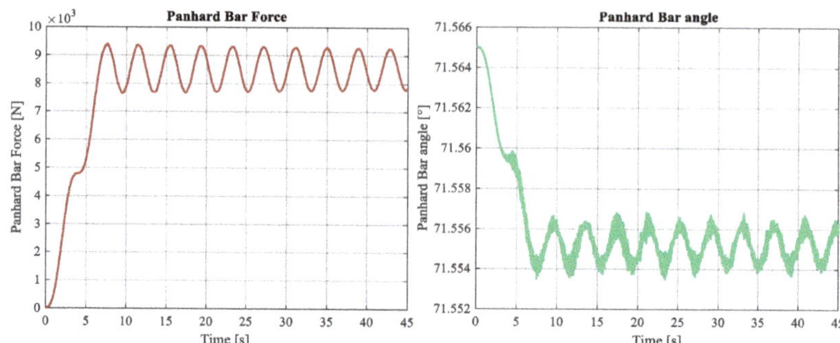

Figure 18. Panhard bar—force and inclination angle.

Those results show the bus dynamic behavior when it is traveling over a transition and circular curve; in fact, it is noticeable that while the bus is over the clothoid its position and rotation state variables are increasing constant and when it arrives at the circular path they become oscillating, consequently, the Panhard bar length and angle have the same behavior.

5.1. Effects of the Panhard Bar

The addition of the Panhard bar has a stabilization function on the mechanical behavior of the bus, the main effect is appreciated on the rear mass rotation angles, which have an effect on the displacements and angles in the front section. It can be said that the Panhard bar shrinks the oscillations of each state variable as can be appreciated, making a comparison between Figures 17 and 19.

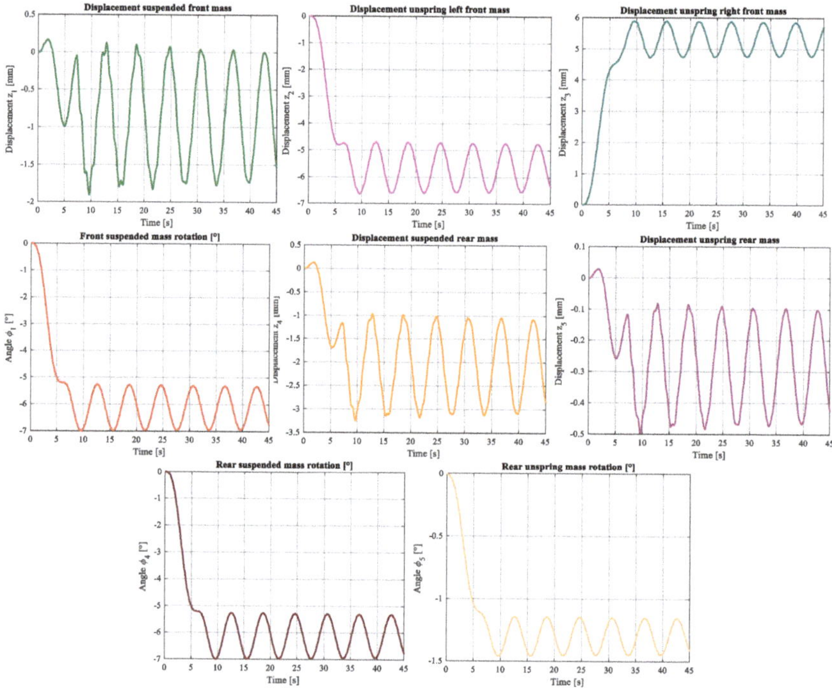

Figure 19. Simulation results without Panhard bar—displacement and rotation angles.

5.2. Panhard Bar Influence over the Structural Torsional Moment

As can be observed in the rotation angles comparison, there is an increment in the values in the system without a Panhard bar compared with the system with a Panhard bar. In fact, this tendency makes a change in the sense of the torsional moment produced along the structure of the bus, as can be seen in Figure 20.

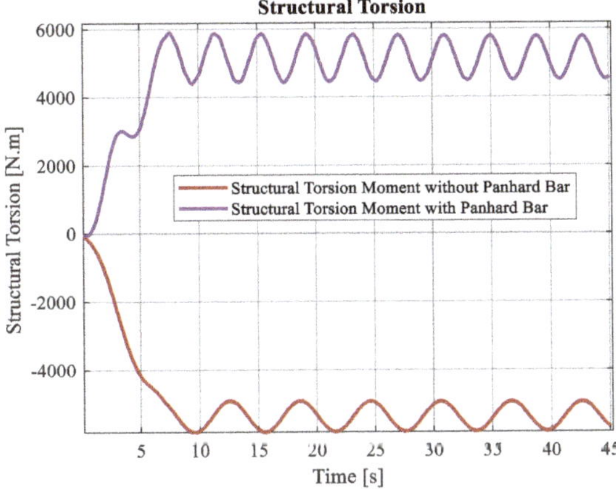

Figure 20. Structural Torsion with and without Panhard bar.

The positive values of the structural moment present in the simulation with the Panhard bar show a better stability due to this moment acts in opposition to the vehicles rollover making the vehicle more stable; on the other hand, the absence of a Panhard bar produces a negative structural moment which contributes to an increased risk of rollover.

5.3. Vehicle Lateral Stability Analysis

Although these results describe how the vehicle will behave, it does not show if the vehicle rolls over or slips; for that reason, it is necessary to handle the obtained results and the states derivatives in a new analysis that evaluates the normal and friction force on each tire. For this purpose, a new free body diagram is performed for the dynamic equilibrium in the front section (Figure 21).

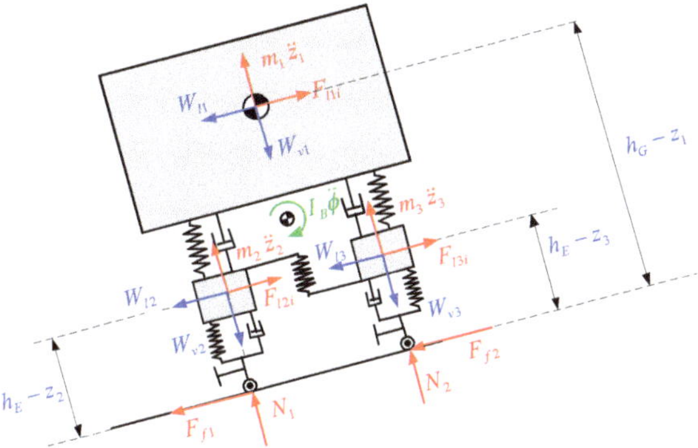

Figure 21. Free Body Diagram for front section dynamic equilibrium.

On the diagram shown, the inertial forces, which are obtained from the acceleration's decomposition used for the motion equations, and weight of each body is already decomposed.

The inertia forces on the vertical direction are defined using the states derivatives, however, the lateral inertia forces and the weight decomposition are calculated as follow:

$$F_{l1i} = \frac{m_1 \, v^2}{\rho_1(t)} \cos[\alpha_1(t) + \phi_1(t)] \tag{33}$$

$$F_{l2i} = \frac{m_2 \, v^2}{\rho_1(t)} \cos[\alpha_1(t)] \tag{34}$$

$$F_{l3i} = \frac{m_3 \, v^2}{\rho_1(t)} \cos[\alpha_1(t)] \tag{35}$$

Analyzing the equilibrium on the lateral direction:

$$F_{f1} + F_{f2} = F_{l1i} + F_{l2i} + F_{l3i} - W_{l1} - W_{l2} - W_{l3} \tag{36}$$

Furthermore, the inertial forces and torques required to be used into the state equations to calculate the normal forces at each tire. Analyzing the equilibrium taking moments from the left frontal wheel and factorizing N_2:

$$\left[\begin{array}{c} [W_{l2} - F_{l2i}](h_{E1} - z_2) + [W_{l3} - F_{l3i}](h_{E1} - z_3) - I_{1B}\,\ddot{\phi}_1 \\ +[W_{l1} - F_{l1i}](h_{G1} - z_1) + (m_1\,\ddot{z}_1 - W_{v1})(B_{via}/2) + (m_3\,\ddot{z}_3 - W_{v3})(a_{via}) \end{array} \right] / (B_t) \tag{37}$$

Analyzing the equilibrium on the vertical direction, N_1 is obtained:

$$N_1 = W_{v1} + W_{v2} + W_{v3} - N_2 - m_1\,\ddot{z}_1 - m_2\,\ddot{z}_2 - m_3\,\ddot{z}_3 \tag{38}$$

The magnitudes of the normal forces indicate if the vehicle rolls over or not when any normal force becomes zero. On the other hand, in the lateral direction there are two unknown variables, which are the friction force and just one equation, so that determines the value of each friction force becomes an unsolvable problem, therefore a new procedure to evaluate if the vehicle slips or not will be performed, using the value of both forces in a conditional way:

$$F_{f1} + F_{f2} \leq F_{f1,\max} + F_{f2,\max} \tag{39}$$

In addition, it can be expressed as:

$$0 \leq N_1\mu_s + N_2\mu_s - F_{f1} - F_{f2} \tag{40}$$

The maximum values of friction force will be determined using the so-called Coulomb's dry friction approach. For that evaluation, it is assumed that both tires are not skidding, so the friction coefficient is the static friction coefficient. The conditional form establishes that when the value becomes negative it means that the vehicle is slipping. This analysis is also applied for the rear section, with a similar free body diagram (Figure 22).

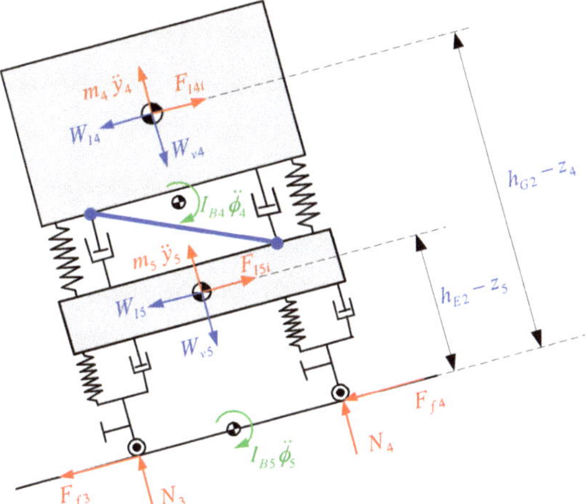

Figure 22. Free Body Diagram for rear section dynamic equilibrium.

The equations for the lateral inertia forces:

$$F_{l4i} = \frac{m_4\, v^2}{\rho_2(t)}\; \cos[\alpha_2(t) + \phi_4(t) + \phi_5(t)] \tag{41}$$

$$F_{l5i} = \frac{m_5\, v^2}{\rho_2(t)}\; \cos[\alpha_2(t) + \phi_5(t)] \tag{42}$$

Analyzing the equilibrium on the lateral direction:

$$F_{f3} + F_{f4} = F_{l4i} + F_{l5i} - W_{l4} - W_{l5} \tag{43}$$

Similar to the front section, analyzing the equilibrium taking moments from the left wheel and factorizing N_4:

$$N_4 = -\left[\begin{array}{l} [W_{l5} - F_{l5i}](h_{E2} - z_5)\ - I_{B4}\,\ddot{\phi}_4 - I_{B5}\,\ddot{\phi}_5 + [W_{l4} - F_{l4i}](h_{G2} - z_4) \\ + (m_4\,\ddot{z}_4 - W_{v4})(B_{via}/2) + (m_5\,\ddot{z}_5 - W_{v5})(B_t/2) \end{array} \right] / B_t \tag{44}$$

Analyzing the equilibrium on the vertical direction, N_1 is obtained:

$$N_3 = W_{V4} + W_{V5} - N_4 - m_4\,\ddot{z}_4 - m_5\,\ddot{z}_5 \tag{45}$$

Finally, the skidding condition is introduced for the rear section:

$$F_{f3} + F_{f4} \le F_{f3,\max} + F_{f4,\max} \tag{46}$$

In other terms:

$$0 \le N_3\mu_s + N_4\mu_s - F_{f3} - F_{f4} \tag{47}$$

For the simulation, a parameter more is necessary:

$\mu_k = 0.4$—Static friction coefficient between the tire and path.

The results for the normal forces at each tire, while the bus is traveling over the simulation path, are shown on the following graphics (Figure 23).

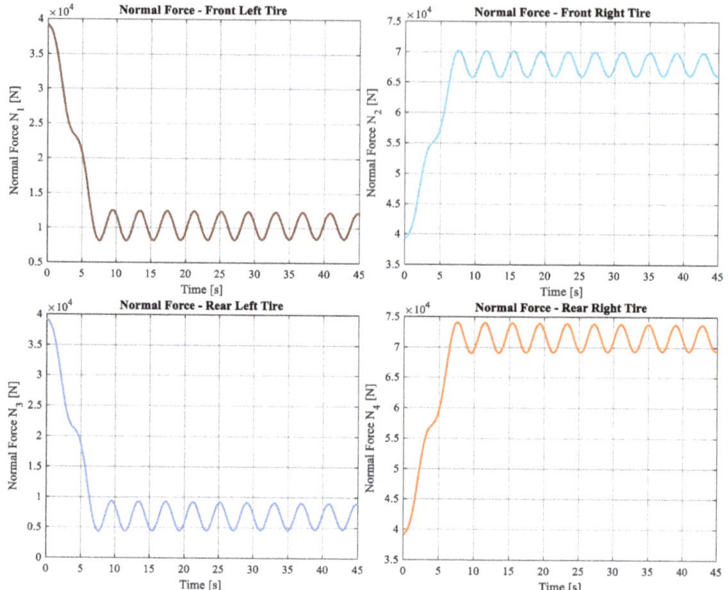

Figure 23. Normal forces on each tire considering Panhard bar.

About the friction force, the skidding condition shows the remaining friction force so that when it becomes negative the vehicle slips. Figure 24 shows the remaining friction force at each section, while the bus is traveling over the simulation:

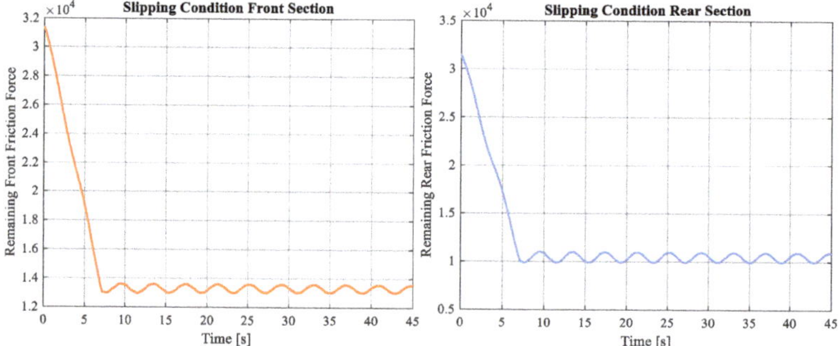

Figure 24. Remaining friction force at each section.

A comparison of the normal force can be made using a model without a Panhard bar to analyze the effect of this element in the lateral stability. The normal force values of the bus without Panhard bar are presented in Figure 25.

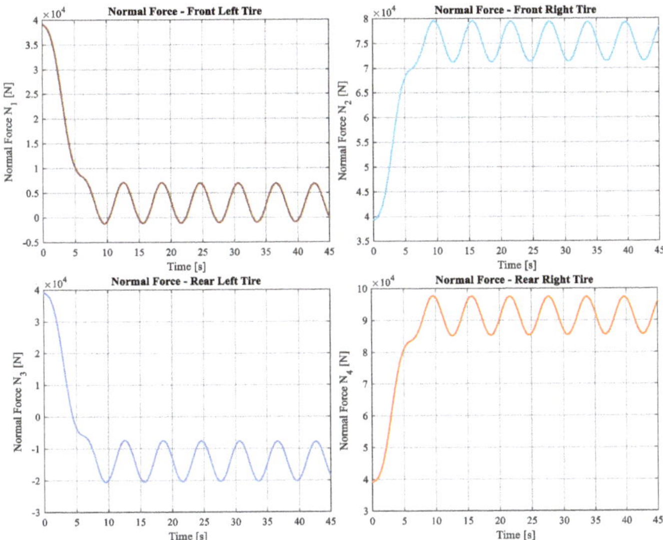

Figure 25. Normal forces on each tire without Panhard bar.

From Figure 25, it is possible to observe that the normal force in the left side reach or even pass negative values, which implies that the vehicle loses adherence with the path. It demonstrates that the Panhard bar improves the adherence between the path and the tires.

6. Conclusions

A model of a bus is presented, elaborated by means of rigid multibody systems with which a virtual or mathematical model (system of ordinary differential equations) has been generated. This model allows the simulation of the lateral dynamics of a bus when it travels on a cambered curve.

Based on the proposed model, a simulation algorithm has been designed to analyze the dynamic behavior of a bus in transit on a cambered curve. This model is adjusted to realistic transport routes since it has been calculated and parameterized with curves of agreement (trajectories that connect a straight and circular trajectory in such a way that it starts with an infinite radius of curvature and ends in a constant radius of curvature) used in road design.

The modeling has been carried out in such a way that the total system representing the bus has been divided into two fundamental parts: the front part of the vehicle, which in turn contemplates several subsystems, such as the unsprung masses, the unsprung masses connected by a stability or sway bar, the flexible wheels, etc. In turn, the rear part has been modeled in an analogous way although, unlike the front part, in this one there is only one unsprung mass and a Panhard type bar is used to improve its vertical and lateral dynamic behavior.

Therefore, the model used to model the bus dynamic behavior integrates two modifiable sections that allow for testing some new models for the calculation of the stiffness of the suspension system or of the tires, etc. For this first approximation, the suspension system and tire stiffness as well as the damping of both bodies have been modeled as linear components. This is responsible for the quasi—linear correlation between the geometric parameters and the tendency shown in the results: the state variables increase as the geometric parameters of the trajectory increase and oscillate, trying to reach a steady state in time while the trajectory parameters are constant.

According to the comparison between the results with and without a Panhard bar, it can be stated that the Panhard bar has a stabilization function, which reduces the oscillations of each state variable. Additionally, the parameters which modify the Panhard bar stiffness can be treated as a control variable on an active suspension system.

From the effects on the rotation angles that the Panhard bar has, it is also possible to highlight that the presence of this element reduces the rotation angles of the mass linked to it, this improves the lateral stability of the vehicle, increasing the value of the normal forces between the tires and the road, as can be concluded by comparing Figures 23 and 25.

The torsional structural stiffness is a parameter that must be considered when the stability of the vehicle is analyzed, as presented in Figure 20, where the direction of the structural moments affects the vehicle lateral stability.

Due to the weak effect of the non-linearities, some state variables have some disturbances which can be treated as noise.

Author Contributions: Conceptualization, E.O., J.R.H. and V.D.; methodology, E.O., E.R.C.L., J.R.H. and V.D.; software, E.R.C.L.; formal analysis, E.O., E.R.C.L., J.R.H. and V.D.; investigation, E.O., E.R.C.L., J.R.H. and V.D.; resources, E.O., E.R.C.L., J.R.H. and V.D.; data curation, E.O., E.R.C.L., J.R.H. and V.D.; writing—original draft preparation, E.O. and E.R.C.L.; writing—review and editing, E.O., E.R.C.L., J.R.H. and V.D.; visualization, E.O., E.R.C.L., J.R.H. and V.D.; supervision, E.O. and V.D.; funding acquisition, V.D. All authors have read and agreed to the published version of the manuscript.

Funding: This research received no external funding.

Institutional Review Board Statement: Not applicable.

Informed Consent Statement: Not applicable.

Data Availability Statement: Not applicable.

Conflicts of Interest: The authors declare no conflict of interest.

References

1. Yang, S.; Chen, L.; Li, S. *Dynamics of Vehicle-Road Coupled System*; Springer: Berlin/Heidelberg, Germany, 2015.
2. Popp, K.; Schiehlen, W. *Ground Vehicle Dynamics*; Springer: Berlin/Heidelberg, Germany, 2010.
3. Rill, G.; Schaeffer, T. *Grundlagen und Methodik der Mehrkörpersimulation—Vertieft in Matlab-Beispielen, Übungen und Anwendungen*; Springer: Berlin/Heidelberg, Germany, 2014.
4. Schramm, D.; Hiller, M.; Bardini, R. *Vehicle Dynamics—Modelling and Simulation*; Springer: Berlin/Heidelberg, Germany, 2014.
5. Jeong, D.; Ko, G.; Choi, S.B. Estimation of sideslip angle and cornering stiffness of an articulated vehicle using a constrained lateral dynamics model. *Mechatronics* **2022**, *85*, 102810. [CrossRef]
6. Seyedi, M.; Jung, S.; Wekezer, J. A comprehensive assessment of bus rollover crashes: Integration of multibody dynamic and finite element simulation methods. *Int. J. Crashworthiness* **2020**, *27*, 273–288. [CrossRef]
7. Niculescu-Faida, O.-C.; Niculescu-Faida, A. Vehicle dynamics modeling during moving along a curved path. Mathematical model usage on studying the robust stability. *UPB Sci. Bull.* **2008**, *4*, 49–60.
8. Gauchia, A.; Olmeda, E.; Aparicio, F.; Díaz, V. Bus mathematical model of acceleration threshold limit estimation in lateral rollover test, Veh. *Syst. Dyn.* **2011**, *49*, 1695–1707. [CrossRef]
9. Gauchia, A.; Díaz, V.; Boada, M.J.L.; Olatunbosun, O.; Boada, B.L. Bus Structure Behaviour under Driving Manoeuvring and Evaluation of the Effect of an Active Roll System. *Int. J. Veh. Struct. Syst.* **2010**, *2*, 14–19. [CrossRef]
10. Eftekharzadeh, S.F.; Khodabakhshi, A. Safety evaluation of highway geometric design criteria in horizontal curves at downgrades. *Int. J. Civ. Eng.* **2014**, *12*, 326–332.
11. Yin, Y.; Wen, H.; Sun, L.; Hou, W. The Influence of Road Geometry on Vehicle Rollover and Skidding. *Int. J. Environ. Res. Public Health* **2020**, *17*, 1648. [CrossRef] [PubMed]
12. Wang, B.; Hallmark, S.; Savolainen, P.; Dong, J. Crashes and near-crashes on horizontal curves along rural two-lane highways: Analysis of naturalistic driving data. *J. Saf. Res.* **2017**, *63*, 163–169. [CrossRef] [PubMed]
13. Geedipally, S.R.; Pratt, M.P.; Lord, D. Effects of geometry and pavement friction on horizontal curve crash frequency. *J. Transp. Saf. Secur.* **2019**, *11*, 167–188. [CrossRef]
14. Bogenreif, C.; Souleyrette, R.R.; Hans, Z. Identifying and measuring horizontal curves and related effects on highway safety. *J. Transp. Saf. Secur.* **2012**, *4*, 179–192. [CrossRef]
15. Abdollahzadeh Nasiri, A.S.; Rahmani, O.; Abdi Kordani, A.; Karballaeezadeh, N.; Mosavi, A. Evaluation of Safety in Horizontal Curves of Roads Using a Multi-Body Dynamic Simulation Process. *Int. J. Environ. Res. Public Health* **2020**, *17*, 5975. [CrossRef] [PubMed]
16. Xu, J.; Xin, T.; Gao, C.; Sun, Z. Study on the Maximum Safe Instantaneous Input of the Steering Wheel against Rollover for Trucks on Horizontal Curves. *Int. J. Environ. Res. Public Health* **2022**, *19*, 2025. [CrossRef] [PubMed]

17. Yin, Q.; Stiharu, I.; Rakheja, S. Kinetostatic analysis of a beam-axle suspension with Panhard rod restraining linkage. *Int. J. Veh. Des.* **1998**, *19*, 108–123.
18. Mayr, S.; Wagner, W. Development of a front axle suspension for special purpose tractors. In Proceedings of the Conference: Agricultural Engineering 2004, Dresden, Germany, 7–8 October 2004.
19. Sonsino, C.M.; Streicher, M. Optimization of cast iron safety components of commercial vehicles by material selection and geometry under consideration of service loadings. *Mater. Test.-Mater. Compon. Technol. Appl.* **2009**, *51*, 428–436.
20. Torbic, D.J.; O'Laughlin, M.K.; Harwood, D.W.; Bauer, K.M.; Bokenkroger, C.D.; Lucas, L.M.; Ronchetto, J.R.; Brennan, S.; Donnell, E.; Brown, A.; et al. *NCHRP 774 Report: Superelevation Criteria for Sharp Horizontal Curves on Steep Grades*; The National Academie Press: Washington, DC, USA, 2014.
21. Esveld, C. *Modern Railway Track*; Delft University, MRT—Productions: Zaltbommel, The Netherlands, 2001.
22. Rajamani, R. *Vehicle Dynamics and Control*; Mechanical Engineering Series; Springer: Berlin/Heidelberg, Germany, 2012.
23. Aparicio, F.; Vera, C.; Díaz, V. *Teoría de los Vehículos Automóviles*; Sección de Publicaciones, ETSII, Universidad Politécnica de Madrid: Madrid, Spain, 1995.
24. Mustafa Siddiqui, O. Dynamic Analysis of a Modern Urban Bus for Assessment of Ride Quality and Dynamic Wheel Loads. Master's Thesis, Concordia University, Montreal, QC, Canada, 2000.

MDPI
St. Alban-Anlage 66
4052 Basel
Switzerland
Tel. +41 61 683 77 34
Fax +41 61 302 89 18
www.mdpi.com

Mathematics Editorial Office
E-mail: mathematics@mdpi.com
www.mdpi.com/journal/mathematics